Abbreviations for Units

A	ampere	H	henry	nm	nanometer (10^{-9} m)
Å	angstrom (10^{-10} m)	h	hour	pt	pint
atm	atmosphere	Hz	hertz	qt	quart
Btu	British thermal unit	in	inch	rev	revolution
Bq	becquerel	J	joule	R	roentgen
C	coulomb	K	kelvin	Sv	seivert
°C	degree Celsius	kg	kilogram	s	second
cal	calorie	km	kilometer	T	tesla
Ci	curie	keV	kilo-electron volt	u	unified mass unit
cm	centimeter	lb	pound	V	volt
dyn	dyne	L	liter	W	watt
eV	electron volt	m	meter	Wb	weber
°F	degree Fahrenheit	MeV	mega-electron volt	y	year
fm	femtometer, fermi (10^{-15} m)	Mm	megameter (10^6 m)	yd	yard
ft	foot	mi	mile	μm	micrometer (10^{-6} m)
Gm	gigameter (10^9 m)	min	minute	μs	microsecond
G	gauss	mm	millimeter	μC	microcoulomb
Gy	gray	ms	millisecond	Ω	ohm
g	gram	N	newton		

Some Conversion Factors

Length

1 m = 39.37 in = 3.281 ft = 1.094 yd
1 m = 10^{15} fm = 10^{10} Å = 10^9 nm
1 km = 0.6215 mi
1 mi = 5280 ft = 1.609 km
1 lightyear = 1 $c \cdot$y = 9.461×10^{15} m
1 in = 2.540 cm

Volume

1 L = 10^3 cm^3 = 10^{-3} m^3 = 1.057 qt

Time

1 h = 3600 s = 3.6 ks
1 y = 365.24 d = 3.156×10^7 s

Speed

1 km/h = 0.278 m/s = 0.6215 mi/h
1 ft/s = 0.3048 m/s = 0.6818 mi/h

Angle–angular speed

1 rev = 2π rad = 360°
1 rad = 57.30°
1 rev/min = 0.1047 rad/s

Force–pressure

1 N = 10^5 dyn = 0.2248 lb
1 lb = 4.448 N
1 atm = 101.3 kPa = 1.013 bar = 76.00 cmHg = 14.70 lb/in^2

Mass

1 u = [(10^{-3} mol^{-1})/N_A] kg = 1.661×10^{-27} kg
1 tonne = 10^3 kg = 1 Mg
1 slug = 14.59 kg
1 kg weighs about 2.205 lb

Energy–power

1 J = 10^7 erg = 0.7373 ft·lb = 9.869×10^{-3} L·atm
1 kW·h = 3.6 MJ
1 cal = 4.184 J = 4.129×10^{-2} L·atm
1 L·atm = 101.325 J = 24.22 cal
1 eV = 1.602×10^{-19} J
1 Btu = 778 ft·lb = 252 cal = 1054 J
1 horsepower = 550 ft·lb/s = 746 W

Thermal conductivity

1 W/(m·K) = 6.938 Btu·in/(h·ft^2·F°)

Magnetic field

1 T = 10^4 G

Viscosity

1 Pa·s = 10 poise

fifth edition

PHYSICS
FOR SCIENTISTS AND ENGINEERS

Volume 2B
Electrodynamics, Light

W. H. Freeman and Company
New York

PT: For Claudia

GM: For Vivian

Publisher:	Susan Finnemore Brennan
Senior Development Editor:	Kathleen Civetta/Jennifer Van Hove
Assistant Editors:	Rebecca Pearce/Amanda McCorquodale/Eileen McGinnis
Marketing Manager:	Mark Santee
Project Editors:	Georgia L. Hadler/Cathy Townsend, PreMediaONE, A Black Dot Group Company
Cover and Text Designers:	Marcia Cohen/Blake Logan
Illustrations:	Network Graphics/PreMediaONE, A Black Dot Group Company
Photo Editors:	Patricia Marx/Dena Betz
Production Manager:	Julia DeRosa
Media and Supplements Editors:	Brian Donnellan
Composition:	PreMediaONE, A Black Dot Group Company
Manufacturing:	RR Donnelley & Sons Company

Cover image: Digital Vision

Library of Congress Cataloging-in-Publication Data
Physics for Scientists and Engineers. - 5th ed.
    p. cm.
    By Paul A. Tipler and Gene Mosca
    Includes index.
    ISBN: 0-7167-0809-4 (Vol. 1 Hardback Ch. 1-20, R)
    ISBN: 0-7167-0900-7 (Vol. 1A Softcover Ch. 1-13, R)
    ISBN: 0-7167-0903-1 (Vol. 1B Softcover Ch. 14-20)
    ISBN: 0-7167-0810-8 (Vol. 2 Hardback Ch. 21-41)
    ISBN: 0-7167-0902-3 (Vol. 2A Softcover Ch. 21-25)
    ISBN: 0-7167-0901-5 (Vol. 2B Softcover Ch. 26-33)
    ISBN: 0-7167-0906-6 (Vol. 2C Softcover Ch. 34-41)
    ISBN: 0-7167-8339-8 (Standard Hardback Ch. 1-33, R)
    ISBN: 0-7167-4389-2 (Extended Hardback Ch. 1-41)

© 2004 by W. H. Freeman and Company

Printed in the United States of America

First printing 2003

CONTENTS

CHAPTER 10

CONSERVATION OF ANGULAR MOMENTUM / 309

CHAPTER R

SPECIAL RELATIVITY / R-1

CHAPTER 11

GRAVITY / 339

CHAPTER 12 *

STATIC EQUILIBRIUM AND ELASTICITY / 370

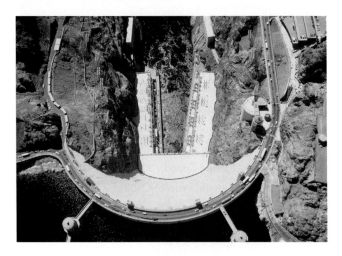

PART II   OSCILLATIONS AND WAVES/425

CHAPTER 14

OSCILLATIONS / 425

CHAPTER 15

TRAVELING WAVES / 465

CHAPTER 16

SUPERPOSITION AND STANDING WAVES / 503

VOLUME 2

PART IV   **ELECTRICITY AND MAGNETISM /651**

CHAPTER 21

THE ELECTRIC FIELD I: DISCRETE CHARGE DISTRIBUTIONS / 651

CHAPTER 22

THE ELECTRIC FIELD II: CONTINUOUS CHARGE DISTRIBUTIONS / 682

CHAPTER 23

ELECTRIC POTENTIAL / 717

CHAPTER 24

ELECTROSTATIC ENERGY AND CAPACITANCE / 748

CHAPTER 25

ELECTRIC CURRENT AND DIRECT-CURRENT CIRCUITS / 786

CHAPTER 26

THE MAGNETIC FIELD / 829

CHAPTER 30

MAXWELL'S EQUATIONS AND ELECTROMAGNETIC WAVES / 971

PART V   LIGHT/997

CHAPTER 31

PROPERTIES OF LIGHT / 997

CHAPTER 32

OPTICAL IMAGES / 1038

CHAPTER 33

INTERFERENCE AND DIFFRACTION / 1084

APPENDIX A

SI UNITS AND CONVERSION FACTORS / AP-1

APPENDIX B

NUMERICAL DATA / AP-3

APPENDIX C

PERIODIC TABLE OF ELEMENTS / AP-6

APPENDIX D

REVIEW OF MATHEMATICS / AP-8

PART VI   MODERN PHYSICS: QUANTUM MECHANICS, RELATIVITY, AND THE STRUCTURE OF MATTER

CHAPTER 34

WAVE PARTICLE DUALITY AND QUANTUM PHYSICS

CHAPTER 35

APPLICATIONS OF THE SCHRÖDINGER EQUATION

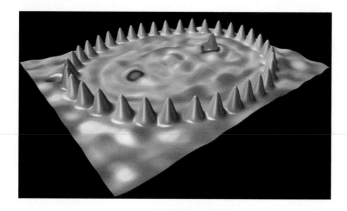

CHAPTER 36

ATOMS

CHAPTER 37

MOLECULES

CHAPTER 38

SOLIDS AND THE THEORY OF CONDUCTION

CHAPTER 39

RELATIVITY

CHAPTER 40

NUCLEAR PHYSICS

CHAPTER 41

ELEMENTARY PARTICLES AND THE BEGINNING OF THE UNIVERSE

The Magnetic Field

THE AURORA BOREALIS APPEARS WHEN "SOLAR WIND," CHARGED PARTICLES PRODUCED BY NUCLEAR FUSION REACTIONS IN THE SUN, BECOME TRAPPED IN THE EARTH'S MAGNETIC FIELD.

? **How does the earth's magnetic field act on subatomic particles? (See Example 26-1.)**

26-1   The Force Exerted by a Magnetic Field

26-2   Motion of a Point Charge in a Magnetic Field

26-3   Torques on Current Loops and Magnets

26-4   The Hall Effect

More than 2000 years ago, the Greeks were aware that a certain type of stone (now called magnetite) attracts pieces of iron, and there are written references to the use of magnets for navigation dating from the twelfth century.

In 1269, Pierre de Maricourt discovered that a needle laid at various positions on a spherical natural magnet orients itself along lines that pass through points at opposite ends of the sphere. He called these points the poles of the magnet. Subsequently, many experimenters noted that every magnet of any shape has two poles, designated the north and south poles, where the force exerted by the magnet is strongest. It was also noted that the like poles of two magnets repel each other and the unlike poles of two magnets attract each other.

In 1600, William Gilbert discovered that the earth is a natural magnet with magnetic poles near the north and south geographic poles. Since the north pole of a compass needle points toward the south pole of a given magnet, what we call the north pole of the earth is actually a south magnetic pole, as illustrated in Figure 26-1.

Although electric charges and magnetic poles are similar in many respects, there is an important difference: Magnetic poles always occur in pairs. When a magnet is broken in half, equal and opposite poles appear at either side of the break point. The result is two magnets, each with a north and south pole. There has long been speculation about the existence of an isolated magnetic pole, and in recent years considerable experimental effort has been made to find such an object. Thus far, there is no conclusive evidence that an isolated magnetic pole exists.

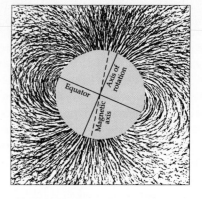

FIGURE 26-1 Magnetic field lines of the earth depicted by iron filings around a uniformly magnetized sphere. The field lines exit from the north magnetic pole, which is near the south geographic pole, and enter the south magnetic pole, which is near the north geographic pole.

➤ In this chapter, we consider the effects of a given magnetic field on moving charges and on wires carrying currents. The sources of magnetic fields are discussed in the next chapter.

26-1 The Force Exerted by a Magnetic Field

The existence of a magnetic field $\vec{B}$ at some point in space can be demonstrated with a compass needle. If there is a magnetic field, the needle will align itself in the direction of the field.

Experimentally it is observed that, when a charge q has velocity $\vec{v}$ in a magnetic field, there is a force on the magnetic field that is proportional to q and to v, and to the sine of the angle between the directions of $\vec{v}$ and $\vec{B}$. Surprisingly, the force is perpendicular to both the velocity and the field. These experimental results can be summarized as follows: When a charge q moves with velocity $\vec{v}$ in a magnetic field $\vec{B}$, the magnetic force $\vec{F}$ on the charge is

$$\vec{F} = q\vec{v} \times \vec{B} \qquad 26\text{-}1$$

MAGNETIC FORCE ON A MOVING CHARGE

Since $\vec{F}$ is perpendicular to both $\vec{v}$ and $\vec{B}$, $\vec{F}$ is perpendicular to the plane defined by these two vectors. The direction of $\vec{v} \times \vec{B}$ is given by the right-hand rule as $\vec{v}$ is rotated into $\vec{B}$, as illustrated in Figure 26-2. If q is positive, then $\vec{F}$ is in the same direction as $\vec{v} \times \vec{B}$.

Examples of the direction of the forces exerted on moving charges when the magnetic field vector $\vec{B}$ is in the vertical direction are shown in Figure 26-3. Note that the direction of any particular magnetic field $\vec{B}$ can be found experimentally by measuring $\vec{F}$ and $\vec{v}$ for several velocities in different directions and then applying Equation 26-1.

Equation 26-1 defines the **magnetic field** $\vec{B}$ in terms of the force exerted on a moving charge. The SI unit of magnetic field is the **tesla** (T). A charge of one coulomb moving with a velocity of one meter per second perpendicular to a magnetic field of one tesla experiences a force of one newton:

$$1\,\text{T} = 1\frac{\text{N}}{\text{Cm/s}} = 1\,\text{N}/(\text{A·m}) \qquad 26\text{-}2$$

This unit is rather large. The magnetic field of the earth has a magnitude somewhat less than 10^{-4} T on the earth's surface. The magnetic fields near powerful

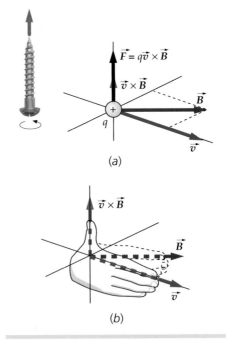

(a)

(b)

FIGURE 26-2 Right-hand rule for determining the direction of a force exerted on a charge moving in a magnetic field. If q is positive, then $\vec{F}$ is in the same direction as $\vec{v} \times \vec{B}$. (a) The cross product $\vec{v} \times \vec{B}$ is perpendicular to both $\vec{v}$ and $\vec{B}$ and is in the direction of the advance of a right-hand-threaded screw if turned in the same direction as to rotate $\vec{v}$ into $\vec{B}$. (b) If the fingers of the right hand are in the direction of $\vec{v}$ so that they can be curled toward $\vec{B}$, the thumb points in the direction of $\vec{v} \times \vec{B}$.

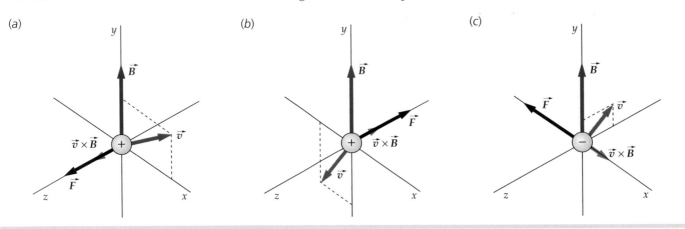

FIGURE 26-3 Direction of the magnetic force on a charged particle moving with velocity $\vec{v}$ in a magnetic field $\vec{B}$.

permanent magnets are about 0.1 T to 0.5 T, and powerful laboratory and industrial electromagnets produce fields of 1 T to 2 T. Fields greater than 10 T are difficult to produce because the resulting magnetic forces will either tear the magnets apart or crush the magnets. A commonly used unit, derived from the cgs system, is the **gauss** (G), which is related to the tesla as follows

$$1\,\text{G} = 10^{-4}\,\text{T} \qquad\qquad\qquad\qquad 26\text{-}3$$

DEFINITION—GAUSS

Since magnetic fields are often given in gauss, which is not an SI unit, remember to convert from gauss to teslas when making calculations.

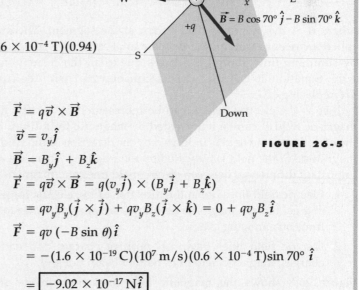

FIGURE 26-4

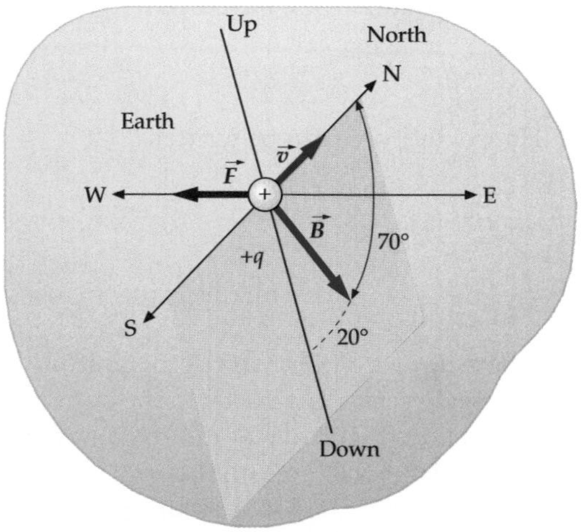

FORCE ON A PROTON GOING NORTH	**EXAMPLE 26-1**

The magnetic field of the earth is measured at a point on the surface to have a magnitude of 0.6 G and is directed downward and, in the northern hemisphere, northward, making an angle of about 70° with the horizontal, as shown in Figure 26-4. (The earth's magnetic field varies from place to place. These data are approximately correct for the central United States.) A proton ($q = +e$) is moving horizontally in the northward direction with speed $v = 10\,\text{Mm/s} = 10^7\,\text{m/s}$. Calculate the magnetic force on the proton (a) using $F = qvB \sin\theta$ and (b) by expressing $\vec{v}$ and $\vec{B}$ in terms of the unit vectors $\hat{i}, \hat{j}, \hat{k}$, and computing $\vec{F} = q\vec{v} \times \vec{B}$.

PICTURE THE PROBLEM Let the x and y directions be east and north, respectively, and let the z direction be upward (Figure 26-5). The velocity vector is then in the y direction.

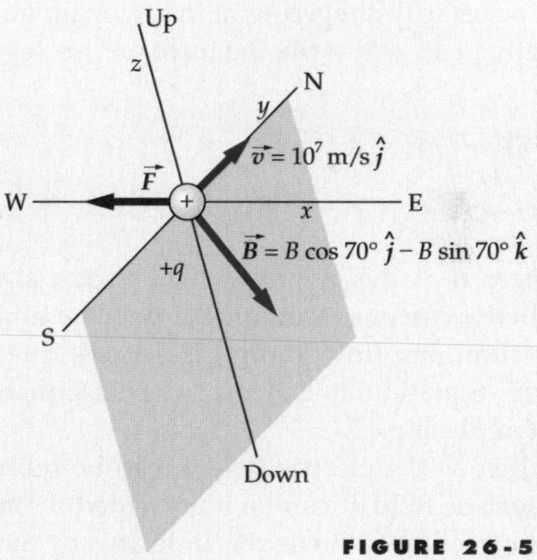

(a) Calculate $F = qvB \sin\theta$ using $\theta = 70°$. From Figure 26-4 we see that the direction of the force is westward:

$$F = qvB \sin 70°$$
$$= (1.6 \times 10^{-19}\,\text{C})(10^7\,\text{m/s})(0.6 \times 10^{-4}\,\text{T})(0.94)$$
$$= \boxed{9.02 \times 10^{-17}\,\text{N}}$$

(b) 1. The magnetic force is the vector product of $q\vec{v}$ and $\vec{B}$: $\vec{F} = q\vec{v} \times \vec{B}$

2. Express $\vec{v}$ and $\vec{B}$ in terms of their components: $\vec{v} = v_y\hat{j}$
$$\vec{B} = B_y\hat{j} + B_z\hat{k}$$

3. Write $\vec{F} = q\vec{v} \times \vec{B}$ in terms of these components: $\vec{F} = q\vec{v} \times \vec{B} = q(v_y\hat{j}) \times (B_y\hat{j} + B_z\hat{k})$
$$= qv_yB_y(\hat{j} \times \hat{j}) + qv_yB_z(\hat{j} \times \hat{k}) = 0 + qv_yB_z\hat{i}$$

4. Evaluate $\vec{F}$: $\vec{F} = qv\,(-B \sin\theta)\hat{i}$
$$= -(1.6 \times 10^{-19}\,\text{C})(10^7\,\text{m/s})(0.6 \times 10^{-4}\,\text{T})\sin 70°\,\hat{i}$$
$$= \boxed{-9.02 \times 10^{-17}\,\text{N}\hat{i}}$$

FIGURE 26-5

REMARKS Note that the direction of $\hat{i}$ is eastward, so the force is directed westward as shown in Figure 26-5.

EXERCISE Find the force on a proton moving with velocity $\vec{v} = 4 \times 10^6\,\text{m/s}\hat{i}$ in a magnetic field $\vec{B} = 2.0\,\text{T}\hat{k}$. (*Answer* $-1.28 \times 10^{-12}\,\text{N}\hat{j}$)

When a wire carries a current in a magnetic field, there is a force on the wire that is equal to the sum of the magnetic forces on the charged particles whose motion produces the current. Figure 26-6 shows a short segment of wire of cross-sectional area A and length L carrying a current I. If the wire is in a magnetic field $\vec{B}$, the magnetic force on each charge is $q\vec{v}_d \times \vec{B}$, where $\vec{v}_d$ is the drift velocity of the charge carriers (the drift velocity is the same as the average velocity). The number of charges in the wire segment is the number n per unit volume times the volume AL. Thus, the total force on the wire segment is

$$\vec{F} = (q\vec{v}_d \times \vec{B})nAL$$

From Equation 25-3, the current in the wire is

$$I = nqv_dA$$

Hence, the force can be written

$$\vec{F} = I\vec{L} \times \vec{B} \qquad \text{26-4}$$

MAGNETIC FORCE ON A SEGMENT OF CURRENT-CARRYING WIRE

where $\vec{L}$ is a vector whose magnitude is the length of the wire and whose direction is parallel to the current. For the current in the positive x direction (Figure 26-7) and the magnetic field vector at the segment in the xy plane, the force on the wire is directed along the z axis.

In Equation 26-4 it is assumed that the wire segment is straight and that the magnetic field does not vary over its length. The equation can be generalized for an arbitrarily shaped wire in any magnetic field. If we choose a very small wire segment $d\vec{\ell}$ and write the force on this segment as $d\vec{F}$, we have

$$d\vec{F} = Id\vec{\ell} \times \vec{B} \qquad \text{26-5}$$

MAGNETIC FORCE ON A CURRENT ELEMENT

where $\vec{B}$ is the magnetic field vector at the segment. The quantity $Id\vec{\ell}$ is called a **current element.** We find the total force on a current-carrying wire by summing (integrating) the forces due to all the current elements in the wire. Equation 26-5 is the same as Equation 26-1 with the current element $Id\vec{\ell}$ replacing $q\vec{v}$.

Just as the electric field $\vec{E}$ can be represented by electric field lines, the magnetic field $\vec{B}$ can be represented by **magnetic field lines.** In both cases, the direction of the field is indicated by the direction of the field lines and the magnitude of the field is indicated by their density. There are, however, two important differences between electric field lines and magnetic field lines:

1. Electric field lines are in the direction of the electric force on a positive charge, but the magnetic field lines are perpendicular to the magnetic force on a moving charge.

2. Electric field lines begin on positive charges and end on negative charges; magnetic field lines neither begin nor end.

Figure 26-8 shows the magnetic field lines both inside and outside a bar magnet.

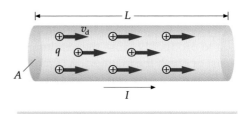

FIGURE 26-6 Wire segment of length L carrying current I. If the wire is in a magnetic field, there will be a force on each charge carrier resulting in a force on the wire.

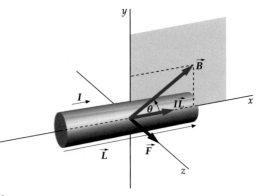

FIGURE 26-7 Magnetic force on a current-carrying segment of wire in a magnetic field. The current is in the x direction, and the magnetic field is in the xy plane and makes an angle θ with the $+x$ direction. The force $\vec{F}$ is in the $+z$ direction, perpendicular to both $\vec{B}$ and $\vec{L}$, and has magnitude $ILB \sin \theta$.

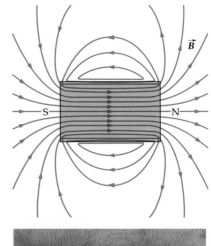

(a)

(b)

FIGURE 26-8 (a) Magnetic field lines inside and outside a bar magnet. The lines emerge from the north pole and enter the south pole, but they have no beginning or end. Instead, they form closed loops. (b) Magnetic field lines outside a bar magnet as indicated by iron filings.

FIGURE 26-9

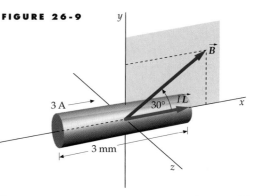

FORCE ON A STRAIGHT WIRE   **EXAMPLE   26-2**

A wire segment 3 mm long carries a current of 3 A in the $+x$ direction. It lies in a magnetic field of magnitude 0.02 T that is in the xy plane and makes an angle of 30° with the $+x$ direction, as shown in Figure 26-9. What is the magnetic force exerted on the wire segment?

PICTURE THE PROBLEM The magnetic force is in the direction of $\vec{L} \times \vec{B}$, which we see from Figure 26-9 is in the positive z direction.

The magnetic force is given by Equation 26-4:

$$\vec{F} = I\vec{L} \times \vec{B} = ILB \sin 30° \, \hat{k}$$

$$= (3 \text{ A})(0.003 \text{ m})(0.02 \text{ T})(\sin 30°)\hat{k}$$

$$= \boxed{9 \times 10^{-5} \text{ N}\hat{k}}$$

FIGURE 26-10

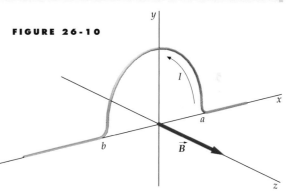

FORCE ON A BENT WIRE   **EXAMPLE   26-3**

A wire bent into a semicircular loop of radius R lies in the xy plane. It carries a current I from point a to point b, as shown in Figure 26-10. There is a uniform magnetic field $\vec{B} = B\hat{k}$ perpendicular to the plane of the loop. Find the force acting on the semicircular loop part of the wire.

PICTURE THE PROBLEM The force $d\vec{F}$ exerted on a segment of the semicircular wire lies in the xy plane, as shown in Figure 26-11. We find the total force by expressing the x and y components of $d\vec{F}$ in terms of θ and integrating them separately from $\theta = 0$ to $\theta = \pi$.

FIGURE 26-11

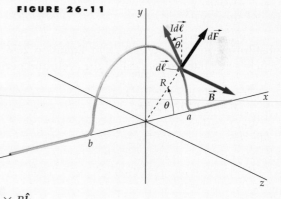

1. Write the force $d\vec{F}$ on a current element $d\vec{\ell}$.

$$d\vec{F} = I d\vec{\ell} \times \vec{B}$$

2. Express $d\vec{\ell}$ in terms of the unit vectors $\hat{i}$ and $\hat{j}$:

$$d\vec{\ell} = -d\ell \sin \theta \hat{i} + d\ell \cos \theta \hat{j}$$

3. Compute $I d\vec{\ell}$ using $d\ell = R d\theta$ and $\vec{B} = B\hat{k}$:

$$d\vec{F} = I d\vec{\ell} \times \vec{B}$$
$$= I(-R \sin \theta d\theta \hat{i} + R \cos \theta d\theta \hat{j}) \times B\hat{k}$$
$$= IRB \sin \theta d\theta \hat{j} + IRB \cos \theta d\theta \hat{i}$$

4. Integrate each component of $d\vec{F}$ from $\theta = 0$ to $\theta = \pi$:

$$\vec{F} = \int d\vec{F} = IRB\hat{i} \int_0^{\pi} \cos \theta d\theta + IRB\hat{j} \int_0^{\pi} \sin \theta d\theta$$
$$= IRB\hat{i} \, (0) + IRB\hat{j} \, (2) = \boxed{2IRB\hat{j}}$$

❶ PLAUSIBILITY CHECK The result that the x component of $\vec{F}$ is zero can be seen from symmetry. For the right half of the loop, $d\vec{F}$ points to the right; for the left half of the loop, $d\vec{F}$ points to the left.

REMARKS The net force on the semicircular wire is the same as if the semicircle were replaced by a straight-line segment of length $2R$ connecting points a and b. (This is a general result, as shown in Problem 30.)

26-2 Motion of a Point Charge in a Magnetic Field

The magnetic force on a charged particle moving through a magnetic field is always perpendicular to the velocity of the particle. The magnetic force thus changes the direction of the velocity but not the velocity's magnitude. Therefore, *magnetic fields do no work on particles and do not change their kinetic energy.*

In the special case where the velocity of a particle is perpendicular to a uniform magnetic field, as shown in Figure 26-12, the particle moves in a circular orbit. The magnetic force provides the centripetal force necessary for the centripetal acceleration v^2/r in circular motion. We can use Newton's second law to relate the radius of the circle to the magnetic field and the speed of the particle. If the velocity is $\vec{v}$, the magnitude of the net force is qvB, since $\vec{v}$ and $\vec{B}$ are perpendicular. Newton's second law gives

$$F = ma$$

$$qvB = m\frac{v^2}{r}$$

or

$$r = \frac{mv}{qB} \qquad\qquad 26\text{-}6$$

The period of the circular motion is the time it takes the particle to travel once around the circumference of the circle. The period is related to the speed by

$$T = \frac{2\pi r}{v}$$

Substituting in $r = mv/(qB)$ from Equation 26-6, we obtain the period of the particle's circular motion, called the **cyclotron period**:

$$T = \frac{2\pi(mv/qB)}{v} = \frac{2\pi m}{qB} \qquad\qquad 26\text{-}7$$

CYCLOTRON PERIOD

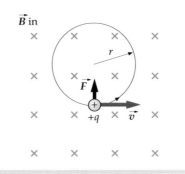

FIGURE 26-12 Charged particle moving in a plane perpendicular to a uniform magnetic field. The magnetic field is into the page as indicated by the crosses. (Each cross represents the tail feathers of an arrow. A field out of the plane of the page would be indicated by dots, each dot representing the point of an arrow.) The magnetic force is perpendicular to the velocity of the particle, causing it to move in a circular orbit.

(*a*) Circular path of electrons moving in the magnetic field produced by two large coils. The electrons ionize the gas in the tube, causing it to give off a bluish glow that indicates the path of the beam. (*b*) False-color photograph showing tracks of a 1.6-MeV proton (red) and a 7-MeV α particle (yellow) in a cloud chamber. The radius of curvature is proportional to the momentum and inversely proportional to the charge of the particle. For these energies, the momentum of the α particle, which has twice the charge of the proton, is about four times that of the proton and so its radius of curvature is greater.

(*b*)

(*a*)

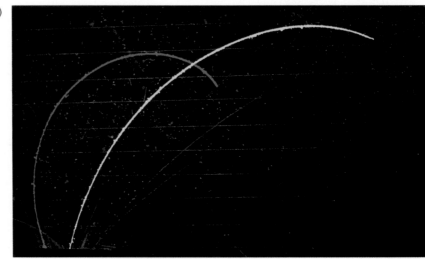

The frequency of the circular motion, called the **cyclotron frequency,** is the reciprocal of the period:

$$f = \frac{1}{T} = \frac{qB}{2\pi m}, \quad \text{so} \quad \omega = 2\pi f = \frac{q}{m}B$$ 26-8

CYCLOTRON FREQUENCY

Note that the period and the frequency given by Equations 26-7 and 26-8 depend on the charge-to-mass ratio q/m, but the period and the frequency are independent of the velocity v or the radius r. Two important applications of the circular motion of charged particles in a uniform magnetic field, the mass spectrometer and the cyclotron, are discussed later in this section.

CYCLOTRON PERIOD   **E X A M P L E   2 6 - 4**

A proton of mass $m = 1.67 \times 10^{-27}$ kg and charge $q = e = 1.6 \times 10^{-19}$ C moves in a circle of radius $r = 21$ cm perpendicular to a magnetic field $B = 4000$ G. Find (a) the period of the motion and (b) the speed of the proton.

1. Calculate the period T from Equation 26-7 with $B = 4000$ G $= 0.4$ T:

$$T = \frac{2\pi m}{qB} = \frac{2\pi(1.67 \times 10^{-27} \text{ kg})}{(1.6 \times 10^{-19} \text{ C})(0.4 \text{ T})}$$

$$= \boxed{1.64 \times 10^{-7} \text{ s} = 164 \text{ ns}}$$

2. Calculate the speed v from Equation 26-6:

$$v = \frac{rqB}{m} = \frac{(0.21 \text{ m})(1.6 \times 10^{-19} \text{ C})(0.4 \text{ T})}{1.67 \times 10^{-27} \text{ kg}}$$

$$= \boxed{8.05 \times 10^6 \text{ m/s} = 8.05 \text{ m/}\mu\text{s}}$$

MASTER the CONCEPT · WEB

REMARKS The radius of the circular motion is proportional to the speed, but the period is independent of both the speed and radius.

PLAUSIBILITY CHECK Note that the product of the speed v and the period T equals the circumference of the circle $2\pi r$ as expected:
$vT = (8.05 \times 10^6 \text{ m/s})(1.64 \times 10^{-7} \text{ s}) = 1.32 \text{ m}; 2\pi r = 2\pi(0.21 \text{ m}) = 1.32 \text{ m}.$

Suppose that a charged particle enters a uniform magnetic field with a velocity that is not perpendicular to $\vec{B}$. There is no force component, and thus no acceleration component, parallel to $\vec{B}$, so the component of the velocity parallel to $\vec{B}$ remains constant. The magnetic force on the particle is perpendicular to $\vec{B}$, so the change in motion of the particle due to this force is the same as that just discussed. The path of the particle is thus a helix, as shown in Figure 26-13.

(a)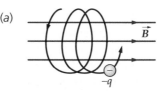

(b)

FIGURE 26-13 (a) When a particle has a velocity component parallel to a magnetic field as well as a velocity component perpendicular to the magnetic field the particle moves in a helical path around the field lines. (b) Cloud-chamber photograph of the helical path of an electron moving in a magnetic field. The path of the electron is made visible by the condensation of water droplets in the cloud chamber.

The motion of charged particles in nonuniform magnetic fields can be quite complex. Figure 26-14 shows a **magnetic bottle,** an interesting magnetic field configuration in which the field is weak at the center and strong at both ends. A detailed analysis of the motion of a charged particle in such a field shows that the particle spirals around the field lines and becomes trapped, oscillating back and forth between points P_1 and P_2 in the figure. Such magnetic field configurations are used to confine dense beams of charged particles, called *plasmas,* in nuclear fusion research. A similar phenomenon is the oscillation of ions back and forth between the earth's magnetic poles in the Van Allen belts (Figure 26-15).

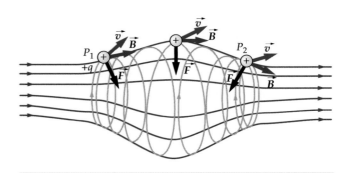

FIGURE 26-14 Magnetic bottle. When a charged particle moves in such a field, which is strong at both ends and weak in the middle, the particle becomes trapped and moves back and forth, spiraling around the field lines.

*The Velocity Selector

The magnetic force on a charged particle moving in a uniform magnetic field can be balanced by an electric force if the magnitudes and directions of the magnetic field and the electric field are properly chosen. Since the electric force is in the direction of the electric field (for positive particles) and the magnetic force is perpendicular to the magnetic field, the electric and magnetic fields in the region through which the particle is moving must be perpendicular to each other if the forces are to balance. Such a region is said to have **crossed fields.**

Figure 26-16 shows a region of space between the plates of a capacitor where there is an electric field and a perpendicular magnetic field (produced by a magnet with poles above and below the paper). Consider a particle of charge q entering this space from the left. The net force on the particle is

$$\vec{F} = q\vec{E} + q\vec{v} \times \vec{B}$$

If q is positive, the electric force of magnitude qE is down and the magnetic force of magnitude qvB is up. If the charge is negative, each of these forces is reversed. The two forces balance if $qE = qvB$ or

$$v = \frac{E}{B} \qquad\qquad 26\text{-}9$$

For given magnitudes of the electric and magnetic fields, the forces balance only for particles with the speed given by Equation 26-9. Any particle with this speed, regardless of its mass or charge, will traverse the space undeflected. A particle with a greater speed will be deflected toward the direction of the magnetic force, and a particle with less speed will be deflected in the direction of the electric force. This arrangement of fields is often used as a **velocity selector,** which is a device that allows only particles with speed, given by Equation 26-9, to pass.

EXERCISE A proton is moving in the x direction in a region of crossed fields where $\vec{E} = 2 \times 10^5$ N/C $\hat{k}$ and $\vec{B} = -3000$ G $\hat{j}$. (*a*) What is the speed of the proton if it is not deflected? (*b*) If the proton moves with twice this speed, in which direction will it be deflected? (*Answer* (*a*) 667 km/s (*b*) in the negative z direction)

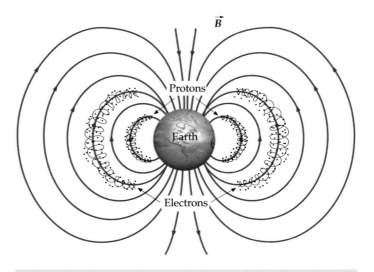

FIGURE 26-15 Van Allen belts. Protons (inner belts) and electrons (outer belts) are trapped in the earth's magnetic field and spiral around the field lines between the north and south poles.

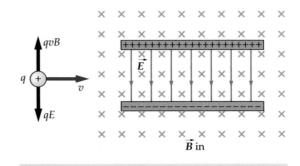

FIGURE 26-16 Crossed electric and magnetic fields. When a positive particle moves to the right, the particle experiences a downward electric force and an upward magnetic force. These forces balance if the speed of the particle is related to the field strengths by $vB = E$.

*Thomson's Measurement of q/m for Electrons

An example of the use of crossed electric and magnetic fields is the famous experiment performed by J. J. Thomson in 1897 where he showed that the rays of a cathode-ray tube can be deflected by electric and magnetic fields, indicating that they must consist of charged particles. By measuring the deflections of these particles Thomson showed that all the particles have the same charge-to-mass ratio q/m. He also showed that particles with this charge-to-mass ratio can be obtained using any material for a source, which means that these particles, now called electrons, are a fundamental constituent of all matter.

Figure 26-17 shows a schematic diagram of the cathode-ray tube Thomson used. Electrons are emitted from the cathode C, which is at a negative potential relative to the slits A and B. An electric field in the direction from A to C accelerates the electrons, and the electrons pass through slits A and B into a field-free region. The electrons then enter the electric field between the capacitor plates D and F that is perpendicular to the velocity of the electrons. This field accelerates the electrons vertically for the short time that they are between the plates. The electrons are deflected and strike the phosphorescent screen S at the far right side of the tube at some deflection Δy from the point at which they strike when there is no field between the plates. The screen glows where the electrons strike the screen, indicating the location of the beam. The initial speed of the electrons v_0 is determined by introducing a magnetic field $\vec{B}$ between the plates in a direction that is perpendicular to both the electric field and the initial velocity of the electrons. The magnitude of $\vec{B}$ is adjusted until the beam is not deflected. The speed is then found from Equation 26-9.

With the magnetic field turned off, the beam is deflected by an amount Δy, which consists of two parts: the deflection Δy_1, which occurs while the electrons are between the plates, and the deflection Δy_2, which occurs after the electrons leave the region between the plates (Figure 26-18).

Let x_1 be the horizontal distance across the deflection plates D and F. If the electron is moving horizontally with speed v_0 when it enters region between the plates, the time spent between the plates is $t_1 = x_1/v_0$, and the vertical velocity when it leaves the plates is

$$v_y = a_y t_1 = \frac{qE_y}{m} t_1 = \frac{qE_y}{m} \frac{x_1}{v_0}$$

where E_y is the upward component of the electric field between the plates. The deflection in this region is

$$\Delta y_1 = \frac{1}{2} a_y t_1^2 = \frac{1}{2} \frac{qE_y}{m} \left(\frac{x_1}{v_0}\right)^2$$

The electron then travels an additional horizontal distance x_2 in the field-free region from the deflection plates to the screen. Since the velocity of the electron is constant in this region, the time to reach the screen is $t_2 = x_2/v_0$, and the additional vertical deflection is

$$\Delta y_2 = v_y t_2 = \frac{qE_y}{m} \frac{x_1}{v_0} \frac{x_2}{v_0}$$

The total deflection at the screen is therefore

$$\Delta y = \Delta y_1 + \Delta y_2 = \frac{1}{2} \frac{qE_y}{mv_0^2} x_1^2 + \frac{qE_y}{mv_0^2} x_1 x_2 \qquad \text{26-10}$$

The measured deflection Δy can be used to determine the charge-to-mass ratio, q/m, from Equation 26-10.

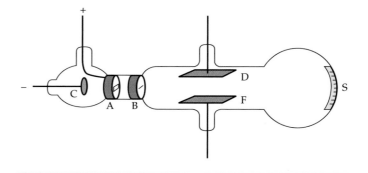

FIGURE 26-17 Thomson's tube for measuring q/m for the particles of cathode rays (electrons). Electrons from the cathode C pass through the slits at A and B and strike a phosphorescent screen S. The beam can be deflected by an electric field between plates D and F or by a magnetic field (not shown).

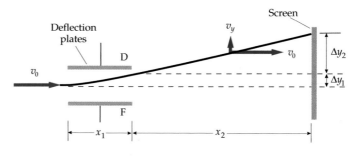

FIGURE 26-18 The total deflection of the beam in the J. J. Thomson experiments consists of the deflection Δy_1 while the electrons are between the plates plus the deflection Δy_2 that occurs in the field-free region between the plates and the screen.

ELECTRON BEAM DEFLECTION   **E X A M P L E   2 6 - 5**

Electrons pass undeflected through the plates of Thomson's apparatus when the electric field is 3000 V/m and there is a crossed magnetic field of 1.40 G. If the plates are 4-cm long and the ends of the plates are 30 cm from the screen, find the deflection on the screen when the magnetic field is turned off.

PICTURE THE PROBLEM The mass and charge of the electron are known: $m = 9.11 \times 10^{-31}$ kg and $q = -e = -1.6 \times 10^{-19}$ C. The speed of the electron can be found from the ratio of the magnetic and electric fields.

1. The total deflection of the electron is given by Equation 26-10:

$$\Delta y = \Delta y_1 + \Delta y_2 = \frac{1}{2}\frac{qE_y}{mv_0^2}x_1^2 + \frac{qE_y}{mv_0^2}x_1 x_2$$

2. The speed v_0 equals E/B:

$$v_0 = \frac{E}{B} = \frac{3000 \text{ V/m}}{1.40 \times 10^{-4} \text{ T}} = 2.14 \times 10^7 \text{ m/s}$$

3. Substitute this value for v_0, the given value of E, and the known values for m and q to find Δy:

$$\Delta y_1 = \frac{1}{2}\frac{(-1.6 \times 10^{-19}\text{ C})(-3000 \text{ V/m})}{(9.11 \times 10^{-31}\text{ kg})(2.14 \times 10^7 \text{ m/s})^2}(0.04 \text{ m})^2$$

$$= 9.20 \times 10^{-4} \text{ m}$$

$$\Delta y_2 = \frac{(-1.6 \times 10^{-19}\text{ C})(-3000 \text{ V/m})}{(9.11 \times 10^{-31}\text{ kg})(2.14 \times 10^7 \text{ m/s})^2}(0.04 \text{ m})(0.30 \text{ m})$$

$$= 1.38 \times 10^{-2} \text{ m}$$

$$\Delta y = \Delta y_1 + \Delta y_2$$

$$= 9.20 \times 10^{-4} \text{ m} + 1.38 \times 10^{-2} \text{ m}$$

$$= 0.92 \text{ mm} + 13.8 \text{ mm} = \boxed{14.7 \text{ mm}}$$

*The Mass Spectrometer

The **mass spectrometer,** first designed by Francis William Aston in 1919, was developed as a means of measuring the masses of isotopes. Such measurements are important in determining both the presence of isotopes and their abundance in nature. For example, natural magnesium has been found to consist of 78.7 percent ^{24}Mg, 10.1 percent ^{25}Mg, and 11.2 percent ^{26}Mg. These isotopes have masses in the approximate ratio 24:25:26.

Figure 26-19 shows a simple schematic drawing of a mass spectrometer. Positive ions are formed by bombarding neutral atoms with X rays or a beam of electrons. (Electrons are knocked out of the atoms by the X rays or bombarding electrons.) These ions are accelerated by an electric field and enter a uniform magnetic field. If the positive ions start from rest and move through a potential difference ΔV, the ions kinetic energy when they enter the magnetic field equals their loss in potential energy, $q|\Delta V|$:

$$\tfrac{1}{2} mv^2 = q|\Delta V| \qquad\qquad 26\text{-}11$$

The ions move in a semicircle of radius r given by Equation 26-6, $r = mv/qB$, and strike a photographic plate at point P_2, a distance $2r$ from the point P_1 where the ions entered the magnetic field.

The speed v can be eliminated from Equations 26-6 and 26-11 to find m/q in terms of the known quantities ΔV, B, and r. We first solve Equation 26-6 for v and square each term, which gives

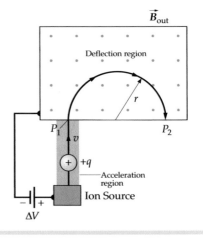

FIGURE 26-19 Schematic drawing of a mass spectrometer. Positive ions from an ion source are accelerated through a potential difference ΔV and enter a uniform magnetic field. The magnetic field is out of the plane of the page as indicated by the dots. The ions are bent into a circular arc and emerge at P_2. The radius of the circle varies with the mass of the ion.

$$v^2 = \frac{r^2 q^2 B^2}{m^2}$$

Substituting this expression for v^2 into Equation 26-11, we obtain

$$\frac{1}{2} m \left(\frac{r^2 q^2 B^2}{m^2} \right) = q |\Delta V|$$

Simplifying this equation and solving for m/q, we obtain

$$\frac{m}{q} = \frac{B^2 r^2}{2 |\Delta V|} \qquad\qquad 26\text{-}12$$

In Aston's original mass spectrometer, mass differences could be measured to a precision of about 1 part in 10,000. The precision has been improved by introducing a velocity selector between the ion source and the magnet, which increases the degree of accuracy with which the velocities of the incoming ions can be determined.

SEPARATING ISOTOPES OF NICKEL **EXAMPLE 26-6**

A ^{58}Ni ion of charge $+e$ and mass 9.62×10^{-26} kg is accelerated through a potential drop of 3 kV and deflected in a magnetic field of 0.12 T. (*a*) Find the radius of curvature of the orbit of the ion. (*b*) Find the difference in the radii of curvature of ^{58}Ni ions and ^{60}Ni ions. (Assume that the mass ratio is 58:60.)

PICTURE THE PROBLEM The radius of curvature r can be found using Equation 26-12. Using the mass dependence of r, we can find the radius for ^{60}Ni ions from the radius for ^{58}Ni ions and then take the difference.

(*a*) Solve Equation 26-12 for r:
$$r = \sqrt{\frac{2m\,|\Delta V|}{qB^2}} = \left[\frac{2(9.62 \times 10^{-26} \text{ kg})(3000 \text{ V})}{(1.6 \times 10^{-19} \text{ C})(0.12 \text{ T})^2} \right]^{1/2}$$

$$= \boxed{0.501 \text{ m}}$$

(*b*) 1. Let r_1 and r_2 be the radius of the orbit of the ^{58}Ni ion and the ^{60}Ni ion, respectively. Use the result in Part (*a*) to find the ratio of r_2 to r_1:
$$\frac{r_2}{r_1} = \sqrt{\frac{m_2}{m_1}} = \sqrt{\frac{60}{58}} = 1.017$$

2. Use the result of the previous step to calculate r_2 for ^{60}Ni:
$$r_2 = 1.017\, r_1 = (1.017)(0.501 \text{ m}) = 0.510 \text{ m}$$

3. The difference in orbital radii is $r_2 - r_1$:
$$r_2 - r_1 = 0.510 \text{ m} - 0.501 \text{ m} = \boxed{9 \text{ mm}}$$

The Cyclotron

The cyclotron was invented by E. O. Lawrence and M. S. Livingston in 1934 to accelerate particles, such as protons or deuterons, to high kinetic energies.† The high-energy particles are used to bombard atomic nuclei, causing nuclear reactions that are then studied to obtain information about the nucleus. High-energy protons and deuterons are also used to produce radioactive materials and for medical purposes.

---

† A deuteron is the nucleus of heavy hydrogen, ^{2}H, which consists of a proton and neutron tightly bound together.

Figure 26-20 is a schematic drawing of a cyclotron. The particles move in two semicircular metal containers called *dees*, after their shape. The dees are housed in a vacuum chamber that is in a uniform magnetic field provided by an electromagnet. The region in which the particles move must be evacuated so that the particles will not be scattered in collisions with air molecules and lose energy. A potential difference ΔV, which alternates in time with a period T, is maintained between the dees. The period is chosen to be the cyclotron period $T = 2\pi m/(qB)$ (Equation 26-7). The potential difference creates an electric field across the gap between the dees. At the same time, there is no electric field within each dee because the metal dees act as shields.

Positively charged particles are initially injected into dee_1 with a small velocity from an ion source S near the center of the dees. They move in a semicircle in dee_1 and arrive at the gap between dee_1 and dee_2 after a time $\frac{1}{2}T$. The potential is adjusted so that dee_1 is at a higher potential than dee_2 when the particles arrive at the gap between them. Each particle is therefore accelerated across the gap by the electric field and gains kinetic energy equal to $q\,\Delta V$.

Because the particle now has more kinetic energy, the particle moves in a semicircle of larger radius in dee_2. It arrives at the gap again after a time $\frac{1}{2}T$, because the period is independent of the particle's speed. By this time, the potential between the dees has been reversed so that dee_2 is now at the higher potential. Once more the particle is accelerated across the gap and gains additional kinetic energy equal to $q\,\Delta V$. Each time the particle arrives at the gap, it is accelerated and gains kinetic energy equal to $q\,\Delta V$. Thus, the particle moves in larger and larger semicircular orbits until it eventually leaves the magnetic field. In the typical cyclotron, each particle may make 50 to 100 revolutions and exit with energies of up to several hundred mega-electron volts.

The kinetic energy of a particle leaving a cyclotron can be calculated by setting r in Equation 26-6 equal to the maximum radius of the dees and solving the equation for v:

$$r = \frac{mv}{qB}, \quad v = \frac{qBr}{m}$$

Then

$$K = \frac{1}{2}mv^2 = \frac{1}{2}\left(\frac{q^2B^2}{m}\right)r^2 \qquad\qquad 26\text{-}13$$

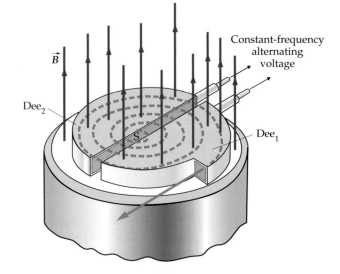

FIGURE 26-20 Schematic drawing of a cyclotron. The upper-pole face of the magnet has been omitted. Charged particles, such as protons, are accelerated from a source at the center by the potential difference across the gap between the dees. When the charged particles arrive at the gap again the potential difference has changed sign so they are again accelerated across the gap and move in a larger circle. The potential difference across the gap alternates with the cyclotron frequency of the particle, which is independent of the radius of the circle.

ENERGY OF ACCELERATED PROTON   **E X A M P L E   2 6 - 7**

A cyclotron for accelerating protons has a magnetic field of 1.5 T and a maximum radius of 0.5 m. (*a*) What is the cyclotron frequency? (*b*) What is the kinetic energy of the protons when they emerge?

(*a*) The cyclotron frequency is given by Equation 26-8:

$$f = \frac{qB}{2\pi m} = \frac{(1.6 \times 10^{-19}\,\text{C})(1.5\,\text{T})}{2\pi(1.67 \times 10^{-27}\,\text{kg})} = 2.29 \times 10^7\,\text{Hz}$$

$$= \boxed{22.9\ \text{MHz}}$$

(b) 1. The kinetic energy of the emerging protons is given by Equation 26-13:

$$K = \frac{1}{2}\left[\frac{(1.6 \times 10^{-19}\text{ C})^2(1.5\text{ T})^2}{1.67 \times 10^{-27}\text{ kg}}\right](0.5\text{ m})^2$$

$$= 4.31 \times 10^{-12}\text{ J}$$

2. The energies of protons and other elementary particles are usually expressed in electron volts. Use $1\text{ eV} = 1.6 \times 10^{-19}\text{ J}$ to convert to eV:

$$K = 4.31 \times 10^{-12}\text{ J} \times \frac{1\text{ eV}}{1.6 \times 10^{-19}\text{ J}} = \boxed{26.9\text{ MeV}}$$

26-3 Torques on Current Loops and Magnets

A current-carrying loop experiences no net force in a uniform magnetic field, but it does experience a torque that tends to twist the current-carrying loop. The orientation of the loop can be described conveniently by a unit vector $\hat{n}$ that is perpendicular to the plane of the loop, as illustrated in Figure 26-21. If the fingers of the right hand curl around the loop in the direction of the current, the thumb points in the direction of $\hat{n}$.

Figure 26-22 shows the forces exerted by a uniform magnetic field on a rectangular loop whose normal unit vector $\hat{n}$ makes an angle θ with the magnetic field $\vec{B}$. The net force on the loop is zero. The forces $\vec{F}_1$ and $\vec{F}_2$ have the magnitude

$$F_1 = F_2 = IaB$$

These forces form a couple, so the torque is the same about any point. Point P in Figure 26-22 is a convenient point about which to compute the torque. The magnitude of the torque is

$$\tau = F_2 b \sin\theta = IaBb \sin\theta = IAB \sin\theta$$

where $A = ab$ is the area of the loop. For a loop with N turns, the torque has the magnitude

$$\tau = NIAB \sin\theta$$

This torque tends to twist the loop so that $\hat{n}$ is in the same direction as $\vec{B}$ (i.e., so that its plane is perpendicular to $\vec{B}$).

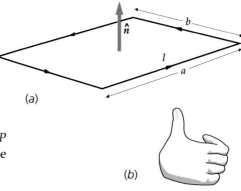

(a)

(b)

FIGURE 26-21 (*a*) The orientation of a current loop is described by the unit vector $\hat{n}$ perpendicular to the plane of the loop. (*b*) Right-hand rule for determining the direction of $\hat{n}$. If the fingers of the right hand curl around the loop in the direction of the current, the thumb points in the direction of $\hat{n}$.

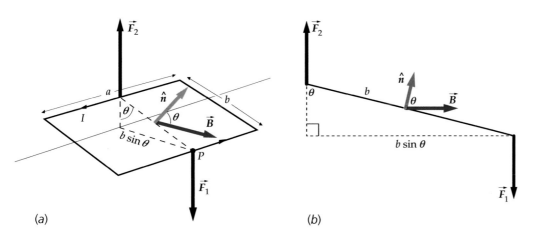

(a)

(b)

FIGURE 26-22 (*a*) Rectangular current loop whose unit normal $\hat{n}$ makes an angle θ with a uniform magnetic field $\vec{B}$. (*b*) An edge-on view of the current loop. The torque on the loop has magnitude $IAB \sin\theta$ and is in the direction such that $\hat{n}$ tends to rotate into $\vec{B}$.

The torque can be written conveniently in terms of the **magnetic dipole moment** $\vec{\mu}$ (also referred to simply as the **magnetic moment**) of the current loop, which is defined as

$$\vec{\mu} = NIA\hat{n} \qquad\qquad 26\text{-}14$$

MAGNETIC DIPOLE MOMENT OF A CURRENT LOOP

The SI unit of magnetic moment is the ampere-meter² (A·m²). In terms of the magnetic dipole moment, the torque on the current loop is given by

$$\vec{\tau} = \vec{\mu} \times \vec{B} \qquad\qquad 26\text{-}15$$

TORQUE ON A CURRENT LOOP

Equation 26-15, which we have derived for a rectangular loop, holds in general for a loop of any shape that is in a single plane. The torque on any loop is the cross product of the magnetic moment $\vec{\mu}$ of the loop and the magnetic field $\vec{B}$, where the magnetic moment is defined as a vector that is perpendicular to the plane of the loop (Figure 26-23), has magnitude equal to NIA, and has the same direction as $\hat{n}$. Comparing Equation 26-15 with Equation 21-11 ($\vec{\tau} = \vec{p} \times \vec{E}$) for the torque on an electric dipole, we see that the expression for the torque on a current loop in a magnetic field has the same form as that for the torque on an electric dipole in an electric field.

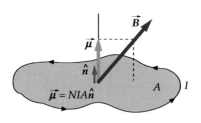

FIGURE 26-23 A flat current loop of arbitrary shape is described by its magnetic moment $\vec{\mu} = NIA\hat{n}$. In a magnetic field $\vec{B}$, the loop experiences a torque $\vec{\mu} \times \vec{B}$.

TORQUE ON A CURRENT LOOP   **EXAMPLE 26-8**

A circular loop of radius 2 cm has 10 turns of wire and carries a current of 3 A. The axis of the loop makes an angle of 30° with a magnetic field of 8000 G. Find the magnitude of the torque on the loop.

The magnitude of the torque is given by Equation 26-15:

$$\tau = |\vec{\mu} \times \vec{B}| = \mu B \sin\theta = NIAB \sin\theta$$

$$= (10)(3\text{ A})\pi(0.02\text{ m})^2(0.8\text{ T})\sin 30°$$

$$= \boxed{1.51 \times 10^{-2}\text{ N·m}^{-2}}$$

TILTING A LOOP   **EXAMPLE 26-9** **Try It Yourself**

A circular wire loop of radius R, mass m, and current I lies on a horizontal surface (Figure 26-24). There is a horizontal magnetic field $\vec{B}$. How large can the current I be before one edge of the loop will lift off the surface?

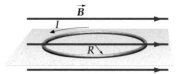

FIGURE 26-24

PICTURE THE PROBLEM The loop (Figure 26-25) will start to rotate when the magnitude of the net torque on the loop is not zero. To eliminate the torque due to the normal force, we calculate torques about the point of contact between the surface and the loop. The magnetic torque is given by $\vec{\tau} = \vec{\mu} \times \vec{B}$. The magnetic torque is the same about any point since the magnetic torque consists of couples. The lever arm for the gravitational torque is the radius of the loop.

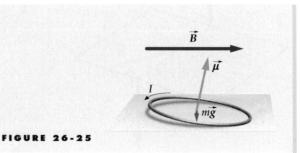

FIGURE 26-25

Cover the column to the right and try these on your own before looking at the answers.

Steps	Answers
1. Find the magnitude of the magnetic torque acting on the loop.	$\tau_m = \mu B = I\pi R^2 B$
2. Find the magnitude of the gravitational torque exerted on the loop.	$\tau_g = mgR$
3. Equate the magnitudes of the torques and solve for the current I.	$I = \dfrac{mg}{\pi RB}$

REMARKS The torque vectors are equal and opposite.

Potential Energy of a Magnetic Dipole in a Magnetic Field

When a torque is exerted through an angle, work is done. When a dipole is rotated through an angle $d\theta$, the work done is

$$dW = -\tau d\theta = -\mu B \sin\theta d\theta$$

The minus sign arises because the torque tends to decrease θ. Setting this work equal to the decrease in potential energy, we have

$$dU = -dW = +\mu B \sin\theta d\theta$$

Integrating, we obtain

$$U = -\mu B \cos\theta + U_0$$

We choose the potential energy to be zero when $\theta = 90°$. Then $U_0 = 0$ and the potential energy of the dipole is

$$U = -\mu B \cos\theta = -\vec{\mu} \cdot \vec{B} \qquad\qquad 26\text{-}16$$

POTENTIAL ENERGY OF A MAGNETIC DIPOLE

Equation 26-16 gives the potential energy of a magnetic dipole at an angle θ to the direction of a magnetic field.

TORQUE ON A COIL   **EXAMPLE   26-10**

A square 12-turn coil with edge-length 40 cm carries a current of 3 A. It lies in the xy plane, as shown in a uniform magnetic field $\vec{B} = 0.3$ T $\hat{i} + 0.4$ T $\hat{k}$. Find (a) the magnetic moment of the coil and (b) the torque exerted on the coil. (c) Find the potential energy of the coil.

PICTURE THE PROBLEM From Figure 26-26, we see that the magnetic moment of the loop is in the positive z direction.

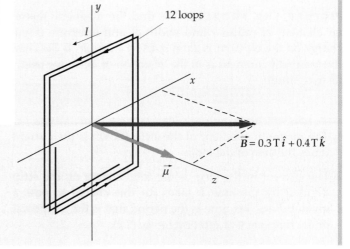

FIGURE 26-26

(a) Calculate the magnetic moment of the loop:

$$\vec{\mu} = NIA\,\hat{k} = (12)(3\text{ A})(0.40\text{ m})^2\,\hat{k}$$

$$= \boxed{5.76\text{ A·m}^2\,\hat{k}}$$

(b) The torque on the current loop is given by Equation 26-15:

$$\vec{\tau} = \vec{\mu} \times \vec{B}$$

$$= (5.76\text{ A·m}^2\,\hat{k}) \times (0.3\text{ T }\hat{i} + 0.4\text{ T }\hat{k})$$

$$= \boxed{1.73\text{ N·m }\hat{j}}$$

(c) The potential energy is the negative dot product of $\vec{\mu}$ and $\vec{B}$:

$$U = -\vec{\mu} \cdot \vec{B}$$

$$= -(5.76\text{ A·m}^2\,\hat{k}) \cdot (0.3\text{ T }\hat{i} + 0.4\text{ T }\hat{k})$$

$$= \boxed{-2.30\text{ J}}$$

REMARKS We have used $\hat{k} \times \hat{k} = 0$ and $\hat{k} \times \hat{i} = \hat{j}$, $\hat{k} \cdot \hat{i} = 0$ and $\hat{k} \cdot \hat{k} = 1$. The torque is in the y direction.

EXERCISE Calculate U if $\vec{B}$ and the magnetic moment $\vec{\mu}$ are in the same direction. (*Answer* $U = -\mu B = -(5.76\text{ A·m}^2)(0.5\text{ T}) = -2.88$ J. Note that this potential energy is lower than that found in the example. The potential energy is lowest when $\vec{\mu}$ and $\vec{B}$ are in the same direction.)

When a small permanent magnet, such as a compass needle, is placed in a magnetic field $\vec{B}$, the field exerts a torque on the magnet that tends to rotate the magnet so that it lines up with the field. This effect also occurs with previously unmagnetized iron filings, which become magnetized in the presence of a $\vec{B}$ field. The bar magnet is characterized by a magnetic moment $\vec{\mu}$, a vector that points in the same direction as an arrow drawn from the south pole to the north pole. A small bar magnet thus behaves like a current loop. This is not a coincidence. The origin of the magnetic moment of a bar magnet is, in fact, microscopic current loops that result from the motion of electrons in the atoms of the magnet.

$\vec{\mu}$ OF A ROTATING DISK　　　　　　　**EXAMPLE　26-11**

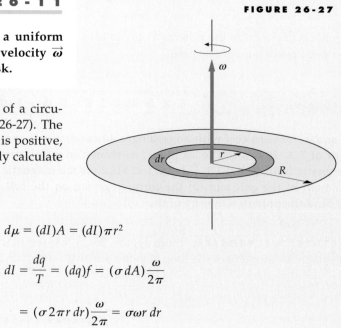

FIGURE 26-27

A thin nonconducting disk of mass m and radius R has a uniform surface charge per unit area σ and rotates with angular velocity $\vec{\omega}$ about its axis. Find the magnetic moment of the rotating disk.

PICTURE THE PROBLEM We find the magnetic moment of a circular element of radius r and width dr and integrate (Figure 26-27). The charge on the element is $dq = \sigma\,dA = \sigma 2\pi r\,dr$. If the charge is positive, the magnetic moment is in the direction of $\vec{\omega}$, so we need only calculate its magnitude.

1. The magnetic moment of the strip shown is the current times the area of the loop:

$$d\mu = (dI)A = (dI)\pi r^2$$

2. The current in the strip is the total charge on the strip divided by the time it takes for this charge to pass a given point. This time is the period that is the reciprocal of the frequency of rotation $f = \omega/(2\pi)$:

$$dI = \frac{dq}{T} = (dq)f = (\sigma\,dA)\frac{\omega}{2\pi}$$

$$= (\sigma 2\pi r\,dr)\frac{\omega}{2\pi} = \sigma\omega r\,dr$$

3. Substitute to obtain the magnetic moment of the strip $d\mu$ in terms of r and dr:

$$d\mu = (dI)\pi r^2 = (\sigma \omega r\, dr)\pi r^2 = \pi\sigma\omega r^3\, dr$$

4. Integrate from $r = 0$ to $r = R$:

$$\mu = \int d\mu = \int_0^R \pi\sigma\omega r^3\, dr = \frac{1}{4}\pi\sigma\omega R^4$$

5. Use the fact that $\vec{\mu}$ is parallel to $\vec{\omega}$ if σ is positive to write the magnetic moment as a vector:

$$\boxed{\vec{\mu} = \tfrac{1}{4}\pi\sigma R^4\,\vec{\omega}}$$

REMARKS In terms of the total charge $Q = \sigma\pi R^2$, the magnetic moment is $\vec{\mu} = \frac{1}{4}QR^2\vec{\omega}$. The angular momentum of the disk is $\vec{L} = (\frac{1}{2}mR^2)\vec{\omega}$, so the magnetic moment can be written $\vec{\mu} = [Q/(2m)]\vec{L}$, which is a more general result. (See Problem 63.)

26-4 The Hall Effect

As we have seen, charges moving in a magnetic field experience a force perpendicular to their motion. When these charges are traveling in a conducting wire, they will be pushed to one side of the wire. This results in a separation of charge in the wire called the **Hall effect.** This phenomenon allows us to determine the sign of the charge on the charge carriers and the number of charge carriers per unit volume n in a conductor. The Hall effect also provides a convenient method for measuring magnetic fields.

Figure 26-28 shows two conducting strips; each conducting strip carries a current I to the right because the left sides of the strips are connected to the positive terminal of a battery and the right sides are connected to the negative terminal. The strips are in a magnetic field that is directed into the paper. Let us assume for the moment that the current in the strip consists of positively charged particles moving to the right, as shown in Figure 26-28a. The magnetic force on these particles is $q\vec{v}_d \times \vec{B}$ (where $\vec{v}_d$ is the drift velocity of the charge carriers). This force is directed upward. The positive particles therefore move up to the top of the strip, leaving the bottom of the strip with an excess negative charge. This separation of charge produces an electric field in the strip that opposes the magnetic force on the charge carriers. When the electric forces and magnetic forces balance, the charge carriers no longer move upward. Since the electric field points in the direction of decreasing potential, the upper part of the strip is at a higher potential than is the lower part of the strip. This potential difference can be measured using a sensitive voltmeter. On the other hand, if the current consists of moving negatively charged particles, as shown in Figure 26-28b, the charge carriers in the strip must move to the left (since the current is still to the right). The magnetic force $q\vec{v}_d \times \vec{B}$ is again up, because the signs of both q and $\vec{v}_d$ have been reversed. Again the carriers are forced to the upper part of the strip, but the upper part of the strip now carries a negative charge (because the charge carriers are negative) and the lower part of the strip now carries a positive charge.

A measurement of the sign of the potential difference between the upper and lower parts of the strip tells us the sign of the charge carriers. In semiconductors, the charge carriers may be negative electrons or positive holes. A measurement of the sign of the potential difference tells us which are dominant for a particular semiconductor. For a normal metallic conductor, we find that the upper part of the strip in Figure 26-28b is at a lower potential than is the lower part of the strip—which means that the upper part must carry a negative charge. Thus, Figure 26-28b is the correct illustration of the current in a normal metallic conductor. It was this type of experiment that led to the discovery that the charge carriers in metallic conductors are negative.

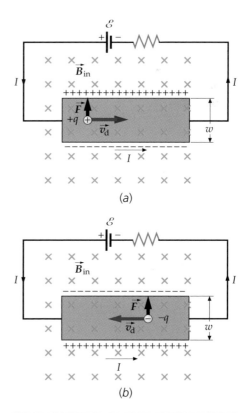

FIGURE 26-28 The Hall effect. The magnetic field is directed into the plane of the page as indicated by the crosses. The magnetic force on a charged particle is upward for a current to the right whether the current is due to (a) positive particles moving to the right or (b) negative particles moving to the left.

The potential difference between the top of the strip and the bottom of the strip is called the **Hall voltage.** We can calculate the magnitude of the Hall voltage in terms of the drift velocity. The magnitude of the magnetic force on the charge carriers in the strip is qv_dB. This magnetic force is balanced by the electrostatic force of magnitude qE_H, where E_H is the electric field due to the charge separation. Thus, we have $E_H = v_dB$. If the width of the strip is w, the potential difference is E_Hw. The Hall voltage is therefore

$$V_H = E_Hw = v_dBw \qquad\qquad 26\text{-}17$$

EXERCISE A conducting strip of width $w = 2.0$ cm is placed in a magnetic field of 0.8 T. The Hall voltage is measured to be 0.64 μV. Calculate the drift velocity of the electrons. (*Answer* 4.0×10^{-5} m/s)

Since the drift velocity for ordinary currents is very small, we can see from Equation 26-17 that the Hall voltage is very small for ordinary-sized strips and magnetic fields. From measurements of the Hall voltage for a strip of a given size, we can determine the number of charge carriers per unit volume in the strip. The current is given by Equation 25-3:

$$I = nqv_dA$$

where A is the cross-sectional area of the strip. For a strip of width w and thickness t, the cross-sectional area is $A = wt$. Since the charge carriers are electrons, the quantity q is the charge on one electron e. The number density of charge carriers n is thus given by

$$n = \frac{I}{Aqv_d} = \frac{I}{wtev_d} \qquad\qquad 26\text{-}18$$

Substituting V_H/B for v_dw (Equation 26-17), we have

$$n = \frac{IB}{teV_H} \qquad\qquad 26\text{-}19$$

CHARGE CARRIER NUMBER DENSITY IN SILVER   **EXAMPLE 26-12**

A silver slab of thickness 1 mm and width 1.5 cm carries a current of 2.5 A in a region in which there is a magnetic field of magnitude 1.25 T perpendicular to the slab. The Hall voltage is measured to be 0.334 μV. (*a*) Calculate the number density of the charge carriers. (*b*) Compare your answer in step 1 to the number density of atoms in silver, which has a mass density of $\rho = 10.5$ g/cm^3 and a molar mass of $M = 107.9$ g/mol.

1. Substitute numerical values into Equation 26-19 to find n:
$$n = \frac{IB}{teV_H} = \frac{(2.5\text{ A})(1.25\text{ T})}{(0.001\text{ m})(1.6 \times 10^{-19}\text{ C})(3.34 \times 10^{-7}\text{ V})}$$

$$= \boxed{5.85 \times 10^{28}\text{ electrons/m}^3}$$

2. The number of atoms per unit volume is $\rho N_A/M$:
$$n_a = \rho\frac{N_A}{M} = (10.5\text{ g/cm}^3)\frac{6.02 \times 10^{23}\text{ atoms/mol}}{107.9\text{ g/mol}}$$

$$= \boxed{5.86 \times 10^{22}\text{ atoms/cm}^3 = 5.86 \times 10^{28}\text{ atoms/m}^3}$$

REMARKS These results indicate that the number of charge carriers in silver is very nearly one per atom.

The Hall voltage provides a convenient method for measuring magnetic fields. If we rearrange Equation 26-19, we can write for the Hall voltage

$$V_{\text{H}} = \frac{I}{nte}B \qquad\qquad\qquad 26\text{-}20$$

A given strip can be calibrated by measuring the Hall voltage for a given current in a known magnetic field. The strip can then be used to measure an unknown magnetic field B by measuring the Hall voltage for a given current.

*The Quantum Hall Effects

According to Equation 26-20, the Hall voltage should increase linearly with magnetic field B for a given current in a given slab. In 1980, while studying the Hall effect in semiconductors at very low temperatures and very large magnetic fields, the German physicist Klaus von Klitzing discovered that a plot of V_{H} versus B resulted in a series of plateaus, as shown in Figure 26-29, rather than a straight line. That is, the Hall voltage is quantized. For the discovery of the integer quantum Hall effect, von Klitzing won the Nobel Prize in physics in 1985.

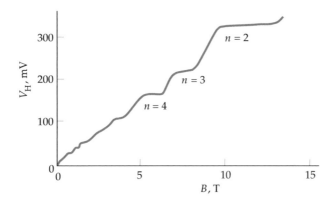

FIGURE 26-29 A plot of the Hall voltage versus applied magnetic field shows plateaus, indicating that the Hall voltage is quantized. These data were taken at a temperature of 1.39 K with the current I held fixed at 25.52 μA.

In the theory of the integer quantum Hall effect, the Hall resistance, defined as $R_{\text{H}} = V_{\text{H}}/I$, can take on only the values

$$R_{\text{H}} = \frac{V_{\text{H}}}{I} = \frac{R_{\text{K}}}{n}, \quad n = 1, 2, 3, \ldots \qquad\qquad 26\text{-}21$$

where n is an integer, and R_{K}, called the **von Klitzing constant,** is related to the fundamental electronic charge e and Planck's constant h by

$$R_{\text{K}} = \frac{h}{e^2} \qquad\qquad\qquad 26\text{-}22$$

Because the von Klitzing constant can be measured to an accuracy of a few parts per billion, the quantum Hall effect is now used to define a standard of resistance. As of January 1990, the **ohm** is defined in terms of the conventional value[†] of the von Klitzing constant $R_{\text{K}-90}$, which has the value

$$R_{\text{K}-90} = 25{,}812.807 \ \Omega \ (\text{exact}) \qquad\qquad 26\text{-}23$$

In 1982 it was observed that under certain special conditions the Hall resistance is given by Equation 26-22, but with the integer n replaced by a series of rational fractions. This is called the fractional quantum Hall effect. For the discovery and explanation of the fractional quantum Hall effect, American professors Laughlin, Stormer, and Tsui won the Nobel Prize in physics in 1998.

---

† The value of $R_{\text{K}-90}$ differs only slightly from that of R_{K}. The currently used value of the von Klitzing constant is $R_{\text{K}} = (25\,812.807\,572 \pm 0.000\,095)\ \Omega$

SUMMARY

1. The magnetic field describes the condition in space in which moving charges experience a force perpendicular to their velocity.

2. The magnetic force is part of the electromagnetic force, one of the four fundamental forces of nature.

3. The magnitude and direction of a magnetic field $\vec{B}$ are defined by the force $\vec{F} = q\vec{v} \times \vec{B}$ exerted on moving charges.

Topic	Relevant Equations and Remarks	
1. Magnetic Force		
On a moving charge	$\vec{F} = q\vec{v} \times \vec{B}$	26-1
On a current element	$d\vec{F} = Id\vec{\ell} \times \vec{B}$	26-5
Unit of the magnetic field	The SI unit of magnetic fields is the tesla (T). A commonly used unit is the gauss (G), which is related to the tesla by	
	$1\,G = 10^{-4}\,T$	26-3
2. Motion of Point Charges	A particle of mass m and charge q moving with speed v in a plane perpendicular to a uniform magnetic field moves in a circular orbit. The period and frequency of this circular motion are independent of the radius of the orbit and of the speed of the particle.	
Newton's second law	$qvB = m\dfrac{v^2}{r}$	26-6
Cyclotron period	$T = \dfrac{2\pi m}{qB}$	26-7
Cyclotron frequency	$f = \dfrac{1}{T} = \dfrac{qB}{2\pi m}$	26-8
*Velocity selector	A velocity selector consists of crossed electric and magnetic fields so that the electric and magnetic forces balance for a particle moving with speed v.	
	$E = vB$	26-9
*Thomson's measurement of q/m	The deflection of a charged particle in an electric field depends on the speed of the particle and is proportional to the charge-to-mass ratio q/m of the particle. J. J. Thomson used crossed electric and magnetic fields to measure the speed of cathode rays and then measured q/m for these particles by deflecting them in an electric field. He showed that all cathode rays consist of particles that all have the same charge-to-mass ratio. These particles are now called electrons.	
*Mass spectrometer	The mass-to-charge ratio of an ion of known speed can be determined by measuring the radius of the circular path taken by the ion in a known magnetic field.	
3. Current Loops		
Magnetic dipole moment	$\vec{\mu} = NIA\hat{n}$	26-14
Torque	$\vec{\tau} = \vec{\mu} \times \vec{B}$	26-15

Potential energy of a magnetic dipole	$U = -\vec{\mu} \cdot \vec{B}$	**26-16**

Net force	The net force on a current loop in a *uniform* magnetic field is zero.

4. The Hall Effect
When a conducting strip carrying a current is placed in a magnetic field, the magnetic force on the charge carriers causes a separation of charge called the Hall effect. This results in a voltage V_H, called the Hall voltage. The sign of the charge carriers can be determined from a measurement of the sign of the Hall voltage, and the number of carriers per unit volume can be determined from the magnitude of V_H.

Hall voltage	$V_H = E_H w = v_d B w = \dfrac{I}{nte} B$	**26-17, 26-20**

*Quantum Hall effects
Measurements at very low temperatures in very large magnetic fields indicate that the Hall resistance $R_H = V_H / I$ is quantized and can take on only the values given by

$$R_H = \frac{V_H}{I} = \frac{R_K}{n}, \quad n = 1, 2, 3, \ldots \qquad \textbf{26-21}$$

*Conventional von Klitzing constant (definition of ohm)	$R_{K-90} = 25{,}812.807 \ \Omega$ (exact)	**26-23**

PROBLEMS

- Single-concept, single-step, relatively easy
- •• Intermediate-level, may require synthesis of concepts
- ••• Challenging
- SSM Solution is in the *Student Solutions Manual*
- iSOLVE Problems available on iSOLVE online homework service
- iSOLVE ✓ These "Checkpoint" online homework service problems ask students additional questions about their confidence level, and how they arrived at their answer.

In a few problems, you are given more data than you actually need; in a few other problems, you are required to supply data from your general knowledge, outside sources, or informed estimates.

Conceptual Problems

1 • SSM When a cathode-ray tube is placed horizontally in a magnetic field that is directed vertically upward, the electrons emitted from the cathode follow one of the dashed paths to the face of the tube in Figure 26-30. The correct path is (a) 1. (b) 2. (c) 3. (d) 4. (e) 5.

FIGURE 26-30
Problem 1

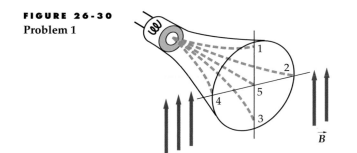

2 • Why not define $\vec{B}$ to be in the direction of $\vec{F}$, as we do for $\vec{E}$?

3 • True or false: The magnetic force does not accelerate a charged particle because the magnetic force is perpendicular to the velocity of the particle.

4 • A beam of positively charged particles passes undeflected from left to right through a velocity selector in which the electric field is up. The beam is then reversed so that it travels from right to left. Will the beam now be deflected in the velocity selector? If so, in which direction?

5 • SSM A *flicker bulb* is a lightbulb with a long, thin filament. When it is plugged in and a magnet is brought near the lightbulb, the filament is seen to oscillate rapidly back and forth. Why does the filament oscillate, and what is the frequency of oscillation?

6 • What orientation of a current loop relative to the direction of the magnetic field gives maximum torque?

7 • True or false:

(a) The magnetic force on a moving charged particle is always perpendicular to the velocity of the particle.

(b) The torque on a magnet by a magnetic field tends to align the magnet's magnetic moment in the direction of the magnetic field.

(c) A current loop in a uniform magnetic field responds to the field in the same manner as a small permanent magnet.

(d) The period of a particle moving in a circle in a magnetic field is proportional to the radius of the circle.

(e) The drift velocity of electrons in a wire can be determined from the Hall effect.

8 • SSM The north-seeking pole of a compass needle located on the magnetic equator is the end of the needle that points toward the north, and the direction of any magnetic field $\vec{B}$ is specified as the direction that the north-seeking pole of a compass needle points when the needle is aligned in the field. Suppose that the direction of the magnetic field $\vec{B}$ were instead specified as the direction of a south-seeking pole of a compass needle aligned in the field. Would the right-hand rule shown in Figure 26-2 then give the direction of the magnetic force on the moving positive charge, or would a left-hand rule be required? Explain.

9 • If the magnetic field is directed toward the north and a positively charged particle is moving toward the east, what is the direction of the magnetic force on the particle?

10 • A positively charged particle is moving northward in a magnetic field. The magnetic force on the particle is toward the northeast. What is the direction of the magnetic field? (a) Up (b) West (c) South (d) Down (e) The force cannot be directed toward the northeast.

11 • A ^{7}Li nucleus with a charge of $+3e$ and a mass of 7 u(1 u = 1.66×10^{-27} kg) and a proton with charge $+e$ and mass 1 u are both moving in a plane perpendicular to a magnetic field $\vec{B}$. The magnitude of the momenta of the two particles are equal. The ratio of the radius of curvature of the path of the proton, R_p, to that of the ^{7}Li nucleus, R_{Li}, is (a) $R_p/R_{Li} = 3$. (b) $R_p/R_{Li} = 1/3$. (c) $R_p/R_{Li} = 1/7$. (d) $R_p/R_{Li} = 3/7$. (e) none of these.

12 • SSM An electron moving with speed v to the right enters a region of uniform magnetic field directed out of the paper. When the electron enters this region, it will be (a) deflected out of the plane of the paper. (b) deflected into the plane of the paper. (c) deflected upward. (d) deflected downward. (e) undeviated in its motion.

13 ••• The theory of relativity tells us that none of the laws of physics can depend on the absolute velocity of an object, which is in fact impossible to define. Instead, the behavior of physical objects can only depend on the relative velocity between the objects. The development of new physical insights can come from this idea. For example, in Figure 26-31, a magnet moving at high speed flies by an electron that is at rest relative to a physicist observing it in a laboratory. Explain why you are sure that a force must be acting on it. What direction will the force point when the north pole of the magnet passes directly underneath the electron? Explain.

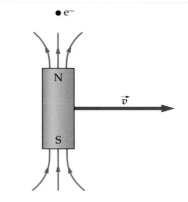

FIGURE 26-31 Problem 13

14 • How are magnetic field lines similar to electric field lines? How are they different?

15 • If a current I in a given wire and a magnetic field $\vec{B}$ are known, the force $\vec{F}$ on the current is uniquely determined. Show that knowing $\vec{F}$ and I does not provide complete knowledge of $\vec{B}$.

Estimation and Approximation

16 •• SSM CRT's used in monitors and televisions commonly use magnetic deflection to steer the electron beams. A schematic diagram is shown in Figure 26-32. The electron beam is accelerated through a potential difference and the electron beam is then accelerated through a magnetic field that deflects the electron beam, as shown in the figure. Given the following parameters, estimate the magnitude of the magnetic field needed for maximum deflection: accelerating voltage, $V = 15$ kV; distance over which electron is in magnetic field, $d = 5$ cm; length, $L = 50$ cm; diagonal of CRT, $r = 19$ in.

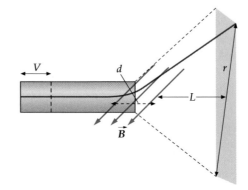

FIGURE 26-32 Problem 16

17 •• (a) Estimate the charge-to-mass ratio of a micrometeorite needed for it to "orbit" the Earth in a low-earth orbit (400 km above the surface of the Earth) under the influence of the Earth's magnetic field alone. Take the magnitude of the Earth's field to be 5×10^{-5} T and assume it perpendicular to the meteorite's velocity. Assume that the speed of the meteorite is about the same as Earth's orbital speed of roughly 30 km/s. (b) If the mass of the micrometeorite is 3×10^{-10} kg, what is its charge?

The Force Exerted by a Magnetic Field

18 • **SOLVE** ✔ Find the magnetic force on a proton moving with velocity 4.46 Mm/s in the positive x direction in a magnetic field of 1.75 T in the positive z direction.

19 • A charge $q = -3.64$ nC moves with a velocity of 2.75×10^6 m/s $\hat{i}$. Find the force on the charge if the magnetic field is (a) $\vec{B} = 0.38$ T $\hat{j}$, (b) $\vec{B} = 0.75$ T $\hat{i} + 0.75$ T $\hat{j}$, (c) $\vec{B} = 0.65$ T $\hat{i}$, and (d) $\vec{B} = 0.75$ T $\hat{i} + 0.75$ T $\hat{k}$.

20 • **SOLVE** A uniform magnetic field of magnitude 1.48 T is in the positive z direction. Find the force exerted by the field on a proton if the proton's velocity is (a) $\vec{v} = 2.7$ Mm/s $\hat{i}$, (b) $\vec{v} = 3.7$ Mm/s $\hat{j}$, (c) $\vec{v} = 6.8$ Mm/s $\hat{k}$, and (d) $\vec{v} = 4.0$ Mm/s $\hat{i} + 3.0$ Mm/s $\hat{j}$.

21 • **SOLVE** A straight wire segment 2 m long makes an angle of 30° with a uniform magnetic field of 0.37 T. Find the magnitude of the force on the wire if it carries a current of 2.6 A.

22 • **SSM** **SOLVE** ✔ A straight wire segment $I\vec{L} = (2.7$ A$)(3$ cm $\hat{i} + 4$ cm $\hat{j})$ is in a uniform magnetic field $\vec{B} = 1.3$ T $\hat{i}$. Find the force on the wire.

23 • What is the force (magnitude and direction) on an electron with velocity $\vec{v} = (2\hat{i} - 3\hat{j} \times 10^6)$ m/s in a magnetic field $\vec{B} = (0.8 \hat{i} + 0.6 \hat{j} - 0.4 \hat{k})$T?

24 •• The wire segment shown in Figure 26-33 carries a current of 1.8 A from a to b. There is a magnetic field $\vec{B} = 1.2$ T $\hat{k}$. Find the total force on the wire and show that the total force is the same as if the wire were a straight segment from a to b.

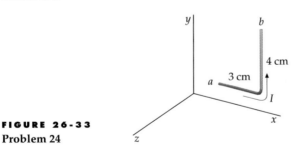

FIGURE 26-33
Problem 24

25 •• **SOLVE** A straight, stiff, horizontal wire of length 25 cm and mass 50 g is connected to a source of emf by light, flexible leads. A magnetic field of 1.33 T is horizontal and perpendicular to the wire. Find the current necessary to float the wire; that is, find the current so the magnetic force balances the weight of the wire.

26 •• **SSM** **SOLVE** ✔ A simple gaussmeter for measuring horizontal magnetic fields consists of a stiff 50-cm wire that hangs vertically from a conducting pivot so that its free end makes contact with a pool of mercury in a dish below. The mercury provides an electrical contact without constraining the movement of the wire. The wire has a mass of 5 g and conducts a current downward. (a) What is the equilibrium angular displacement of the wire from vertical if the horizontal magnetic field is 0.04 T and the current is 0.20 A? (b) If the current is 20 A and a displacement from vertical of 0.5 mm can be detected for the free end, what is the horizontal magnetic field sensitivity of this gaussmeter?

27 •• A current-carrying wire is bent into a semicircular loop of radius R that lies in the xy plane. There is a uniform magnetic field $\vec{B} = B\hat{k}$ perpendicular to the plane of the loop (Figure 26-34). Verify that the force acting on the loop is 0.

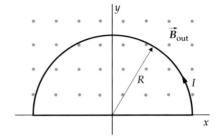

FIGURE 26-34 Problem 27

28 •• A 10-cm length of wire carries a current of 4.0 A in the positive z direction. The force on this wire due to a uniform magnetic field $\vec{B}$ is $\vec{F} = (-0.2 \hat{i} + 0.2 \hat{j})$N. If this wire is rotated so that the current flows in the positive x direction, the force on the wire is $\vec{F} = 0.2 \hat{k}$N. Find the magnetic field $\vec{B}$.

29 •• **SOLVE** A 10-cm length of wire carries a current of 2.0 A in the positive x direction. The force on this wire due to the presence of a magnetic field $\vec{B}$ is $\vec{F} = (3.0 \hat{j} + 2.0 \hat{k})$N. If this wire is now rotated so that the current flows in the positive y direction, the force on the wire is $\vec{F} = (-3.0 \hat{i} - 2.0 \hat{k})$N. Determine the magnetic field $\vec{B}$.

30 ••• A wire bent in some arbitrary shape carries a current I in a uniform magnetic field $\vec{B}$. Show explicitly that the total force on the part of the wire from some point a to some point b is $\vec{F} = I\vec{L} \times \vec{B}$, where $\vec{L}$ is the vector from point a to point b.

Motion of a Point Charge in a Magnetic Field

31 • **SSM** A proton moves in a circular orbit of radius 65 cm perpendicular to a uniform magnetic field of magnitude 0.75 T. (a) What is the period for this motion? (b) Find the speed of the proton. (c) Find the kinetic energy of the proton.

32 • **SOLVE** ✔ An electron of kinetic energy 45 keV moves in a circular orbit perpendicular to a magnetic field of 0.325 T. (a) Find the radius of the orbit. (b) Find the frequency and period of the motion.

33 • **SOLVE** An electron from the sun with a speed of 1×10^7 m/s enters the earth's magnetic field high above the equator where the magnetic field is 4×10^{-7} T. The electron moves nearly in a circle, except for a small drift along the direction of the earth's magnetic field that will take the electron toward the north pole. (a) What is the radius of the circular motion? (b) What is the radius of the circular motion near the north pole where the magnetic field is 2×10^{-5} T?

34 •• Protons and deuterons (each with charge $+e$) and alpha particles (with charge $+2e$) of the same kinetic energy enter a uniform magnetic field $\vec{B}$ that is perpendicular to their velocities. Let R_p, R_d, and R_α be the radii of their circular orbits. Find the ratios R_d/R_p and R_α/R_p. Assume that $m_\alpha = 2m_d = 4m_p$.

35 •• A proton and an alpha particle move in a uniform magnetic field in circles of the same radii. Compare (a) their velocities, (b) their kinetic energies, and (c) their angular momenta. (See Problem 34.)

36 •• A particle of charge q and mass m has momentum $p = mv$ and kinetic energy $K = p^2/2m$. If the particle moves in a circular orbit of radius R perpendicular to a uniform magnetic field $\vec{B}$, show that (a) $p = BqR$ and (b) $K = \frac{1}{2}B^2q^2R^2/m$.

37 •• [SSM] A beam of particles with velocity $\vec{v}$ enters a region of uniform magnetic field $\vec{B}$ that makes a small angle θ with $\vec{v}$. Show that after a particle moves a distance $2\pi(m/qB)v \cos\theta$, measured along the direction of $\vec{B}$, the velocity of the particle is in the same direction as it was when the particle entered the field.

38 •• [SOLVE] A proton with speed $v = 10^7$ m/s enters a region of uniform magnetic field $B = 0.8$ T, which is into the page, as shown in Figure 26-35. The angle $\theta = 60°$. Find the angle ϕ and the distance d.

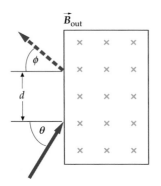

FIGURE 26-35 Problems 38 and 39

39 •• Suppose that in Figure 26-35 $B = 0.6$ T, the distance $d = 0.4$ m, and $\theta = 24°$. Find the speed v and the angle ϕ if the particles are (a) protons and (b) deuterons.

40 •• The galactic magnetic field in some region of interstellar space has a magnitude of 10^{-9} T. A particle of interstellar dust has a mass of 10 μg and a total charge of 0.3 nC. How many years does it take to complete a circular orbit in the magnetic field?

The Velocity Selector

41 • [SSM] [SOLVE] A velocity selector has a magnetic field of magnitude 0.28 T perpendicular to an electric field of magnitude 0.46 MV/m. (a) What must the speed of a particle be for the particle to pass through undeflected? What energy must (b) protons and (c) electrons have to pass through undeflected?

42 • [SOLVE✓] A beam of protons moves along the x axis in the positive x direction with a speed of 12.4 km/s through a region of crossed fields balanced for zero deflection. (a) If there is a magnetic field of magnitude 0.85 T in the positive y direction, find the magnitude and direction of the electric field. (b) Would electrons of the same velocity be deflected by these fields? If so, in what direction?

Thomson's Measurement of q/m for Electrons and the Mass Spectrometer

43 •• [SSM] The plates of a Thomson q/m apparatus are 6.0 cm long and are separated by 1.2 cm. The end of the plates is 30.0 cm from the tube screen. The kinetic energy of the electrons is 2.8 keV. (a) If a potential of 25 V is applied across the deflection plates, by how much will the beam deflect? (b) Find the magnitude of the crossed magnetic field that will allow the beam to pass between the plates undeflected.

44 •• Chlorine has two stable isotopes, ^{35}Cl and ^{37}Cl, whose natural abundances are about 76 percent and 24 percent, respectively. Singly ionized chlorine gas is to be separated into its isotopic components using a mass spectrometer. The magnetic field in the spectrometer is 1.2 T. What is the minimum value of the potential through which these ions must be accelerated so that the separation between them is 1.4 cm?

45 •• [SOLVE] A singly ionized ^{24}Mg ion (mass 3.983 × 10^{-26} kg) is accelerated through a 2.5-kV potential difference and deflected in a magnetic field of 557 G in a mass spectrometer. (a) Find the radius of curvature of the orbit for the ion. (b) What is the difference in radius for ^{26}Mg ions and for ^{24}Mg ions? (Assume that their mass ratio is 26:24.)

46 •• [SSM] A beam of ^{6}Li and ^{7}Li ions passes through a velocity selector and enters a magnetic spectrometer. If the diameter of the orbit of the ^{6}Li ions is 15 cm, what is the diameter of the orbit for ^{7}Li ions?

The Cyclotron

47 •• In Example 26-6, determine the time required for a ^{58}Ni ion and a ^{60}Ni ion to complete the semicircular path.

48 •• Before entering a mass spectrometer, ions pass through a velocity selector consisting of parallel plates separated by 2.0 mm and having a potential difference of 160 V. The magnetic field between the plates is 0.42 T. The magnetic field in the mass spectrometer is 1.2 T. Find (a) the speed of the ions entering the mass spectrometer and (b) the difference in the diameters of the orbits of singly ionized ^{238}U and ^{235}U. (The mass of a ^{235}U ion is 3.903 × 10^{-25} kg.)

49 •• [SSM] [SOLVE] A cyclotron for accelerating protons has a magnetic field of 1.4 T and a radius of 0.7 m. (a) What is the cyclotron frequency? (b) Find the maximum energy of the protons when they emerge. (c) How will your answers change if deuterons, which have the same charge but twice the mass, are used instead of protons?

50 •• A certain cyclotron with a magnetic field of 1.8 T is designed to accelerate protons to 25 MeV. (a) What is the cyclotron frequency? (b) What must the minimum radius of the magnet be to achieve a 25-MeV emergence energy? (c) If the alternating potential applied to the dees has a maximum value of 50 kV, how many revolutions must the protons make before emerging with an energy of 25 MeV?

51 •• [SOLVE] Show that for a certain cyclotron the cyclotron frequencies of deuterons and alpha particles are the same and are half that of a proton in the same magnetic field. (See Problem 34.)

52 •• Show that the radius of the orbit of a charged parti-cle in a cyclotron is proportional to the square root of the number of orbits completed.

Torques on Current Loops and Magnets

53 • **ISOLVE**✓ A small circular coil of 20 turns of wire lies in a uniform magnetic field of 0.5 T, so that the normal to the plane of the coil makes an angle of 60° with the direction of $\vec{B}$. The radius of the coil is 4 cm, and it carries a current of 3 A. (*a*) What is the magnitude of the magnetic moment of the coil? (*b*) What is the magnitude of the torque exerted on the coil?

54 • **ISOLVE**✓ What is the maximum torque on a 400-turn circular coil of radius 0.75 cm that carries a current of 1.6 mA and resides in a uniform magnetic field of 0.25 T?

55 • **SSM** **ISOLVE** A current-carrying wire is bent into the shape of a square of edge-length $L = 6$ cm and is placed in the xy plane. It carries a current $I = 2.5$ A. What is the magni-tude of the torque on the wire if there is a uniform magnetic field of 0.3 T (*a*) in the z direction and (*b*) in the x direction?

56 • Repeat Problem 55 if the wire is bent into an equi-lateral triangle of edge-length 8 cm.

57 •• **ISOLVE**✓ A rigid, circular loop of radius R and mass m carries a current I and lies in the xy plane on a rough, flat table. There is a horizontal magnetic field of magnitude B. What is the minimum value of B so that one edge of the loop will lift off the table?

58 •• A rectangular, 50-turn coil has sides 6-cm long and 8-cm long and carries a current I of 1.75 A. It is oriented and pivoted about the z axis, as shown in Figure 26-36. (*a*) If the wire in the xy plane makes an angle $\theta = 37°$ with the y axis as shown, what angle does the unit normal $\hat{n}$ make with the x axis? (*b*) Write an expression for $\hat{n}$ in terms of the unit vectors $\hat{i}$ and $\hat{j}$. (*c*) What is the magnetic moment of the coil? (*d*) Find the torque on the coil when there is a uniform mag-netic field $\vec{B} = 1.5$ T $\hat{j}$. (*e*) Find the potential energy of the coil in this field.

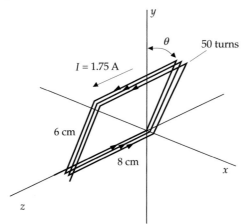

FIGURE 26-36 Problems 58 and 59

59 •• The coil in Problem 58 is pivoted about the z axis and held at various positions in a uniform magnetic field $\vec{B} = 2.0$ T $\hat{j}$. Sketch the position of the coil and find the torque exerted when the unit normal is (*a*) $\hat{n} = \hat{i}$, (*b*) $\hat{n} = \hat{j}$, (*c*) $\hat{n} = -\hat{j}$, and (*d*) $\hat{n} = (\hat{i} + \hat{j})/\sqrt{2}$.

Magnetic Moments

60 •• **SSM** **ISOLVE**✓ A small magnet of length 6.8 cm is placed at an angle of 60° to the direction of a uniform magnetic field of magnitude 0.04 T. The observed torque has a magni-tude of 0.10 N·m. Find the magnetic moment of the magnet.

61 •• **ISOLVE** A wire loop consists of two semicircles connected by straight segments (Figure 26-37). The inner and outer radii are 0.3 m and 0.5 m, respectively. A current I of 1.5 A flows in this loop with the current in the outer semicircle in the clockwise direction. What is the magnetic moment of this current loop?

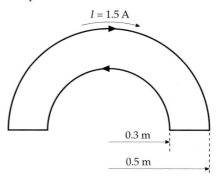

$I = 1.5$ A

0.3 m

0.5 m

FIGURE 26-37 Problem 61

62 •• A wire of length L is wound into a circular coil of N loops. Show that when this coil carries a current I, its magnetic moment has the magnitude $IL^2/4\pi N$.

63 •• A particle of charge q and mass m moves in a circle of radius R and with angular velocity ω. (*a*) Show that the average current is $I = q\omega/(2\pi)$ and that the magnetic moment has the magnitude $\mu = \frac{1}{2}q\omega r^2$. (*b*) Show that the angular momentum of this particle has the magnitude $L = mr^2\omega$ and that the magnetic moment and angular momentum vectors are related by $\vec{\mu} = (\frac{1}{2}q/m)\vec{L}$.

64 ••• **SSM** A hollow cylinder has length L and inner and outer radii R_i and R_o, respectively (Figure 26-38). The cylinder carries a uniform charge density ρ. Derive an expression for the magnetic moment as a function of ω, the angular velocity of rotation of the cylinder about its axis.

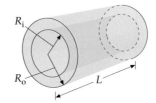

R_i

R_o

L

FIGURE 26-38 Problem 64

65 ••• A nonconducting rod of mass m and length L has a uniform charge per unit length λ and rotates with angular velocity ω about an axis through one end and perpendic-ular to the rod. (*a*) Consider a small segment of the rod of length dx and charge $dq = \lambda\ dx$ at a distance x from the pivot (Figure 26-39). Show that the magnetic moment of this seg-ment is $\frac{1}{2}\lambda\omega x^2 dx$. (*b*) Integrate your result to show that the total magnetic moment of the rod is $\mu = \frac{1}{6}\lambda\omega L^3$. (*c*) Show that the magnetic moment $\vec{\mu}$ and angular momentum $\vec{L}$ are re-lated by $\vec{\mu} = (\frac{1}{2}Q/m)\vec{L}$, where Q is the total charge on the rod.

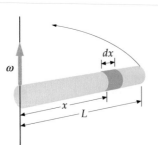

FIGURE 26-39
Problem 65

66 ••• A nonuniform, nonconducting disk of mass m, radius R, and total charge Q has a surface charge density $\sigma = \sigma_0 r/R$ and a mass per unit area $\sigma_m = (m/Q)\sigma$. The disk rotates with angular velocity ω about its axis. (a) Show that the magnetic moment of the disk has a magnitude $\mu = \frac{1}{5}\pi\omega\sigma_0 R^4 = \frac{3}{10}Q\omega R^2$. (b) Show that the magnetic moment $\vec{\mu}$ and angular momentum $\vec{L}$ are related by $\vec{\mu} = (\frac{1}{2}Q/m)\,\vec{L}$.

67 ••• A spherical shell of radius R carries a surface charge density σ. The sphere rotates about its diameter with angular velocity ω. Find the magnetic moment of the rotating sphere.

68 ••• A solid sphere of radius R carries a uniform volume charge density ρ. The sphere rotates about its diameter with angular velocity ω. Find the magnetic moment of this rotating sphere.

69 ••• **SSM** A uniform disk of mass m, radius R, and surface charge σ rotates about its center with angular velocity ω in Figure 26-40. A uniform magnetic field of magnitude $\vec{B}$ threads the disk, making an angle θ with respect to the rotation axis of the disk. Calculate (a) the net torque acting on the disk and (b) the precession frequency of the disk in the magnetic field. (See pp. 316–317 for a discussion of precession.)

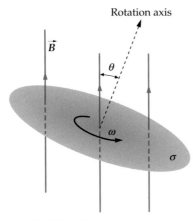

FIGURE 26-40 Problem 69

The Hall Effect

70 • **SOLVE✔** A metal strip 2-cm wide and 0.1-cm thick carries a current of 20 A in a uniform magnetic field of 2 T, as shown in Figure 26-41. The Hall voltage is measured to be 4.27 μV. (a) Calculate the drift velocity of the electrons in the strip. (b) Find the number density of the charge carriers in the strip. (c) Is point a or point b at the higher potential?

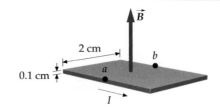

FIGURE 26-41 Problems 70 and 71

71 •• The number density of free electrons in copper is 8.47×10^{22} electrons per cubic centimeter. If the metal strip in Figure 26-41 is copper and the current is 10 A, find (a) the drift velocity v_d and (b) the Hall voltage. (Assume that the magnetic field is 2.0 T.)

72 •• **SSM** **SOLVE** A copper strip ($n = 8.47 \times 10^{22}$ electrons per cubic centimeter) 2-cm wide and 0.1-cm thick is used to measure the magnitudes of unknown magnetic fields that are perpendicular to the strip. Find the magnitude of B when $I = 20$ A and the Hall voltage is (a) 2.00 μV, (b) 5.25 μV, and (c) 8.00 μV.

73 •• **SOLVE** Because blood contains charged ions, moving blood develops a Hall voltage across the diameter of an artery. A large artery with a diameter of 0.85 cm has a flow speed of 0.6 m/s. If a section of this artery is in a magnetic field of 0.2 T, what is the maximum possible potential difference across the diameter of the artery?

74 •• **SOLVE** The Hall coefficient R is defined as $R = E_y/(J_x B_z)$, where J_x is the current per unit area in the x direction in the slab, B_z is the magnetic field in the z direction, and E_y is the resulting Hall field in the y direction. Show that the Hall coefficient is $1/(nq)$, where q is the charge of the charge carriers, -1.6×10^{-19} C if they are electrons. (The Hall coefficients of monovalent metals, such as copper, silver, and sodium are therefore negative.)

75 •• **SSM** Aluminum has a density of 2.7×10^3 kg/m^3 and a molar mass of 27 g/mol. The Hall coefficient of aluminum is $R = -0.3 \times 10^{-10}$ m^3/C. (See Problem 74 for the definition of R.) Find the number of conduction electrons per aluminum atom.

General Problems

76 • **SOLVE✔** A long wire parallel to the x axis carries a current of 6.5 A in the positive x direction. There is a uniform magnetic field $\vec{B} = 1.35$ T $\hat{j}$. Find the force per unit length on the wire.

77 • **SOLVE** An alpha particle (charge $+2e$) travels in a circular path of radius 0.5 m in a magnetic field of 1 T. Find (a) the period, (b) the speed, and (c) the kinetic energy (in electron volts) of the alpha particle. Take $m = 6.65 \times 10^{-27}$ kg for the mass of the alpha particle.

78 •• The pole strength q_m of a bar magnet is defined by $q_m = |\vec{\mu}|/L$, where L is the length of the magnet. Show that the torque exerted on a bar magnet in a uniform magnetic field $\vec{B}$ is the same as if a force $+q_m\vec{B}$ is exerted on the north pole and a force $-q_m\vec{B}$ is exerted on the south pole.

79 •• SSM A particle of mass m and charge q enters a region where there is a uniform magnetic field $\vec{B}$ along the x axis. The initial velocity of the particle is $\vec{v} = v_{0x}\hat{i} + v_{0y}\hat{j}$, so the particle moves in a helix. (a) Show that the radius of the helix is $r = mv_{0y}/qB$. (b) Show that the particle takes a time $t = 2\,\pi m/qB$ to make one orbit around the helix.

80 •• SSM ISOLVE A metal crossbar of mass m rides on a pair of long, horizontal conducting rails separated by a distance L and connected to a device that supplies constant current I to the circuit, as shown in Figure 26-42. A uniform magnetic field $\vec{B}$ is established, as shown. (a) If there is no friction and the bar starts from rest at $t = 0$, show that at time t the bar has velocity $v = (BIL/m)t$. (b) In which direction will the bar move? (c) If the coefficient of static friction is μ_S, find the minimum field B necessary to start the bar moving.

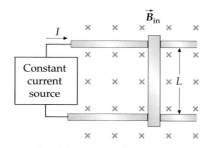

FIGURE 26-42 Problems 80 and 81

81 •• Assume that the rails in Figure 26-42 are frictionless but tilted upward so that they make an angle θ with the horizontal. (a) What vertical magnetic field $\vec{B}$ is needed to keep the bar from sliding down the rails? (b) What is the acceleration of the bar if B has twice the value found in Part (a)?

82 •• A long, narrow bar magnet that has magnetic moment $\vec{\mu}$ parallel to its long axis is suspended at its center as a frictionless compass needle. When placed in a horizontal magnetic field $\vec{B}$, the needle lines up with the field. If it is displaced by a small angle θ, show that the needle will oscillate about its equilibrium position with frequency $f = \frac{1}{2\pi}\sqrt{\mu B/I}$, where I is the moment of inertia about the point of suspension.

83 •• ISOLVE A conducting wire is parallel to the y axis. It moves in the positive x direction with a speed of 20 m/s in a magnetic field $\vec{B} = 0.5$ T $\hat{k}$. (a) What are the magnitude and direction of the magnetic force on an electron in the conductor? (b) Because of this magnetic force, electrons move to one end of the wire leaving the other end positively charged, until the electric field due to this charge separation exerts a force on the electrons that balances the magnetic force. Find the magnitude and direction of this electric field in the steady state. (c) Suppose the moving wire is 2-m long. What is the potential difference between its two ends due to this electric field?

84 ••• The rectangular frame shown in Figure 26-43 is free to rotate about the axis A–A on the horizontal shaft. The frame is 10-cm long and 6-cm wide, and the rods that make up the frame have a mass per unit length of 20 g/cm. A uniform magnetic field $B = 0.2$ T is directed, as shown. A current may be sent around the frame by means of the wires attached at the top. (a) If no current passes through the frame, what is the period of this physical pendulum for small oscillations? (b) If a current of 8 A passes through the frame in the direction indicated by the arrow, what is then the period of this physical

pendulum? (c) Suppose the direction of the current is opposite to the direction shown. The frame is displaced from the vertical by some angle θ. What must be the magnitude of the current so that this frame will be in equilibrium?

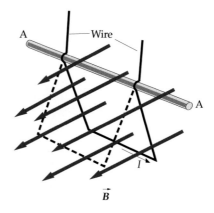

FIGURE 26-43 Problem 84

85 ••• SSM A stiff, straight horizontal wire of length 25 cm and mass 20 g is supported by electrical contacts at its ends, but is otherwise free to move vertically upward. The wire is in a uniform, horizontal magnetic field of magnitude 0.4 T perpendicular to the wire. A switch connecting the wire to a battery is closed and the wire flies upward, rising to a maximum height h. The battery delivers a total charge of 2 C during the short time it makes contact with the wire. Find the height h.

86 ••• A circular loop of wire with mass m carries a current I in a uniform magnetic field. It is initially in equilibrium with its magnetic moment vector aligned with the magnetic field. The loop is given a small twist about a diameter and then released. What is the period of the motion? (Assume that the only torque exerted on the loop is due to the magnetic field.)

87 ••• A small bar magnet has a magnetic moment $\vec{\mu}$ that makes an angle θ with the x axis and lies in a nonuniform magnetic field given by $\vec{B} = B_x(x)\hat{i} + B_y(y)\hat{j}$. Use $F_x = -dU/dx$ and $F_y = -dU/dy$ to show that there is a net force on the magnet that is given by

$$\vec{F} = \mu_x \frac{\partial B_x}{\partial x}\hat{i} + \mu_y \frac{\partial B_y}{\partial y}\hat{j}$$

88 ••• SSM The special theory of relativity tells us that a particle's mass depends on its speed through the formula:

$$m(v) = \frac{m_0}{\sqrt{1 - \dfrac{v^2}{c^2}}} = \gamma(v)m_0$$

where m_0 is the particle's rest mass and $\gamma(v) = 1/\sqrt{1 - (v^2/c^2)}$ (a) *Taking into account the special theory of relativity*, what is the radius and period of a particle's orbit if it has speed v and is moving in a magnetic field with magnitude B that is perpendicular to the direction of the velocity? Assume the force on the particle is given by $\vec{F} = q(\vec{v} \times \vec{B})$. The particle has rest mass m_0 and charge q. (b) Using a spreadsheet program, make graphs of the radius and period of the orbit of an electron in a 10-T magnetic field versus $\gamma(v)$ for speeds between $v = 0.1c$ and $v = 0.999c$. Use a logarithmic scale to display $\gamma(v)$.

Sources of the Magnetic Field

THESE COILS AT THE KETTERING MAGNETICS LABORATORY AT OAKLAND UNIVERSITY ARE CALLED HELMHOLTZ COILS. THEY ARE USED TO CANCEL THE EARTH'S MAGNETIC FIELD AND TO PROVIDE A UNIFORM MAGNETIC FIELD IN A SMALL REGION OF SPACE FOR STUDYING THE MAGNETIC PROPERTIES OF MATTER.

? **Have you any idea what the magnetic field of a current-carrying coil looks like? There are illustrations of the magnetic field of a coil in Section 27-2.**

The earliest known sources of magnetism were permanent magnets. One month after Oersted announced his discovery that a compass needle is deflected by an electric current, Jean-Baptiste Biot and Félix Savart announced the results of their measurements of the torque on a magnet near a long, current-carrying wire and they analyzed these results in terms of the magnetic field produced by each element of the current. André-Marie Ampère extended these experiments and showed that current elements also experience a force in the presence of a magnetic field and that two currents exert forces on each other.

➤ **In this chapter, we begin by considering the magnetic field produced by a single moving charge and by the moving charges in a current element. We then calculate the magnetic fields produced by some common current configurations, such as a straight wire segment; a long, straight wire; a current loop; and a solenoid. Next we discuss Ampère's law, which relates the line integral of the magnetic field around a closed loop to the total current that passes through the loop. Finally, we consider the magnetic properties of matter.**

27-1 The Magnetic Field of Moving Point Charges

When a point charge q moves with velocity $\vec{v}$, the moving point charge produces a magnetic field $\vec{B}$ in space, given by[†]

$$\vec{B} = \frac{\mu_0}{4\pi}\frac{q\vec{v} \times \hat{r}}{r^2}$$ 27-1

MAGNETIC FIELD OF A MOVING POINT CHARGE

where $\hat{r}$ is a unit vector (see Figure 27-1) that points to the field point P from the charge q moving with velocity $\vec{v}$, and μ_0 is a constant of proportionality called the **permeability of free space**,[‡] which has the exact value

$$\mu_0 = 4\pi \times 10^{-7}\,\text{T}\cdot\text{m/A} = 4\pi \times 10^{-7}\,\text{N/A}^2$$ 27-2

The units of μ_0 are such that B is in teslas when q is in coulombs, v is in meters per second, and r is in meters. The unit N/A^2 comes from the fact that $1\,\text{T} = 1\,\text{N/(A·m)}$. The constant $1/(4\pi)$ is arbitrarily included in Equation 27-1 so that the factor 4π will not appear in Ampère's law (Equation 27-15), which we will study in Section 27-4.

FIGURE 27-1 A positive point charge q moving with velocity $\vec{v}$ produces a magnetic field $\vec{B}$ at a field point P that is in the direction $\vec{v} \times \hat{r}$, where $\hat{r}$ is the unit vector pointing from the charge to the field point. The field varies inversely as the square of the distance from the charge to the field point and is proportional to the sine of the angle between $\vec{v}$ and $\hat{r}$. (The blue × at the field point indicates that the direction of the field is into the page.)

MAGNETIC FIELD OF A MOVING POINT CHARGE   **EXAMPLE 27-1**

A point particle with charge $q = 4.5$ nC is moving with velocity $\vec{v} = 3 \times 10^3$ m/s$\hat{i}$ parallel to the x axis along the line $y = 3$ m. Find the magnetic field at the origin produced by this charge when the charge is at the point $x = -4$ m, $y = 3$ m, as shown in Figure 27-2.

FIGURE 27-2

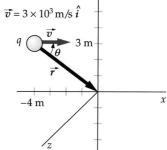

1. The magnetic field is given by Equation 27-1:

$$\vec{B} = \frac{\mu_0}{4\pi}\frac{q\vec{v} \times \hat{r}}{r^2}, \quad \text{with } \vec{v} = v\hat{i}$$

2. Find $\vec{r}$ and r from Figure 27-2 and write $\hat{r}$ in terms of $\hat{i}$ and $\hat{j}$:

$$\vec{r} = 4\,\text{m}\hat{i} - 3\,\text{m}\hat{j}$$

$$r = \sqrt{4^2 + 3^2}\,\text{m} = 5\,\text{m}$$

$$\hat{r} = \frac{\vec{r}}{r} = \frac{4\,\text{m}\hat{i} - 3\,\text{m}\hat{j}}{5\,\text{m}} = 0.8\,\hat{i} - 0.6\,\hat{j}$$

3. Substitute the above results in Equation 27-1 to obtain:

$$\vec{B} = \frac{\mu_0}{4\pi}\frac{q\vec{v} \times \hat{r}}{r^2} = \frac{\mu_0}{4\pi}\frac{q(v\hat{i}) \times (0.8\,\hat{i} - 0.6\,\hat{j})}{r^2} = \frac{\mu_0}{4\pi}\frac{q(-0.6\,v\hat{k})}{r^2}$$

$$= -(10^{-7}\,\text{T·m/A})\frac{(4.5 \times 10^{-9}\,\text{C})(0.6)(3 \times 10^3\,\text{m/s})}{(5\,\text{m})^2}\hat{k}$$

$$= \boxed{-3.24 \times 10^{-14}\,\text{T}\hat{k}}$$

[†] This expression is used for speeds much less than the speed of light.
[‡] Some care must be taken not to confuse the constant μ_0 with the magnitude of the magnetic moment vector $\vec{\mu}$.

REMARKS It is also possible to obtain $\vec{B}$ without finding an explicit expression for the unit vector $\hat{r}$. From Figure 27-2 we note that $\vec{v} \times \hat{r}$ is in the negative z direction. In addition, the magnitude of $\vec{v} \times \hat{r}$ is $v \sin \theta$, where $\sin \theta = (3 \text{ m})/(5 \text{ m}) = 0.6$. Combining these results, we have $\vec{v} \times \hat{r} = v \sin \theta(-\hat{k}) = -v(0.6)\hat{k}$, in agreement with our result in line 1 of step 3. Finally, this example shows that the magnetic field due to a moving charge is quite small. For comparison, the earth's magnetic field near its surface has a magnitude of about 10^{-4} T.

EXERCISE At the same instant, find the magnetic field on the y axis both at $y = 3$ m and at $y = 6$ m. (*Answer* $\vec{B} = 0, \vec{B} = 3.24 \times 10^{-14}$ T$\hat{k}$)

(a)   (b)

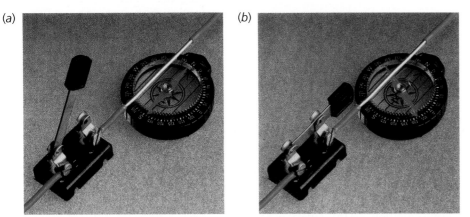

Oersted's experiment. (*a*) With no current in the wire, the compass needle points north. (*b*) When the wire carries a current, the compass needle is deflected in the direction of the resultant magnetic field. The current in the wire is directed upward, from left to right. The insulation has been stripped from the wire to improve the contrast of the photograph.

27-2 The Magnetic Field of Currents: The Biot–Savart Law

In the previous chapter we extended our discussion of forces on point charges to forces on current elements by replacing $q\vec{v}$ with the current element $I \, d\vec{\ell}$. We do the same for the magnetic field produced by a current element. The magnetic field $d\vec{B}$ produced by a current element $I \, d\vec{\ell}$ is given by Equation 27-1, with $q\vec{v}$ replaced by $I \, d\vec{\ell}$:

$$d\vec{B} = \frac{\mu_0}{4\pi} \frac{I \, d\vec{\ell} \times \hat{r}}{r^2}$$

27-3

BIOT–SAVART LAW

Equation 27-3, known as the **Biot–Savart law,** was also deduced by Ampère. The Biot–Savart law and Equation 27-1 are analogous to Coulomb's law for the electric field of a point charge. The source of the magnetic field is a moving charge $q\vec{v}$ or a current element $I \, d\vec{\ell}$, just as the charge q is the source of the electrostatic field. The magnetic field decreases with the square of the distance from the moving charge or current element, just as the electric field decreases with the square of the distance from a point charge. However, the directional aspects of the electric and magnetic fields are quite different. Whereas the electric field points in the radial direction $\hat{r}$ from the point charge to the field point (for a positive charge), the magnetic field is perpendicular to both $\hat{r}$ and to $\vec{v}$, in the case of a point charge, or to $d\vec{\ell}$ in the case of a current element. At a point along the line of a current element, such as point P_2 in Figure 27-3, the magnetic field due to that element is zero. (Equation 27-3 gives $d\vec{B} = 0$ if $d\vec{\ell}$ and $\hat{r}$ are either parallel or antiparallel.)

The magnetic field due to the total current in a circuit can be calculated by using the Biot–Savart law to find the field due to each current element, and then summing (integrating) over all the current elements in the circuit. This calculation is difficult for all but the simplest circuit geometries.

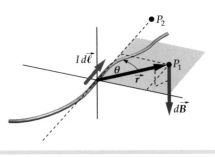

FIGURE 27-3 The current element $I \, d\vec{\ell}$ produces a magnetic field at point P_1 that is perpendicular to both $d\vec{\ell}$ and $\vec{r}$. The current element produces no magnetic field at point P_2, which is along the line of $d\vec{\ell}$.

$\vec{B}$ Due to a Current Loop

Figure 27-4 shows a current element $I\, d\vec{\ell}$ of a current loop of radius R and the unit vector $\hat{r}$ that is directed from the element to the center of the loop. The magnetic field at the center of the loop due to this element is directed along the axis of the loop, and its magnitude is given by

$$dB = \frac{\mu_0}{4\pi}\frac{I\, d\ell \sin\theta}{R^2}$$

where θ is the angle between $d\vec{\ell}$ and $\hat{r}$, which is $90°$ for each current element, so $\sin\theta = 1$. The magnetic field due to the entire current is found by integrating over all the current elements in the loop. Since R is the same for all elements, we obtain

$$B = \int dB = \frac{\mu_0}{4\pi}\frac{I}{R^2}\oint d\ell$$

The integral of $d\ell$ around the complete loop gives the total length $2\pi R$, the circumference of the loop. The magnetic field due to the entire loop is thus

$$B = \frac{\mu_0}{4\pi}\frac{I}{R^2}2\pi R = \frac{\mu_0 I}{2R} \qquad\qquad 27\text{-}4$$

$$B \text{ AT THE CENTER OF A CURRENT LOOP}$$

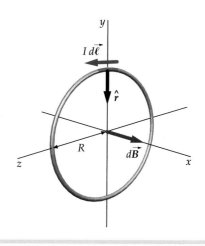

FIGURE 27-4 Current element for calculating the magnetic field at the center of a circular current loop. Each element produces a magnetic field that is directed along the axis of the loop.

EXERCISE Find the current in a circular loop of radius 8 cm that will give a magnetic field of 2 G at the center of the loop. (*Answer*   25.5 A)

Figure 27-5 shows the geometry for calculating the magnetic field at a point on the axis of a circular current loop a distance x from the circular loop's center. We first consider the current element at the top of the loop. Here, as everywhere on the loop, $I\, d\vec{\ell}$ is tangent to the loop and perpendicular to the vector $\vec{r}$ from the current element to the field point P. The magnetic field $d\vec{B}$ due to this element is in the direction shown in the figure, perpendicular to $\hat{r}$ and also perpendicular to $I\, d\vec{\ell}$. The magnitude of $d\vec{B}$ is

$$|d\vec{B}| = \frac{\mu_0}{4\pi}\frac{I|d\vec{\ell}\times\hat{r}|}{r^2} = \frac{\mu_0}{4\pi}\frac{I\, d\ell}{(x^2 + R^2)}$$

where we have used the facts that $r^2 = x^2 + R^2$ and that $d\vec{\ell}$ and $\hat{r}$ are perpendicular, so $|d\vec{\ell}\times\hat{r}| = d\ell$.

When we sum around all the current elements in the loop, the components of $d\vec{B}$ perpendicular to the axis of the loop, such as dB_y in Figure 27-5, sum to zero, which leave only the components dB_x that are parallel to the axis. We thus compute only the x component of the field. From Figure 27-5, we have

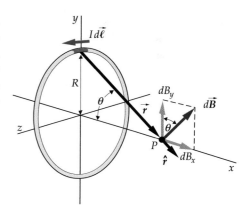

FIGURE 27-5 Geometry for calculating the magnetic field at a point on the axis of a circular current loop.

$$dB_x = dB\sin\theta = \left(\frac{\mu_0}{4\pi}\frac{I\, d\ell}{(x^2 + R^2)}\right)\left(\frac{R}{\sqrt{x^2 + R^2}}\right) = \frac{\mu_0}{4\pi}\frac{IR\, d\ell}{(x^2 + R^2)^{3/2}}$$

To find the field due to the entire loop of current, we integrate dB_x around the loop:

$$B_x = \oint dB_x = \oint \frac{\mu_0}{4\pi}\frac{IR}{(x^2 + R^2)^{3/2}}\, d\ell$$

Since neither x nor R varies as we sum over the elements in the loop, we can remove these quantities from the integral. Then,

$$B_x = \frac{\mu_0}{4\pi} \frac{IR}{(x^2 + R^2)^{3/2}} \oint d\ell$$

The integral of $d\ell$ around the loop gives $2\pi R$. Thus,

$$B_x = \frac{\mu_0}{4\pi} \frac{IR}{(x^2 + R^2)^{3/2}} 2\pi R = \frac{\mu_0}{4\pi} \frac{2\pi R^2 I}{(x^2 + R^2)^{3/2}} \qquad 27\text{-}5$$

B ON THE AXIS OF A CURRENT LOOP

EXERCISE Show that Equation 27-5 reduces to $B_x = \mu_0 I/2R$ (Equation 27-4) at the center of the loop.

At great distances from the loop, $|x|$ is much greater than R, so $(x^2 + R^2)^{3/2} \approx (x^2)^{3/2} = |x|^3$. Then,

$$B_x = \frac{\mu_0}{4\pi} \frac{2I\pi R^2}{|x|^3}$$

or

$$B_x = \frac{\mu_0}{4\pi} \frac{2\mu}{|x|^3} \qquad 27\text{-}6$$

MAGNETIC-DIPOLE FIELD ON THE AXIS OF THE DIPOLE

where $\mu = I\pi R^2$ is the magnitude of the magnetic moment of the loop. Note the similarity of this expression and the electric field on the axis of an electric dipole of moment p (Equation 21-10):

$$E_x = \frac{1}{4\pi \epsilon_0} \frac{2p}{|x|^3}$$

Although it has not been demonstrated, our result that a current loop produces a magnetic dipole field far away holds in general for any point whether it is on the axis of the loop or off of the axis of the loop. Thus, a current loop behaves as a magnetic dipole because it experiences a torque $\vec{\mu} \times \vec{B}$ when placed in an external magnetic field (as was shown in Chapter 26) and it also produces a magnetic dipole field at a great distance from the current loop. Figure 27-6 shows the magnetic field lines for a current loop.

(a)

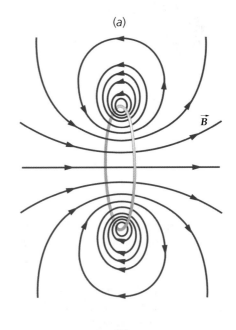

(b)

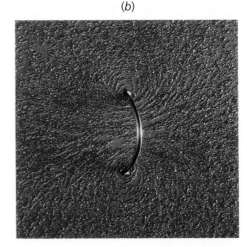

FIGURE 27-6 (a) The magnetic field lines of a circular current loop. (b) The magnetic field lines of a circular current loop indicated by iron filings.

EXAMPLE 27-2

A circular coil of radius 5.0 cm has 12 turns and lies in the $x = 0$ plane and is centered at the origin. It carries a current of 4 A so that the direction of the magnetic moment of the coil is along the x axis. Using Equation 27-5, find the magnetic field on the x axis at (a) $x = 0$, (b) $x = 15$ cm, and (c) $x = 3$ m. (d) Using Equation 27-6, find the magnetic field on the x axis at $x = 3$ m.

PICTURE THE PROBLEM The magnetic field due to a loop with N turns is N times that due to a single turn. (a) At $x = 0$ (center of the loops) $B = \mu_0 N/(2R)$ (from Equation 27-4). Equation 27-5 gives the magnetic field on axis due to the current in a single turn. Far from the loop, as in Part (c), the field can be found using Equation 27-6. In this case, since we have N loops, the magnetic moment is $\mu = NI\pi R^2$.

(a) B_x at the center is N times that given by Equation 27-4 for a single loop:

$$B_x = \frac{\mu_0 NI}{2R}$$

$$= (4\pi \times 10^{-7}\,\text{T·m/A})\frac{(12)(4\,\text{A})}{2(0.05\,\text{m})} = \boxed{6.03 \times 10^{-4}\,\text{T}}$$

(b) B_x on the axis is N times that given by Equation 27-5:

$$B_x = \frac{\mu_0}{4\pi}\frac{2\pi R^2 NI}{(x^2 + R^2)^{3/2}}$$

$$= (10^{-7}\,\text{T·m/A})\frac{2\pi(0.05\,\text{m})^2(12)(4\,\text{A})}{\left[(0.15\,\text{m})^2 + (0.05\,\text{m})^2\right]^{3/2}}$$

$$= \boxed{1.91 \times 10^{-5}\,\text{T}}$$

(c) Use Equation 27-5 again:

$$B_x = \frac{\mu_0}{4\pi}\frac{2\pi R^2 NI}{(x^2 + R^2)^{3/2}}$$

$$= (10^{-7}\,\text{T·m/A})\frac{2\pi(0.05\,\text{m})^2(12)(4\,\text{A})}{\left[(3\,\text{m})^2 + (0.05\,\text{m})^2\right]^{3/2}}$$

$$= \boxed{2.791 \times 10^{-9}\,\text{T}}$$

(d) 1. Since 3 m is much greater than the radius $R = 0.05$ m, we can use Equation 27-6 for the magnetic field far from the loop:

$$B_x = \frac{\mu_0}{4\pi}\frac{2\mu}{|x|^3}$$

2. The magnitude of the magnetic moment of the loop is N/A:

$$\mu = NI\pi R^2 = (12)(4\,\text{A})\pi(0.05\,\text{m})^2 = 0.377\,\text{A·m}^2$$

3. Substitute μ and $x = 3$ m into B_x in step 1:

$$B_x = \frac{\mu_0}{4\pi}\frac{2\mu}{|x|^3} = (10^{-7}\,\text{T·m/A})\frac{2(0.377\,\text{A·m}^2)}{(3\,\text{m})^3}$$

$$= \boxed{2.793 \times 10^{-9}\,\text{T}}$$

REMARKS In Part (d) $x = 60R$, so we were able to use an approximation that is valid for $x \gg R$. The result differs from the exact value, calculated in Part (c), by less than one tenth of one percent.

CIRCULATING THE AMOUNT OF MOBILE CHARGE **EXAMPLE 27-3**

In the coil described in Example 27-2 the current is 4 A. Assuming the drift speed is 1.4×10^{-4} m/s, find the number of coulombs of mobile charge in the wire. (The drift speed for a wire carrying a current of 1 A was found to be 3.4×10^{-5} m/s in Example 25-1.)

PICTURE THE PROBLEM The amount of moving charge Q in the wire is the product of the rate at which charge enters one end of the wire and the time it takes the charge to travel the length of the wire. The rate at which charge enters one end of the wire is the current I, and the time for the charge to travel the length L of the wire is L/v_d, which is the drift speed.

1. The amount of moving charge is the product of the current and the time for a charge carrier to travel the length of the wire:

$$Q = I\,\Delta t$$

2. The drift speed is the length of the wire divided by the time:

$$v_d = \frac{L}{\Delta t}$$

3. The length L is the number of turns times the length per turn. Also, we solve the step 2 result for the time:

$$L = N2\pi R = (12)2\pi(.05\text{ m}) = 3.77\text{ m}$$

and

$$\Delta t = \frac{L}{v_d} = \frac{3.77\text{ m}}{1.4 \times 10^{-4}\text{ m/s}} = 2.69 \times 10^4\text{ s}$$

4. Solve the step 1 result for the amount of moving charge in the wire:

$$Q = I\,\Delta t = (4\text{ A})(2.69 \times 10^4\text{ s})$$

$$= \boxed{1.08 \times 10^5\text{ C}}$$

REMARKS The current consists of more than 10^5 C of moving charges. This is an enormous amount of charge, in comparison to the amount of charge stored in an ordinary capacitor.

TORQUE ON A BAR MAGNET **EXAMPLE 27-4** **Try It Yourself**

A small bar magnet of magnetic moment $\mu = 0.03$ A·m² is placed at the center of the loop of Example 27-2 so that its magnetic moment vector lies in the xy plane and makes an angle of 30° with the x axis. Neglecting any variation in $\vec{B}$ over the region of the magnet, find the torque on the magnet.

PICTURE THE PROBLEM The torque on a magnetic moment is given by $\vec{\tau} = \vec{\mu} \times \vec{B}$. Since $\vec{B}$ is in the positive x direction, you can see from Figure 27-7 that $\vec{\mu} \times \vec{B}$ is in the negative z direction.

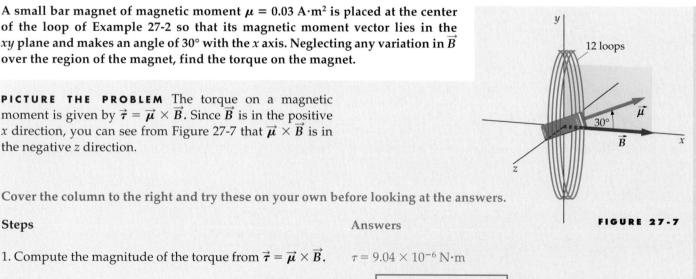

FIGURE 27-7

Cover the column to the right and try these on your own before looking at the answers.

Steps	Answers
1. Compute the magnitude of the torque from $\vec{\tau} = \vec{\mu} \times \vec{B}$.	$\tau = 9.04 \times 10^{-6}$ N·m
2. Indicate the direction with a unit vector.	$\vec{\tau} = \boxed{-(9.04 \times 10^{-6}\text{ N·m})\hat{k}}$

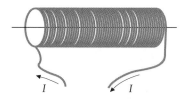

FIGURE 27-8 A tightly wound solenoid can be considered as a set of circular current loops placed side by side that carry the same current. The solenoid produces a uniform magnetic field inside the loops.

$\vec{B}$ Due to a Current in a Solenoid

A **solenoid** is a wire tightly wound into a helix of closely spaced turns, as illustrated in Figure 27-8. A solenoid is used to produce a strong, uniform magnetic field in the region surrounded by its loops. The solenoid's role in magnetism is analogous to that of the parallel-plate capacitor, which produces a strong, uniform electric field between its plates. The magnetic field of a solenoid is essentially that of a set of N identical current loops placed side by side. Figure 27-9 shows the magnetic field lines for two such loops.

Figure 27-10 shows the magnetic field lines for a long, tightly wound solenoid. Inside the solenoid, the field lines are approximately parallel to the axis and are closely and uniformly spaced, indicating a strong, uniform magnetic field. Outside the solenoid, the lines are much less dense. The field lines diverge from one end and converge at the other end. Comparing this figure with Figure 27-8, we see that the field lines of a solenoid, both inside and outside the solenoid, are identical to those of a bar magnet of the same shape as the solenoid.

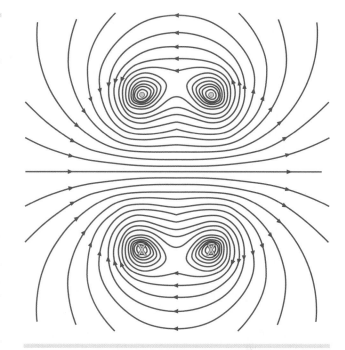

FIGURE 27-9 Magnetic field lines due to two coaxial loops carrying the same current. The points where the loops intersect the plane of the page are each marked by an × where the current enters and by a dot where the current emerges. In the region between the loops near the axis the magnetic fields of the individual loops superpose, so the resultant field is strong and surprisingly uniform. In the regions away from the loops, the resultant field is relatively weak.

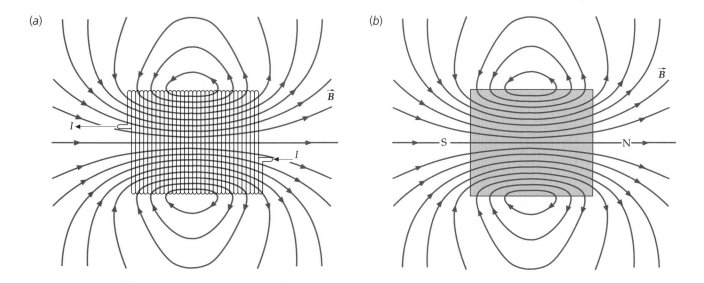

(a)

(b)

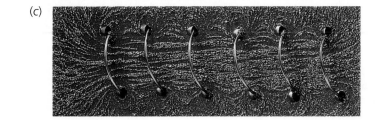

(c)

FIGURE 27-10 (a) Magnetic field lines of a solenoid. The lines are identical to those of a bar magnet of the same shape, as in Figure 27-10 (b). (c) Magnetic field lines of a solenoid shown by iron filings.

Consider a solenoid of length L, which consists of N turns of wire carrying a current I. We choose the axis of the solenoid to be the x axis, with the left end at $x = x_1$ and the right end at $x = x_2$, as shown in Figure 27-11. We will calculate the magnetic field at the origin. The figure shows an element of the solenoid of length dx at a distance x from the origin. If $n = N/L$ is the number of turns per unit length, there are $n\,dx$ turns of wire in this element, with each turn carrying a current I. The element is thus equivalent to a single loop carrying a current $di = nI\,dx$. The magnetic field at a point on the x axis due to a loop at the origin carrying a current $nI\,dx$ is given by Equation 27-5 with I replaced by $di = nI\,dx$:

$$dB_x = \frac{\mu_0}{4\pi} \frac{2\pi R^2 nI\,dx}{(x^2 + R^2)^{3/2}}$$

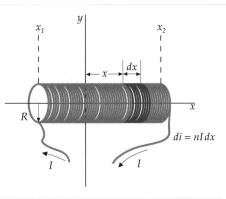

FIGURE 27-11 Geometry for calculating the magnetic field inside a solenoid on its axis. The number of turns in the element dx is $n\,dx$, where $n = N/L$ is the number of turns per unit length. The element dx is treated as a current loop carrying a current $di = nI\,dx$.

This expression also gives the magnetic field at the origin due to a current loop at x. We find the magnetic field at the origin due to the entire solenoid by integrating this expression from $x = x_1$ to $x = x_2$:

$$B_x = \frac{\mu_0}{4\pi} 2\pi R^2 nI \int_{x_1}^{x_2} \frac{dx}{(x^2 + R^2)^{3/2}} \qquad 27\text{-}7$$

The integral in Equation 27-7 can be evaluated using trigonometric substitution with $x = R \tan \theta$. Also, the integral can be looked up in standard tables of integrals. The integral's value is

$$\int_{x_1}^{x_2} \frac{dx}{(x^2 + R^2)^{3/2}} = \frac{x}{R^2\sqrt{x^2 + R^2}} \Big|_{x_1}^{x_2} = \frac{1}{R^2}\left(\frac{x_2}{\sqrt{x_2^2 + R^2}} - \frac{x_1}{\sqrt{x_1^2 + R^2}} \right)$$

Substituting this into Equation 27-7, we obtain

$$B_x = \frac{1}{2} \mu_0 nI \left(\frac{x_2}{\sqrt{x_2^2 + R^2}} - \frac{x_1}{\sqrt{x_1^2 + R^2}} \right) \qquad 27\text{-}8$$

B_x ON THE AXIS OF A SOLENOID AT $X = 0$

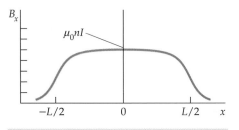

FIGURE 27-12 Graph of the magnetic field on the axis inside a solenoid versus the position x on the axis. The field inside the solenoid is nearly constant except near the ends. The length L of the solenoid is ten times longer than the radius.

A solenoid is called a long solenoid if its length L is much greater than its radius R. Inside and far from the ends of a long solenoid, the left term in the parentheses tends toward $+1$ and the right term tends toward -1. In the region satisfying these conditions, the magnetic field is

$$B_x = \mu_0 nI \qquad 27\text{-}9$$

B_x INSIDE A LONG SOLENOID

If the origin is at the left end of the solenoid, $x_1 = 0$ and $x_2 = L$. Then, if $L \gg R$, the right term in the parentheses of Equation 27-8 is zero and the left term approaches 1, so $B \approx \frac{1}{2}\mu_0 nI$. Thus, the magnitude of $\vec{B}$ at either end of a long solenoid is half the magnitude at points within the solenoid that are distant from either end. Figure 27-12 gives a plot of the magnetic field on the axis of a solenoid versus position x on the axis (with the origin at the center of the solenoid). The approximation that the field is uniform (independent of the position) along the axis is good, except for very near the ends.

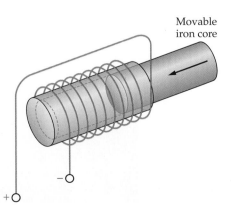

FIGURE 27-13 An automotive starter solenoid. When the solenoid is energized, its magnetic field pulls in the iron core. This engages gears that connect the starter motor to the flywheel of the engine. Once the current to the solenoid is interrupted, a spring disengages the gears and pushes the iron core to the right.

B AT CENTER OF A SOLENOID   **EXAMPLE   2 7 - 5**

Find the magnetic field at the center of a solenoid of length 20 cm, radius 1.4 cm, and 600 turns that carries a current of 4 A.

PICTURE THE PROBLEM

1. We will calculate the field exactly, using Equation 27-8:

$$B_x = \frac{1}{2} \mu_0 n I \left(\frac{x_2}{\sqrt{x_2^2 + R^2}} - \frac{x_1}{\sqrt{x_1^2 + R^2}} \right)$$

2. For a point at the center of the solenoid, $x_1 = -10$ cm and $x_2 = +10$ cm. Thus, the terms in the parentheses in Equation 27-8 have values of:

$$\frac{x_2}{\sqrt{x_2^2 + R^2}} = \frac{10 \text{ cm}}{\sqrt{(10 \text{ cm})^2 + (1.4 \text{ cm})^2}} = 0.990$$

$$\frac{x_1}{\sqrt{x_1^2 + R^2}} = \frac{-10 \text{ cm}}{\sqrt{(-10 \text{ cm})^2 + (1.4 \text{ cm})^2}} = 0.990$$

3. Substitute these results into B_x in step 1:

$$B_x = \frac{1}{2} (4\pi \times 10^{-7} \text{ T·m/A})[(600 \text{ turns})/(0.2 \text{ m})](4 \text{ A})(0.990 + 0.990)$$

$$= \boxed{1.50 \times 10^{-2} \text{ T}}$$

REMARKS Note that the approximation obtained using Equation 27-9 amounts to replacing 0.99 by 1.00, which differs by only one percent. Note also that the magnitude of the magnetic field inside this solenoid is fairly large—about 250 times the magnetic field of the earth.

EXERCISE Calculate B_x using the long-solenoid approximation. (*Answer* 1.51×10^{-2} T)

$\vec{B}$ Due to a Current in a Straight Wire

Figure 27-14 shows the geometry for calculating the magnetic field $\vec{B}$ at a point P due to the current in the straight wire segment shown. We choose R to be the perpendicular distance from the wire to point P, and we choose the x axis to be along the wire with $x = 0$ at the projection of P onto the x axis.

A typical current element $I\,d\vec{\ell}$ at a distance x from the origin is shown. The vector $\vec{r}$ points from the element to the field point P. The direction of the magnetic field at P due to this element is the direction of $I\,d\vec{\ell} \times \hat{r}$, which is out of the paper. Note that the magnetic fields due to all the current elements of the wire are in this same direction. Thus, we need to compute only the magnitude of the field. The field due to the current element shown has the magnitude (Equation 27-3)

$$dB = \frac{\mu_0}{4\pi} \frac{I\,dx}{r^2} \sin \phi$$

It is more convenient to write this in terms of θ rather than ϕ:

$$dB = \frac{\mu_0}{4\pi} \frac{I\,dx}{r^2} \cos \theta \qquad\qquad 27\text{-}10$$

A cross section of a doorbell. When the solenoid is energized, its magnetic field pulls on the plunger, causing it to strike the bell (not shown). The spring returns the plunger to its normal position.

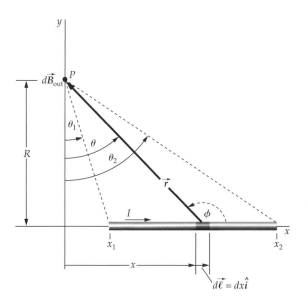

FIGURE 27-14 Geometry for calculating the magnetic field at point P due to a straight current segment. Each element of the segment contributes to the total magnetic field at point P, which is directed out of the paper. The result is expressed in terms of the angles θ_1 and θ_2.

To sum over all the current elements, we need to relate the variables θ, r, and x. It turns out to be easiest to express x and r in terms of θ. We have

$$x = R \tan \theta$$

Then, taking the differential of each side with R as a constant gives

$$dx = R \sec^2 \theta \, d\theta = R \frac{r^2}{R^2} d\theta = \frac{r^2}{R} d\theta$$

where we have used $\sec \theta = r/R$. Substituting this expression for dx into Equation 27-10, we obtain

$$dB = \frac{\mu_0}{4\pi} \frac{I}{r^2} \frac{r^2 \, d\theta}{R} \cos \theta = \frac{\mu_0}{4\pi} \frac{I}{R} \cos \theta \, d\theta$$

We sum over these elements by integrating from $\theta = \theta_1$ to $\theta = \theta_2$, where θ_1 and θ_2 are shown in Figure 27-14. This gives

$$B = \int_{\theta_1}^{\theta_2} \frac{\mu_0}{4\pi} \frac{I}{R} \cos \theta \, d\theta = \frac{\mu_0}{4\pi} \frac{I}{R} \int_{\theta_1}^{\theta_2} \cos \theta \, d\theta$$

Evaluating the integral, we obtain

$$B = \frac{\mu_0}{4\pi} \frac{I}{R} (\sin \theta_2 - \sin \theta_1) \qquad \text{27-11}$$

B DUE TO A STRAIGHT WIRE SEGMENT

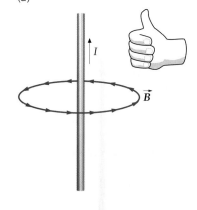

(a)

This result gives the magnetic field due to any wire segment in terms of the perpendicular distance R and θ_1 and θ_2 are the angles subtended at the field point by the ends of the wire. If the length of the wire approaches infinity in both directions, θ_2 approaches $+90°$ and θ_1 approaches $-90°$. The result for such a very long wire is obtained from Equation 27-11, by setting $\theta_1 = -90°$ and $\theta_2 = +90°$:

$$B = \frac{\mu_0}{4\pi} \frac{2I}{R} \qquad \text{27-12}$$

B DUE TO AN INFINITELY LONG, STRAIGHT WIRE

At any point in space, the magnetic field lines of a long, straight, current-carrying wire are tangent to a circle of radius R about the wire, where R is the perpendicular distance from the wire to the field point. The direction of $\vec{B}$ can be determined by applying the right-hand rule, as shown in Figure 27-15a. The magnetic field lines thus encircle the wire, as shown in Figure 27-15b.

The result expressed by Equation 27-12 was found experimentally by Biot and Savart in 1820. From their analysis, Biot and Savart were able to discover the expression given in Equation 27-3 for the magnetic field due to a current element.

(b)

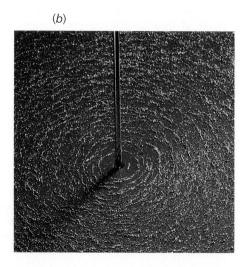

FIGURE 27-15 (a) Right-hand rule for determining the direction of the magnetic field due to a long, straight, current-carrying wire. The magnetic field lines encircle the wire in the direction of the fingers of the right hand when the thumb points in the direction of the current. (b) Magnetic field lines due to a long wire, which is indicated by iron filings.

$\vec{B}$ AT CENTER OF SQUARE CURRENT LOOP   **EXAMPLE   27-6**

Find the magnetic field at the center of a square current loop of edge length $L = 50$ cm, which carries a current of 1.5 A.

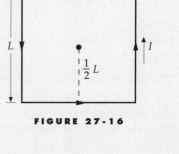

FIGURE 27-16

PICTURE THE PROBLEM The magnetic field at the center of the loop is the sum of the contributions from each of the four sides of the loop. From Figure 27-16, we can see that each side of the loop produces a field of equal magnitude pointing out of the page. Thus, we use Equation 27-11 for a given side, then multiply by 4 for the total field.

1. The total field is 4 times the field B_s due to a side:

$$B = 4B_s$$

2. Calculate the magnetic field B_s due to a given side of the loop. Note from the figure that $R = \frac{1}{2}L$ and $\theta_1 = -45°$ and $\theta_2 = +45°$:

$$B_s = \frac{\mu_0}{4\pi}\frac{I}{R}(\sin\theta_2 - \sin\theta_1) = \frac{\mu_0}{4\pi}\frac{I}{\frac{1}{2}L}\left[\sin(+45°) - \sin(-45°)\right]$$

$$= (10^{-7}\,\text{T·m/A})\frac{1.5\,\text{A}}{0.25\,\text{m}}\,2\sin 45° = 8.49 \times 10^{-7}\,\text{T}$$

3. Multiply this value by 4 to find the total field:

$$B = 4B_s = 4(8.49 \times 10^{-7}\,\text{T}) = \boxed{3.39 \times 19^{-6}\,\text{T}}$$

EXERCISE Compare the magnetic field at the center of a circular current loop of radius R with the magnetic field at the center of a square current loop of side $L = 2R$ carrying the same current. Which is larger? (*Answer* B at the center is larger for the circle, by about 10 percent)

EXERCISE Find the distance from a long, straight wire carrying a current of 12 A, where the magnetic field due to the current in the wire is equal in magnitude to 0.6 G (the magnitude of the earth's magnetic field). (*Answer* $R = 4.00$ cm)

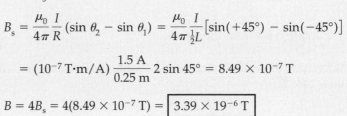

A current gun used to measure electric current. The jaws of the current gun clamp around a current-carrying wire without touching the wire. The magnetic field produced by the wire is measured with a Hall-effect device mounted in the current gun. The Hall-effect device puts out a voltage proportional to the magnetic field, which in turn is proportional to the current in the wire.

$\vec{B}$ DUE TO TWO PARALLEL WIRES   **EXAMPLE   27-7**

A long, straight wire carrying a current of 1.7 A in the positive z direction lies along the line $x = -3$ cm, $y = 0$. A second such wire carrying a current of 1.7 A in the positive z direction lies along the line $x = +3$ cm, $y = 0$, as shown in Figure 27-17. Find the magnetic field at a point P on the y axis at $y = 6$ cm.

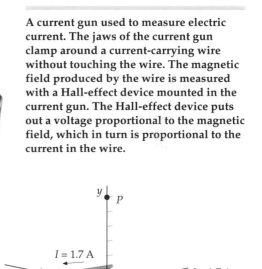

FIGURE 27-17

PICTURE THE PROBLEM The magnetic field at point P is the vector sum of the field $\vec{B}_L$ due to the wire on the left in Figure 27-18, and the field $\vec{B}_R$ due to the wire on the right. Since each wire carries the same current, and each wire is the same distance from point P, the magnitudes B_L and B_R are equal. $\vec{B}_L$ is perpendicular to the radius from the left wire to point P, and $\vec{B}_R$ is perpendicular to the radius from the right wire to the point P.

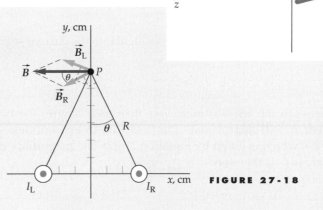

FIGURE 27-18

1. The field at P is the vector sum of the fields $\vec{B}_L$ and $\vec{B}_R$:

$$\vec{B} = \vec{B}_L + \vec{B}_R$$

2. From Figure 27-18, we see that the resultant magnetic field is in the negative x direction and has the magnitude $2B_L \cos \theta$.

$$\vec{B} = -2B_L \cos \theta \, \hat{i}$$

3. The magnitudes of $\vec{B}_L$ and $\vec{B}_R$ are given by Equation 27-12:

$$B_L = B_R = \frac{\mu_0}{4\pi} \frac{2I}{R}$$

4. R is the distance from each wire to the point P. We find R from the figure and substitute R into the expression for B_L and B_R:

$$R = \sqrt{(3 \text{ cm})^2 + (6 \text{ cm})^2} = 6.71 \text{ cm}$$

so

$$B_L = B_R = (10^{-7} \text{ T·m/A}) \frac{2(1.7 \text{ A})}{0.0671 \text{ m}} = 5.07 \times 10^{-6} \text{ T}$$

5. We obtain $\cos \theta$ from the figure:

$$\cos \theta = \frac{6 \text{ cm}}{R} = \frac{6 \text{ cm}}{6.71 \text{ cm}} = 0.894$$

6. Substitute the values of $\cos \theta$ and B_L into the equation in step 2 for $\vec{B}$:

$$\vec{B} = -2(5.07 \times 10^{-6} \text{ T})(0.894)\hat{i} = \boxed{-9.07 \times 10^{-6} \text{ T } \hat{i}}$$

EXERCISE Find $\vec{B}$ at the origin. (*Answer* 0)

EXERCISE Find $\vec{B}$ at the origin assuming that I_R goes into the page. (*Answer* $\vec{B} = 2.27 \times 10^{-5} \text{ T } \hat{j}$)

Magnetic Force Between Parallel Wires

We can use Equation 27-12 for the magnetic field due to a long, straight, current-carrying wire and $d\vec{F} = I \, d\vec{\ell} \times \vec{B}$ (Equation 26-5) for the force exerted by a magnetic field on a segment of a current-carrying wire to find the force exerted by one long straight current on another. Figure 27-19 shows two long parallel wires carrying currents in the same direction. We consider the force on a segment $d\vec{\ell}_2$ carrying current I_2, as shown. The magnetic field $\vec{B}_1$ at this segment due to current I_1 is perpendicular to the segment $I_2 \, d\vec{\ell}_2$, as shown. This is true for all current elements along the wire. The magnetic force $d\vec{F}_2$ on current segment $I_2 \, d\vec{\ell}_2$ is directed toward current I_1, since $d\vec{F}_2 = I_2 \, d\vec{\ell}_2 \times \vec{B}_2$. Similarly, a current segment $I_1 \, d\vec{\ell}_1$ will experience a magnetic force directed toward current I_2 due to a magnetic field arising from current I_2. Thus, two parallel currents attract each other. If one of the currents is reversed the force will be reversed, so two antiparallel currents will repel each other. The attraction or repulsion of parallel or antiparallel currents was discovered experimentally by Ampère one week after he heard of Oersted's discovery of the effect of a current on a compass needle.

The magnitude of the magnetic force on the segment $I_2 \, d\vec{\ell}_2$ is

$$dF_2 = |I_2 \, d\vec{\ell}_2 \times \vec{B}_1|$$

Since the magnetic field at segment $I_2 \, d\vec{\ell}_2$ is perpendicular to the current segment, we have

$$dF_2 = I_2 \, d\ell_2 \, B_1$$

If the distance R between the wires is much less than their length, the field at $I_2 \, d\vec{\ell}_2$ due to current I_1 will approximate the field due to an infinitely long, current-carrying wire, which is given by Equation 27-12. The magnitude of the force on the segment $I_2 \, d\vec{\ell}_2$ is therefore

$$dF_2 = I_2 \, d\ell_2 \frac{\mu_0 I_1}{2\pi R}$$

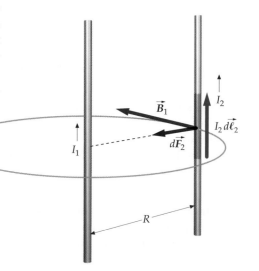

FIGURE 27-19 Two long straight wires carrying parallel currents. The magnetic field $\vec{B}_1$ due to current I_1 is perpendicular to current I_2. The force on current I_2 is toward current I_1. There is an equal and opposite force exerted by current I_2 on I_1. The current-carrying wires thus attract each other.

The force per unit length is

$$\frac{dF_2}{d\ell_2} = I_2 \frac{\mu_0 I_1}{2\pi R} = 2\frac{\mu_0}{4\pi}\frac{I_1 I_2}{R}$$ 27-13

In Chapter 21, the coulomb was defined in terms of the ampere, but the definition of the ampere was deferred. The ampere is defined as follows:

> The ampere is the constant electric current that, when maintained in two straight parallel conductors of infinite length and of negligible circular cross sections placed one meter apart in a vacuum, would produce a force between the conductors equal to 2×10^{-7} newtons per meter of length.

DEFINITION—AMPERE

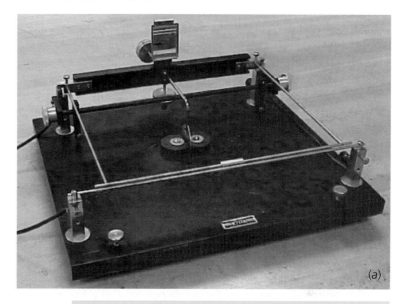

(a)

This definition of the ampere makes the permeability of free space μ_0 equal to exactly $4\pi \times 10^{-7}\,\text{N/A}^2$. It also allows the unit of current (and therefore the unit of electric charge) to be determined by a mechanical measurement. In practice, currents much closer together than 1 m are used so that the force can be measured accurately with long but finite wires.

Figure 27-20 shows a **current balance,** which is a device that can be used to calibrate an ammeter from the definition of the ampere. The upper conductor, directly above the lower conductor, is free to rotate about knife-edge contacts and is balanced so that the wires (or conducting rods) are a small distance apart. The conductors are connected in series to carry the same current but in opposite directions so that the currents will repel each other. Weights are placed on the upper conductor until it balances again at the original separation. The force of repulsion is thus determined by measuring the total weight required to balance the upper conductor.

FIGURE 27-20 (a) A picture of a current balance used in a general physics lab. (b) A schematic diagram of a current balance. The two parallel rods in front carry equal but oppositely directed currents and therefore repel each other. The force of repulsion is balanced by weights placed on the upper rod, which is part of a rectangle that is balanced on knife edges at the back. The mirror on top is used to reflect a beam of laser light to accurately determine the position of the upper rod.

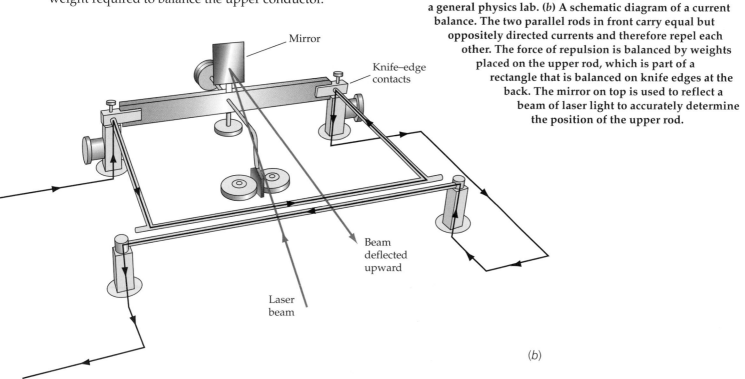

(b)

BALANCING THE MAGNETIC FORCE **EXAMPLE 27-8** Try It Yourself

Two straight rods 50-cm long with axes 1.5-mm apart in a current balance carry currents of 15 A each in opposite directions. What mass must be placed on the upper rod to balance the magnetic force of repulsion?

PICTURE THE PROBLEM Equation 27-13 gives the magnitude of the magnetic force per unit length exerted by the lower rod on the upper rod. Find this force for a rod of length L and set it equal to the weight mg.

Cover the column to the right and try these on your own before looking at the answers.

Steps   Answers

1. Set the weight mg equal to the magnetic force of repul- $mg = 2 \dfrac{\mu_0}{4\pi} \dfrac{I_1 I_2}{R} L$
   sion of the rods.

2. Solve for the mass m. $m = 1.53 \times 10^{-3} \text{ kg} = \boxed{1.53 \text{ g}}$

REMARKS Since only 1.53 g are required to balance the system, we see that the magnetic force between two straight current-carrying wires is relatively small, even for currents as large as 15 A separated by only 1.5 mm.

27-3 Gauss's Law for Magnetism

The magnetic field lines shown in Figure 27-6, Figure 27-9, and Figure 27-10 differ from electric field lines because the lines of $\vec{B}$ form closed curves, whereas lines of $\vec{E}$ begin and end on electric charges. The magnetic equivalent of an electric charge is a magnetic pole, such as appears to be at the ends of a bar magnet. Magnetic field lines appear to diverge from the north-pole end of a bar magnet (Figure 27-10b) and appear to converge on the south-pole end. However, inside the magnet the magnetic field lines neither diverge from a point near the north-pole end, nor do they converge on a point near the south-pole end. Instead, the magnetic field lines pass through the bar magnet from the south-pole end to the north-pole end, as shown in Figure 27-10b. If a Gaussian surface encloses one end of a bar magnet, the number of magnetic field lines that leave through the surface is exactly equal to the number of magnetic field lines that enter through the surface. That is, the net flux $\phi_{m,\,net}$ of the magnetic field through any closed surface S is always zero.[†]

$$\phi_{m,\,net} = \oint_S B_n \, dA = 0 \qquad\qquad 27\text{-}14$$

GAUSS'S LAW FOR MAGNETISM

where B_n is the component of $\vec{B}$ normal to surface S at area element dA. The definition of the magnetic flux ϕ_m is exactly analogous to the electric flux, with $\vec{B}$ replacing $\vec{E}$. This result is called Gauss's law for magnetism. It is the mathematical statement that there exist no points in space from which magnetic field lines diverge, or to which magnetic field lines converge. That is, isolated magnetic poles do not exist.[‡] The fundamental unit of magnetism is the magnetic

---

[†] Recall that the net flux of the electric field is a measure of the net number of field lines that leave a closed surface and is equal to Q_{inside}/ϵ_0.

[‡] The existence of magnetic monopoles is a subject of great debate, and the search for magnetic monopoles remains active. To date, however, none have been discovered.

dipole. Figure 27-21 compares the field lines of $\vec{B}$ for a magnetic dipole with the field lines of $\vec{E}$ for an electric dipole. Note that far from the dipoles the field lines are identical. But inside the dipole, the field lines of $\vec{E}$ are directed opposite to the field lines of $\vec{B}$. The field lines of $\vec{E}$ diverge from the positive charge and converge to the negative charge, whereas the field lines of $\vec{B}$ are continuous loops.

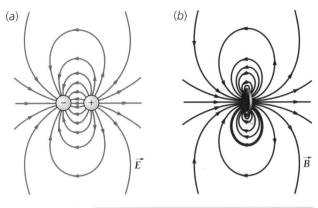

27-4 Ampère's Law

In Chapter 22, we found that for highly symmetric charge distributions we could calculate the electric field more easily using Gauss's law than Coulomb's law. A similar situation exists in magnetism. Ampère's law relates the tangential component B_t of the magnetic field summed (integrated) around a closed curve C to the current I_C that passes through any surface bounded by C. It can be used to obtain an expression for the magnetic field in situations that have a high degree of symmetry. In mathematical form, **Ampère's law** is

$$\oint_C B_t \, d\ell = \oint_C \vec{B} \cdot d\vec{\ell} = \mu_0 I_C, \quad C \text{ is any closed curve} \qquad 27\text{-}15$$

AMPÈRE'S LAW

FIGURE 27-21 (a) Electric field lines of an electric dipole. (b) Magnetic field lines of a magnetic dipole. Far from the dipoles, the field lines are identical. In the region between the charges in Figure 27-21(a), the electric field lines are opposite the direction of the dipole moment, whereas inside the loop in Figure 27-21(b), the magnetic field lines are parallel to the direction of the dipole moment.

where I_C is the net current that penetrates any surface S bounded by the curve C. The positive tangential direction for the path integral is related to the choice for the positive direction for the current I_C through S by the right-hand rule shown in Figure 27-22. Ampère's law holds for any curve C, as long as the currents are steady and continuous. This means the current does not change in time and that charge is not accumulating anywhere. Ampère's law is useful in calculating the magnetic field $\vec{B}$ in situations that have a high degree of symmetry so that the line integral $\oint_C \vec{B} \cdot d\vec{\ell}$ can be written as $B \oint_C d\ell$ (the product of B and some distance). The integral $\oint_C \vec{B} \cdot d\vec{\ell}$ is called a **circulation integral**. More specifically, $\oint_C \vec{B} \cdot d\vec{\ell}$ is called the circulation of $\vec{B}$ around curve C. Ampère's law and Gauss's law are both of considerable theoretical importance, and both laws hold whether there is symmetry or there is no symmetry. If there is no symmetry, neither law is very useful in calculating electric or magnetic fields.

The simplest application of Ampère's law is to find the magnetic field of an infinitely long, straight, current-carrying wire. Figure 27-23 shows a circular curve around a long wire with its center at the wire. We know the direction of the magnetic field due to each current element is tangent to this circle from the Biot–Savart law. Assuming that the magnetic field is tangent to this circle, that the magnetic field is in the same direction as $d\vec{\ell}$, and that the magnetic field has the same magnitude B at any point on the circle, Ampère's law ($\oint_C B_t \, d\vec{\ell} = \mu_0 I_C$) then gives

$$B \oint_C d\ell = \mu_0 I_C$$

where $B = B_t$. We can factor B out of the integral because B has the same value everywhere on the circle. The integral of $d\ell$ around the circle equals $2\pi R$ (the circumference of the circle). The current I_C is the current I in the wire. We thus obtain $B 2\pi R = \mu_0 I$

$$B = \frac{\mu_0 I}{2\pi R}$$

which is Equation 27-12.

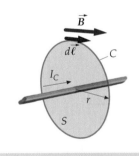

FIGURE 27-22 The positive direction for the path integral for Ampère's law is related to the positive direction for the current passing through the surface by a right-hand rule.

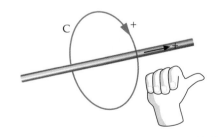

FIGURE 27-23 Geometry for calculating the magnetic field of a long, straight, current-carrying wire using Ampère's law. On a circle around the wire, the magnetic field is constant and tangent to the circle.

$\vec{B}$ INSIDE AND OUTSIDE A WIRE   **EXAMPLE 27-9**

A long, straight wire of radius R carries a current I that is uniformly distributed over the circular cross section of the wire. Find the magnetic field both outside the wire and inside the wire.

PICTURE THE PROBLEM We can use Ampère's law to calculate $\vec{B}$ because of the high degree of symmetry. At a distance r (Figure 27-24), we know that $\vec{B}$ is tangent to the circle of radius r about the wire and $\vec{B}$ is constant in magnitude everywhere on the circle. The current through the surface S bounded by C depends on whether r is less than or greater than the radius of the wire R.

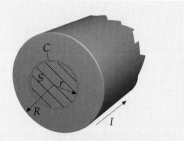

FIGURE 27-24

1. Ampère's law is used to relate the circulation of $\vec{B}$ around curve C to the current passing through the surface S bounded by C:

$$\oint_C \vec{B} \cdot d\vec{\ell} = \mu_0 I_C$$

2. Evaluate the circulation of $\vec{B}$ around a circle of radius r that is coaxial with the wire:

$$\oint_C \vec{B} \cdot d\vec{\ell} = B \oint_C d\ell = B2\pi r$$

3. Substitute into Ampère's law and solve for B:

$$B2\pi r = \mu_0 I_C$$

so

$$B = \frac{\mu_0 I_C}{2\pi r}$$

4. Outside the wire, $r > R$, and the total current passes through the surface bounded by C:

$$I_C = I$$

$$\boxed{B = \frac{\mu_0 I}{2\pi r} \quad (r \geq R)}$$

5. Inside the wire, $r < R$. Assume that the current is distributed uniformly to solve for I_C. Solve for B:

$$\frac{I_C}{\pi r^2} = \frac{I}{\pi R^2}$$

or

$$\left(I_C = \frac{r^2}{R^2} I\right)$$

so

$$B = \frac{\mu_0}{2\pi}\frac{I_C}{r} = \frac{\mu_0}{2\pi}\frac{(r^2/R^2)I}{r} = \boxed{\frac{\mu_0}{2\pi}\frac{I}{R^2}r \quad r \leq R}$$

REMARKS Inside the wire, the field increases with distance from the center of the wire. Figure 27-25 shows the graph of B versus r for this example.

We see from Example 27-9 that the magnetic field due to a current uniformly distributed over a wire of radius R is given by

$$B = \frac{\mu_0 I}{2\pi R^2}r \quad (r \leq R)$$

$$B = \frac{\mu_0 I}{2\pi r} \quad (r \geq R)$$ 27-16

For the next application of Ampère's law, we calculate the magnetic field of a tightly wound **toroid,** which consists of loops of wire wound around a doughnut-

FIGURE 27-25

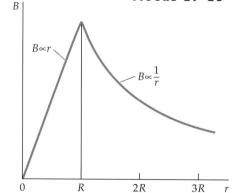

shaped form, as shown in Figure 27-26. There are N turns of wire, each carrying a current I. To calculate B, we evaluate the line integral $\oint_C \vec{B} \cdot d\vec{\ell}$ around a circle of radius r centered in the middle of the toroid. By symmetry, $\vec{B}$ is tangent to this circle and constant in magnitude at every point on the circle. Then,

$$\oint_C \vec{B} \cdot d\vec{\ell} = B2\pi r = \mu_0 I_C$$

Let a and b be the inner and outer radii of the toroid, respectively. The total current through the surface S bounded by a circle of radius r for $a < r < b$ is NI. Ampère's law then gives

$$\oint_C \vec{B} \cdot d\vec{\ell} = \mu_0 I_C, \quad \text{or} \quad (B2\pi r = \mu_0 NI)$$

or

$$B = \frac{\mu_0 NI}{2\pi r}, \quad a < r < b \qquad\qquad 27\text{-}17$$

B INSIDE A TIGHTLY WOUND TOROID

If r is less than a, there is no current through the surface S. If r is greater than b, the total current through S is zero because for each turn of the wire the current penetrates the surface twice (Figure 27-27), once going into the page and once coming out of the page. Thus, the magnetic field is zero for both $r < a$ and $r > b$:

$$B = 0, \quad r < a \quad \text{or} \quad r > b$$

The magnetic field intensity inside the toroid is not uniform but decreases with increasing r. However, if the radius of the loops of the coil, $\frac{1}{2}(b - a)$ is much less than the radius $\frac{1}{2}(b + a)$ of the center of the loops, the variation in r from $r = a$ to $r = b$ is small and B is approximately uniform, as it is in a solenoid.

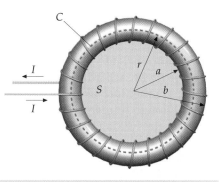

FIGURE 27-26 A toroid consists of loops of wire wound around a doughnut-shaped form. The magnetic field at any distance r can be found by applying Ampère's law to the circle of radius r. The surface S is bounded by curve C. The wire penetrates S once for each turn.

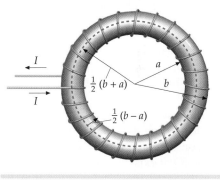

FIGURE 27-27 The toroid has mean radius $r = \frac{1}{2}(b + a)$, where a and b are the inner and outer radii of the toroid. Each turn of the wire is a circle of radius $\frac{1}{2}(b - a)$.

(a)   (b)

(a) The Tokamak fusion-test reactor is a large toroid that produces a magnetic field for confining charged particles. Coils containing over 10 km of water-cooled copper wire carry a pulsed current, which has a peak value of 73,000 A and produces a magnetic field of 5.2 T for about 3 s. (b) Inspection of the assembly of the Tokamak reactor from inside the toroid.

Limitations of Ampère's Law

Ampère's law is useful for calculating the magnetic field only when there is both a steady current and a high degree of symmetry. Consider the current loop shown in Figure 27-28. According to Ampère's law, the line integral $\oint_C \vec{B} \cdot d\vec{\ell} = \oint_C B_t \, d\ell$ around a curve, such as curve C in the figure, equals μ_0 times the current I in the loop. Although Ampère's law is valid for this curve, the tangential component of magnetic field B_t is not constant along any curve encircling the current. Thus, there is not enough symmetry in this situation to allow us to evaluate the integral $\oint_C B_t \, d\ell$ and solve for B_t.

Figure 27-29 shows a finite current segment of length ℓ. We wish to find the magnetic field at point P, which is equidistant from the ends of the segment and at a distance r from the center of the segment. A direct application of Ampère's law gives

$$B = \frac{\mu_0}{2\pi} \frac{I}{r}$$

This result is the same as for an infinitely long wire, since the same symmetry arguments apply. It does not agree with the result obtained from the Biot–Savart law, which depends on the length of the current segment and which agrees with experiment. If the current segment is just one part of a continuous circuit carrying a current, as shown in Figure 27-30, Ampère's law for curve C is valid, but it cannot be used to find the magnetic field at point P because there is insufficient symmetry.

In Figure 27-31, the current in the segment arises from a small spherical conductor with initial charge $+Q$ at the left of the segment and another small spherical conductor at the right with charge $-Q$. When they are connected, a current $I = -dQ/dt$ exists in the segment for a short time, until the spheres are uncharged. For this case, we *do* have the symmetry needed to assume that $\vec{B}$ is tangential to the curve and $\vec{B}$ is constant in magnitude along the curve. For a situation like this, in which the current is discontinuous in space, Ampère's law is not valid. In Chapter 30, we will see how Maxwell was able to modify Ampère's law so that it holds for all currents. When Maxwell's generalized form of Ampère's law is used to calculate the magnetic field for a current segment, such as the current segment shown in Figure 27-31, the result agrees with the result found from the Biot–Savart law.

27-5 Magnetism in Matter

Atoms have magnetic dipole moments due to the motion of their electrons and due to the intrinsic magnetic dipole moment associated with the spin of the electrons. Unlike the situation with electric dipoles, the alignment of magnetic dipoles parallel to an external magnetic field tends to *increase* the field. We can see this difference by comparing the electric field lines of an electric dipole with the magnetic field lines of a magnetic dipole, such as a small current loop, as was shown in Figure 27-21. Far from the dipoles, the field lines are identical. However, between the charges of the electric dipole, the electric field lines are opposite the direction of the dipole moment, whereas inside the current loop, the magnetic field lines are parallel to the magnetic dipole moment. Thus, inside a magnetically polarized material, the magnetic dipoles create a magnetic field that is parallel to the magnetic dipole moment vectors.

Materials fall into three categories—**paramagnetic, diamagnetic,** and **ferromagnetic**—according to the behavior of their magnetic moments in an external magnetic field. Paramagnetism arises from the partial alignment of the electron spins (in metals) or from the atomic or molecular magnetic moments by an

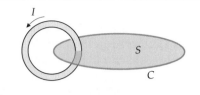

FIGURE 27-28 Ampère's law holds for the curve C encircling the current in the circular loop, but it is not useful for finding B_t, because B_t cannot be factored out of the circulation integral.

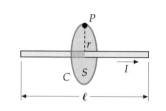

FIGURE 27-29 The application of Ampère's law to find the magnetic field on the bisector of a finite current segment gives an incorrect result.

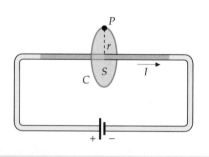

FIGURE 27-30 If the current segment in Figure 27-28 is part of a complete circuit, Ampère's law for the curve C is valid, but there is not enough symmetry to use Ampère's law to find the magnetic field at point P.

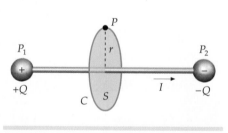

FIGURE 27-31 If the current segment in Figure 27-29 is due to a momentary flow of charge from a small conductor on the left to a small conductor on the right, there is enough symmetry to use Ampère's law to compute the magnetic field at P, but Ampère's law is not valid because the current is not continuous in space.

applied magnetic field in the direction of the field. In paramagnetic materials, the magnetic dipoles do not interact strongly with each other and are normally randomly oriented. In the presence of an applied magnetic field, the dipoles are partially aligned in the direction of the field, thereby increasing the field. However, in external magnetic fields of ordinary strength at ordinary temperatures, only a very small fraction of the molecules are aligned because thermal motion tends to randomize their orientation. The increase in the total magnetic field is therefore very small. Ferromagnetism is much more complicated. Because of a strong interaction between neighboring magnetic dipoles, a high degree of alignment occurs even in weak external magnetic fields, which causes a very large increase in the total field. Even when there is no external magnetic field, a ferromagnetic material may have its magnetic dipoles aligned, as in permanent magnets. Diamagnetism arises from the orbital magnetic dipole moments induced by an applied magnetic field. These magnetic moments are opposite the direction of the applied magnetic field, thereby decreasing the field. This effect actually occurs in all materials; however, because the induced magnetic moments are very small compared to the permanent magnetic moments, diamagnetism is often masked by paramagnetic or ferromagnetic effects. Diamagnetism is thus observed only in materials whose molecules have no permanent magnetic moments.

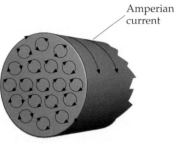

FIGURE 27-32 A model of atomic current loops in which all the atomic dipoles are parallel to the axis of the cylinder. The net current at any point inside the material is zero due to cancellation of neighboring atoms. The result is a surface current similar to that of a solenoid.

Magnetization and Magnetic Susceptibility

When some material is placed in a strong magnetic field, such as that of a solenoid, the magnetic field of the solenoid tends to align the magnetic dipole moments (either permanent or induced) inside the material and the material is said to be magnetized. We describe a magnetized material by its **magnetization** $\vec{M}$, which is defined as the net magnetic dipole moment per unit volume of the material:

$$\vec{M} = \frac{d\vec{\mu}}{dV} \qquad\qquad 27\text{-}18$$

Long before we had any understanding of atomic or molecular structure, Ampère proposed a model of magnetism in which the magnetization of materials is due to microscopic current loops inside the magnetized material. We now know that these current loops are a classical model for the orbital motion and spin of the electrons in atoms. Consider a cylinder of magnetized material. Figure 27-32 shows atomic current loops in the cylinder aligned with their magnetic moments along the axis of the cylinder. Because of cancellation of neighboring current loops, the net current at any point inside the material is zero, leaving a net current on the surface of the material (Figure 27-33). This surface current, called an **amperian current,** is similar to the real current in the windings of the solenoid.

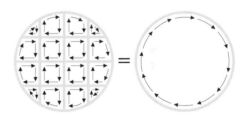

FIGURE 27-33 The currents in the adjacent current loops in the interior of a uniformly magnetized material cancel, leaving only a surface current. Cancellation occurs at every interior point independent of the shape of the loops.

Figure 27-34 shows a small disk of cross-sectional area A, length $d\ell$, and volume $dV = A\,d\ell$. Let di be the amperian current on the surface of the disk. The magnitude of the magnetic dipole moment of the disk is the same as that of a current loop of area A carrying a current di:

$$d\mu = A\,di$$

The magnitude of the magnetization of the disk is the magnetic moment per unit volume:

$$M = \frac{d\mu}{dV} = \frac{A\,di}{A\,d\ell} = \frac{di}{d\ell} \qquad\qquad 27\text{-}19$$

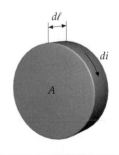

FIGURE 27-34 A disk element for relating the magnetization M to the surface current per unit length.

Thus, the magnitude of the magnetization vector is the amperian current per unit length along the surface of the magnetized material. We see from this result that the units of M are amperes per meter.

Consider a cylinder that has a uniform magnetization $\vec{M}$ parallel to its axis. The effect of the magnetization is the same as if the cylinder carried a surface current per unit length of magnitude M. This current is similar to the current carried by a tightly wound solenoid. For a solenoid, the current per unit length is nI, where n is the number of turns per unit length and I is the current in each turn. The magnitude of the magnetic field B_m inside the cylinder and far from its ends is thus given by Equation 27-9 for a solenoid with nI replaced by M:

$$B_m = \mu_0 M \tag{27-20}$$

Suppose we place a cylinder of magnetic material inside a long solenoid with n turns per unit length that carries a current I. The applied field of the solenoid $\vec{B}_{app}$ ($B_{app} = \mu_0 nI$) magnetizes the material so that it has a magnetization $\vec{M}$. The resultant magnetic field at a point inside the solenoid and far from its ends due to the current in the solenoid plus the magnetized material is

$$\vec{B} = \vec{B}_{app} + \mu_0\vec{M} \tag{27-21}$$

For paramagnetic and ferromagnetic materials, $\vec{M}$ is in the same direction as $\vec{B}_{app}$; for diamagnetic materials, $\vec{M}$ is opposite to $\vec{B}_{app}$. For paramagnetic and diamagnetic materials, the magnetization is found to be proportional to the applied magnetic field that produces the alignment of the magnetic dipoles in the material. We can thus write

$$\vec{M} = \chi_m \frac{\vec{B}_{app}}{\mu_0} \tag{27-22}$$

where χ_m is a dimensionless number called the **magnetic susceptibility.** Equation 27-21 is then

$$\vec{B} = \vec{B}_{app} + \mu_0\vec{M} = \vec{B}_{app}(1 + \chi_m) = K_m\vec{B}_{app} \tag{27-23}$$

where

$$K_m = 1 + \chi_m \tag{27-24}$$

is called the **relative permeability** of the material. For paramagnetic materials, χ_m is a small positive number that depends on temperature. For diamagnetic materials (other than superconductors), it is a small negative constant independent of temperature. Table 27-1 lists the magnetic susceptibility of various paramagnetic and diamagnetic materials. We see that the magnetic susceptibility for the solids listed is of the order of 10^{-5}, and $K_m \approx 1$.

The magnetization of ferromagnetic materials, which we discuss shortly, is much more complicated. The relative permeability K_m defined as the ratio B/B_{app} is not constant and has maximum values ranging from 5000 to 100,000. In the case of permanent magnets, K_m is not even defined since such materials exhibit magnetization even in the absence of an applied field.

Atomic Magnetic Moments

The magnetization of a paramagnetic or ferromagnetic material can be related to the permanent magnetic moments of the individual atoms or electrons of the material. The orbital magnetic moment of an atomic electron can be derived semiclassically, even though it is quantum mechanical in origin. Consider a particle of mass m and charge q moving with speed v in a circle of radius r, as shown in Figure 27-35. The magnitude of the angular momentum of the particle is

$$L = mvr \tag{27-25}$$

TABLE 27-1

Magnetic Susceptibility of Various Materials at 20°C

Material	χ_m
Aluminum	2.3×10^{-5}
Bismuth	-1.66×10^{-5}
Copper	-0.98×10^{-5}
Diamond	-2.2×10^{-5}
Gold	-3.6×10^{-5}
Magnesium	1.2×10^{-5}
Mercury	-3.2×10^{-5}
Silver	-2.6×10^{-5}
Sodium	-0.24×10^{-5}
Titanium	7.06×10^{-5}
Tungsten	6.8×10^{-5}
Hydrogen (1 atm)	-9.9×10^{-9}
Carbon dioxide (1 atm)	-2.3×10^{-9}
Nitrogen (1 atm)	-5.0×10^{-9}
Oxygen (1 atm)	2090×10^{-9}

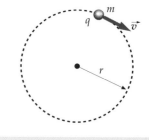

FIGURE 27-35 A particle of charge q and mass m moving with speed v in a circle of radius r. The angular momentum is into the paper and has a magnitude mvr and the magnetic moment is into the paper (if q is positive) and has a magnitude $\frac{1}{2}qvr$.

The magnitude of the magnetic moment is the product of the current and the area of the circle:

$$\mu = IA = I\pi r^2$$

If T is the time for the charge to complete one revolution, the current (charge passing a point per unit time) is q/T. Since the period T is the distance $2\pi r$ divided by the velocity v, the current is

$$I = \frac{q}{T} = \frac{qv}{2\pi r}$$

The magnetic moment is then

$$\mu = IA = \frac{qv}{2\pi r}\pi r^2 = \frac{1}{2}qvr \qquad\qquad 27\text{-}26$$

Using $vr = L/m$ from Equation 27-25, we have for the magnetic moment

$$\mu = \frac{q}{2m}L$$

If the charge q is positive, the angular momentum and magnetic moment are in the same direction. We can therefore write

$$\vec{\boldsymbol{\mu}} = \frac{q}{2m}\vec{\boldsymbol{L}} \qquad\qquad 27\text{-}27$$

CLASSICAL RELATION BETWEEN MAGNETIC MOMENT AND ANGULAR MOMENTUM

Equation 27-27 is the general classical relation between magnetic moment and angular momentum. It also holds in the quantum theory of the atom for orbital angular momentum, but the equation does not hold for the intrinsic spin angular momentum of the electron. For electron spin, the magnetic moment is twice that predicted by this equation.[†] The extra factor of 2 is a result from quantum theory that has no analog in classical mechanics.

Since angular momentum is quantized, the magnetic moment of an atom is also quantized. The quantum of angular momentum is $\hbar = h/(2\pi)$, where h is Planck's constant, so we express the magnetic moment in terms of $\vec{L}/\hbar$

$$\vec{\boldsymbol{\mu}} = \frac{q\hbar}{2m}\frac{\vec{\boldsymbol{L}}}{\hbar}$$

For an electron, $m = m_e$ and $q = -e$, so the magnetic moment of the electron due to its orbital motion is

$$\vec{\boldsymbol{\mu}}_\ell = -\frac{e\hbar}{2m_e}\frac{\vec{\boldsymbol{L}}}{\hbar} = -\mu_B\frac{\vec{\boldsymbol{L}}}{\hbar} \qquad\qquad 27\text{-}28$$

MAGNETIC MOMENT DUE TO THE ORBITAL MOTION OF AN ELECTRON

---

† This result, and the phenomenon of electron spin itself, was predicted in 1927 by Paul Dirac, who combined special relativity and quantum mechanics into a relativistic wave equation called the Dirac equation. Precise measurements indicate that the magnetic moment of the electron due to its spin is 2.00232 times that predicted by Equation 27-27. The fact that the intrinsic magnetic moment of the electron is approximately twice what we would expect makes it clear that the simple model of the electron as a spinning ball is not to be taken literally.

where

$$\mu_B = \frac{e\hbar}{2m_e} = 9.27 \times 10^{-24} \text{ A·m}^2 = 9.27 \times 10^{-24} \text{ J/T}$$

$$= 5.79 \times 10^{-5} \text{ eV/T} \qquad\qquad 27\text{-}29$$

BOHR MAGNETON

is the quantum unit of magnetic moment called a **Bohr magneton.** The magnetic moment of an electron due to its intrinsic spin angular momentum $\vec{S}$ is

$$\vec{\mu}_s = -2 \times \frac{e\hbar}{2m_e}\frac{\vec{S}}{\hbar} = -2\mu_B\frac{\vec{S}}{\hbar} \qquad\qquad 27\text{-}30$$

MAGNETIC MOMENT DUE TO ELECTRON SPIN

Although the calculation of the magnetic moment of any atom is a complicated problem in quantum theory, the result for all electrons, according to both theory and experiment, is that the magnetic moment is of the order of a few Bohr magnetons. For atoms with zero net angular momentum, the net magnetic moment is zero. (The shell structure of atoms is discussed in Chapter 36.)

If all the atoms or molecules in some material have their magnetic moments aligned, the magnetic moment per unit volume of the material is the product of the number of molecules per unit volume n and the magnetic moment μ of each molecule. For this extreme case, the **saturation magnetization** M_s is

$$M_s = n\mu \qquad\qquad 27\text{-}31$$

The number of molecules per unit volume can be found from the molecular mass M, the density ρ of the material, and Avogadro's number N_A:

$$n = \frac{N_A \text{ (atoms/mol)}}{M \text{ (kg/mol)}} \rho(\text{kg/m}^3) \qquad\qquad 27\text{-}32$$

SATURIZATION MAGNETIZATION FOR IRON   **EXAMPLE 27-10**

Find the saturation magnetization and the magnetic field it produces for iron, assuming that each iron atom has a magnetic moment of 1 Bohr magneton.

PICTURE THE PROBLEM We find the number of molecules per unit volume from the density of iron, $\rho = 7.9 \times 10^3 \text{ kg/m}^3$, and its molecular mass $M = 55.8 \times 10^{-3} \text{ kg/mol}$.

1. The saturation magnetization is the product of the number of molecules per unit volume and the magnetic moment of each molecule:

$$M_s = n\mu$$

2. Calculate the number of molecules per unit volume from Avogadro's number, the molecular mass, and the density:

$$n = \frac{N_A}{M}\rho = \frac{6.02 \times 10^{23} \text{ atoms/mol}}{55.8 \times 10^{-3} \text{ kg/mol}} (7.9 \times 10^3 \text{ kg/m}^3)$$

$$= 8.52 \times 10^{28} \text{ atoms/m}^3$$

3. Substitute this result and $\mu = 1$ Bohr magneton to calculate the saturation magnetization:

$$M_s = n\mu$$
$$= (8.52 \times 10^{28} \text{ atoms/m}^3)(9.27 \times 10^{-24} \text{ A·m}^2)$$
$$= \boxed{7.90 \times 10^5 \text{ A/m}}$$

4. The magnetic field on the axis inside a long iron cylinder resulting from this maximum magnetization is given by $B = \mu_0 M_s$:

$$B = \mu_0 M_s$$
$$= (4\pi \times 10^{-7} \text{ T·A})(7.90 \times 10^5 \text{ A/m})$$
$$= \boxed{0.993 \text{ T} \approx 1 \text{ T}}$$

REMARKS The measured saturation magnetic field of annealed iron is about 2.16 T, indicating that the magnetic moment of an iron atom is slightly greater than 2 Bohr magnetons. This magnetic moment is due mainly to the spins of two unpaired electrons in the iron atom.

*Paramagnetism

Paramagnetism occurs in materials whose atoms have permanent magnetic moments that interact with each other only very weakly, resulting in a very small, positive magnetic susceptibility χ_m. When there is no external magnetic field, these magnetic moments are randomly oriented. In the presence of an external magnetic field, the magnetic moments tend to line up parallel to the field, but this is counteracted by the tendency for the magnetic moments to be randomly oriented due to thermal motion. The degree to which the moments line up with the field depends on the strength of the field and on the temperature. This degree of alignment usually is small because the energy of a magnetic moment in an external magnetic field is typically much smaller than the thermal energy of an atom of the material, which is of the order of kT, where k is Boltzmann's constant and T is the absolute temperature.

The potential energy of a magnetic dipole of moment $\vec{\mu}$ in an external magnetic field $\vec{B}$ is given by Equation 26-16:

$$U = -\mu B \cos \theta = -\vec{\mu} \cdot \vec{B}$$

The potential energy when the moment is parallel with the field ($\theta = 0$) is thus lower than when the moment is antiparallel ($\theta = 180°$) by the amount $2\mu B$. For a typical atomic magnetic moment of 1 Bohr magneton and a typical strong magnetic field of 1 T, the difference in potential energy is

$$\Delta U = 2\mu_B B = 2(5.79 \times 10^{-5} \text{ eV/T})(1 \text{ T}) = 1.16 \times 10^{-4} \text{ eV}$$

At a normal temperature of $T = 300$ K, the typical thermal energy kT is

$$kT = (8.62 \times 10^{-5} \text{ eV/K})(300 \text{ K}) = 2.59 \times 10^{-2} \text{ eV}$$

which is more than 200 times greater than $2\mu_B B$. Thus, even in a very strong magnetic field of 1 T, most of the magnetic moments will be randomly oriented because of thermal motions (unless the temperature is very low).

Figure 27-36 shows a plot of the magnetization M versus an applied external magnetic field B_{app} at a given temperature. In very strong fields, nearly all the magnetic moments are aligned with the field and $M \approx M_s$. (For magnetic fields attainable in the laboratory, this can occur only for very low temperatures.) When $B_{app} = 0$, $M = 0$, indicating that the orientation of the moments is completely random. In weak fields, the magnetization is approximately proportional to the

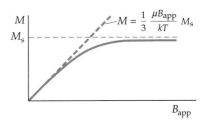

FIGURE 27-36 Plot of magnetization M versus an applied magnetic field B_{app}. In very strong fields, the magnetization approaches the saturation value M_s. This can be achieved only at very low temperatures. In weak fields, the magnetization is approximately proportional to B_{app}, a result known as Curie's law.

applied field, as indicated by the orange dashed line in the figure. In this region, the magnetization is given by

$$M = \frac{1}{3}\frac{\mu B_{app}}{kT}M_s \qquad\qquad 27\text{-}33$$

CURIE'S LAW

Note that $\mu B_{app}/(kT)$ is the ratio of the maximum energy of a dipole in the magnetic field to the characteristic thermal energy. The result that the magnetization varies inversely with the absolute temperature was discovered experimentally by Pierre Curie and is known as **Curie's law.**

Liquid oxygen, which is paramagnetic, is attracted by the magnetic field of a permanent magnet. A net force is exerted on the magnetic dipoles because the magnetic field is not uniform.

APPLYING CURIE'S LAW   **E X A M P L E   2 7 - 1 1**

If $\mu = \mu_B$, at what temperature will the magnetization be 1 percent of the saturation magnetization in an applied magnetic field of 1 T?

PICTURE THE PROBLEM

1. Curie's law relates M, T, M_s, and B_{app}:

$$M = \frac{1}{3}\frac{\mu B_{app}}{kT}M_s$$

2. Solve for T using $\mu = \mu_B$ and $M/M_s = 0.01$:

$$T = \frac{\mu_B B_{app}}{3k}\frac{M_s}{M} = \frac{(5.79 \times 10^{-5}\ \text{eV/T})(1\ \text{T})}{3(8.62 \times 10^{-5}\ \text{eV/K})}100$$

$$= \boxed{22.4\ \text{K}}$$

REMARKS From this example, we see that even in a strong applied magnetic field of 1 T, the magnetization is less than 1 percent of saturation at temperatures above 22.4 K.

EXERCISE If $\mu = \mu_B$, what fraction of the saturation magnetization is M at 300 K for an external magnetic field of 1.5 T? (*Answer* $M/M_s = 1.12 \times 10^{-3}$)

*Ferromagnetism

Ferromagnetism occurs in pure iron, cobalt, and nickel as well as in alloys of these metals with each other. It also occurs in gadolinium, dysprosium, and a few compounds. Ferromagnetism arises from a strong interaction between the electrons in a partially full band in a metal or between the localized electrons that form magnetic moments on neighboring atoms or molecules. This interaction, called the **exchange interaction,** lowers the energy of a pair of electrons with parallel spins.

Ferromagnetic materials have very large positive values of magnetic susceptibility χ_m (as measured under conditions described, which follow). In these substances, a small external magnetic field can produce a very large degree of alignment of the atomic magnetic dipole moments. In some cases, the alignment can persist even when the external magnetizing field is removed. This alignment persists because the magnetic dipole moments exert strong forces on their neighbors so that over a small region of space the moments are aligned with each other even when there is no external field. The region of space over which the magnetic dipole moments are aligned is called a **magnetic domain.** The size of a domain is usually microscopic. Within the domain, all the permanent atomic magnetic moments are aligned, but the direction of alignment varies from domain to domain so that the net magnetic moment of a macroscopic piece of ferromagnetic

A Canadian quarter that is attracted by a magnet. Canadian coins often contain significant amounts of nickel, which is ferromagnetic.

(a)

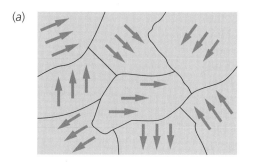

(b)

FIGURE 27-37 (*a*) Schematic illustration of ferromagnetic domains. Within a domain, the magnetic dipoles are aligned, but the direction of alignment varies from domain to domain so that the net magnetic moment is zero. A small external magnetic field may cause the enlargement of those domains that are aligned parallel to the field, or it may cause the alignment within a domain to rotate. In either case, the result is a net magnetic moment parallel to the field. (*b*) Magnetic domains on the surface of an FE−3 percent Si crystal observed using a scanning electron microscope with polarization analysis. The four colors indicate four possible domain orientations.

material is zero in the normal state. Figure 27-37 illustrates this situation. The dipole forces that produce this alignment are predicted by quantum theory but cannot be explained with classical physics. At temperatures above a critical temperature, called the **Curie temperature,** thermal agitation is great enough to break up this alignment and ferromagnetic materials become paramagnetic.

When an external magnetic field is applied, the boundaries of the domains may shift or the direction of alignment within a domain may change so that there is a net macroscopic magnetic moment in the direction of the applied field. Since the degree of alignment is large for even a small external field, the magnetic field produced in the material by the dipoles is often much greater than the external field.

Let us consider what happens when we magnetize a long iron rod by placing it inside a solenoid and gradually increase the current in the solenoid windings. We assume that the rod and the solenoid are long enough to permit us to neglect end effects. Since the induced magnetic moments are in the same direction as the applied field, $\vec{B}_{app}$ and $\vec{M}$ are in the same direction. Then,

$$B = B_{app} + \mu_0 M = \mu_0 n I + \mu_0 M \qquad \text{27-34}$$

In ferromagnetic materials, the magnetic field $\mu_0 M$ due to the magnetic moments is often greater than the magnetizing field B_{app} by a factor of several thousand.

A chunk of magnetite (lodestone) attracts the needle of a compass.

(a)

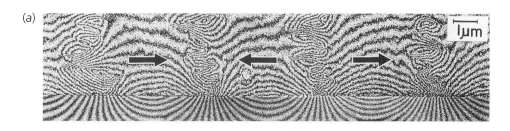

(b)

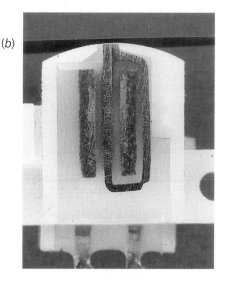

(*a*) Magnetic field lines on a cobalt magnetic recording tape. The solid arrows indicate the encoded magnetic bits. (*b*) Cross section of a magnetic tape recording head. Current from an audio amplifier is sent to wires around a magnetic core in the recording head where it produces a magnetic field. When the tape passes over a gap in the core of the recording head, the fringing magnetic field encodes information on the tape.

Figure 27-38 shows a plot of B versus the magnetizing field B_{app}. As the current is gradually increased from zero, B increases from zero along the part of the curve from the origin O to point P_1. The flattening of this curve near point P_1 indicates that the magnetization M is approaching its saturation value M_s, at which all the atomic magnetic moments are aligned. Above saturation, B increases only because the magnetizing field $B_{app} = \mu_0 nI$ increases. When B_{app} is gradually decreased from point P_1, there is not a corresponding decrease in the magnetization. The shift of the domains in a ferromagnetic material is not completely reversible, and some magnetization remains even when B_{app} is reduced to zero, as indicated in the figure. This effect is called **hysteresis,** from the Greek word *hysteros* meaning later or behind, and the curve in Figure 27-38 is called a **hysteresis curve.** The value of the magnetic field at point r when B_{app} is zero is called the **remnant field** B_{rem}. At this point, the iron rod is a permanent magnet. If the current in the solenoid is now reversed so that B_{app} is in the opposite direction, the magnetic field B is gradually brought to zero at point c. The remaining part of the hysteresis curve is obtained by further increasing the current in the opposite direction until point P_2 is reached, which corresponds to saturation in the opposite direction, and then decreasing the current to zero at point P_3 and increasing it again in its original direction.

Since the magnetization M depends on the previous history of the material, and since it can have a large value even when the applied field is zero, it is not simply related to the applied field B_{app}. However, if we confined ourselves to that part of the magnetization curve from the origin to point P_1 in Figure 27-38, $\vec{B}_{app}$ and $\vec{M}$ are parallel and M is zero when B_{app} is zero. We can then define the magnetic susceptibility as in Equation 27-22,

$$M = \chi_m \frac{B_{app}}{\mu_0}$$

and

$$B = B_{app} + \mu_0 M = B_{app}(1 + \chi_m) = K_m \mu_0 nI = \mu nI \qquad \text{27-35}$$

where

$$\mu = (1 + \chi_m)\mu_0 = K_m \mu_0 \qquad \text{27-36}$$

is called the **permeability** of the material. (For paramagnetic and diamagnetic materials, χ_m is much less than 1 so the permeability μ and the permeability of free space μ_0 are very nearly equal.)

Since B does not vary linearly with B_{app}, as can be seen from Figure 27-38, the relative permeability is not constant. The maximum value of K_m occurs at a magnetization that is considerably less than the saturation magnetization. Table 27-2 lists the saturation magnetic field $\mu_0 M_s$ and the maximum values of K_m for some ferromagnetic materials. Note that the maximum values of K_m are much greater than 1.

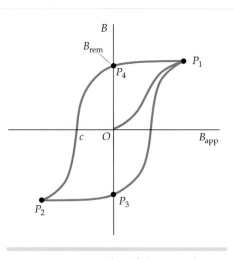

FIGURE 27-38 Plot of B versus the applied magnetizing field B_{app}. The outer curve is called a hysteresis curve. The field B_{rem} is called the remnant field. It remains when the applied field returns to zero.

TABLE 27-2

Maximum Values of $\mu_0 M$ and K_m for Some Ferromagnetic Materials

Material	$\mu_0 M_s$, T	K_m
Iron (annealed)	2.16	5,500
Iron-silicon (96 percent Fe, 4 percent Si)	1.95	7,000
Permalloy (55 percent Fe, 45 percent Ni)	1.60	25,000
Mu-metal (77 percent Ni, 16 percent Fe, 5 percent Cu, 2 percent Cr)	0.65	100,000

The area enclosed by the hysteresis curve is proportional to the energy dissipated as heat in the irreversible process of magnetizing and demagnetizing. If the hysteresis effect is small, so that the area inside the curve is small, indicating a small energy loss, the material is called **magnetically soft.** Soft iron is an example. The hysteresis curve for a magnetically soft material is shown in Figure 27-39. Here the remnant field B_{rem} is nearly zero, and the energy loss per cycle is small. Magnetically soft materials are used for transformer cores to allow the magnetic field B to change without incurring large energy losses as the field alternates. On the other hand, a large remnant field is desirable in a permanent magnet. **Magnetically hard** materials, such as carbon steel and the alloy Alnico 5, are used for permanent magnets.

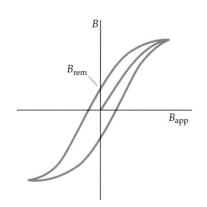

FIGURE 27-39 Hysteresis curve for a magnetically soft material. The remnant field is very small compared with the remnant field for a magnetically hard material such as that shown in Figure 27-38.

(*a*) An extremely high-capacity, hard-disk drive for magnetic storage of information, capable of storing over 250 gigabytes of information. (*b*) A magnetic test pattern on a hard disk, magnified 2400 times. The light and dark regions correspond to oppositely directed magnetic fields. The smooth region just outside the pattern is a region of the disk that has been erased just prior to writing.

(*a*)

(*b*)

SOLENOID WITH IRON CORE   **EXAMPLE 27-12**

A long solenoid with 12 turns per centimeter has a core of annealed iron. When the current is 0.50 A, the magnetic field inside the iron core is 1.36 T. Find (a) the applied field B_{app}, (b) the relative permeability K_m, and (c) the magnetization M.

PICTURE THE PROBLEM The applied field is just that of a long solenoid given by $B_{app} = \mu_0 nI$. Since the total magnetic field is given, we can find the relative permeability from its definition ($K_m = B/B_{app}$) and we can find M from $B = B_{app} + \mu_0 M$.

1. The applied field is given by Equation 27-9:

$$B_{app} = \mu_0 nI$$
$$= (4\pi \times 10^{-7}\,\text{T·m/A})(1200\,\text{turns/m})(0.5\,\text{A})$$
$$= 7.54 \times 10^{-4}\,\text{T}$$

2. The relative permeability is the ratio of B to B_{app}:

$$K_m = \frac{B}{B_{app}} = \frac{1.36\,\text{T}}{7.54 \times 10^{-4}\,\text{T}} = 1.80 \times 10^3$$

3. The magnetization M is found from Equation 27-34:

$$\mu_0 M = B - B_{app}$$
$$= 1.36\,\text{T} - 7.54 \times 10^{-4}\,\text{T} \approx B = 1.36\,\text{T}$$
$$M = \frac{B}{\mu_0} = \frac{1.36\,\text{T}}{4\pi \times 10^{-7}\,\text{T·m/A}} = \boxed{1.08 \times 10^6\,\text{A/m}}$$

REMARKS The applied magnetic field of 7.54×10^{-4} T is a negligible fraction of the total field of 1.36 T. Note that the value for K_m of 1800 is considerably smaller than the maximum value of 5500 in Table 27-2. Note also that the susceptibility $\chi_m = K_m - 1 \approx K_m$ to the three-place accuracy with which we calculated K_m.

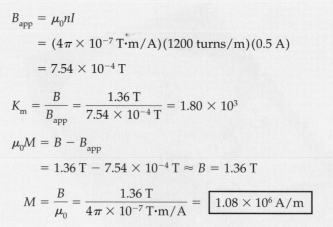

*Diamagnetism

Diamagnetic materials are those materials that have very small negative values of magnetic susceptibility χ_m. Diamagnetism was discovered by Michael Faraday in 1845 when Faraday found that a piece of bismuth is repelled by either pole of a magnet, indicating that the external field of the magnet induces a magnetic moment in bismuth in the direction opposite the field.

We can understand this effect qualitatively from Figure 27-40, which shows two positive charges moving in circular orbits with the same speed but in opposite directions. Their magnetic moments are in opposite directions and therefore cancel.[†] In the presence of an external magnetic field $\vec{B}$ directed into the paper, the charges experience an extra force $q\vec{v} \times \vec{B}$, which is along the radial direction. For the charge on the left, this extra force is inward, increasing the centripetal force. If the charge is to remain in the same circular orbit, it must speed up so that mv^2/r equals the total centripetal force.[‡] Its magnetic moment, which is outward, is thus increased. For the charge on the right, the additional force is outward, so the particle must slow down to maintain its circular orbit. Its magnetic moment, which is

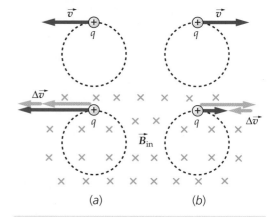

FIGURE 27-40 (a) A positive charge moving counterclockwise in a circle has its magnetic moment directed out of the page. When an external, magnetic field directed into the page is turned on, the magnetic force increases the centripetal force so the speed of the particle must increase. The change in the magnetic moment is out of the page. (b) A positive charge moving clockwise in a circle has its magnetic moment directed into the page. When an external, magnetic field directed into the page is turned on, the magnetic force decreases the centripetal force so the speed of the particle must decrease. As in (a), the change in the magnetic moment is directed out of the page.

---

† It is simpler to consider positive charges even though it is the negatively charged electrons that provide the magnetic moments in matter.

‡ The electron speeds up because of an electric field induced by the changing magnetic field, an effect called induction, which we discuss in Chapter 28.

inward, is decreased. In each case, the *change* in the magnetic moment of the charges is in the direction out of the page, opposite that of the external applied field. Since the permanent magnetic moments of the two charges are equal and oppositely directed they add to zero, leaving only the induced magnetic moments which are both opposite the direction of the applied magnetic field.

A material will be diamagnetic if its atoms have zero net angular momentum and therefore no permanent magnetic moment. (The net angular momentum of an atom depends on the electronic structure of the atom, which is a subject that we will study in Chapter 35.) The induced magnetic moments that cause diamagnetism have magnitudes of the order of 10^{-5} Bohr magnetons. Since this is much smaller than the permanent magnetic moments of the atoms of paramagnetic or ferromagnetic materials, the diamagnetic effect in these atoms is masked by the alignment of their permanent magnetic moments. However, since this alignment decreases with temperature, all materials are theoretically diamagnetic at sufficiently high temperatures.

When a superconductor is placed in an external magnetic field, electric currents are induced on the superconductor's surface so that the net magnetic field in the superconductor is zero. Consider a superconducting rod inside a solenoid of n turns per unit length. When the solenoid is connected to a source of emf so that it carries a current I, the magnetic field due to the solenoid is $\mu_0 nI$. A surface current of $-nI$ per unit length is induced on the superconducting rod that cancels out the field due to the solenoid so that the net field inside the superconductor is zero. From Equation 27-23,

$$\vec{B} = \vec{B}_{app}(1 + \chi_m) = 0$$

so

$$\chi_m = -1$$

A superconductor is thus a perfect diamagnet with a magnetic susceptibility of -1.

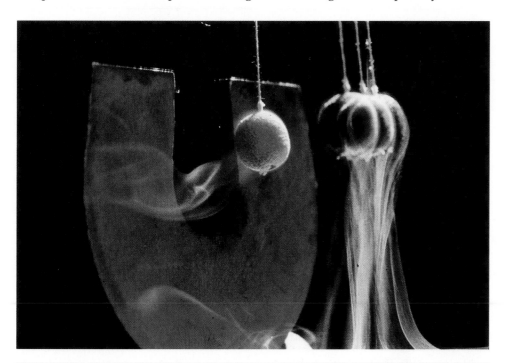

A superconductor is a perfect diamagnet. Here, the superconducting pendulum bob is repelled by the permanent magnet.

SUMMARY

1. Magnetic fields arise from moving charges, and therefore from currents.
2. The Biot–Savart law describes the magnetic field produced by a current element.
3. Ampère's law relates the line integral of the magnetic field along some closed curve to the current that passes through any surface bounded by the curve.
4. The magnetization vector $\vec{M}$ describes the magnetic moment per unit volume of matter.
5. The classical relation $\vec{\mu} = [q/(2m)]\,\vec{L}$ is derived from the definitions of angular momentum and magnetic moment.
6. The Bohr magneton is a convenient unit for atomic and nuclear magnetic moments.

Topic	Relevant Equations and Remarks	

1. Magnetic Field $\vec{B}$

Due to a moving point charge	$$\vec{B} = \frac{\mu_0}{4\pi}\frac{q\vec{v} \times \hat{r}}{r^2}$$	27-1
	where $\hat{r}$ is a unit vector that points to the field point P from the charge q moving with velocity $\vec{v}$, and μ_0 is a constant of proportionality called the permeability of free space:	
	$$\mu_0 = 4\pi \times 10^{-7}\ \text{T·m/A} = 4\pi \times 10^{-7}\ \text{N/A}^2$$	27-2
Due to a current element (Biot–Savart law)	$$d\vec{B} = \frac{\mu_0}{4\pi}\frac{I\,d\vec{\ell} \times \hat{r}}{r^2}$$	27-3
On the axis of a current loop	$$B_x = \frac{\mu_0}{4\pi}\frac{2\pi R^2 I}{(x^2 + R^2)^{3/2}}$$	27-5
On the axis of a current loop, at great distances from the loop	$$B_x = \frac{\mu_0}{4\pi}\frac{2\mu}{|x|^3}$$ where μ is the magnitude of the magnetic moment of the loop.	27-6
Inside a long solenoid, far from the ends	$B_x = \mu_0 n I$ where n is the number of turns per unit length.	27-9
Due to a straight wire segment	$$B = \frac{\mu_0}{4\pi}\frac{I}{R}(\sin\theta_2 - \sin\theta_1)$$ where R is the perpendicular distance to the wire and θ_1 and θ_2 are the angles subtended at the field point by the ends of the wire.	27-11
Due to a long, straight wire	$$B = \frac{\mu_0}{4\pi}\frac{2I}{R}$$ The direction of $\vec{B}$ is such that the magnetic field lines of $\vec{B}$ encircle the wire in the direction of the fingers of the right hand if the thumb points in the direction of the current.	27-12
Inside a tightly wound toroid	$$B = \frac{\mu_0}{2\pi}\frac{NI}{r}, \quad a < r < b$$	27-17

2. Magnetic Field Lines

The magnetic field is indicated by lines parallel to $\vec{B}$ at any point whose density is proportional to the magnitude of $\vec{B}$. Magnetic lines do not begin or end at any point in space. Instead, they form continuous loops.

3. Gauss's Law for Magnetism

$$\phi_{m,\,net} = \oint_S B_n \, dA = 0$$

27-14

4. Magnetic Poles

Magnetic poles always occur in pairs. Isolated magnetic poles have not been found.

5. Ampère's Law

$$\oint_C \vec{B} \cdot d\vec{\ell} = \mu_0 I_C$$

where C is any closed curve.

27-15

Validity of Ampère's law

Ampère's law is valid only if the currents are steady and continuous. It can be used to derive expressions for the magnetic field for situations with a high degree of symmetry, such as a long, straight, current-carrying wire or a long, tightly wound solenoid.

6. Magnetism in Matter

Matter can be classified as paramagnetic, ferromagnetic, or diamagnetic.

Magnetization

A magnetized material is described by its magnetization vector $\vec{M}$, which is defined as the net magnetic dipole moment per unit volume of the material:

$$\vec{M} = \frac{d\vec{\mu}}{dV}$$

27-18

The magnetic field due to a uniformly magnetized cylinder is the same as if the cylinder carried a current per unit length of magnitude M on its surface. This current, which is due to the intrinsic motion of the atomic charges in the cylinder, is called an amperian current.

7. $\vec{B}$ in Magnetic Materials

$$\vec{B} = \vec{B}_{app} + \mu_0 \vec{M}$$

27-21

Magnetic susceptibility χ_m

$$\vec{M} = \chi_m \frac{\vec{B}_{app}}{\mu_0}$$

27-22

For paramagnetic materials, χ_m is a small positive number that depends on temperature. For diamagnetic materials (other than superconductors), it is a small negative constant independent of temperature. For superconductors, $\chi_m = -1$. For ferromagnetic materials, the magnetization depends not only on the magnetizing current but also on the past history of the material.

Relative permeability

$$\vec{B} = K_m \vec{B}_{app}$$

27-23

where

$$K_m = 1 + \chi_m$$

27-24

8. Atomic Magnetic Moments

$$\vec{\mu} = \frac{q}{2m} \vec{L}$$

27-27

where $\vec{L}$ is the orbital angular momentum of the particle.

Due to the orbital motion of an electron

$$\vec{\mu}_\ell = -\frac{e\hbar}{2m_e} \frac{\vec{L}}{\hbar} = -\mu_B \frac{\vec{L}}{\hbar}$$

27-28

Due to electron spin

$$\vec{\mu}_s = -2 \times \frac{e\hbar}{2m_e} \frac{\vec{S}}{\hbar} = -2\mu_B \frac{\vec{S}}{\hbar}$$

27-30

Bohr magneton

$$\mu_B = \frac{e\hbar}{2m_e} = 9.27 \times 10^{-24} \, \text{A·m}^2$$

$$= 9.27 \times 10^{-24} \, \text{J/T} = 5.79 \times 10^{-5} \, \text{eV/T}$$

27-29

where

$$\hbar = \frac{h}{2\pi} = 1.05 \times 10^{-34}\,\text{J·s}$$

and $h = 6.626 \times 10^{-34}$ J·s is Planck's constant.

---

*9. Paramagnetism

Paramagnetic materials have permanent atomic magnetic moments that have random directions in the absence of an applied magnetic field. In an applied field these dipoles are aligned with the field to some degree, producing a small contribution to the total field that adds to the applied field. The degree of alignment is small except in very strong fields and at very low temperatures. At ordinary temperatures, thermal motion tends to maintain the random directions of the magnetic moments.

Curie's law

In weak fields, the magnetization is approximately proportional to the applied field and inversely proportional to the absolute temperature.

$$M = \frac{1}{3}\frac{\mu B_{app}}{kT}M_s \qquad\qquad \textbf{27-33}$$

---

*10. Ferromagnetism

Ferromagnetic materials have small regions of space called magnetic domains in which all the permanent atomic magnetic moments are aligned. When the material is unmagnetized, the direction of alignment in one domain is independent of that in another domain so that no net magnetic field is produced. When the material is magnetized, the domains of a ferromagnetic material are aligned, producing a very strong contribution to the magnetic field. This alignment can persist even when the external field is removed, thus leading to permanent magnetism.

---

*11. Diamagnetism

Diamagnetic materials are those materials in which the magnetic moments of all electrons in each atom cancel, leaving each atom with zero magnetic moment in the absence of an external field. In an external field, a very small magnetic moment is induced that tends to weaken the field. This effect is independent of temperature. Superconductors are diamagnetic with a magnetic susceptibility equal to -1.

PROBLEMS

- Single-concept, single-step, relatively easy
- •• Intermediate-level, may require synthesis of concepts
- ••• Challenging
- **SSM** Solution is in the *Student Solutions Manual*
- **iSOLVE** Problems available on iSOLVE online homework service
- **iSOLVE** ✓ These "Checkpoint" online homework service problems ask students additional questions about their confidence level, and how they arrived at their answer.

In a few problems, you are given more data than you actually need; in a few other problems, you are required to supply data from your general knowledge, outside sources, or informed estimates.

Conceptual Problems

1 • **SSM** Compare the directions of the electric force and the magnetic force between two positive charges, which move along parallel paths (*a*) in the same direction and (*b*) in opposite directions.

2 • Is $\vec{B}$ uniform everywhere within a current loop? Explain.

3 • Sketch the field lines for the electric dipole and the magnetic dipole shown in Figure 27-41. How do they differ in appearance close to the center of each dipole?

Electric dipole   Magnetic dipole

FIGURE 27-41 Problem 3

4 • Two wires lie in the plane of the paper and carry equal currents in opposite directions, as shown in Figure 27-42. At a point midway between the wires, the magnetic field is (*a*) zero. (*b*) into the page. (*c*) out of the page. (*d*) toward the top or bottom of the page. (*e*) toward one of the two wires.

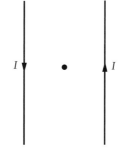

FIGURE 27-42 Problem 4

5 • Two parallel wires carry currents I_1 and $I_2 = 2I_1$ in the same direction. The forces F_1 and F_2 on the wires are related by (*a*) $F_1 = F_2$. (*b*) $F_1 = 2F_2$. (*c*) $2F_1 = F_2$. (*d*) $F_1 = 4F_2$. (*e*) $4F_1 = F_2$.

6 • **SSM** A wire carries an electrical current straight up. What is the direction of the magnetic field due to the wire a distance of 2 m north of the wire? (*a*) North (*b*) East (*c*) West (*d*) South (*e*) Upward

7 • Two current-carrying wires are perpendicular to each other. The current in one wire flows vertically upward and the current in the other wire flows horizontally toward the east. The horizontal wire is 1 m south of the vertical wire. What is the direction of the net magnetic force on the horizontal wire? (*a*) North (*b*) East (*c*) West (*d*) South (*e*) There is no net magnetic force on the horizontal wire.

8 • Make a field-line sketch of the magnetic field due to the currents in the pair of coaxial coils (Figure 27-43). The currents in the coils have the same magnitude and are in the same direction in each coil.

FIGURE 27-43 Problems 8 and 9

9 • **SSM** Make a field-line sketch of the magnetic field due to the currents in the pair of coaxial coils (Figure 27-43). The currents in the coils have the same magnitude but are opposite in direction in each coil.

10 • Ampère's law is valid (*a*) when there is a high degree of symmetry. (*b*) when there is no symmetry. (*c*) when the current is constant. (*d*) when the magnetic field is constant. (*e*) in all of these situations if the current is continuous.

11 • True or false:

(*a*) Diamagnetism is the result of induced magnetic dipole moments.

(*b*) Paramagnetism is the result of the partial alignment of permanent magnetic dipole moments.

12 • **SSM** If the magnetic susceptibility is positive, (*a*) paramagnetic effects or ferromagnetic effects must be greater than diamagnetic effects. (*b*) diamagnetic effects must be greater than paramagnetic effects. (*c*) diamagnetic effects must be greater than ferromagnetic effects. (*d*) ferromagnetic effects must be greater than paramagnetic effects. (*e*) paramagnetic effects must be greater than ferromagnetic effects.

13 • True or false:

(*a*) The magnetic field due to a current element is parallel to the current element.

(*b*) The magnetic field due to a current element varies inversely with the square of the distance from the element.

(*c*) The magnetic field due to a long wire varies inversely with the square of the distance from the wire.

(*d*) Ampère's law is valid only if there is a high degree of symmetry.

(*e*) Ampère's law is valid only for continuous currents.

14 • Can a particle have angular momentum and not have a magnetic moment?

15 • Can a particle have a magnetic moment and not have angular momentum?

16 • A circular loop of wire carries a current I. Is there angular momentum associated with the magnetic moment of the loop? If so, why is it not noticed?

17 • A hollow tube carries a current. Inside the tube, $\vec{B} = 0$. Why is this the case, because $\vec{B}$ is strong inside a solenoid?

18 • **SSM** When a current is passed through the wire in Figure 27-44, will the wire tend to bunch up or form a circle?

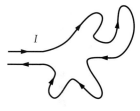

FIGURE 27-44 Problem 18

19 • Which of the four gases listed in Table 27-1 are diamagnetic and which of the four gases are paramagnetic?

Estimation and Approximation

20 •• The magnetic moment of the earth is about 9×10^{22} A·m². (*a*) If the magnetization of the earth's core were 1.5×10^9 A/m, what is the core volume? (*b*) What is the radius of such a core if it were spherical and centered with the earth?

21 •• **SSM** Estimate the transient magnetic field 100 m away from a lightning bolt if a charge of about 30 C is transferred from cloud to ground and the average speed of the charges is 10^6 m/s.

22 •• **SSM** The rotating disk of Problem 125 (page 896) can be used as a model for the magnetic field due to a sunspot. If the sunspot radius is approximately 10^7 m rotating at an angular velocity of about 10^{-2} rad/s, calculate the total charge Q on the sunspot needed to create a magnetic field of order 0.1 T at the center of the sunspot. What is the electrical field magnitude just above the center of the sunspot due to this charge?

The Magnetic Field of Moving Point Charges

23 • At time $t = 0$, a particle with charge $q = 12$ μC is located at $x = 0$, $y = 2$ m; the particle's velocity at that time is $\vec{v} = 30$ m/s$\hat{i}$. Find the magnetic field at (*a*) the origin; (*b*) $x = 0$, $y = 1$ m; (*c*) $x = 0$, $y = 3$ m; and (*d*) $x = 0$, $y = 4$ m.

24 • For the particle in Problem 23, find the magnetic field at (*a*) $x = 1$ m, $y = 3$ m; (*b*) $x = 2$ m, $y = 2$ m; and (*c*) $x = 2$ m, $y = 3$ m.

25 • A proton (charge $+e$) traveling with a velocity of $\vec{v} = 1 \times 10^4$ m/s$\hat{i} + 2 \times 10^4$ m/s$\hat{j}$ is located at $x = 3$ m, $y = 4$ m at some time t. Find the magnetic field at (*a*) $x = 2$ m, $y = 2$ m; (*b*) $x = 6$ m, $y = 4$ m; and (*c*) $x = 3$ m, $y = 6$ m.

26 • SOLVE An electron orbits a proton at a radius of 5.29×10^{-11} m. What is the magnetic field at the proton due to the orbital motion of the electron?

27 •• SSM Two equal charges q located at $(0, 0, 0)$ and at $(0, b, 0)$ at time zero are moving with speed v in the positive x direction ($v \ll c$). Find the ratio of the magnitudes of the magnetic force and electrostatic force on each charge.

The Magnetic Field of Currents: The Biot–Savart Law

28 • A small current element $I\,d\vec{\ell}$ with $d\vec{\ell} = 2$ mm$\hat{k}$ and $I = 2$ A, is centered at the origin. Find the magnetic field $d\vec{B}$ at the following points: (*a*) on the x axis at $x = 3$ m, (*b*) on the x axis at $x = -6$ m, (*c*) on the z axis at $z = 3$ m, and (*d*) on the y axis at $y = 3$ m.

29 • For the current element in Problem 28, find the magnitude and direction of $d\vec{B}$ at $x = 0, y = 3$ m, $z = 4$ m.

30 • SSM For the current element in Problem 28, find the magnitude of $d\vec{B}$ and indicate its direction on a diagram at (*a*) $x = 2$ m, $y = 4$ m, $z = 0$ and (*b*) $x = 2$ m, $y = 0$, $z = 4$ m.

$\vec{B}$ Due to a Current Loop

31 • A single loop of wire with radius 3 cm carries a current of 2.6 A. What is the magnitude of B on the axis of the loop at (*a*) the center of the loop, (*b*) 1 cm from the center, (*c*) 2 cm from the center, and (*d*) 35 cm from the center?

32 • SSM SOLVE A single-turn circular loop of radius 10.0 cm is to produce a field at its center that will just cancel the earth's magnetic field at the equator, which is 0.7 G directed north. Find the current in the loop and make a sketch that shows the orientation of the loop and the current.

33 •• For the loop of wire in Problem 32, at what point along the axis of the loop is the magnetic field (*a*) 10 percent of the field at the center, (*b*) 1 percent of the field at the center, and (*c*) 0.1 percent of the field at the center?

34 •• SOLVE A single-turn circular loop of radius 8.5 cm is to produce a field at its center that will just cancel the earth's field of magnitude 0.7 G directed at 70° below the horizontal north direction. Find the current in the loop and make a sketch that shows the orientation of the loop and the current in the loop.

35 •• A circular current loop of radius R carrying a current $I = 10$ A is centered at the origin with its axis along the x axis. Its current is such that it produces a magnetic field in the positive x direction. (*a*) Using a spreadsheet program or graphing calculator, construct a graph of B_x versus x/R for points on the x/R axis $-5 < x/R < +5$. Compare this graph with that for E_x due to a charged ring of the same size. (*b*) A second, identical current loop, carrying an equal current in the same sense, is in a plane parallel to the yz plane with its center at $x = R$. Make separate graphs of B_x on the x axis due to each loop and also graph the resultant field due to the two loops. Show from your sketch that dB_x/dx is zero midway between the two loops.

36 •• A pair of identical coils, each of radius r, are separated by a distance r. Called *Helmholtz coils,* the coils are coaxial and carry equal currents such that their axial fields add. A feature of Helmholtz coils is that the resultant magnetic field in the region between the coils is very uniform. Let $r = 30$ cm, $I = 15$ A and $N = 250$ turns for each coil. Using a spreadsheet program calculate and graph the magnetic field as a function of x, the distance from the center of the coils along the common axis, for $-r < x < r$. Over what range of x does the field vary by less than 20%?

37 ••• Two Helmholtz coils with radii R have their axes along the x axis (see Problem 36). One coil is in the yz plane and the second coil is in a parallel plane at $x = R$. Show that at the midpoint of the coils $dB_x/dx = 0$, $d^2B_x/dx^2 = 0$, and $d^3B_x/dx^3 = 0$. (*Note:* This shows that the magnetic field at points near the midpoint is approximately equal to that at the midpoint.)

38 ••• SSM *Anti-Helmholtz* coils are used in many physics applications, such as laser cooling and trapping, where a spatially inhomogeneous field with a uniform gradient is desired. These coils have the same construction as a Helmholtz coil, except that the currents flow in opposite directions, so that the axial fields subtract, and the coil separation is $r\sqrt{3}$ rather than r. Graph the magnetic field as a function of x, the axial distance from the center of the coils, for an anti-Helmholtz coil using the same parameters as in Problem 36.

39 •• Two concentric coplaner conducting circular loops have radii $r_1 = 10$ cm and $r_2 > r_1$ are in a horizontal plane. A current $I = 1$ A flows in each coil, but in opposite directions, with the current in the inner coil being counterclockwise as viewed from above. Using a spreadsheet program, calculate and graph the magnetic field as a function of the height x above the center of the coils for $r_2 = $ (*a*) 10.1 cm, (*b*) 11 cm, (*c*) 15 cm and (*d*) 20 cm.

40 ••• Two concentric circular loops of wire in the same plane have radii $r_1 = 10$ cm and $r_2 > r_1$. A current $I = 1$ A flows in each loop but in the opposite direction. Using a spreadsheet program, calculate and graph the magnetic field component B_x on the axis of the loops as a function of the distance x from the center of the coils. Construct a separate curve for $r_2 = $ (*a*) 10.1 cm, (*b*) 11 cm, (*c*) 15 cm, and (*d*) 20 cm.

41 ••• For the coils considered in Problem 40, show that if $r_2 = r_1 + \Delta r$, where $\Delta r \ll r_1$, then

$$B(x) \approx \left(\frac{\mu_0 I \Delta r}{2}\right)\left(\frac{2rx^2 - r^3}{(x^2 + r_1^2)^{5/2}}\right).$$

Straight-Line Current Segments

42 •• For the coils considered in Problem 41, show that when $x \gg r_1$, then

$$B(x) \approx -\left(\frac{\mu_0 I \Delta r}{2}\right)\left(\frac{2r_1}{x^3}\right).$$

Compare this to the results of Problem 39 (*a*).

Problems 43 to 48 refer to Figure 27-45, which shows two long straight wires in the xy plane and parallel to the x axis. One wire is at y = −6 cm and the other wire is at y = +6 cm. The current in each wire is 20 A.

FIGURE 27-45
Problems 43–48

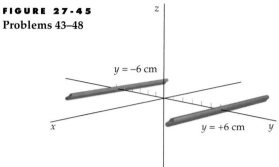

43 • **SSM** If the currents in Figure 27-45 are in the negative x direction, find $\vec{B}$ at the points on the y axis at (a) y = −3 cm, (b) y = 0, (c) y = +3 cm, and (d) y = +9 cm.

44 •• Using a spreadsheet program or graphing calculator, graph B_z versus y for points on the y axis when both currents are in the negative x direction.

45 • Find $\vec{B}$ at points on the y axis, as in Problem 43, when the current in the wire at y = −6 cm is in the negative x direction and the current in the wire at y = +6 cm is in the positive x direction.

46 •• Using a spreadsheet program or graphing calculator, graph B_z versus y for points on the y axis when the directions of the currents are opposite to those in Problem 45.

47 • Find $\vec{B}$ on the z axis at z = +8 cm if (a) the currents are parallel, as in Problem 43 and (b) the currents are antiparallel, as in Problem 45.

48 • **iSOLVE** Find the magnitude of the force per unit length exerted by one wire on the other.

49 • **iSOLVE** Two long, straight parallel wires 8.6 cm apart carry currents of equal magnitude I. The parallel wires repel each other with a force per unit length of 3.6 nN/m. (a) Are the currents parallel or antiparallel? (b) Find I.

50 •• **iSOLVE** The current in the wire shown in Figure 27-46 is 8 A. Find B at point P due to each wire segment and sum to find the resultant B.

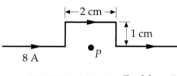

FIGURE 27-46 Problem 50

51 •• **iSOLVE** A wire of length 16 cm is suspended by flexible leads above a long straight wire. Equal but opposite currents are established in the wires so that the 16-cm wire floats 1.5 mm above the long wire with no tension in its suspension leads. If the mass of the 16-cm wire is 14 g, what is the current?

52 •• **SSM** Three long, parallel straight wires pass through the corners of an equilateral triangle of sides 10 cm, as shown in Figure 27-47, where a dot means that the current is out of the paper and a cross means that the current is into the paper. If each current is 15 A, find

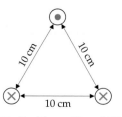

FIGURE 27-47 Problems 52 and 53

(a) the force per unit length on the upper wire and (b) the magnetic field B at the upper wire due to the two lower wires.

53 •• Rework Problem 52, with the current in the lower right corner of Figure 27-47 reversed.

54 •• An infinitely long insulated wire lies along the x axis and carries current I in the positive x direction. A second infinitely long insulated wire lies along the y axis and carries current I in the positive y direction. Where in the xy plane is the resultant magnetic field zero?

55 •• An infinitely long wire lies along the z axis and carries a current of 20 A in the positive z direction. A second infinitely long wire is parallel to the z axis at x = 10 cm. (a) Find the current in the second wire if the magnetic field at x = 2 cm is zero. (b) What is the magnetic field at x = 5 cm?

56 •• Three very long parallel wires are at the corners of a square, as shown in Figure 27-48. The wires each carry a current of magnitude I. Find the magnetic field B at the unoccupied corner of the square when (a) all the currents are into the paper, (b) I_1 and I_3 are into the paper and I_2 is out, and (c) I_1 and I_2 are into the paper and I_3 is out.

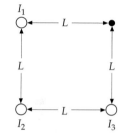

FIGURE 27-48 Problem 56

57 •• **SSM** Four long, straight parallel wires each carry current I. In a plane perpendicular to the wires, the wires are at the corners of a square of side a. Find the force per unit length on one of the wires if (a) all the currents are in the same direction and (b) the currents in the wires at adjacent corners are oppositely directed.

58 •• An infinitely long nonconducting cylinder of radius R lies along the z axis. Five long conducting wires are parallel to the cylinder and are spaced equally on the upper half of the cylinder's surface. Each wire carries a current I in the positive z direction. Find the magnetic field on the z axis.

$\vec{B}$ Due to a Current in a Solenoid

59 • A solenoid with length 30 cm, radius 1.2 cm, and 300 turns carries a current of 2.6 A. Find B on the axis of the solenoid (a) at the center, (b) inside the solenoid at a point 10 cm from one end, and (c) at one end.

60 • **SSM** **iSOLVE** A solenoid 2.7-m long has a radius of 0.85 cm and 600 turns. It carries a current I of 2.5 A. What is the approximate magnetic field B on the axis of the solenoid?

61 ••• A solenoid has n turns per unit length and radius R and carries a current I. Its axis is along the x axis with one end at $x = -\frac{1}{2}\ell$ and the other end at $x = +\frac{1}{2}\ell$, where ℓ is the total length of the solenoid. Show that the magnetic field B at a point on the axis outside the solenoid is given by

$$B = \tfrac{1}{2}\mu_0 nI(\cos\theta_1 - \cos\theta_2)$$

where $\cos\theta_1 = \dfrac{x + \frac{1}{2}\ell}{[R^2 + (x + \frac{1}{2}\ell)^2]^{1/2}}$

and $\cos\theta_2 = \dfrac{x - \frac{1}{2}\ell}{[R^2 + (x - \frac{1}{2}\ell)^2]^{1/2}}$

62 ••• In Problem 61, a formula for the magnetic field along the axis of a solenoid is given. For $x \gg \ell$ and $\ell > R$, the angles θ_1 and θ_2 are very small, so the small-angle approximation $\cos \theta \approx 1 - \theta^2/2$ is valid. (a) Draw a diagram and show that

$$\theta_1 \approx \frac{R}{x + \frac{1}{2}\ell}$$

and

$$\theta_2 \approx \frac{R}{x - \frac{1}{2}\ell}$$

(b) Show that the magnetic field at a point far from either end of the solenoid can be written

$$B = \frac{\mu_0}{4\pi}\left(\frac{q_m}{r_1^2} - \frac{q_m}{r_2^2}\right)$$

where $r_1 = x - \frac{1}{2}\ell$ is the distance to the near end of the solenoid, $r_1 = x + \frac{1}{2}\ell$ is the distance to the far end, and $q_m = nI\pi R^2 = \mu/\ell$, where $\mu = NI\pi R^2$ is the magnetic moment of the solenoid.

Ampère's Law

63 • SSM iSOLVE A long, straight, thin-walled cylindrical shell of radius R carries a current I. Find B inside the cylinder and outside the cylinder.

64 • In Figure 27-49, one current is 8 A into the paper, the other current is 8 A out of the paper, and each curve is a circular path. (a) Find $\oint_C \vec{B} \cdot d\vec{\ell}$ for each path indicated, where each integral is taken with $d\vec{\ell}$ counterclockwise. (b) Which path, if any, can be used to find B at some point due to these currents?

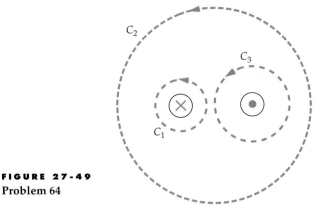

FIGURE 27-49
Problem 64

65 • A very long coaxial cable consists of an inner wire and a concentric outer cylindrical conducting shell of radius R. At one end, the wire is connected to the shell. At the other end, the wire and shell are connected to opposite terminals of a battery, so there is a current down the wire and back up the shell. Assume that the cable is straight. Find B (a) at points between the wire and the shell far from the ends and (b) outside the cable.

66 •• iSOLVE A wire of radius 0.5 cm carries a current of 100 A that is uniformly distributed over its cross-sectional area. Find B (a) 0.1 cm from the center of the wire, (b) at the surface of the wire, and (c) at a point outside the wire 0.2 cm from the surface of the wire. (d) Sketch a graph of B versus the distance from the center of the wire.

67 •• SSM Show that a uniform magnetic field with no fringing field, such as that shown in Figure 27-50, is impossible because it violates Ampère's law. Do this by applying Ampère's law to the rectangular curve shown by the dashed lines.

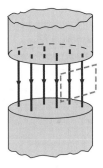

FIGURE 27-50 Problem 67

68 •• A coaxial cable consists of a solid inner cylindrical conductor of radius 1.00 mm and an outer cylindrical shell conductor of inner radius 2.00 mm with an outer radius of 3.00 mm. There is a current of 15 A down the inner wire and an equal return current in the outer wire. The currents are uniform over the cross section of each conductor. Using a spreadsheet program or graphing calculator, graph the magnitude of the magnetic field B as a function of the distance r from the cable axis for $0 \text{ mm} < r < 3.00 \text{ mm}$. What is the field outside the wire?

69 •• An infinitely long, thick cylindrical shell of inner radius a and outer radius b carries a current I uniformly distributed across a cross section of the shell. Find the magnetic field for (a) $r < a$, (b) $a < r < b$, and (c) $r > b$.

70 •• Figure 27-51 shows a solenoid carrying a current I with n turns per unit length. Apply Ampère's law to the rectangular curve shown in the figure to derive an expression for B, assuming that B is uniform inside the solenoid and that B is zero outside the solenoid.

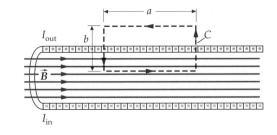

FIGURE 27-51 Problem 70

71 •• iSOLVE A tightly wound toroid of inner radius 1 cm and outer radius 2 cm has 1000 turns of wire and carries a current of 1.5 A. (a) What is the magnetic field at a distance of 1.1 cm from the center? (b) What is the magnetic field at a distance of 1.5 cm from the center?

72 •• SSM The xz plane contains an infinite sheet of current in the positive z direction. The current per unit length (along the x direction) is λ. Figure 27-52a shows a point P above the sheet ($y > 0$) and two portions of the current sheet labeled I_1 and I_2. (a) What is the direction of the magnetic field $\vec{B}$ at point P due to the two portions of the current shown? (b) What is the direction of the magnetic field $\vec{B}$ at point P due to the entire sheet? (c) What is the direction of $\vec{B}$ at a point below the sheet ($y < 0$)? (d) Apply Ampère's law to the rectangular curve shown in Figure 27-52b to show that the magnetic field at any point above the sheet is given by $\vec{B} = -\frac{1}{2}\mu_0\lambda\hat{i}$

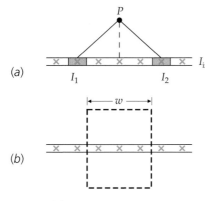

(a)

(b)

FIGURE 27-52 Problem 72

Magnetization and Magnetic Susceptibility

73 • A tightly wound solenoid 20-cm long has 400 turns and carries a current of 4 A so that its axial field is in the z direction. Neglecting end effects, find B and B_{app} at the center when (a) there is no core in the solenoid and (b) there is an iron core with a magnetization $M = 1.2 \times 10^6$ A/m.

74 • If the solenoid of Problem 73 has an aluminum core, find B_{app}, M, and B at the center, neglecting end effects.

75 • Repeat Problem 74 for a tungsten core.

76 • A long solenoid is wound around a tungsten core and carries a current. (a) If the core is removed while the current is held constant, does the magnetic field inside the solenoid decrease or increase? (b) By what percentage does the magnetic field inside the solenoid decrease or increase?

77 • **SOLVE** When a sample of liquid is inserted into a solenoid carrying a constant current, the magnetic field inside the solenoid decreases by 0.004 percent. What is the magnetic susceptibility of the liquid?

78 • **SOLVE** A long solenoid carrying a current of 10 A has 50 turns/cm. What is the magnetic field in the interior of the solenoid when the interior is (a) a vacuum, (b) filled with aluminum, and (c) filled with silver?

79 •• **SSM** A cylinder of magnetic material is placed in a long solenoid of n turns per unit length and current I. The values for magnetic field B within the material versus nI is given below. Use these values to plot B versus B_{app} and K_m versus nI.

nI, A/m	0	50	100	150	200	500	1000	10,000
B, T	0	0.04	0.67	1.00	1.2	1.4	1.6	1.7

80 •• A small magnetic sample is in the form of a disk that has a radius of 1.4 cm, a thickness of 0.3 cm, and a uniform magnetization along its axis throughout its volume. The magnetic moment of the sample is 1.5×10^{-2} A·m². (a) What is the magnetization $\vec{M}$ of the sample? (b) If this magnetization is due to the alignment of N electrons, each with a magnetic moment of 1 μ_B, what is N? (c) If the magnetization is along the axis of the disk, what is the magnitude of the amperian surface current?

81 •• A cylindrical shell in the shape of a flat washer has inner radius r, outer radius R, and thickness (length) t, where $t << R$. The material of the shell has a uniform magnetization of magnitude M parallel to its axis. Show that the magnetic field due to the cylinder can be modeled using the concentric conducting loops model of Problem 39. What is the amperian current I which we must use to model this field?

Atomic Magnetic Moments

82 •• **SSM** **SOLVE** Nickel has a density of 8.7 g/cm³ and a molecular mass of 58.7 g/mol. Nickel's saturation magnetization is given by $\mu_0 M_s = 0.61$ T. Calculate the magnetic moment of a nickel atom in Bohr magnetons.

83 •• **SOLVE** Repeat Problem 82 for cobalt, which has a density of 8.9 g/cm³, a molecular mass of 58.9 g/mol, and a saturation magnetization given by $\mu_0 M_s = 1.79$ T.

*Paramagnetism

84 • Show that Curie's law predicts that the magnetic susceptibility of a paramagnetic substance is $\chi_m = \mu_0 M_s/3kT$.

85 •• In a simple model of paramagnetism, we can consider that some fraction f of the molecules have their magnetic moments aligned with the external magnetic field and that the rest of the molecules are randomly oriented and therefore do not contribute to the magnetic field. (a) Use this model and Curie's law to show that at temperature T and external magnetic field B the fraction of aligned molecules is $f = \mu B/3kT$. (b) Calculate this fraction for $T = 300$ K, $B = 1$ T, assuming μ to be 1 Bohr magneton.

86 •• **SSM** Assume that the magnetic moment of an aluminum atom is 1 Bohr magneton. The density of aluminum is 2.7 g/cm³, and its molecular mass is 27 g/mol. (a) Calculate M_s and $\mu_0 M_s$ for aluminum. (b) Use the results of Problem 84 to calculate χ_m at $T = 300$ K. (c) Explain why the result for Part (b) is larger than the value listed in Table 27-1.

87 •• A toroid with N turns carrying a current I has a mean radius R and a cross-sectional radius r, where $r << R$ (Figure 27-53). When the toroid is filled with material, it is called a *Rowland ring*. Find B_{app} and B in such a ring, assuming a magnetization $\vec{M}$ everywhere parallel to $\vec{B}_{app}$.

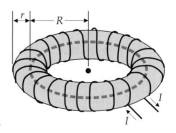

FIGURE 27-53 Problem 87

88 •• A toroid is filled with liquid oxygen that has a susceptibility of 4×10^{-3}. The toroid has 2000 turns and carries a current of 15 A. Its mean radius is 20 cm, and the radius of its cross section is 0.8 cm. (a) What is the magnetization M? (b) What is the magnetic field B? (c) What is the percentage increase in B produced by the liquid oxygen?

89 •• A toroid has an average radius of 14 cm and a cross-sectional area of 3 cm². It is wound with fine wire, 60 turns/cm measured along its mean circumference, and the wire carries a current of 4 A. The core is filled with a paramagnetic material of magnetic susceptibility 2.9×10^{-4}. (a) What is the magnitude of the magnetic field within the substance? (b) What is the magnitude of the magnetization? (c) What would the magnitude of the magnetic field be if there were no paramagnetic core present?

*Ferromagnetism

90 • **SSM** For annealed iron, the relative permeability K_m has its maximum value of approximately 5500 at $B_{app} = 1.57 \times 10^{-4}$ T. Find M and B when K_m is maximum.

91 •• The saturation magnetization for annealed iron occurs when $B_{app} = 0.201$ T. Find the permeability μ and the relative permeability K_m of annealed iron at saturation. (See Table 27-2.)

92 •• **iSOLVE** The coercive force is defined to be the applied magnetic field needed to bring B back to zero along the hysteresis curve (which is point c in Figure 27-38). For a certain permanent bar magnet, the coercive force $B_{app} = 5.53 \times 10^{-2}$ T. The bar magnet is to be demagnetized by placing it inside a 15-cm-long solenoid with 600 turns. What minimum current is needed in the solenoid to demagnetize the magnet?

93 •• A long solenoid with 50 turns/cm carries a current of 2 A. The solenoid is filled with iron and B is measured to be 1.72 T. (a) Neglecting end effects, what is B_{app}? (b) What is M? (c) What is the relative permeability K_m?

94 •• When the current in Problem 93 is 0.2 A, the magnetic field is measured to be 1.58 T. (a) Neglecting end effects, what is B_{app}? (b) What is M? (c) What is the relative permeability K_m?

95 •• A long, iron-core solenoid with 2000 turns/m carries a current of 20 mA. At this current, the relative permeability of the iron core is 1200. (a) What is the magnetic field within the solenoid? (b) With the iron core removed, what current will produce the same field within the solenoid?

96 •• **SSM** Two long straight wires 4-cm apart are embedded in a uniform insulator that has a relative permeability of $K_m = 120$. The wires carry 40 A in opposite directions. (a) What is the magnetic field at the midpoint of the plane of the wires? (b) What is the force per unit length on the wires?

97 •• The toroid of Problem 88 has its core filled with iron. When the current is 10 A, the magnetic field in the toroid is 1.8 T. (a) What is the magnetization M? (b) Find the values for K_m, μ, and χ_m for the iron sample.

98 •• Find the magnetic field in the toroid of Problem 89 if the current in the wire is 0.2 A and soft iron, which has a relative permeability of 500, is substituted for the paramagnetic core?

99 •• A long straight wire with a radius of 1.0 mm is coated with an insulating ferromagnetic material that has a thickness of 3.0 mm and a relative magnetic permeability of $K_m = 400$. The coated wire is in air and the wire itself is nonmagnetic. The wire carries a current of 40 A. (a) Find the magnetic field inside the wire as a function of radius R. (b) Find the magnetic field inside the ferromagnetic material as a function of radius R. (c) Find the magnetic field outside the ferromagnetic material as a function of R. (d) What must the magnitudes and directions of the amperian currents be on the surfaces of the ferromagnetic material to account for the magnetic fields observed?

General Problems

100 • Find the magnetic field at point P in Figure 27-54.

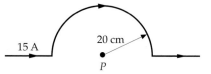

FIGURE 27-54 Problem 100

101 • **SSM** In Figure 27-55, find the magnetic field at point P, which is at the common center of the two semicircular arcs.

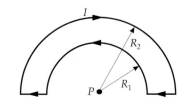

FIGURE 27-55 Problem 101

102 •• A wire of length ℓ is wound into a circular coil of N loops and carries a current I. Show that the magnetic field at the center of the coil is given by $B = \mu_0 \pi N^2 I / \ell$.

103 •• A very long wire carrying a current I is bent into the shape shown in Figure 27-56. Find the magnetic field at point P.

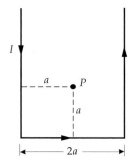

FIGURE 27-56 Problem 103

104 •• **SSM** A power cable carrying 50 A is 2 m below the earth's surface, but the cable's direction and precise position are unknown. Show how you could locate the cable using a compass. Assume that you are at the equator, where the earth's magnetic field is 0.7 G north.

105 •• A long straight wire carries a current of 20 A, as shown in Figure 27-57. A rectangular coil with two sides parallel to the straight wire has sides 5 cm and 10 cm with the near side a distance 2 cm from the wire. The coil carries a current of 5 A. (a) Find the force on each segment of the rectangular coil due to the current in the long straight wire.

(b) What is the net force on the coil?

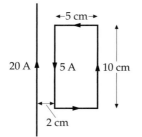

FIGURE 27-57 Problem 105

106 •• [iSOLVE] The closed loop shown in Figure 27-58 carries a current of 8 A in the counterclockwise direction. The radius of the outer arc is 60 cm, that of the inner arc is 40 cm. Find the magnetic field at point P.

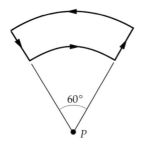

FIGURE 27-58 Problem 106

107 •• [iSOLVE] A closed circuit consists of two semicircles of radii 40 cm and 20 cm that are connected by straight segments, as shown in Figure 27-59. A current of 3 A flows around this circuit in the clockwise direction. Find the magnetic field at point P.

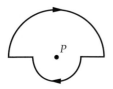

FIGURE 27-59 Problem 107

108 •• [SSM] [iSOLVE] A very long straight wire carries a current of 20 A. An electron 1 cm from the center of the wire is moving with a speed of 5.0×10^6 m/s. Find the force on the electron when it moves (a) directly away from the wire, (b) parallel to the wire in the direction of the current, and (c) perpendicular to the wire and tangent to a circle around the wire.

109 •• A current I of 5 A is uniformly distributed over the cross section of a long straight wire of radius $r_0 = 2.55$ mm. Using a spreadsheet program, graph the magnitude of the magnetic field as a function of r and the distance from the center of the wire for $0 \le r \le 10r_0$.

110 •• A large, 50-turn circular coil of radius 10 cm carries a current of 4 A. At the center of the large coil is a small 20-turn coil of radius 0.5 cm carrying a current of 1 A. The planes of the two coils are perpendicular. Find the torque exerted by the large coil on the small coil. (Neglect any variation in B due to the large coil over the region occupied by the small coil.)

111 •• [SSM] Figure 27-60 shows a bar magnet suspended by a thin wire that provides a restoring torque $-\kappa\theta$. The magnet is 16-cm long, has a mass of 0.8 kg, a dipole moment of $\mu = 0.12$ A·m², and it is located in a region where a uniform magnetic field B can be established. When the external magnetic field is 0.2 T and the magnet is given a small angular displacement $\Delta\theta$, the bar magnet oscillates about its equilibrium position with a period of 0.500 s. Determine the constant κ and the period of this torsional pendulum when $B = 0$.

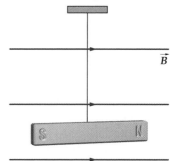

FIGURE 27-60 Problem 111

112 •• A long, narrow bar magnet that has magnetic moment μ parallel to its long axis is suspended at its center as a frictionless compass needle. When placed in a magnetic field $\vec{B}$, the needle lines up with the field. If it is displaced by a small angle θ, show that the needle will oscillate about its equilibrium position with frequency $f = (\frac{1}{2})\pi\sqrt{\mu B/I}$, where I is the moment of inertia about the point of suspension.

113 •• A small bar magnet of mass 0.1 kg, length 1 cm, and magnetic moment $\mu = 0.04$ A·m² is located at the center of a 100-turn loop of 0.2 m diameter. The loop carries a current of 5.0 A. At equilibrium, the bar magnet is aligned with the field due to the current loop. The bar magnet is given a displacement along the axis of the loop and released. Show that if the displacement is small, the bar magnet executes simple harmonic motion, and find the period of this motion.

114 •• Suppose the needle in Problem 112 is a uniformly magnetized iron rod that is 8 cm long and has a cross-sectional area of 3 mm². Assume that the magnetic dipole moment for each iron atom is 2.2 μ_B and that all the iron atoms have their dipole moments aligned. Calculate the frequency of small oscillations about the equilibrium position when the magnetic field is 0.5 G.

115 •• The needle of a magnetic compass has a length of 3 cm, a radius of 0.85 mm, and a density of 7.96×10^3 kg/m³. The needle is free to rotate in a horizontal plane, where the horizontal component of the earth's magnetic field is 0.6 G. When disturbed slightly, the compass executes simple harmonic motion about its midpoint with a frequency of 1.4 Hz. (a) What is the magnetic dipole moment of the needle? (b) What is the magnetization M? (c) What is the amperian current on the surface of the needle? (See Problem 112.)

116 •• [SSM] An iron bar of length 1.4 m has a diameter of 2 cm and a uniform magnetization of 1.72×10^6 A/m directed along the bar's length. The bar is stationary in space and is suddenly demagnetized so that its magnetization disappears. What is the rotational angular velocity of the bar if its angular momentum is conserved? (Assume that Equation 27-27 holds where m is the mass of an electron and $q = -e$.)

117 •• The magnetic dipole moment of an iron atom is 2.219 μ_B. (a) If all the atoms in an iron bar of length 20 cm and cross-sectional area 2 cm² have their dipole moments aligned, what is the dipole moment of the bar? (b) What torque must be supplied to hold the iron bar perpendicular to a magnetic field of 0.25 T?

118 •• **SSM** A relatively inexpensive ammeter, called a *tangent galvanometer*, can be made using the earth's field. A plane circular coil of N turns and radius R is oriented so the field B_c it produces in the center of the coil is either east or west. A compass is placed at the center of the coil. When there is no current in the coil, the compass needle points north. When there is a current I, the compass needle points in the direction of the resultant magnetic field $\vec{B}$ at an angle θ to the north. Show that the current I is related to θ and to the horizontal component of the earth's field B_e by

$$I = \frac{2RB_c}{\mu_0 N} \tan \theta$$

119 •• **iSOLVE** An infinitely long straight wire is bent, as shown in Figure 27-61. The circular portion has a radius of 10 cm with its center a distance r from the straight part. Find r so that the magnetic field at the center of the circular portion is zero.

FIGURE 27-61
Problem 119

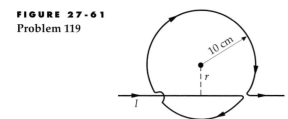

120 •• (a) Find the magnetic field at point P for the wire carrying current I, as shown in Figure 27-62. (b) Use your result from Part (a) to find the field at the center of a polygon of N sides. Show that when N is very large, your result approaches that for the magnetic field at the center of a circle.

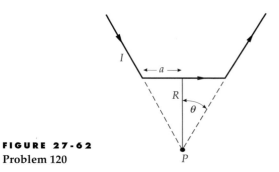

FIGURE 27-62
Problem 120

121 •• The current in a long cylindrical conductor of radius R = 10 cm varies with distance from the axis of the cylinder according to the relation I(r) = (50 A/m)r. Find the magnetic field at (a) r = 5 cm, (b) r = 10 cm, and (c) r = 20 cm.

122 •• Figure 27-63 shows a square loop, 20 cm per side, in the xy plane with its center at the origin. The loop carries a current of 5 A. Above it at y = 0, z = 10 cm is an infinitely long wire parallel to the x axis carrying a current of 10 A. (a) Find the torque on the loop. (b) Find the net force on the loop.

FIGURE 27-63
Problem 122

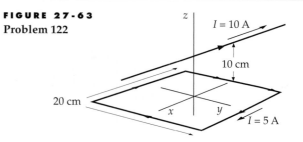

123 •• A current balance is constructed in the following way: A 10-cm-long section of wire is placed on top of the pan of an electronic balance used in a chemistry lab. Leads are clipped to it running into a power supply and through the supply to another segment of wire that is suspended directly above it, parallel with it. (See figure below.) The distance between the two wires is L = 2.0 cm. The power supply provides a current I running through the wires. When the power supply is switched on, the reading on the balance increases by 5.0 mg (1 mg = 10^{-6} kg). What is the current running through the wire?

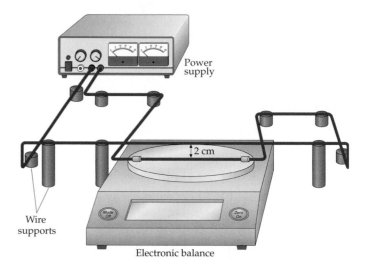

FIGURE 27-64 **Problem 123**

124 •• Consider the current balance of Problem 123. If the sensitivity of the balance is 0.1 mg, what is the minimum current detectable using this current balance? Discuss any advantages or disadvantages of this type of current balance versus the "standard" current balance discussed in the chapter.

125 ••• **SSM** A disk of radius R carries a fixed charge density σ and rotates with angular velocity ω. (a) Consider a circular strip of radius r and width dr with charge dq. Show that the current produced by this strip $dI = (\omega/2\pi) \, dq = \omega \sigma r \, dr$. (b) Use your result from Part (a) to show that the magnetic field at the center of the disk is $B = \frac{1}{2}\mu_0 \sigma \omega R$. (c) Use your result from Part (a) to find the magnetic field at a point on the axis of the disk a distance x from the center.

126 ••• A square loop of side ℓ lies in the yz plane with its center at the origin. It carries a current I. Find the magnetic field B at any point on the x axis and show from your expression that for x much larger than ℓ,

$$B \approx \frac{\mu_0 2\mu}{4\pi x^3}$$

where $\mu = I\ell^2$ is the magnetic moment of the loop.

Magnetic Induction

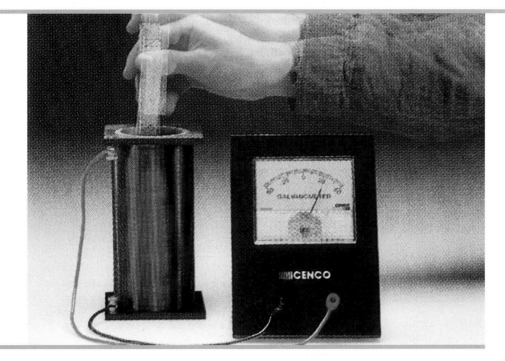

DEMONSTRATION OF INDUCED EMF.
WHEN THE MAGNET IS MOVING
TOWARD OR AWAY FROM THE COIL,
AN EMF IS INDUCED IN THE COIL, AS
SHOWN BY THE GALVANOMETER'S
DEFLECTION. NO DEFLECTION IS
OBSERVED WHEN THE MAGNET
IS STATIONARY.

? **What causes the current
when the magnet moves? This
is discussed in Section 28-2.**

n the early 1830s, Michael Faraday in England and Joseph Henry in America independently discovered that in a *changing* magnetic field a changing magnetic flux through a surface bounded by a closed stationary loop of wire induces a current in the wire. The emfs and currents caused by such changing magnetic fluxes are called **induced emfs** and **induced currents.** The process itself is referred to as **induction.** Faraday and Henry also discovered that in a *static* magnetic field a changing magnetic flux through a surface bounded by a moving loop of wire induces an emf in the wire. An emf caused by the motion of a conductor in a region with a magnetic field is called a **motional emf.**

When you pull the plug of an electric cord from its socket, you sometimes observe a small spark. Before the cord is disconnected, the cord carries a current that produces a magnetic field encircling the current. When the cord is disconnected, the current abruptly ceases and the magnetic field around the cord collapses. This changing magnetic field induces an emf that tends to maintain the original current, resulting in a spark at the points of the disconnect. Once the magnetic field collapses to zero it is no longer changing, and the induced emf is zero.

Changing magnetic fields can result from changing currents or from moving magnets. The chapter-opening photo illustrates a simple classroom demonstration of emf induced by a changing magnetic field. The ends of a coil are attached to a galvanometer and a strong magnet is moved toward or away from the coil. The momentary deflection shown by the galvanometer *during* the motion indicates that there is an induced electric current in the coil–galvanometer circuit. A current is also induced if the coil is moved toward a stationary magnet, away from a stationary magnet, or if the coil is rotated in a region with a static magnetic field. A coil rotating in a static magnetic field is the basic element of a generator, which converts mechanical energy into electrical energy.

➤ **This chapter will explore the various methods of magnetic induction, all of which can be summarized by a single relation known as Faraday's law. Faraday's law relates the induced emf in a circuit to the rate of change in magnetic flux through the circuit. (The *magnetic flux through the circuit* refers to the flux of the magnetic field through any surface bounded by the circuit.)**

28-1 Magnetic Flux

The flux of any vector field through a surface is calculated in the same way as the flux of an electric field through a surface (Section 22-2). Let dA be an element of area on the surface S, and let $\hat{n}$ be a unit normal, a unit vector normal to the area element (Figure 28-1). There are two directions normal to any area element, and which of the two directions is selected for the direction of $\hat{n}$ is a matter of choice. However, the sign of the flux does depend on this choice. The magnetic flux ϕ_m through S is

$$\phi_m = \int_S \vec{B} \cdot \hat{n}\, dA = \int_S B_n\, dA \qquad 28\text{-}1$$

MAGNETIC FLUX

The unit of magnetic flux is that of magnetic field intensity times area, teslameter squared, which is called a **weber** (Wb):

$$1\,\text{Wb} = 1\,\text{T·m}^2 \qquad 28\text{-}2$$

Since B is proportional to the number of field lines per unit area, the magnetic flux is proportional to the number of lines through an element of area.

EXERCISE Show that a weber per second is a volt.

If the surface is flat with area A, and if $\vec{B}$ is uniform (has the same magnitude and direction) over the surface, the magnetic flux through the surface is

$$\phi_m = \vec{B} \cdot \hat{n}\, A = BA \cos\theta = B_n A$$

where θ is the angle between the direction of $\vec{B}$ and the positive normal direction. We are often interested in the flux through a surface bounded by a coil that contains several turns of wire. If the coil contains N turns, the flux through the surface is N times the flux through each turn (Figure 28-2). That is,

$$\phi_m = NBA \cos\theta \qquad 28\text{-}3$$

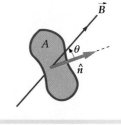

FIGURE 28-1 When $\vec{B}$ makes an angle θ with the normal to the area of a loop, the flux through the loop is $\vec{B}\cdot\hat{n}A = BA \cos\theta$.

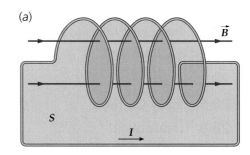

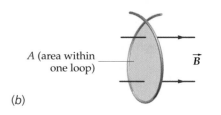

FIGURE 28-2 (*a*) The flux through the surface S bounded by a coil with N turns is proportional to the number of field lines penetrating the surface. The coil shown has 4 turns. For the two field lines shown, each line penetrates the surface four times, once for each turn, so the flux through S is four times greater than the flux through the surface "bounded" by a single turn of the coil. The coil shown is not tightly wound so the surface S can better be observed. (*b*) The area A of the flat surface that is (almost) bounded by a single turn.

where A is the area of the flat surface bounded by a single turn. (*Note:* Only a closed curve can actually bound a surface.) A single turn of a multiturn coil is not closed, so a single turn can not actually bound a surface. However, if a coil is tightly wound a single turn is almost closed, and A is the area of the flat surface that it (almost) bounds.

FLUX THROUGH A SOLENOID   **EXAMPLE 28-1**

Find the magnetic flux through a solenoid that is 40-cm long, has a radius of 2.5 cm, has 600 turns, and carries a current of 7.5 A.

PICTURE THE PROBLEM The magnetic field $\vec{B}$ inside the solenoid is uniform and parallel with the axis of the solenoid. It is therefore perpendicular to the plane of each coil. Therefore, we need to find B inside the solenoid and then multiply B by NA.

1. The magnetic flux is the product of the number of turns, the magnetic field strength, and the area bounded by one turn:

$$\phi_m = NBA$$

2. The magnetic field inside the solenoid is given by $B = \mu_0 nI$, where $n = N/\ell$ is the number of turns per unit length:

$$\phi_m = N\mu_0 nIA = N\mu_0 \frac{N}{\ell}IA = \frac{\mu_0 N^2 IA}{\ell}$$

3. Express the area A in terms of its radius:

$$A = \pi r^2$$

4. Substitute the given values to calculate the flux:

$$\phi_m = \frac{\mu_0 N^2 I \pi r^2}{\ell}$$

$$= \frac{(4\pi \times 10^{-7} \text{ T·m/A})(600 \text{ turns})^2 (7.5 \text{ A}) \pi (0.025 \text{ m})^2}{0.40 \text{ m}}$$

$$= \boxed{1.66 \times 10^{-2} \text{ Wb}}$$

REMARKS Note that since $\phi_m = NBA$ and B is proportional to the number of turns N, ϕ_m is proportional to N^2.

28-2 Induced EMF and Faraday's Law

Experiments by Faraday, Henry, and others showed that if the magnetic flux through a surface bounded by a circuit is changed by any means, an emf equal in magnitude to the rate of change of the flux is induced in the circuit. We usually detect the emf by observing a current in the circuit, but the emf is present even if the circuit is nonexistent or incomplete (not closed) and there is no current. Previously we considered emfs that were localized in a specific part of the circuit, such as between the terminals of the battery. However, induced emfs can be distributed throughout the circuit.

The magnetic flux through a surface bounded by a circuit can be changed in several ways. The current producing the magnetic field may be increased or decreased, permanent magnets may be moved toward the surface or away from the surface, the circuit itself may be rotated in a region with a static magnetic field or translated in a region with a nonuniform static magnetic field $\vec{B}$, the orientation of the circuit may be changed, or the area of the surface in a region with a uniform static magnetic field may be increased or decreased. In every case, an emf $\mathcal{E}$ is induced in the circuit that is equal in magnitude to the rate of change of the magnetic flux through (a surface bounded by) the circuit. That is

$$\mathcal{E} = -\frac{d\phi_m}{dt}$$

28-4

FARADAY'S LAW

This result is known as **Faraday's law.** The negative sign in Faraday's law has to do with the direction of the induced emf, which is addressed shortly.

Figure 28-3 shows a single stationary loop of wire in a magnetic field. The flux through the loop is changing because the magnetic field strength is increasing, so an emf is induced in the loop. Since emf is the work done per unit charge, we know there must be forces exerted on the mobile charges doing work on them. Magnetic forces can do no work, therefore, we cannot attribute the emf to the work done by magnetic forces. It is electric forces associated with a nonconservative electric field $\vec{E}_{nc}$ doing the work on the mobile charges. The line integral of this electric field around a complete circuit equals the work done per unit charge, which is the induced emf in the circuit.

The electric fields that we studied in earlier chapters resulted from static electric charges. Such electric fields are conservative, meaning that their circulation about any curve C is zero. (The circulation of a vector field $\vec{A}$ about a closed curve C is defined as $\oint_C \vec{A} \cdot d\vec{\ell}$.) However, the electric field associated with a changing magnetic field is nonconservative. Its circulation about C is an induced emf, equal to the negative of the rate of change of the magnetic flux through any surface S bounded by C:

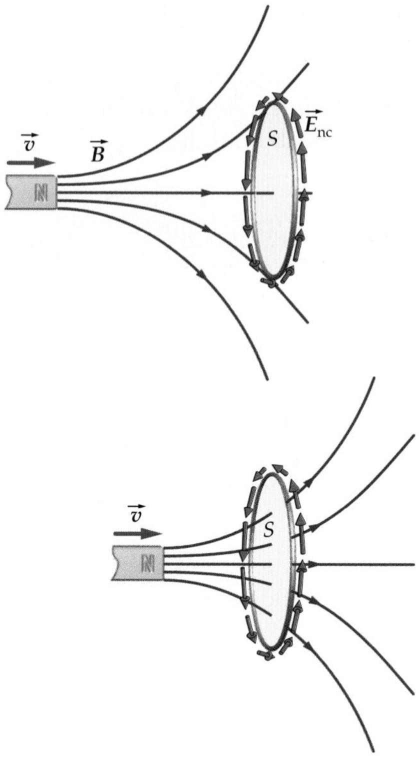

$$\mathcal{E} = \oint_C \vec{E}_{nc} \cdot d\vec{\ell} = -\frac{d}{dt} \int_S \vec{B} \cdot \hat{n}\, dA = -\frac{d\phi_m}{dt}$$

28-5

INDUCED EMF IN A STATIONARY CIRCUIT IN A CHANGING MAGNETIC FIELD

FIGURE 28-3 If the magnetic flux through the stationary wire loop is changing, an emf is induced in the loop. The emf is distributed throughout the loop, which is due to a nonconservative electric field $\vec{E}_{nc}$ tangent to the wire.

INDUCED EMF IN A CIRCULAR COIL I | **EXAMPLE 28-2**

A uniform magnetic field makes an angle of 30° with the axis of a circular coil of 300 turns and a radius of 4 cm. The magnitude of the magnetic field increases at a rate of 85 T/s while its direction remains fixed. Find the magnitude of the induced emf in the coil.

PICTURE THE PROBLEM The induced emf equals N times the rate of change of the flux through a single turn. Since $\vec{B}$ is uniform, the flux through each turn is simply $\phi_m = BA \cos \theta$, where $A = \pi r^2$ is the area of the coil.

1. The magnitude of the induced emf is given by Faraday's law:

$$\mathcal{E} = -\frac{d\phi_m}{dt}$$

2. For a uniform field, the flux is:

$$\phi_m = N\vec{B} \cdot \hat{n}A = NBA \cos \theta$$

3. Substitute this expression for ϕ_m and calculate $\mathcal{E}$:

$$\mathcal{E} = -\frac{d\phi_m}{dt} = -\frac{d}{dt}(NBA \cos \theta) = -N\pi r^2 \cos \theta \frac{dB}{dt}$$

$$= -(300)\pi(0.04\text{ m})^2 \cos 30°(85\text{ T/s}) = -111\text{ V}$$

$$|\mathcal{E}| = \boxed{111\text{ V}}$$

EXERCISE If the resistance of the coil is 200 Ω, what is the induced current? (*Answer* 0.555 A)

INDUCED EMF IN A CIRCULAR COIL II   **EXAMPLE 28-3**   Try It Yourself

An 80-turn coil of radius 5 cm and resistance of 30 Ω sits in a region with a uniform magnetic field normal to the plane of the coil. At what rate must the magnitude of the magnetic field change to produce a current of 4 A in the coil?

PICTURE THE PROBLEM The rate of change of the magnetic field is related to the rate of change of the flux, which is related to the induced emf by Faraday's law. The emf in the coil equals IR.

Cover the column to the right and try these on your own before looking at the answers.

Steps	Answers
1. Write the magnetic flux in terms of B, N, and the radius r, and solve for B.	$\phi_m = NBA = NB\pi r^2$ $B = \dfrac{\phi_m}{N\pi r^2}$
2. Take the time derivative of B.	$\dfrac{dB}{dt} = \dfrac{1}{N\pi r^2}\dfrac{d\phi_m}{dt}$
3. Use Faraday's law to relate the rate of change of the flux to the emf.	$\mathcal{E} = -\dfrac{d\phi_m}{dt}$
4. Calculate the magnitude of the emf in the coil from the current and resistance of the coil.	$\lvert\mathcal{E}\rvert = IR = 120 \text{ V}$
5. Substitute numerical values of E, N, and r to calculate dB/dt.	$\left\lvert\dfrac{dB}{dt}\right\rvert = \dfrac{1}{N\pi r^2}\lvert\mathcal{E}\rvert = \boxed{191 \text{ T/s}}$

A sign convention allows us to use Equation 28-5 to find the direction of both the induced electric field and the induced emf. According to this convention, the positive tangential direction along the integration path C is related to the direction of the unit normal $\hat{n}$ on the surface S bounded by C by a right-hand rule (Figure 28-4). By placing your right thumb in the direction of $\hat{n}$, the fingers of your hand curl in the positive tangential direction on C. If $d\phi_m/dt$ is positive, then in accord with Faraday's law (Equation 28-5), both $\vec{E}_{nc}$ and $\mathcal{E}$ are in the negative tangential direction. (The direction of both $\vec{E}_{nc}$ and $\mathcal{E}$ can be determined via Lenz's law, which is discussed in Section 28-3.)

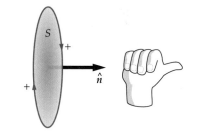

FIGURE 28-4 By placing your right thumb in the direction of $\hat{n}$ on the surface S, the fingers of your hand curl in the positive tangential direction on C.

INDUCED NONCONSERVATIVE ELECTRIC FIELD   **EXAMPLE 28-4**

A magnetic field $\vec{B}$ is perpendicular to the plane of the page. $\vec{B}$ is uniform throughout a circular region of radius R, as shown in Figure 28-5. Outside this region, B equals zero. The direction of $\vec{B}$ remains fixed and rate of change of B is dB/dt. What are the magnitude and direction of the induced electric field in the plane of the page (a) a distance $r < R$ from the center of the circular region and (b) a distance $r > R$ from the center, where $B = 0$.

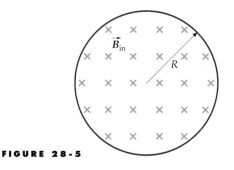

FIGURE 28-5

PICTURE THE PROBLEM The magnetic field $\vec{B}$ is into the page and uniform over a circular region of radius R, as shown in Figure 28-6. As B increases or decreases, the magnetic flux through a surface bounded by closed curve C also changes, and an emf $\mathcal{E} = \oint_C \vec{E} \cdot d\vec{\ell}$ is induced around C. The induced electric field is found by applying $\oint_C \vec{E} \cdot d\vec{\ell} = -d\phi_m/dt$ (Equation 28-5). To take advantage of the system's symmetry, we choose C to be a circular curve of radius r and then evaluate the line integral. By symmetry, $\vec{E}$ is tangent to circle C and has the same magnitude at any point on the circle. We will assign into the page as the direction of $\hat{n}$. The sign convention then tells us that the positive tangential direction is clockwise. We then calculate the magnetic flux ϕ_m, take its time derivative, and solve for E_t.

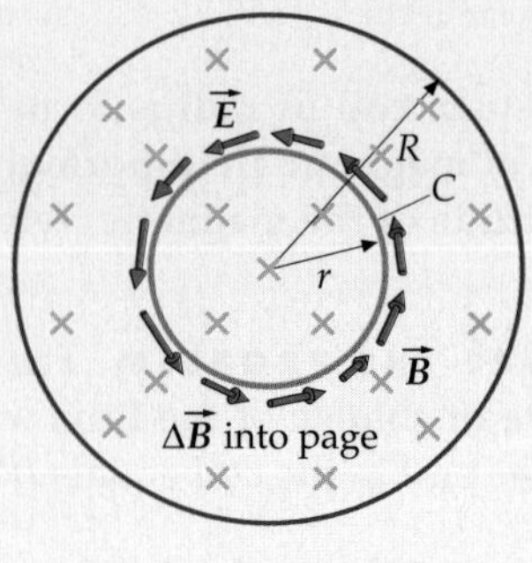

FIGURE 28-6

(a) 1. The $\vec{E}$ and $\vec{B}$ fields are related by Equation 28-5:

$$\oint_C \vec{E} \cdot d\vec{\ell} = -\frac{d\phi_m}{dt}$$

where

$$\phi_m = \int_S \vec{B} \cdot \hat{n} \, dA$$

2. E_t (the tangential component of $\vec{E}$) is found from the line integral for a circle of radius $r < R$. $\vec{E}$ is tangent to the circle and has a constant magnitude:

$$\oint_C \vec{E} \cdot d\vec{\ell} = \oint_C E_t \, d\ell = E_t \oint_C d\ell = E_t \, 2\pi r$$

3. For $r < R$, $\vec{B}$ is uniform on the flat surface S bounded by the circle C. We choose into the page as the direction of $\hat{n}$. Because $\vec{B}$ is also into the page, the flux through S is simply BA:

$$\phi_m = \int_S \vec{B} \cdot \hat{n} \, dA = \int_S B_n \, dA = B_n \int_S dA$$
$$= BA = B\pi r^2$$

4. Calculate the time derivative of ϕ_m:

$$\frac{d\phi_m}{dt} = \frac{d}{dt}(B\pi r^2) = \frac{dB}{dt}\pi r^2$$

5. Substitute the step 2 and step 4 results into the step 1 result and solve for E_t. The positive tangential direction is clockwise.

$$E_t \, 2\pi r = -\frac{dB}{dt}\pi r^2$$

so

$$\boxed{E_t = -\frac{r}{2}\frac{dB}{dt}, \quad r < R}$$

6. For the choice for the direction of $\hat{n}$ in step 3, the positive tangential direction is clockwise:

E_t is negative, so $\vec{E}$ is $\boxed{\text{counterclockwise}}$.

(b) 1. For a circle of radius $r > R$ (the region where the magnetic field is zero), the line integral is the same as before:

$$\oint_C \vec{E} \cdot d\vec{\ell} = E_t 2\pi r$$

2. Since $B = 0$ for $r > R$, the magnetic flux through S is $B\pi R^2$:

$$\phi_m = B\pi R^2$$

3. Apply Faraday's law to find E_t:

$$E_t 2\pi r = -\frac{dB}{dt}\pi R^2$$

$$E_t = -\frac{R^2}{2r}\frac{dB}{dt}, \quad r > R$$

E_t is negative, so $\vec{E}$ is $\boxed{\text{counterclockwise}}$.

REMARKS The positive tangential direction is clockwise. When $d\phi_m/dt$ is positive, E_t is negative and the electric field direction is counterclockwise, as shown in Figure 28-7. Note that the electric field in this example is produced by a changing magnetic field rather than by electric charges. Note also that $\vec{E}$, and thus the emf, exists along any closed curve bounding the area through which the magnetic flux is changing, whether there is a wire or circuit along the curve or there is not.

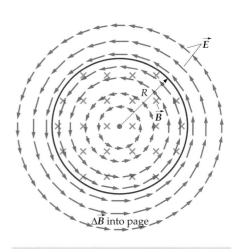

FIGURE 28-7 The magnetic field is into the page and increasing in magnitude. The induced electric field is counterclockwise.

28-3 Lenz's Law

The negative sign in Faraday's law has to do with the direction of the induced emf. This can be obtained by the sign convention described in the previous section, or by a general physical principle known as **Lenz's law**:

> The induced emf is in such a direction as to oppose, or tend to oppose, the change that produces it.

LENZ'S LAW

Note that Lenz's law does not specify just what kind of change causes the induced emf and current. The statement of Lenz's law is purposely left vague to cover a variety of conditions, which we will now illustrate.

Figure 28-8 shows a bar magnet moving toward a loop that has a resistance R. It is the motion of the bar magnetic to the right that induces an emf and current in the loop. Lenz's law tells us that this induced emf and current must be in a direction to oppose the motion of the bar magnet. That is, the current induced in the loop produces a magnetic field of its own, and this magnetic field must exert a force to the left on the approaching bar magnet. Figure 28-9 shows the induced magnetic moment of the current loop when the magnet is moving toward it. The loop acts like a small magnet with its north pole to the left and its south pole to the right. Since like poles repel, the induced magnetic moment of the loop repels the bar magnet; that is, it opposes its motion toward the loop. This means the direction of the induced current in the loop must be as shown in Figure 28-9.

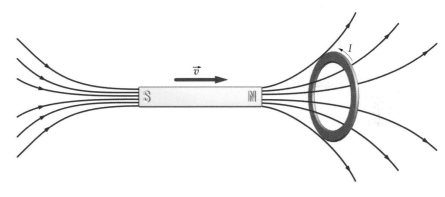

FIGURE 28-8 When the bar magnet is moving to the right, toward the loop, the emf induced in the loop produces an induced current in the direction shown. The magnetic field due to this induced current in the loop produces a magnetic field that exerts a force on the bar magnet opposing its motion to the right.

Suppose the induced current in the loop shown in Figure 28-9 was opposite to the direction shown. Then there would be a magnetic force on the approaching bar magnet to the right, causing it to gain speed. This gain in speed would cause an increase in the induced current, which in turn would cause the force on the bar magnet to increase, and so forth. This is too good to be true. Any time we nudge a bar magnetic toward a conducting loop it would move toward the loop with ever increasing speed and with no significant effort on our part. Were this to occur, it would be a violation of energy conservation. However, the reality is that energy is conserved, and the statement called Lenz's law is consistent with this reality.

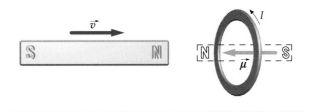

FIGURE 28-9 The magnetic moment of the loop $\vec{\mu}$ (shown in outline as if it were a bar magnet) due to the induced current is such as to oppose the motion of the bar magnet. The bar magnet is moving toward the loop, so the induced magnetic moment repels the bar magnet.

An alternative statement of Lenz's law in terms of magnetic flux is frequently of use. This statement is:

When a magnetic flux through a surface changes, the magnetic field due to any induced current produces a flux of its own—through the same surface and in opposition to the change.

ALTERNATIVE STATEMENT OF LENZ'S LAW

For an example of how this alternative statement is applied, see Example 28-5.

LENZ'S LAW AND INDUCED CURRENT   **E X A M P L E   2 8 - 5**

Using the alternative statement of Lenz's law, find the direction of the induced current in the loop shown in Figure 28-8.

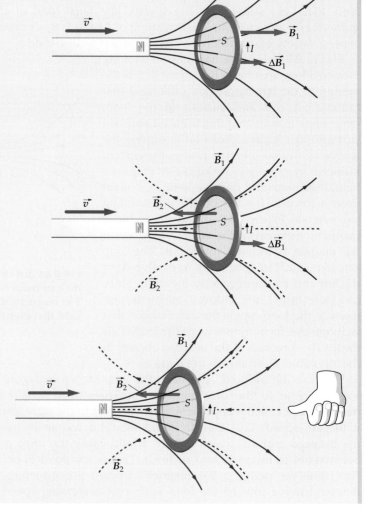

PICTURE THE PROBLEM Use the alternative statement of Lenz's law to determine the direction of the magnetic field due to the current induced in the loop. Then use a right-hand rule to determine the direction of the induced current.

1. Draw a sketch of the loop bounding the flat surface S (Figure 28-10). On surface S draw the vector $\Delta \vec{B}_1$, which is the change in the magnetic field $\vec{B}_1$ of the approaching bar magnet:

FIGURE 28-10

2. On the sketch draw the vector $\vec{B}_2$, which is the magnetic field of the current induced in the loop (Figure 28-11). Use the alternative statement of Lenz's law to determine the direction of $\vec{B}_2$:

FIGURE 28-11

3. Using the right-hand rule and the direction of $\vec{B}_2$, determine the direction of the current induced in the loop (Figure 28-12):

FIGURE 28-12

EXERCISE Using the alternative statement of Lenz's law, find the direction of the induced current in the loop shown in Figure 28-8 if the magnet is moving to the left (away from the loop). (*Answer* Opposite to the direction shown in Figure 28-12)

In Figure 28-13, the bar magnet is at rest and the loop is moving away from the magnet. The induced current and magnetic moment are shown in the figure. In this case, the bar magnet attracts the loop, thus opposing the motion of the loop as required by Lenz's law.

In Figure 28-14, when the current in circuit 1 is changing, there is a changing flux through circuit 2. Suppose that the switch S in circuit 1 is initially open so that there is no current in the circuit (Figure 28-14a). When we close the switch (Figure 28-14b), the current in circuit 1 does not reach its steady value $\mathcal{E}_1/R_1$ instantaneously but takes some time to change from zero to this value. During the time the current is increasing, the flux through circuit 2 is changing and a current is induced in circuit 2 in the direction shown. When the current in circuit 1 reaches its steady value, the flux through circuit 2 is no longer changing, so there is no longer an induced current in circuit 2. An induced current in circuit 2 in the opposite direction appears momentarily when the switch in circuit 1 is opened (Figure 28-14c) and the current in circuit 1 is decreasing to zero. It is important to understand that there is an induced emf *only while the flux is changing*. The emf does not depend on the magnitude of the flux itself, but only on its rate of change. If there is a large steady flux through a circuit, there is no induced emf.

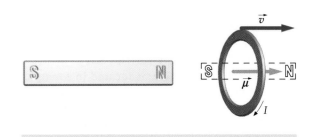

FIGURE 28-13 When the loop is moving away from the stationary bar magnet, the bar magnet attracts the magnetic moment of the loop, again opposing the relative motion.

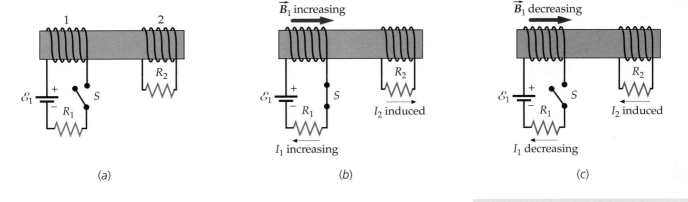

(a) (b) (c)

For our next example, we consider the single isolated circuit shown in Figure 28-15. If there is a current in the circuit, there is a magnetic flux through the coil due to its own current. If the current is changing, the flux in the coil is changing and there is an induced emf in the circuit while the flux is changing. This *self-induced emf* opposes the change in the current. It is therefore called a **back emf**. Because of this self-induced emf, the current in a circuit cannot jump instantaneously from zero to some finite value or from some finite value to zero. Henry first noticed this effect when he was experimenting with a circuit consisting of many turns of a wire like that in Figure 28-15. This arrangement gives a large flux through the circuit for even a small current. Joseph Henry noticed a spark across the switch when he tried to break the circuit. Such a spark is due to the large induced emf that occurs when the current varies rapidly, as during the opening of the switch. In this case, the induced emf is directed so as to maintain the original current. The large induced emf produces a large potential difference across the switch as it is opened. The electric field between the contacts of the switch is large enough to produce dielectric breakdown in the surrounding air. When dielectric breakdown occurs, the air conducts electric current in the form of a spark.

FIGURE 28-14 (a) Two adjacent circuits. (b) Just after the switch is closed, I_1 is increasing in the direction shown. The changing flux through circuit 2 induces the current I_2. The flux through circuit 2 due to I_2 opposes the change in flux due to I_1. (c) As the switch is opened, I_1 decreases and the flux through circuit 2 changes. The induced current I_2 then tends to maintain the flux through circuit 2.

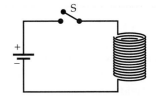

FIGURE 28-15 The coil with many turns of wire gives a large flux for a given current in the circuit. Thus, when the current changes, there is a large emf induced in the coil opposing the change.

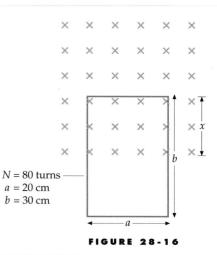

FIGURE 28-16

EXAMPLE 2 8 - 6

A rectangular coil of N turns, each of width a and length b; where $N = 80$, $a = 20$ cm, and $b = 30$ cm; is located in a magnetic field of magnitude $B = 0.8$ T directed into the page (Figure 28-16), with only half of the coil in the region of the magnetic field. The resistance R of the coil is 30 Ω. Find the magnitude and direction of the induced current if the coil is moved with a speed of 2 m/s (*a*) to the right, (*b*) up, and (*c*) down.

PICTURE THE PROBLEM The induced current equals the induced emf divided by the resistance. We can calculate the emf induced in the circuit as the coil moves by calculating the rate of change of the flux through the coil. The flux is proportional to the distance x. The direction of the induced current is found from Lenz's law.

(*a*) 1. The induced current equals the emf divided by the resistance:

$$I = \frac{\mathcal{E}}{R}$$

2. The induced emf and the magnetic flux are related by Faraday's law:

$$\mathcal{E} = -\frac{d\phi_m}{dt}$$

3. The flux through the coil is N times the flux through each turn of the coil. We choose into the page as the direction of $\hat{n}$. The flux through the surface S bounded by a single turn is Bax:

$$\phi_m = N\vec{B} \cdot \hat{n}A = NBax$$

4. When the coil is moving to the right (or to the left), the flux does not change (until the coil leaves the region of magnetic field). The current is therefore zero:

$$\mathcal{E} = -\frac{d\phi_m}{dt} = 0$$

so

$$I = \boxed{0}$$

(*b*) 1. Compute the rate of change of the flux when the coil is moving up. In this case x is increasing, so dx/dt is positive:

$$\frac{d\phi_m}{dt} = \frac{d}{dt}(NBax) = NBa\frac{dx}{dt}$$

2. Calculate the magnitude of the current:

$$I = \frac{\mathcal{E}}{R} = \frac{NBa(dx/dt)}{R}$$

$$= \frac{(80)(0.8\text{ T})(0.20\text{ m})(2\text{ m/s})}{30\ \Omega} = 0.853\text{ A}$$

3. As the coil moves upward, the flux of $\vec{B}$ through S is increasing. The induced current must produce a magnetic field whose flux through S decreases as x increases. That would be a magnetic field whose dot product with $\hat{n}$ is negative. Such a magnetic field is directed out of the page on S. To produce a magnetic field in this direction the induced current must be counterclockwise:

$$\boxed{I = 0.853\text{ A, counterclockwise}}$$

(*c*) As the coil moves downward, the flux of $\vec{B}$ through S is decreasing. The induced current must produce a magnetic field whose flux through S increases as x decreases. That would be a magnetic field whose dot product with $\hat{n}$ is positive. Such a magnetic field is directed into the page on S. To produce a magnetic field in this direction the induced current must be clockwise:

$$\boxed{I = 0.853\text{ A, clockwise}}$$

REMARKS In this example the magnetic field is static, so there is no nonconservative electric field. Thus, the emf is not the work done by a nonconservative electric field. This issue is examined in the next section.

28-4 Motional EMF

The emf induced in a conductor moving through a magnetic field is called **motional emf.** More generally,

> Motional emf is any emf induced by the motion of a conductor in a magnetic field.

DEFINITION—MOTIONAL EMF

TOTAL CHARGE THROUGH A FLIPPED COIL | **EXAMPLE 28-7**

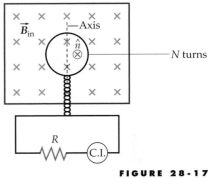

FIGURE 28-17

A small coil of N turns has its plane perpendicular to a uniform static magnetic field $\vec{B}$, as shown in Figure 28-17. The coil is connected to a current integrator (C.I.), which is a device used to measure the total charge passing through the coil. Find the charge passing through the coil if the coil is rotated through 180° about the axis shown.

PICTURE THE PROBLEM When the coil in Figure 28-17 is rotated, the magnetic flux through the coil changes, causing an induced emf $\mathcal{E}$. The emf in turn causes a current $I = \mathcal{E}/R$, where R is the total resistance of the circuit. Since $I = dq/dt$, we can find the charge Q passing through the integrator by integrating I; that is, $Q = \int dq = \int I\, dt$.

1. The increment of charge dq equals the current I times the increment of time dt:

$$dq = I\, dt$$

2. The emf $\mathcal{E}$ is related to I by Ohm's law:

$$\mathcal{E} = RI$$

so

$$\mathcal{E}\, dt = RI\, dt$$

3. The emf is related to the flux ϕ_m by Faraday's law:

$$\mathcal{E} = -\frac{d\phi_m}{dt}$$

or

$$\mathcal{E}\, dt = -d\phi_m$$

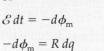

Before rotation   After rotation

FIGURE 28-18

4. Substitute $-d\phi_m$ for $\mathcal{E}\, dt$ and dq for $I\, dt$ in the step 2 result and solve for dq:

$$-d\phi_m = R\, dq$$

so

$$dq = -\frac{1}{R}\, d\phi_m$$

5. Integrate to find the total charge Q:

$$Q = \int_0^Q dq = -\frac{1}{R}\int_{\phi_{m,i}}^{\phi_{m,f}} d\phi_m = -\frac{1}{R}(\phi_{m,f} - \phi_{m,i}) = -\frac{\Delta\phi_m}{R}$$

6. The flux through the coil is $\phi_m = N\vec{B} \cdot \hat{n} A$, where $\hat{n}$ is the normal to the flat surface bounded by the coil (Figure 28-18). Initially, the normal is directed into the page. When the coil rotates, so does the surface and its normal. Find the change in ϕ_m when the coil rotates 180°:

$$\Delta\phi_m = \phi_{m,f} - \phi_{m,i} = N\vec{B} \cdot \hat{n}_f A - N\vec{B} \cdot \hat{n}_i A$$
$$= NA(\vec{B} \cdot \hat{n}_f - \vec{B} \cdot \hat{n}_i) = NA[(-B) - (+B)] = -2NBA$$

7. Combining the previous two results yields Q:

$$\boxed{Q = \frac{2NBA}{R}}$$

REMARKS Note that the charge Q does not depend on whether or not the coil is rotated slowly or quickly—all that matters is the change in magnetic flux through the coil. A coil used in this way is called a *flip coil.* It is used to measure magnetic fields. For example, if the current integrator (C.I.) measures a total charge Q passing through the coil when it is flipped, the magnetic field strength can be found from $B = RQ/(2NA)$.

EXERCISE A flip coil of 40 turns has a radius of 3 cm, a resistance of 16 Ω, and the plane of the coil is initially perpendicular to a static, uniform 0.50-T magnetic field. If the coil is flipped through 90°, how much charge passes through the coil? (*Answer* 3.53 mC)

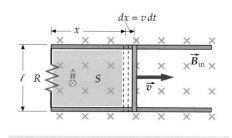

FIGURE 28-19 A conducting rod sliding on conducting rails in a magnetic field. As the rod moves to the right, the area of the surface S increases, so the magnetic flux through S into the paper increases. An emf of magnitude $B\ell v$ is induced in the circuit, inducing a counterclockwise current that produces flux through the surface S directed out of the paper opposing the change in flux due to the motion of the rod.

Figure 28-19 shows a thin conducting rod sliding to the right along conducting rails that are connected by a resistor. A uniform magnetic field $\vec{B}$ is directed into the page.

Consider the magnetic flux through the flat surface S bounded by the circuit. Let the normal $\hat{n}$ to the surface be into the page. As the rod moves to the right the surface S increases, as does the magnetic flux through the surface S. Thus, an emf is induced in the circuit. Let ℓ be the separation of the rails and x be the distance from the left end of the rails to the rod. The area of surface S is then ℓx, and the magnetic flux through S is

$$\phi_m = \vec{B} \cdot \hat{n} A = B_n A = B\ell x$$

When x increases by dx, the area of surface S increases by $dA = \ell\,dx$ and the flux ϕ_m increases by $d\phi_m = B\ell\,dx$. The rate of change of the flux is

$$\frac{d\phi_m}{dt} = B\ell\frac{dx}{dt} = B\ell v$$

where $v = dx/dt$ is the speed of the rod. The emf induced in this circuit is therefore

$$\mathcal{E} = -\frac{d\phi_m}{dt} = -B\ell v$$

where the negative sign tells us that the emf is in the negative tangential direction. Put your right thumb in the direction of $\hat{n}$ (into the page) and your fingers will curl in the positive tangential direction (clockwise). Thus, the induced emf is counterclockwise.

We can check this result (the direction of the induced emf) using Lenz's law. It is the motion of the rod to the right that produces the induced current, so the magnetic force on this rod due to the induced current must be to the left. The magnetic force on a current-carrying conductor is given by $I\vec{L} \times \vec{B}$ (Equation 26-4), where $\vec{L}$ is in the direction of the current. If $\vec{L}$ is upward the force is to the left, which affirms our previous result (that the induced emf is counterclockwise). If the rod is given some initial velocity $\vec{v}$ to the right and is then released, the force due to the induced current slows the rod until it stops. To maintain the motion of the rod, an external force pushing the rod to the right must be maintained.

A second check on the direction of the induced emf and current is implemented by considering the direction of the magnetic force on the charge carriers moving to the right with the rod. The charge carriers move rightward with the same velocity $\vec{v}$ as the rod, so the charge carriers experience a magnetic force $\vec{F} = q\vec{v} \times \vec{B}$. If q is positive this force is upward, which means the induced emf is counterclockwise.

Emf is the work per unit charge on the charge carriers, but what is the force that is doing this work in the circuit shown in Figure 28-19? It turns out this work is done by the superposition of a magnetic force and an electric force (Figure 28-20). To see how this comes about, consider that the current in the rod is upward, so the drift velocity $\vec{v}_d$ of the assumed positive charge carriers is upward. Thus, a magnetic force $(\vec{F}_L = q\vec{v}_d \times \vec{B})$ toward the left acts on the charge carriers and, as a result, the rod becomes polarized—its left side positively charged and its right side negatively charged. These surface charges produce an electric field $\vec{E}_\perp$ inside the rod toward the right, and this field exerts an electric force $(\vec{F}_R = q\vec{E}_\perp)$ toward the right on the charge carriers. The sum $\vec{F}_L + \vec{E}_\perp = 0$, since the net horizontal force on the charge carriers is zero. In addition, an upward magnetic force $\vec{F}_U = q\vec{v} \times \vec{B}$, where $\vec{v}$ is the velocity of both the charge carriers and the rod to the right. The total work done by all three of these forces on a charge carrier traversing the rod is just the work done by $\vec{F}_U$, and this work is $F_U\ell = qvB\ell$. Thus, the work per unit charge is $vB\ell$, which is obtained by dividing the total work by the charge q. The magnitude of the emf equals this work.

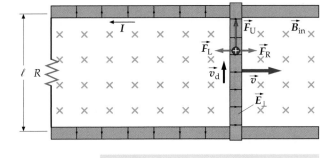

FIGURE 28-20 As a positive charge carrier moves along the moving rod, electric forces and magnetic forces act on the charge carrier. The net electromagnetic force on the charge carrier is directed upward, in the direction of the drift velocity. The work per unit charge done by this force on the charge carrier as it transverses the rod is the motional emf.

$$\mathcal{E} = vB\ell \qquad\qquad 28\text{-}6$$

MAGNITUDE OF EMF FOR A ROD MOVING PERPENDICULAR TO BOTH THE ROD AND $\vec{B}$

The magnitude of the emf is the total work per unit charge done by all three forces $\vec{F}_L$, $\vec{F}_R$, and $\vec{F}_U$. Taken together, $\vec{F}_L$ and $\vec{F}_U$ constitute the total magnetic force. The total magnetic force, however, is perpendicular to the velocity of the charge carriers and thus does no work. Therefore, the total work done by all three forces is done solely by the electric force $\vec{F}_R$.

Figure 28-21 shows a positive charge carrier in a conducting rod that is moving at constant speed through a uniform magnetic field directed into the paper. Because the charge carrier is moving horizontally with the rod, there is an upward magnetic force on the charge carrier of magnitude qvB. Responding to this force, the charge carriers in the rod move upward, producing a net positive charge at the top of the rod and leaving a net negative charge at the bottom of the rod. The charge carriers continue to move upward until the electric field $\vec{E}_\parallel$ produced by the separated charges exerts a downward force of magnitude $qE_\parallel$ on the separated charges, which balances the upward magnetic force qvB. In equilibrium, the magnitude of this electric field in the rod is

$$E_\parallel = vB$$

The direction of this electric field is parallel to the rod, directed downward. The associated potential difference across the length ℓ of the rod is

$$\Delta V = E_\parallel \ell = vB\ell$$

with the potential being higher at the top. That is, when there is no current through the rod, the potential difference across the rod equals $vB\ell$ (the motional emf). When there is a current I through the rod, the potential difference is

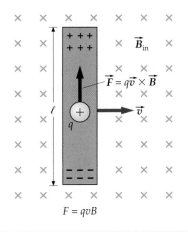

FIGURE 28-21 A positive charge carrier in a conducting rod that is moving through a magnetic field experiences a magnetic force that has an upward component. Some of these charge carriers move to the top of the rod, leaving the bottom of the rod negative. The charge separation produces a downward electric field of magnitude $E_\parallel = vB$ in the rod. Thus, the potential at the top of the rod is greater than the potential at the bottom of the rod by $E_\parallel \ell = vB\ell$.

$$\Delta V = vB\ell - Ir \qquad\qquad 28\text{-}7$$

POTENTIAL DIFFERENCE ACROSS A MOVING ROD

where r is the resistance of the rod.

EXERCISE A rod 40-cm long moves at 12 m/s in a plane perpendicular to a magnetic field of 0.30 T. The rod's velocity is perpendicular to its length. Find the emf induced in the rod. (*Answer* 1.44 V)

A U-SHAPED CONDUCTOR AND A SLIDING ROD   **EXAMPLE 28-8**   Try It Yourself

Using Figure 28-19, let $B = 0.6$ T, $v = 8$ m/s, $\ell = 15$ cm, and $R = 25\ \Omega$; assume that the resistances of the rod and the rails are negligible. Find (*a*) the induced emf in the circuit, (*b*) the current in the circuit, (*c*) the force needed to move the rod with constant velocity, and (*d*) the power dissipated in the resistor.

PICTURE THE PROBLEM

Cover the column to the right and try these on your own before looking at the answers.

Steps	Answers
1. Calculate the induced emf from Equation 28-6.	$\mathcal{E} = Bv\ell =$ [0.720 V]
2. Find the current from Ohm's law.	$I = \dfrac{\mathcal{E}}{R} =$ [28.8 mA]
3. The force needed to move the rod with constant velocity is equal and opposite to the force exerted by the magnetic field on the rod, which has the magnitude $I\ell B$ (Equation 26-4). Calculate the magnitude of this force.	$F = IB\ell =$ [2.59 mN]
4. Find the power dissipated in the resistor.	$P = I^2 R =$ [20.7 mW]

⬤ PLAUSIBILITY CHECK Using $P = Fv$, we confirm that the power is 20.7 mW.

REMARKS The potential at the top of the rod is greater than the potential at the bottom of the rod by the emf.

MAGNETIC DRAG   **EXAMPLE 28-9**

A rod of mass m slides on frictionless conducting rails in a region of static uniform magnetic field $\vec{B}$ directed into the page (Figure 28-22). An external agent is pushing the rod, maintaining its motion to the right at constant speed v_0. At time $t = 0$, the agent abruptly stops pushing and the rod continues forward, being slowed by the magnetic force. Find the speed v of the rod as a function of time.

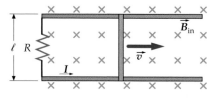

FIGURE 28-22

PICTURE THE PROBLEM The speed of the rod changes because a magnetic force acts on the induced current. The motion of the rod through a magnetic field induces an emf $\mathcal{E} = B\ell v$ and, therefore, a current in the rod, $I = \mathcal{E}/R$. This causes a magnetic force to act on the rod, $F = IB\ell$. With the force known, we apply Newton's second law to find the speed as a function of time. Take the positive x direction as being to the right.

1. Apply Newton's second law to the rod:

$$F_x = ma_x = m\frac{dv}{dt}$$

2. The force exerted on the rod is the magnetic force, which is proportional to the current and in the negative x direction, as shown in Figure 28-22:

$$F_x = -IB\ell$$

3. The current equals the motional emf divided by the resistance of the rod:

$$I = \frac{\mathcal{E}}{R} = \frac{B\ell v}{R}$$

4. Combining these results, we find the magnitude of the magnetic force exerted on the rod:

$$F_x = -IB\ell = -\frac{B\ell v}{R}B\ell = -\frac{B^2\ell^2 v}{R}$$

5. Newton's second law then gives:

$$-\frac{B^2\ell^2 v}{R} = m\frac{dv}{dt}$$

6. Separate the variables, then integrate the velocity from v_0 to v_f and integrate the time from 0 to t_f:

$$\frac{dv}{v} = -\frac{B^2\ell^2}{mR}\,dt$$

$$\int_{v_0}^{v_f}\frac{dv}{v} = -\frac{B^2\ell^2}{mR}\int_0^{t_f}dt$$

$$\ln\frac{v_f}{v_0} = -\frac{B^2\ell^2}{mR}\,t_f$$

7. Let $v = v_f$ and $t = t_f$, then solve for v:

$$\boxed{v = v_0 e^{-t/\tau}, \text{ where } \tau = \frac{mR}{B^2\ell^2}}$$

REMARKS If the force were constant, the rod's speed would decrease linearly with time. However, because the force is proportional to the rod's speed, as found in step 4, the force is large initially but the force decreases as the speed decreases. In principle, the rod never stops moving. Even so, the rod travels only a finite distance. (See Problem 37.)

The general equation for motional emf is

$$\mathcal{E} = \oint_C (\vec{v} \times \vec{B}) \cdot d\vec{\ell} = -\frac{d\phi_m}{dt} \qquad \text{28-8}$$

GENERAL EQUATION FOR MOTIONAL EMF

where $\vec{v}$ is the velocity of the wire at the element $d\vec{\ell}$. The integral is taken at an instant in time.

FIGURE 28-23 The positive x, y, and z directions are to the right, into the page, and up the page respectively. The rod moves to the right with the velocity $\vec{v}_r$, and there is a uniform static magnetic field directed into the page.

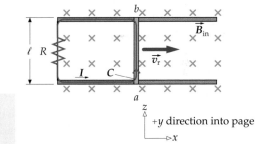

+y direction into page

VERIFYING $\mathcal{E} = vB\ell$ **EXAMPLE 28-10**

Integrate $\mathcal{E} = \oint_C(\vec{v} \times \vec{B})\cdot d\vec{\ell}$ to show that the emf in the circuit in Figure 28-23 is given by Equation 28-6.

PICTURE THE PROBLEM The circuit C can be divided into two parts: part C_1, which is moving, and part C_2, which is stationary.

1. Divide the circuit into two parts, C_1 and C_2. On C_1, $\vec{v} = \vec{v}_r$ and on C_2, $\vec{v} = 0$:

$$\oint_C (\vec{v} \times \vec{B}) \cdot d\vec{\ell} = \int_a^b{}_{C_1} (\vec{v} \times \vec{B}) \cdot d\vec{\ell} + \int_b^a{}_{C_2} (\vec{v} \times \vec{B}) \cdot d\vec{\ell}$$

$$= \int_a^b{}_{C_1} (\vec{v}_r \times \vec{B}) \cdot d\vec{\ell} + 0$$

2. Evaluate $(\vec{v}_r \times \vec{B})\cdot d\vec{\ell}$ on C_1:

$$\vec{v}_r \times \vec{B} = v_r\hat{i} \times B\hat{j} = v_rB\,\hat{k}$$

and

$$d\vec{\ell} = d\ell\,\hat{k}$$

so

$$(\vec{v}_r \times \vec{B}) \cdot d\vec{\ell} = v_rB\hat{k} \cdot d\ell\,\hat{k} = v_rB\,d\ell$$

3. Evaluate the integral and find the emf:

$$\mathcal{E} = \int_a^b{}_{C_1} (\vec{v}_r \times \vec{B}) \cdot d\vec{\ell} = \int_a^b{}_{C_1} v_rB\,d\ell = v_rB\int_a^b{}_{C_1} d\ell$$

$$= \boxed{v_rB\ell}$$

28-5 Eddy Currents

In the examples we have discussed, currents were induced in thin wires or rods. Often a changing flux sets up circulating currents, which are called *eddy currents,* in a piece of bulk metal like the core of a transformer. The heat produced by such current constitutes a power loss in the transformer. Consider a conducting slab between the pole faces of an electromagnet (Figure 28-24). If the magnetic field $\vec{B}$ between the pole faces is changing with time (as it will if the current in the magnet windings is alternating current), the flux through any closed loop in the slab, such as through the curve C indicated in the figure, will be changing. Since path C is in a conductor, there will be an induced emf around C.

The existence of eddy currents can be demonstrated by pulling a copper or aluminum sheet between the poles of a strong permanent magnet (Figure 28-25). Part of the area enclosed by curve C in this figure is in the magnetic field, and part of the area enclosed by curve C is outside the magnetic field. As the sheet is pulled to the right, the flux through this curve decreases (assuming that into the paper is the positive normal direction). A clockwise emf is induced around this curve. This emf drives a current that is directed upward in the region between the pole faces, and the magnetic field exerts a force on this current to the left opposing motion of the sheet. You can feel this drag force on the sheet if you pull a conducting sheet rapidly through a strong magnetic field.

Eddy currents are usually unwanted because power is lost due to joule heating by the current, and this dissipated energy must be transferred to the environment. The power loss can be reduced by increasing the resistance of the possible paths for the eddy currents, as shown in Figure 28-26a. Here the conducting slab is laminated; that is, the conducting slab is made up of small strips glued together. Because insulating glue separates the strips, the eddy currents are essentially confined to the strips. The large eddy-current loops are broken up, and the power loss is greatly reduced. Similarly, if there are cuts in the sheet, as shown in Figure 28-26b, the eddy currents are lessened and the magnetic force is greatly reduced.

Eddy currents are not always undesirable. For example, eddy currents are often used to damp unwanted oscillations. With no damping present, sensitive mechanical balance scales that are used to weigh small masses might oscillate back and forth around their equilibrium reading many times. Such scales are usually designed so that a small sheet of aluminum (or some other metal) moves between the poles of a magnet as the scales oscillate. The resulting eddy currents dampen the oscillations so that equilibrium is quickly reached. Eddy currents also play a role in the magnetic braking systems of some rapid transit cars. A large electromagnet is positioned in the vehicle over the rails. If the magnet is energized by a current in its windings, eddy currents are induced in the rails by the motion of the magnet and the magnetic forces provide a drag force on the magnet that slows the car.

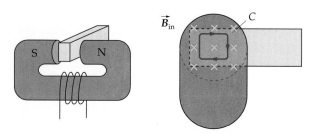

FIGURE 28-24 Eddy currents. When the magnetic field through a metal slab is changing, and emf is induced in any closed loop in the metal, such as loop C. The induced emfs drive currents, which are called eddy currents.

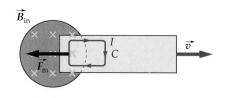

FIGURE 28-25 Demonstration of eddy currents. When the metal sheet is pulled to the right, there is a magnetic force to the left on the induced current opposing the motion.

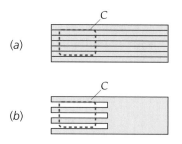

FIGURE 28-26 Disrupting the conduction paths in the metal slab can reduce the eddy current. (*a*) If the slab is constructed from strips of metal glued together, the insulating glue between the slabs increases the resistance of the closed loop C. (*b*) Slots cut into the metal slab also reduce the eddy current.

28-6 Inductance

Self-Inductance

The magnetic flux through a circuit is related to the current in that circuit and the currents in other nearby circuits.[†] Consider a coil carrying a current I. The current in the coil produces a magnetic field $\vec{B}$ that varies from point to point, but the value of $\vec{B}$ at each point is proportional to I. The magnetic flux through the coil is therefore also proportional to I:

[†] We are assuming that there are no permanent magnets around.

$$\phi_{\mathrm{m}} = LI \qquad\qquad 28\text{-}9$$

DEFINITION—SELF-INDUCTANCE

where L, the proportionality constant, is called the self-inductance of the coil. The self-inductance depends on the geometric shape of the coil. The SI unit of inductance is the **henry** (H). From Equation 28-9, we can see that the unit of inductance equals the unit of flux divided by the unit of current:

$$1\,\mathrm{H} = 1\,\frac{\mathrm{Wb}}{\mathrm{A}} = 1\,\frac{\mathrm{T\cdot m^2}}{\mathrm{A}}$$

In principle, the self-inductance of any coil or circuit can be calculated by assuming a current I, calculating $\vec{B}$ at every point on a surface bounded by the coil, calculating the flux ϕ_{m}, and using $L = \phi_{\mathrm{m}}/I$. In actual practice, the calculation is often very challenging. However, the self-inductance of a long, tightly wound solenoid can be calculated directly. The magnetic flux through a solenoid of length ℓ and N turns carrying a current I was calculated in Example 28-1:

$$\phi_{\mathrm{m}} = \frac{\mu_0 N^2 I A}{\ell} = \mu_0 n^2 I A \ell \qquad\qquad 28\text{-}10$$

where $n = N/\ell$ is the number of turns per unit length. As expected, the flux is proportional to the current I. The proportionality constant is the self-inductance:

$$L = \frac{\phi_{\mathrm{m}}}{I} = \mu_0 n^2 A \ell \qquad\qquad 28\text{-}11$$

SELF-INDUCTANCE OF A SOLENOID

The self-inductance of a solenoid is proportional to the square of the number of turns per unit length n and to the volume $A\ell$. Thus, like capacitance, self-inductance depends only on geometric factors.[†] From the dimensions of Equation 28-11, we can see that μ_0 can be expressed in henrys per meter:

$$\mu_0 = 4\pi \times 10^{-7}\,\mathrm{H/m}$$

[†] If the inductor has an iron core, the self-inductance also depends on properties of the core.

SELF-INDUCTANCE OF A SOLENOID   **EXAMPLE 28-11**

Find the self-inductance of a solenoid of length 10 cm, area 5 cm², and 100 turns.

PICTURE THE PROBLEM We can calculate the self-inductance in henrys from Equation 28-11.

1. L is given by Equation 28-11:

$$L = \mu_0 n^2 A \ell$$

2. Convert the given quantities to SI units:

$$\ell = 10\,\mathrm{cm} = 0.1\,\mathrm{m}$$

$$A = 5\,\mathrm{cm^2} = 5 \times 10^{-4}\,\mathrm{m^2}$$

$$n = N/\ell = (100\,\mathrm{turns})/(0.1\,\mathrm{m}) = 1000\,\mathrm{turns/m}$$

$$\mu_0 = 4\pi \times 10^{-7}\,\mathrm{H/m}$$

3. Substitute the given quantities:

$$L = \mu_0 n^2 A \ell$$

$$= (4\pi \times 10^{-7}\,\mathrm{H/m})(10^3\,\mathrm{turns/m})^2(5 \times 10^{-4}\,\mathrm{m^2})(0.1\,\mathrm{m})$$

$$= \boxed{6.28 \times 10^{-5}\,\mathrm{H}}$$

When the current in a circuit is changing, the magnetic flux due to the current is also changing, so an emf is induced in the circuit. Because the self-inductance of a circuit is constant, the change in flux is related to the change in current by

$$\frac{d\phi_m}{dt} = \frac{d(LI)}{dt} = L\frac{dI}{dt}$$

According to Faraday's law, we have

$$\mathcal{E} = -\frac{d\phi_m}{dt} = -L\frac{dI}{dt} \qquad\qquad 28\text{-}12$$

Thus, the self-induced emf is proportional to the rate of change of the current. A coil or solenoid with many turns has a large self-inductance and is called an **inductor**. In circuits, it is denoted by the symbol ⎍⎍⎍. Typically, we can neglect the self-inductance of the rest of the circuit compared with that of an inductor. The potential difference across an inductor is given by

$$\Delta V = \mathcal{E} - Ir = -L\frac{dI}{dt} - Ir \qquad\qquad 28\text{-}13$$

POTENTIAL DIFFERENCE ACROSS AN INDUCTOR

where r is the internal resistance of the inductor.[†] For an ideal inductor, $r = 0$.

EXERCISE At what rate must the current in the solenoid of Example 28-11 change to induce a back emf of 20 V? (*Answer* 3.18×10^5 A/s)

Mutual Inductance

When two or more circuits are close to each other, as in Figure 28-27, the magnetic flux through one circuit depends not only on the current in that circuit but also on the current in the nearby circuits. Let I_1 be the current in circuit 1, on the left in Figure 28-27, and let I_2 be the current in circuit 2, on the right in Figure 28-27. The magnetic field $\vec{B}$ at surface S_2 is the superposition of $\vec{B}_1$ due to I_1, and $\vec{B}_2$ due to I_2, where $\vec{B}_1$ is proportional to I_1 (and $\vec{B}_2$ is proportional to I_2). We can therefore write the flux of $\vec{B}_1$ through circuit 2, $\phi_{m2,1}$ as:

$$\phi_{m2,1} = M_{2,1}I_1 \qquad\qquad 28\text{-}14a$$

DEFINITION—MUTUAL INDUCTANCE

where $M_{2,1}$ is called the **mutual inductance** of the two circuits. The mutual inductance depends on the geometrical arrangement of the two circuits. For instance, if the circuits are far apart, the flux of $\vec{B}_1$ through circuit 2 will be small and the mutual inductance will be small. (The net flux ϕ_{m2} of $\vec{B} = \vec{B}_1 + \vec{B}_2$ through circuit 2 is given by $\phi_{m2} = \phi_{m2,2} + \phi_{m2,1}$.) An equation similar to Equation 28-14a can be written for the flux of $\vec{B}_2$ through circuit 1:

$$\phi_{m1,2} = M_{1,2}I_2 \qquad\qquad 28\text{-}14b$$

We can calculate the mutual inductance for two tightly wound concentric solenoids like the solenoids shown in Figure 28-28. Let ℓ be the length of both solenoids, and let the inner solenoid have N_1 turns and radius r_1 and the outer

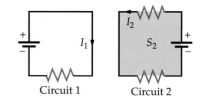

FIGURE 28-27 Two adjacent circuits. The magnetic field on S_2 is partly due to current I_1 and partly due to current I_2. The flux through the magnetic field is the sum of two terms, one proportional to I_1 and the other to I_2.

[†] If the inductor has an iron core, the internal resistance includes properties of the core.

solenoid have N_2 turns and radius r_2. We will first calculate the mutual inductance $M_{2,1}$ by assuming that the inner solenoid carries a current I_1 and finding the magnetic flux ϕ_{m2} due to this current through the outer solenoid.

The magnetic field $\vec{B}_1$ due to the current in the inner solenoid is constant in the space within the inner solenoid and has magnitude

$$B_1 = \mu_0(N_1/\ell)I_1 = \mu_0 n_1 I_1, \quad r < r_1 \qquad \text{28-15}$$

and outside the inner solenoid this magnetic field B_1 is negligible. The flux of $\vec{B}_1$ through the outer solenoid is therefore

$$\phi_{m2} = N_2 B_1(\pi r_1^2) = n_2 \ell B_1(\pi r_1^2) = \mu_0 n_2 n_1 \ell(\pi r_1^2) I_1$$

Note that the area used to compute the flux through the outer solenoid is not the area of that solenoid, πr_2^2, but rather is the area of the inner solenoid, πr_1^2, because the magnetic field due to the inner solenoid is zero outside the inner solenoid. The mutual inductance $M_{1,2}$ is thus

$$M_{2,1} = \frac{\phi_{m2,1}}{I_1} = \mu_0 n_2 n_1 \ell \pi r_1^2 \qquad \text{28-16}$$

EXERCISE Calculate the mutual inductance $M_{1,2}$ of the concentric solenoids of Figure 28-28 by finding the flux through the inner solenoid due to a current I_2 in the outer solenoid. (*Answer* $M_{1,2} = M_{2,1} = \mu_0 n_2 n_1 \ell \pi r_1^2$)

Note from the exercise that $M_{1,2} = M_{2,1}$. It can be shown that this is a general result. We will therefore drop the subscripts for mutual inductance and simply write M.

(a)

(b)

FIGURE 28-28 (*a*) A long narrow solenoid inside a second solenoid of the same length. A current in either solenoid produces magnetic flux in the other. (*b*) A tesla coil illustrating the geometry of the wires in Figure 28-28*a*. Such a device functions as a transformer.[†] Here, low-voltage alternating current in the outer winding is transformed into a higher-voltage alternating current in the inner winding. The emf induced in the inner coil by the field of the charging current in the outer coil is high enough to light the bulb above the coils.

28-7 Magnetic Energy

An inductor stores magnetic energy, just as a capacitor stores electrical energy. Consider the circuit shown in Figure 28-29, which consists of an inductance L and a resistance R in series with a battery of emf $\mathcal{E}_0$ and a switch S. We assume that R and L are the resistance and inductance of the entire circuit. The switch is initially open, so there is no current in the circuit. A short time after the switch is closed there is a current I in the circuit, a potential difference $-IR$ across the resistor, and a potential difference $-L\,dI/dt$ across the inductor. (For an inductor with negligible resistance, the difference in potential across the inductor equals the back emf, which was given in Equation 28-12.) Applying Kirchhoff's loop rule to this circuit gives

$$\mathcal{E}_0 - IR - L\frac{dI}{dt} = 0 \qquad \text{28-17}$$

If we multiply each term by the current I and rearrange, we obtain

$$\mathcal{E}_0 I = I^2 R + LI\frac{dI}{dt} \qquad \text{28-18}$$

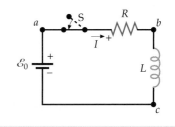

FIGURE 28-29 Just after the switch S is closed in this circuit, the current begins to increase and a back emf of magnitude $L\,dI/dt$ is induced in the inductor. The potential drop across the resistor IR plus the potential drop across the inductor $L\,dI/dt$ equals the emf of the battery.

---

† The transformer is discussed in Chapter 29.

The term $\mathcal{E}_0 I$ is the rate at which electrical potential energy is delivered by the battery. The term I^2R is the rate at which potential energy is delivered to the resistor. (It is also the rate at which potential energy is dissipated by the resistance in the circuit.) The term $LI\,dI/dt$ is the rate at which potential energy is delivered to the inductor. If U_m is the energy in the inductor, then

$$\frac{dU_m}{dt} = LI\frac{dI}{dt}$$

which implies

$$dU_m = LI\,dI$$

Integrating this equation from time $t = 0$, when the current is zero, to $t = \infty$, when the current has reached its final value I_f, we obtain

$$U_m = \int dU_m = \int_0^{I_f} LI\,dI = \frac{1}{2}LI_f^2$$

The energy stored in an inductor carrying a current I is thus given by

$$U_m = \frac{1}{2}LI^2 \qquad\qquad\qquad 28\text{-}19$$

ENERGY STORED IN AN INDUCTOR

When a current is produced in an inductor, a magnetic field is created in the space within the inductor coil. We can think of the energy stored in an inductor as energy stored in the magnetic field. For the special case of a long solenoid, the magnetic field is related to the current I and the number of turns per unit length n by

$$B = \mu_0 n I$$

and the self-inductance is given by Equation 28-11:

$$L = \mu_0 n^2 A\ell$$

where A is the cross-sectional area and ℓ is the length. Substituting $B/(\mu_0 n)$ for I and $\mu_0 n^2 A\ell$ for L in Equation 28-19, we obtain

$$U_m = \frac{1}{2}LI^2 = \frac{1}{2}\mu_0 n^2 A\ell\left(\frac{B}{\mu_0 n}\right)^2 = \frac{B^2}{2\mu_0}A\ell$$

The quantity $A\ell$ is the volume of the space within the solenoid containing the magnetic field. The energy per unit volume is the **magnetic energy density** u_m:

$$u_m = \frac{B^2}{2\mu_0} \qquad\qquad\qquad 28\text{-}20$$

MAGNETIC ENERGY DENSITY

Although we derived this by considering the special case of the magnetic field in a long solenoid, it is a general result. Whenever there is a magnetic field in space, the magnetic energy per unit volume is given by Equation 28-20. Note the similarity to the energy density in an electric field (Equation 24-13):

$$u_e = \frac{1}{2}\epsilon_0 E^2$$

ELECTROMAGNETIC ENERGY DENSITY   **EXAMPLE 28-12**

A certain region of space contains a uniform magnetic field of 0.020 T and a uniform electric field of 2.5 × 10⁶ N/C. Find (a) the total electromagnetic energy density and (b) the energy in a cubical box of edge length ℓ = 12 cm.

PICTURE THE PROBLEM   The total energy density u is the sum of the electrical and magnetic energy densities, $u = u_e + u_m$. The energy in a volume $\mathcal{V}$ is given by $U = u\mathcal{V}$.

(a) 1. Calculate the electrical energy density:

$$u_e = \frac{1}{2}\,\epsilon_0\,E^2$$

$$= \frac{1}{2}\,(8.85 \times 10^{-12}\,\text{C}^2/\text{N·m}^2)(2.5 \times 10^6\,\text{N/C})^2$$

$$= 27.7\,\text{J/m}^3$$

2. Calculate the magnetic energy density:

$$u_m = \frac{B^2}{2\mu_0} = \frac{(0.02\,\text{T})^2}{2(4\pi \times 10^{-7}\,\text{N/A}^2)} = 159\,\text{J/m}^3$$

3. The total energy density is the sum of the above two contributions:

$$u = u_e + u_m = 27.7\,\text{J/m}^3 + 159\,\text{J/m}^3 = \boxed{187\,\text{J/m}^3}$$

(b) The total energy in the box is $U = u\mathcal{V}$, where $\mathcal{V} = \ell^3$ is the volume of the box:

$$U = u\mathcal{V} = u\ell^3 = (187\,\text{J/m}^3)(0.12\,\text{m})^3 = \boxed{0.323\,\text{J}}$$

*28-8   RL Circuits

A circuit containing a resistor and an inductor, such as that shown in Figure 28-29, is called an **RL circuit.** Because all circuits have resistance and self-inductance at room temperature, the analysis of an *RL* circuit can be applied to some extent to all circuits.[†]

For the circuit shown in Figure 28-29, application of Kirchhoff's loop rule (Equation 28-17) gave us

$$\mathcal{E}_0 - IR - L\,\frac{dI}{dt} = 0$$

Let us look at some general features of the current before we solve this equation. Just after we close the switch in the circuit the current is still zero, so *IR* is zero, and *L dI/dt* equals the emf of the battery, $\mathcal{E}_0$. Setting *I* = 0 in Equation 28-17, we get

$$\left.\frac{dI}{dt}\right|_{I=0} = \frac{\mathcal{E}_0}{L} \qquad\qquad\qquad 28\text{-}21$$

As the current increases *IR* increases, and *dI/dt* decreases. Note that the current cannot abruptly jump from zero to some finite value as it would if there were no inductance. When the inductance *L* is not negligible *dI/dt* is finite, and therefore the current must be continuous in time. After a short time, the current has reached a positive value *I*, and the rate of change of the current is

$$\frac{dI}{dt} = \frac{\mathcal{E}_0}{L} - \frac{IR}{L}$$

[†] All circuits also have some capacitance between parts of the circuits at different potentials. We will consider the effects of capacitance in Chapter 29 when we study ac circuits. Here we will neglect capacitance to simplify the analysis and to focus on the effects of inductance.

At this time the current is still increasing, but its rate of increase is less than at $t = 0$. The final value of the current can be obtained by setting dI/dt equal to zero:

$$I_f = \frac{\mathcal{E}_0}{R}$$ 28-22

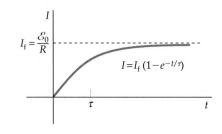

Figure 28-30 shows the current in this circuit as a function of time. This figure is the same as that for the charge on a capacitor as a function of time when the capacitor is charged in an RC circuit (Figure 25-41).

Equation 28-17 is of the same form as Equation 25-36 for the charging of a capacitor and can be solved in the same way—by separating variables and integrating. The result is

$$I = \frac{\mathcal{E}_0}{R}(1 - e^{-(R/L)t}) = I_f(1 - e^{-t/\tau})$$ 28-23

FIGURE 28-30 Current versus time in an RL circuit. At a time $t = \tau = L/R$, the current is at 63 percent of its maximum value $\mathcal{E}_0/R$.

where $I_f = \mathcal{E}_0/R$ is the current as $t \to \infty$, and

$$\tau = \frac{L}{R}$$ 28-24

is the **time constant** of the circuit. The larger the self-inductance L or the smaller the resistance R, the longer it takes for the current to reach any specified fraction of its final current I_f.

ENERGIZING A COIL **EXAMPLE 28-13**

A coil of self-inductance 5 mH and a resistance of 15 Ω is placed across the terminals of a 12-V battery of negligible internal resistance. (*a*) What is the final current? (*b*) What is the time constant? (*c*) How many time constants does it take for the current to reach 99 percent of its final value?

PICTURE THE PROBLEM The final current is the current when $dI/dt = 0$, as given in Equation 28-22. The current as a function of time is given by Equation 28-23, $I = I_f(1 - e^{-t/\tau})$, where $\tau = L/R$.

1. Use Equation 28-22 to find the final current, I_f:

$$I_f = \frac{\mathcal{E}_0}{R} = \frac{12 \text{ V}}{15 \text{ }\Omega} = \boxed{0.800 \text{ A}}$$

2. Calculate the time constant τ.

$$\tau = \frac{L}{R} = \frac{5 \times 10^{-3} \text{ H}}{15 \text{ }\Omega} = \boxed{333 \text{ } \mu s}$$

3. Use Equation 28-23 and calculate the time t for $I = 0.99I_f$:

$$I = I_f(1 - e^{-t/\tau})$$

so

$$e^{-t/\tau} = \left(1 - \frac{I}{I_f}\right)$$

and

$$-\frac{t}{\tau} = \ln\left(1 - \frac{I}{I_f}\right)$$

Thus,

$$t = -\tau \ln\left(1 - \frac{I}{I_f}\right) = -\tau \ln(1 - 0.99)$$

$$= -\tau \ln(0.01) = +\tau \ln 100 = \boxed{4.61 \text{ } \tau}$$

REMARKS In five time constants the current is within one percent of its final value.

EXERCISE How much energy is stored in this inductor when the final current has been attained? (*Answer* $U_m = \frac{1}{2}LI_f^2 = 1.6 \times 10^{-3}$ J)

In Figure 28-31, the circuit has a make-before-break switch (shown in Figure 28-32) that allows us to remove the battery from the circuit without interrupting the current through the inductor. The resistor R_1 protects the battery so that the battery is not shorted when the switch is thrown. If the switch pole is in position e, the battery, the inductor, and the two resistors are connected in series and the current builds up in the circuit as just discussed, except that the total resistance is now $R_1 + R$ and the final current is $\mathcal{E}_0/(R + R_1)$. Suppose that the pole has been in position e for a long time, so that the current remains at its final value, which we will call I_0. At time $t = 0$ we rapidly move the pole to position f (to remove the battery from consideration completely). We now have a circuit (loop $abcda$) with just a resistor and an inductor carrying an initial current I_0. Applying Kirchoff's loop rule to this circuit gives

$$-IR - L\frac{dI}{dt} = 0$$

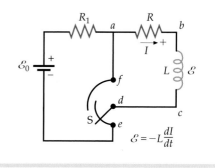

$$\mathcal{E} = -L\frac{dI}{dt}$$

FIGURE 28-31 An *RL* circuit with a make-before-break switch so that the battery can be removed from the circuit without interrupting the current through the inductor. The current in the inductor reaches its maximum value with the switch pole in position e. The pole is then rapidly moved to position f.

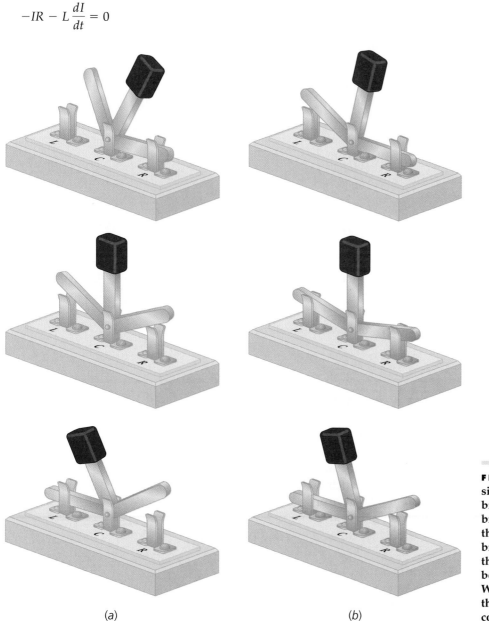

(a)　　　　　　　　(b)

FIGURE 28-32 (*a*) The standard single-pole, double-throw switch is a break-before-make switch. That is, it breaks the first contact before making the second contact. (*b*) In a make-before-break, single-pole, double-throw switch the throw makes the second contact before breaking the first contact. With the throw in the middle position, the throw is in electrical contact with contact *L* and contact *R*.

Rearranging this equation to separate the variables I and t gives

$$\frac{dI}{I} = -\frac{R}{L}\,dt \qquad\qquad\qquad 28\text{-}25$$

Equation 28-25 is of the same form as Equation 25-31 for the discharge of a capacitor. Integrating and then solving for I gives

$$I = I_0 e^{-t/\tau} \qquad\qquad\qquad 28\text{-}26$$

where $\tau = L/R$ is the time constant. Figure 28-33 shows the current as a function of time.

EXERCISE What is the time constant of a circuit of resistance 85 Ω and inductance 6 mH? (*Answer* 70.6 μs)

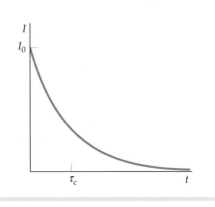

FIGURE 28-33 Current versus time for the circuit in Figure 28-31. The current decreases exponentially with time.

ENERGY DISSIPATED   **EXAMPLE 28-14**

Find the total energy dissipated in the resistor R, as shown in Figure 28-31, when the current in the inductor decreases from its initial value of I_0 to 0.

PICTURE THE PROBLEM The rate of energy dissipation I^2R varies with time so to calculate the total energy dissipated requires that we integrate.

1. The rate of heat production is I^2R:
$$P = I^2R$$

2. The total energy U dissipated in the resistor is the integral of $P\,dt$ from $t = 0$ to $t = \infty$:
$$U = \int_0^\infty I^2R\,dt$$

3. The current I is given by Equation 28-26:
$$I = I_0 e^{-(R/L)t}$$

4. Substitute this current into the integral:
$$U = \int_0^\infty I^2R\,dt = \int_0^\infty I_0^2 e^{-2(R/L)t}R\,dt = I_0^2 R\int_0^\infty e^{-2(R/L)t}\,dt$$

5. The integration can be done by substituting $x = 2Rt/L$:
$$U = I_0^2 R\,\frac{e^{-2(R/L)t}}{-2(R/L)}\bigg|_0^\infty = I_0^2 R\frac{-L}{2R}(0-1) = \boxed{\frac{1}{2}LI_0^2}$$

PLAUSIBILITY CHECK The total amount of energy dissipated equals the energy $\frac{1}{2}LI_0^2$ originally stored in the inductor.

INITIAL CURRENTS AND FINAL CURRENTS   **EXAMPLE 28-15**

For the circuit shown in Figure 28-34, find the currents I_1, I_2, and I_3 (*a*) immediately after switch S is closed and (*b*) a long time after switch S has been closed. After the switch has been closed for a long time the switch is opened. Immediately after the switch is opened (*c*) find the three currents and (*d*) find the potential drop across the 20-Ω resistor. (*e*) Find all three currents a long time after switch S was opened.

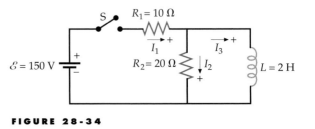

FIGURE 28-34

PICTURE THE PROBLEM (*a*) We simplify our calculations by using the fact that the current in an inductor cannot change abruptly. Thus, the current in the inductor must be zero just after the switch is closed, because the current is zero before the switch is closed. (*b*) When the current reaches its final value dI/dt equals zero, so there is no potential drop across the inductor. The inductor thus acts like a short circuit; that is, the inductor acts like a wire with zero resistance. (*c*) Immediately after the switch is opened, the current in the inductor is the same as it was before. (*d*) A long time after the switch is opened, all the currents must be zero.

(*a*) 1. The switch is just opened. The current through the inductor is zero, just as it was before the switch was closed. Apply the junction rule to relate I_1 and I_2:

$$I_3 = \boxed{0}$$
$$I_1 = I_2 + I_3$$

so

$$I_1 = I_2$$

   2. The current in the left loop is obtained by applying the loop rule to the loop on the left:

$$\mathcal{E} - I_1 R_1 - I_2 R_2 = 0$$

so

$$I_1 = \frac{\mathcal{E}}{R_1 + R_2} = \frac{150\ \text{V}}{10\ \Omega + 20\ \Omega} = \boxed{5\ \text{A}} = I_2$$

(*b*) 1. After a long time, the currents are steady and the inductor acts like a short circuit, so the potential drop across R_2 is zero. Apply the loop rule to the right loop and solve for I_2:

$$-L\frac{dI_3}{dt} + I_2 R_2 = 0$$
$$0 + I_2 R_2 = 0 \quad \Rightarrow \quad I_2 = \boxed{0}$$

   2. Apply the loop rule to the left loop and solve for I_1:

$$\mathcal{E} - I_1 R_1 - I_2 R_2 = 0$$
$$\mathcal{E} - I_1 R_1 - 0 = 0$$

so

$$I_1 = \frac{\mathcal{E}}{R_1} = \frac{150\ \text{V}}{10\ \Omega} = \boxed{15\ \text{A}}$$

   3. Apply the junction rule and solve for I_3:

$$I_1 = I_2 + I_3$$
$$15\ \text{A} = 0 + I_3$$

so

$$I_3 = \boxed{15\ \text{A}}$$

(*c*) When the switch is reopened, I_1 *instantly* becomes zero. The current I_3 in the inductor changes continuously, so at that instant $I_3 = 15$ A. Apply the junction rule and solve for I_2:

$$I_3 = \boxed{15\ \text{A}}$$
$$I_1 = I_2 + I_3$$

so

$$I_2 = I_1 - I_3 = 0 - 15\ \text{A} = \boxed{-15\ \text{A}}$$

(*d*) Apply Ohm's law to find the potential drop across R_2:

$$V = I_2 R_2 = (15\ \text{A})(20\ \Omega) = \boxed{300\ \text{V}}$$

(*e*) A long time after the switch is opened, all the currents must equal zero.

$$I_1 = I_2 = I_3 = \boxed{0}$$

REMARKS Were you surprised to find the potential drop across R_2 in Part (*d*) to be larger than the emf of the battery? This potential drop is equal to the emf of the inductor.

EXERCISE Suppose $R_2 = 200\ \Omega$ and the switch has been closed for a long time. What is the potential drop across it immediately after the switch is then opened? (*Answer*   3000 V)

*28-9 Magnetic Properties of Superconductors

Superconductors have resistivities of zero below a critical temperature T_c, which varies from material to material. In the presence of a magnetic field $\vec{B}$, the critical temperature is lower than the critical temperature is when there is no field. As the magnetic field increases, the critical temperature decreases. If the magnetic field magnitude is greater than some critical field B_c, superconductivity does not exist at any temperature.

*Meissner Effect

As a superconductor is cooled below the critical temperature in an applied magnetic field, the magnetic field inside the superconductor becomes zero (Figure 28-35). This effect was discovered by Walter Meissner and Robert Ochsenfeld in 1933 and is now known as the **Meissner effect.** The magnetic field becomes zero because superconducting currents induced on the surface of the superconductor produce a second magnetic field that cancels out the applied one. The magnetic levitation (see the following photo) results from the repulsion between the permanent magnet producing the applied field and the magnetic field produced by the currents induced in the superconductor. Only certain superconductors, called **type I superconductors,** exhibit the complete Meissner effect. Figure 28-36a shows a plot of the magnetization M times μ_0 versus the applied magnetic field B_{app} for a type I superconductor. For a magnetic field less than the critical field B_c, the magnetic field $\mu_0 M$ induced in the superconductor is equal and opposite to the applied magnetic field. The values of B_c for type I superconductors are always too small for such materials to be useful in the coils of a superconducting magnet.

Other materials, known as **type II superconductors,** have a magnetization curve similar to that in Figure 28-36b. Such materials are usually alloys or metals that have large resistivities in the normal state. Type II superconductors exhibit the electrical properties of superconductors except for the Meissner effect up to the critical field B_{c2}, which may be several hundred times the typical values of critical fields for type I superconductors. For example, the alloy Nb_3Ge has a critical field $B_{c2} = 34$ T. Such materials can be used for high-field superconducting magnets. Below the critical field B_{c1}, the behavior of a type II superconductor is the same as that of a type I superconductor. In the region between fields B_{c1} and B_{c2}, the superconductor is said to be in a vortex state.

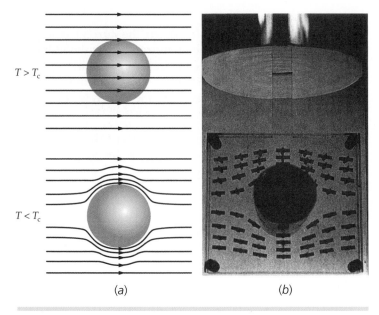

(a)　　　　　　　　(b)

FIGURE 28-35 (a) The Meissner effect in a superconducting solid sphere cooled in a constant applied magnetic field. As the temperature drops below the critical temperature T_c, the magnetic field inside the sphere becomes zero. (b) Demonstration of the Meissner effect. A superconducting tin cylinder is situated with its axis perpendicular to a horizontal magnetic field. The directions of the field lines are indicated by weakly magnetized compass needles mounted in a Lucite sandwich so that they are free to turn.

The cube is a superconductor. The magnetic levitation results from the repulsion between the permanent magnet producing the applied field and the magnetic field produced by the currents induced in the superconductor.

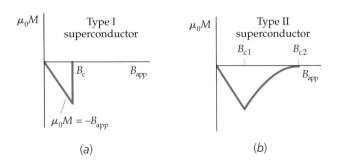

(a)　　　　　　　　(b)

FIGURE 28-36 Plots of μ_0 times the magnetization M versus applied magnetic field for type I and type II superconductors. (a) In a type I superconductor, the resultant magnetic field is zero below a critical applied field B_c because the field due to induced currents on the surface of the superconductor exactly cancels the applied field. Above the critical field, the material is a normal conductor and the magnetization is too small to be seen on this scale. (b) In a type II superconductor, the magnetic field starts to penetrate the superconductor at a field B_{c1}, but the material remains superconducting up to a field B_{c2}, after which the material becomes a normal conductor.

*Flux Quantization

Consider a superconducting ring of area A carrying a current. There can be a magnetic flux $\phi_m = B_n A$ through the flat surface S bounded by the ring due to the current in the ring and due also perhaps to other currents external to the ring. According to Equation 28-5, if the flux through S changes, an electric field will be induced in the ring whose circulation is proportional to the rate of change of the flux. But there can be no electric field in a superconducting ring because it has no resistance, so a finite electric field would drive an infinite current. The flux through the ring is thus frozen and cannot change.

Another effect, which results from the quantum-mechanical treatment of superconductivity, is that the total flux through surface S is quantized and is given by

$$\phi_m = n\frac{h}{2e}, \quad n = 1, 2, 3, \ldots \qquad 28\text{-}27$$

The smallest unit of flux, called a **fluxon,** is

$$\phi_0 = \frac{h}{2e} = 2.0678 \times 10^{-15} \text{ T·m}^2 \qquad 28\text{-}28$$

SUMMARY

1. Faraday's law and Lenz's law are fundamental laws of physics.
2. Self-inductance is a property of a circuit element that relates the flux through the element to the current.

Topic	Relevant Equations and Remarks	
1. Magnetic Flux ϕ_m		
General definition	$\phi_m = \int_S \vec{B} \cdot \hat{n}\, dA$	28-1
Uniform field, flat surface bounded by coil of N turns	$\phi_m = NBA \cos\theta$ where A is the area of the flat surface bounded by a single turn.	28-3
Units	$1 \text{ Wb} = 1 \text{ T·m}^2$	28-2
Due to current in a circuit	$\phi_m = LI$	28-9
Due to current in two circuits	$\phi_{m1} = L_1 I_1 + MI_2$ $\phi_{m2} = L_2 I_2 + MI_1$	28-14
*Quantization	$\phi_m = n\dfrac{h}{2e}, \quad n = 1, 2, 3, \cdots$	28-27
*Fluxon	$\phi_0 = \dfrac{h}{2e} = 2.0678 \times 10^{-15} \text{ T·m}^2$	28-28

---

2. EMF

Faraday's law (includes both induction and motional emf)	$\mathcal{E} = -\dfrac{d\phi_m}{dt}$	28-4
Induction (time varying magnetic field, C stationary)	$\mathcal{E} = \oint_C \vec{E} \cdot d\vec{\ell}$	28-5
Motional (static magnetic field, C not stationary)	$\mathcal{E} = \oint_C (\vec{v} \times \vec{B}) \cdot d\vec{\ell}$ where $\vec{v}$ is the velocity of the conducting path.	28-8
Rod moving perpendicular to both itself and $\vec{B}$	$\mathcal{E} = vB\ell$	28-6
Self-induced (back emf)	$\mathcal{E} = -L\dfrac{dI}{dt}$	28-12

---

3. Faraday's Law $\mathcal{E} = -\dfrac{d\phi_m}{dt}$ 28-4

---

4. Lenz's Law   The induced emf and induced current are in such a direction as to oppose, or tend to oppose, the change that produces them.

Alternative statement	When a magnetic flux through a surface changes, the magnetic field due to any induced current produces a flux of its own—through the same surface and in opposition to the change.

---

5. Inductance

Self-inductance	$L = \dfrac{\phi_m}{I}$	28-9
Self-inductance of a solenoid	$L = \mu_0 n^2 A\ell$	28-11
Mutual inductance	$M = \dfrac{\phi_{m2,1}}{I_1} = \dfrac{\phi_{m1,2}}{I_2}$	28-16
Units	$1\,\text{H} = 1\dfrac{\text{Wb}}{\text{A}} = 1\dfrac{\text{T}\cdot\text{m}^2}{\text{A}}$ $\mu_0 = 4\pi \times 10^{-7}\,\text{H/m}$	

---

6. Magnetic Energy

Energy stored in an inductor	$U_m = \dfrac{1}{2}LI^2$	28-19
Energy density in a magnetic field	$u_m = \dfrac{B^2}{2\mu_0} = \dfrac{1}{2}\mu_0^{-1}B^2$	28-20

***7. RL Circuits**

Potential difference across an inductor	$\Delta V = \mathcal{E} - Ir = -L\dfrac{dI}{dt} - Ir$ where r is the internal resistance of the inductor. For an ideal inductor $r = 0$.	28-13

Energizing an inductor with a battery	In a circuit consisting of a resistance R, an inductance L, and a battery of emf $\mathcal{E}_0$ in series, the current does not reach its maximum value I_f instantaneously, but rather takes some time to build up. If the current is initially zero, its value at some later time t is given by	
	$$I = \frac{\mathcal{E}_0}{R}(1 - e^{-t/\tau}) = I_f(1 - e^{-t/\tau})$$	28-23
Time constant τ	$$\tau = \frac{L}{R}$$	28-24
De-energizing an inductor through a resistor	In a circuit consisting of a resistance R and an inductance L, the current does not drop to zero instantaneously, but rather takes some time to decrease. If the current is initially I_0, its value at some later time t is given by	
	$$I = I_0 e^{-t/\tau}$$	28-26

PROBLEMS

- Single-concept, single-step, relatively easy
- •• Intermediate-level, may require synthesis of concepts
- ••• Challenging
- **SSM** Solution is in the *Student Solutions Manual*
- **iSOLVE** Problems available on iSOLVE online homework service
- **iSOLVE ✓** These "Checkpoint" online homework service problems ask students additional questions about their confidence level, and how they arrived at their answer.

In a few problems, you are given more data than you actually need; in a few other problems, you are required to supply data from your general knowledge, outside sources, or informed estimates.

Conceptual Problems

1 • **SSM** **iSOLVE** A conducting loop lies in the plane of this page and carries a clockwise induced current. Which of the following statements could be true? (*a*) A constant magnetic field is directed into the page. (*b*) A constant magnetic field is directed out of the page. (*c*) An increasing magnetic field is directed into the page. (*d*) A decreasing magnetic field is directed into the page. (*e*) A decreasing magnetic field is directed out of the page.

2 • Give the direction of the induced current in the circuit, shown on the right in Figure 28-37, when the resistance in the circuit on the left is suddenly (*a*) increased and (*b*) decreased.

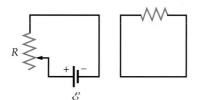

FIGURE 28-37 Problem 2

3 •• The two circular loops in Figure 28-38 have their planes parallel to each other. As viewed from the left, there is a counterclockwise current in loop A. Give the direction of the current in loop B and state whether the loops attract or repel each other if the current in loop A is (*a*) increasing and (*b*) decreasing.

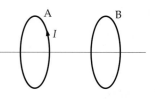

FIGURE 28-38 Problem 3

4 •• A bar magnet moves with constant velocity along the axis of a loop, as shown in Figure 28-39. (*a*) Make a qualitative graph of the flux ϕ_m through the loop as a function of time. Indicate the time t_1 when the magnet is halfway through the loop. (*b*) Sketch a graph of the current I in the loop versus time, choosing I to be positive when it is clockwise as viewed from the left.

FIGURE 28-39 Problem 4

5 •• A bar magnet is mounted on the end of a coiled spring in such a way that it moves with simple harmonic motion along the axis of a loop, as shown in Figure 28-40. (a) Make a qualitative graph of the flux ϕ_m through the loop as a function of time. Indicate the time t_1 when the magnet is halfway through the loop. (b) Sketch the current I in the loop versus time, choosing I to be positive when it is clockwise as viewed from above.

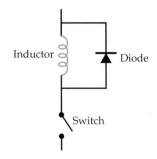

FIGURE 28-40 Problem 5

6 • [SSM] If the current through an inductor were doubled, the energy stored in the inductor would be (a) the same. (b) doubled. (c) quadrupled. (d) halved. (e) quartered.

7 • Inductors in circuits that are switched on and off are often protected by being placed in parallel with a diode, as shown in Figure 28-41. (A diode is a one-way valve for current; current can flow in the direction of the arrow, but not opposite to the direction of the arrow.) Why is such protection needed? Explain how the voltage across the inductor changes with no diode protection if a switch is opened suddenly while current is flowing through the inductor.

Inductor — Diode

Switch

FIGURE 28-41 Problem 7

8 • Two inductors are made from identical lengths of wire wrapped around identical circular cores of the same radius. However, one inductor has three times the number of coils per unit length as the other. Which coil has the higher self-inductance? What is the ratio of the self-inductance of the two coils?

9 • True or false:

(a) The induced emf in a circuit is proportional to the magnetic flux through the circuit.
(b) There can be an induced emf at an instant when the flux through the circuit is zero.
(c) Lenz's law is related to the conservation of energy.
(d) The inductance of a solenoid is proportional to the rate of change of the current in the solenoid.
(e) The magnetic energy density at some point in space is proportional to the square of the magnetic field at that point.

10 • [SSM] A pendulum is fabricated from a thin, flat piece of aluminum. At the bottom of its arc, it passes between the poles of a strong permanent magnet. In Figure 28-42a, the metal sheet is continuous, whereas in Figure 28-42b, there are slots in it. The pendulum with slots swings back and forth many times, but the pendulum without slots comes to a stop in no more than one complete oscillation. Explain why.

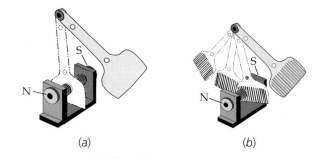

(a)   (b)

FIGURE 28-42 Problem 10

11 • A bar magnet is dropped inside a long vertical tube. If the tube is made of metal, the magnet quickly approaches a terminal speed, but if the tube is made of cardboard, the magnet does not. Explain.

12 •• An experimental setup for a jumping ring demonstration is shown in Figure 28-43. A metal ring is placed on top of a large coil, with an iron rod threading the center of the ring and the coil. When a current is suddenly started in the coil, the ring jumps several feet into the air. Explain how the demonstration works. Will the demonstration work if a slot is cut into the ring?

Ring

FIGURE 28-43 Problem 12

Estimation and Approximation

13 •• [SSM] A physics teacher attempts the following emf demonstration. She has two of her students hold a long wire connected to a voltmeter. The wire is held slack, so that there is a large arc in it. When she says "start," the students begin rotating the wire in a large vertical arc, as if they were playing jump rope. The students stand 3.0 m apart, and the sag in the wire is about 1.5 m. (You may idealize the shape of the wire as a perfect semicircular arc of diameter $d = 1.5$ m.) The induced emf from the jump rope is then measured on the voltmeter. (a) Estimate a reasonable value for the maximum angular velocity that the students can rotate the wire. (b) From this, estimate the maximum emf induced in the wire. The magnitude of the Earth's magnetic field is approximately 0.7 G. (c) Can the students rotate the jump rope fast enough to generate an emf of 1 V? (d) Suggest modifications to the demonstration that would allow higher emfs to be generated.

14 • Compare the energy density stored in the earth's electric field, which has a value $E \sim 100$ V/m at the surface of the earth, to that of the earth's magnetic field, where $B \sim 5 \times 10^{-5}$ T.

15 •• A lightning strike transfers roughly 30 C of charge from the sky to the ground in approximately 1 μs. Estimate the maximum emf induced by the lightning strike in an antenna consisting of a single loop of wire with cross-sectional area 0.1 m^2 a distance 300 m away from the lightning strike.

Magnetic Flux

16 • A uniform magnetic field of magnitude 2000 G is parallel to the x axis. A square coil of side 5 cm has a single turn and makes an angle θ with the z axis, as shown in Figure 28-44. Find the magnetic flux through the coil when (*a*) $\theta = 0°$, (*b*) $\theta = 30°$, (*c*) $\theta = 60°$, and (*d*) $\theta = 90°$.

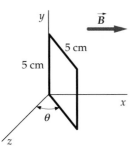

FIGURE 28-44 Problem 16

17 • SSM A circular coil has 25 turns and a radius of 5 cm. It is at the equator, where the earth's magnetic field is 0.7 G north. Find the magnetic flux through the coil when its plane is (*a*) horizontal, (*b*) vertical with its axis pointing north, (*c*) vertical with its axis pointing east, and (*d*) vertical with its axis making an angle of 30° with north.

18 • ISOLVE✓ A magnetic field of 1.2 T is perpendicular to a square coil of 14 turns. The length of each side of the coil is 5 cm. (*a*) Find the magnetic flux through the coil. (*b*) Find the magnetic flux through the coil if the magnetic field makes an angle of 60° with the normal to the plane of the coil.

19 • ISOLVE✓ A uniform magnetic field $\vec{B}$ is perpendicular to the base of a hemisphere of radius R. Calculate the magnetic flux through the spherical surface of the hemisphere.

20 •• Find the magnetic flux through a 400-turn solenoid of length 25 cm and radius 1 cm that carries a current of 3 A.

21 •• ISOLVE Rework Problem 20 for an 800-turn solenoid of length 30 cm and radius 2 cm that carries a current of 2 A.

22 •• A circular coil of 15 turns of radius 4 cm is in a uniform magnetic field of 4000 G in the positive x direction. Find the flux through the coil when the unit vector perpendicular to the plane of the coil is (*a*) $\hat{n} = \hat{i}$, (*b*) $\hat{n} = \hat{j}$, (*c*) $\hat{n} = (\hat{i} + \hat{j})/\sqrt{2}$, (*d*) $\hat{n} = \hat{k}$, and (*e*) $\hat{n} = 0.6\hat{i} + 0.8\hat{j}$.

23 •• A solenoid has n turns per unit length, radius R_1, and carries a current I. (*a*) A large circular loop of radius $R_2 > R_1$ and N turns encircles the solenoid at a point far away from the ends of the solenoid. Find the magnetic flux through the loop. (*b*) A small circular loop of N turns and radius $R_3 < R_1$ is completely inside the solenoid, far from its ends, with its axis parallel to that of the solenoid. Find the magnetic flux through this small loop.

24 •• SSM ISOLVE✓ A long straight wire carries a current I. A rectangular loop with two sides parallel to the straight wire has sides a and b, with its near side a distance d from the straight wire, as shown in Figure 28-45. (*a*) Compute the magnetic flux through the rectangular loop. (*Hint:* Calculate the flux through a strip of area $dA = b\ dx$ and integrate from $x = d$ to $x = d + a$.) (*b*) Evaluate your answer for $a = 5$ cm, $b = 10$ cm, $d = 2$ cm, and $I = 20$ A.

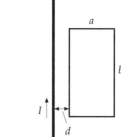

FIGURE 28-45 Problem 24

25 ••• A long cylindrical conductor of radius R carries a current I that is uniformly distributed over its cross-sectional area. Find the magnetic flux per unit length through the area indicated in Figure 28-46.

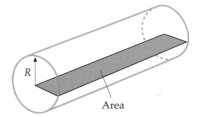

FIGURE 28-46 Problem 25

26 ••• A rectangular coil in the plane of the page has dimensions a and b. A long wire that carries a current I is placed directly above the coil (Figure 28-47). (*a*) Obtain an expression for the magnetic flux through the coil as a function of x for $0 \le x \le 2b$. (*b*) For what value of x is the flux through the coil a maximum? For what value of x is the flux a minimum?

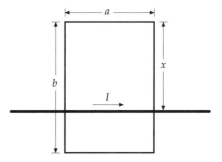

FIGURE 28-47 Problem 26

Induced EMF and Faraday's Law

27 • SSM A uniform magnetic field $\vec{B}$ is established perpendicular to the plane of a loop of radius 5 cm, resistance 0.4 Ω, and negligible self-inductance. The magnitude of $\vec{B}$ is increasing at a rate of 40 mT/s. Find (*a*) the induced emf $\mathcal{E}$ in the loop, (*b*) the induced current in the loop, and (*c*) the rate of joule heating in the loop.

28 • The flux through a loop is given by $\phi_m = (t^2 - 4t) \times 10^{-1}$ Wb, where t is in seconds. (a) Find the induced emf $\mathcal{E}$ as a function of time. (b) Find both ϕ_m and $\mathcal{E}$ at $t = 0$, $t = 2$ s, $t = 4$ s, and $t = 6$ s.

29 • (a) For the flux given in Problem 28, sketch graphs of ϕ_m and $\mathcal{E}$ versus t. (b) At what time is the flux minimum? What is the emf at this time? (c) At what times is the flux zero? What is the emf at these times?

30 • ✓ A solenoid of length 25 cm and radius 0.8 cm with 400 turns is in an external magnetic field of 600 G that makes an angle of 50° with the axis of the solenoid. (a) Find the magnetic flux through the solenoid. (b) Find the magnitude of the emf induced in the solenoid if the external magnetic field is reduced to zero in 1.4 s.

31 •• SSM A 100-turn circular coil has a diameter of 2 cm and resistance of 50 Ω. The plane of the coil is perpendicular to a uniform magnetic field of magnitude 1 T. The direction of the field is suddenly reversed. (a) Find the total charge that passes through the coil. If the reversal takes 0.1 s, find (b) the average current in the coil and (c) the average emf in the coil.

32 •• ✓ At the equator, a 1000-turn coil with a cross-sectional area of 300 cm² and a resistance of 15 Ω is aligned with its plane perpendicular to the earth's magnetic field of 0.7 G. If the coil is flipped over, how much charge flows through the coil?

33 •• ✓ A circular coil of 300 turns and radius 5 cm is connected to a current integrator. The total resistance of the circuit is 20 Ω. The plane of the coil is originally aligned perpendicular to the earth's magnetic field at some point. When the coil is rotated through 90°, the charge that passes through the current integrator is measured to be 9.4 μC. Calculate the magnitude of the earth's magnetic field at that point.

34 •• The wire in Problem 26 is placed at $x = b/4$. (a) Obtain an expression for the emf induced in the coil if the current varies with time according to $I = 2t$. (b) If $a = 1.5$ m and $b = 2.5$ m, what should be the resistance of the coil so that the induced current is 0.1 A? What is the direction of this current?

35 •• Repeat Problem 34 if the wire is placed at $x = b/3$.

Motional EMF

36 • SSM ✓ A rod 30 cm long moves at 8 m/s in a plane perpendicular to a magnetic field of 500 G. The velocity of the rod is perpendicular to its length. Find (a) the magnetic force on an electron in the rod, (b) the electrostatic field $\vec{E}$ in the rod, and (c) the potential difference V between the ends of the rod.

37 • ✓ Find the speed of the rod in Problem 36 if the potential difference between the ends is 6 V.

38 • In Figure 28-22, let B be 0.8 T, $v = 10$ m/s, $\ell = 20$ cm, and $R = 2$ Ω. Find (a) the induced emf in the circuit, (b) the current in the circuit, and (c) the force needed to move the rod with constant velocity assuming negligible friction. Find (d) the power input by the force found in Part (c), and (e) the rate of joule heat production I^2R.

FIGURE 28-22
Problem 38

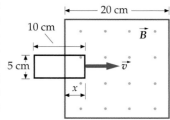

39 •• A 10-cm by 5-cm rectangular loop with resistance 2.5 Ω is pulled through a region of uniform magnetic field $B = 1.7$ T (Figure 28-48) with constant speed $v = 2.4$ cm/s. The front of the loop enters the region of the magnetic field at time $t = 0$. (a) Find and graph the flux through the loop as a function of time. (b) Find and graph the induced emf and the current in the loop as functions of time. Neglect any self-inductance of the loop and extend your graphs from $t = 0$ to $t = 16$ s.

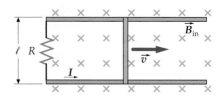

FIGURE 28-48 Problem 39

40 •• ✓ A uniform magnetic field of magnitude 1.2 T is in the z direction. A conducting rod of length 15 cm lies parallel to the y axis and oscillates in the x direction with displacement given by $x = (2$ cm$)\cos 120\,\pi t$. What is the emf induced in the rod?

41 •• In Figure 28-49, the rod has a resistance R and the rails are horizontal and have negligible resistance. A battery of emf $\mathcal{E}$ and negligible internal resistance is connected between points a and b so that the current in the rod is downward. The rod is placed at rest at $t = 0$. (a) Find the force on the rod as a function of the speed v and write Newton's second law for the rod when it has speed v. (b) Show that the rod moves at a terminal speed and find an expression for it. (c) What is the current when the rod will approach its terminal speed?

FIGURE 28-49
Problems 41 and 44

42 •• SSM In Example 28-9, find the total energy dissipated in the resistance and show that it is equal to mv_0^2.

43 •• Find the total distance traveled by the rod in Example 28-9.

44 •• In Figure 28-49, the rod has a resistance R and the rails have negligible resistance. A capacitor with charge Q_0 and capacitance C is connected between points a and b so that the current in the rod is downward. The rod is placed at rest at $t = 0$. (a) Write the equation of motion for the rod on the rails. (b) Show that the terminal speed of the rod down the rails is related to the final charge on the capacitor.

45 •• SSM In Figure 28-50, a conducting rod of mass m and negligible resistance is free to slide without friction along two parallel rails of negligible resistance separated by a distance ℓ and connected by a resistance R. The rails are attached to a long inclined plane that makes an angle θ with the horizontal. There is a magnetic field B directed upward. (a) Show that there is a retarding force directed up the incline given by $F = (B^2\ell^2 v \cos^2 \theta)/R$. (b) Show that the terminal speed of the rod is $v_t = (mgR \sin \theta)/(B^2\ell^2 \cos^2 \theta)$.

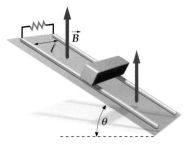

FIGURE 28-50 Problems 45 and 49

46 •• A square loop of a conducting wire (area A) is pulled out of a region of constant, very high magnetic field B that is directed perpendicular to the plane of the wire. Half of the wire is in the field and half of the wire is out of the field when the wire is pulled out. A constant force F is exerted on the wire to pull the wire out. The wire is pulled out in time t. All else being equal, if the force were doubled, approximately how long would it take to pull the wire out? (a) t (b) $t/\sqrt{2}$ (c) $t/2$ (d) $t/4$

47 •• If instead of doubling the force the resistance of the wire in Problem 46 were halved (all else being equal), what would the new time be? (a) t (b) $2t$ (c) $t/2$ (d) $t\sqrt{2}$

48 •• A wire lies along the z axis and carries current $I = 20$ A in the positive z direction. A small conducting sphere of radius $R = 2$ cm is initially at rest on the y axis at a distance $h = 45$ m above the wire. The sphere is dropped at time $t = 0$. (a) What is the electric field at the center of the sphere at $t = 3$ s? Assume that the only magnetic field is the magnetic field produced by the wire. (b) What is the voltage across the sphere at $t = 3$ s?

49 •• **iSOLVE** In Figure 28-50, let $\theta = 30°$; $m = 0.4$ kg, $\ell = 15$ m, and $R = 2 \, \Omega$. The rod starts from rest at the top of the inclined plane at $t = 0$. The rails have negligible resistance. There is a constant, vertically directed magnetic field of magnitude $B = 1.2$ T. (a) Find the emf induced in the rod as a function of its velocity down the rails. (b) Write Newton's law of motion for the rod; show that the rod will approach a terminal speed and determine its value.

50 ••• A solid conducting cylinder of radius 0.1 m and mass 4 kg rests on horizontal conducting rails (Figure 28-51). The rails, separated by a distance $a = 0.4$ m, have a rough surface, so the cylinder rolls rather than slides. A 12-V battery is connected to the rails as shown. The only significant resistance in the circuit is the contact resistance of 6 Ω between the cylinder and rails. The system is in a uniform vertical magnetic field. The cylinder is initially at rest next to the battery. (a) What must be the magnitude and the direction of $\vec{B}$ so that the cylinder has an initial acceleration of 0.1 m/s² to the right? (b) Find the force on the cylinder as a function of its speed v. (c) Find the terminal velocity of the cylinder. (d) What is the kinetic energy of the cylinder when it has reached its terminal velocity? (Neglect the magnetic field due to the current in

the battery–rails–cylinder loop, and assume that the current density in the cylinder is uniform.)

FIGURE 28-51 Problem 50

51 ••• **SSM** The loop in Problem 24 moves away from the wire with a constant speed v. At time $t = 0$, the left side of the loop is a distance d from the long straight wire. (a) Compute the emf in the loop by computing the motional emf in each segment of the loop that is parallel to the long wire. Explain why you can neglect the emf in the segments that are perpendicular to the wire. (b) Compute the emf in the loop by first computing the flux through the loop as a function of time and then using $\mathcal{E} = -d\phi_m/dt$. Compare your answer with that obtained in Part (a).

52 ••• A conducting rod of length ℓ rotates at constant angular velocity about one end, in a plane perpendicular to a uniform magnetic field B (Figure 28-52). (a) Show that the magnetic force on a body whose charge is q at a distance r from the pivot is $Bqr\omega$. (b) Show that the potential difference between the ends of the rod is $V = \frac{1}{2}B\omega\ell^2$. (c) Draw any radial line in the plane from which to measure $\theta = \omega t$. Show that the area of the pie-shaped region between the reference line and the rod is $A = \frac{1}{2}\ell^2 \theta$. Compute the flux through this area, and show that $\mathcal{E} = \frac{1}{2}B\omega\ell^2$ follows when Faraday's law is applied to this area.

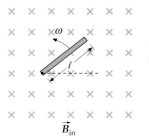

FIGURE 28-52 Problem 52

Inductance

53 • A coil with a self-inductance of 8 H carries a current of 3 A that is changing at a rate of 200 A/s. Find (a) the magnetic flux through the coil and (b) the induced emf in the coil.

54 • **SSM** A coil with self-inductance L carries a current I, given by $I = I_0 \sin 2\pi f t$. Find and graph the flux ϕ_m and the self-induced emf as functions of time.

55 •• **iSOLVE** ✓ A solenoid has a length of 25 cm, a radius of 1 cm, 400 turns, and carries a 3-A current. Find (a) B on the axis at the center of the solenoid; (b) the flux through the solenoid, assuming B to be uniform; (c) the self-inductance of the solenoid; and (d) the induced emf in the solenoid when the current changes at 150 A/s.

56 •• **iSOLVE** Two solenoids of radii 2 cm and 5 cm are coaxial. They are each 25 cm long and have 300 turns and 1000 turns, respectively. Find their mutual inductance.

57 •• [SSM] [ISOLVE] ✔ A long insulated wire with a resistance of 18 Ω/m is to be used to construct a resistor. First, the wire is bent in half and then the doubled wire is wound in a cylindrical form, as shown in Figure 28-53. The diameter of the cylindrical form is 2 cm, its length is 25 cm, and the total length of wire is 9 m. Find the resistance and inductance of this wire-wound resistor.

FIGURE 28-53 Problem 57

58 ••• In Figure 28-54, circuit 2 has a total resistance of 300 Ω. A total charge of 2×10^{-4} C flows through the galvanometer in circuit 2 when switch S in circuit 1 is closed. After a long time, the current in circuit 1 is 5 A. What is the mutual inductance between the two coils?

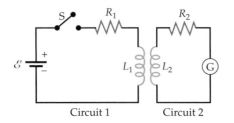

FIGURE 28-54 Problem 58

59 ••• Show that the inductance of a toroid of rectangular cross section, as shown in Figure 28-55, is given by

$$L = \frac{\mu_0 N^2 H \ln(b/a)}{2\pi}$$

where N is the total number of turns, a is the inside radius, b is the outside radius, and H is the height of the toroid.

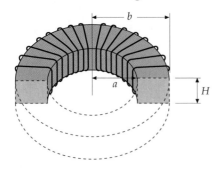

FIGURE 28-55 Problem 59

Magnetic Energy

60 • A coil with a self-inductance of 2 H and a resistance of 12 Ω is connected across a 24-V battery of negligible internal resistance. (*a*) What is the final current? (*b*) How much energy is stored in the inductor when the final current is attained?

61 •• [SSM] In a plane electromagnetic wave, such as a light wave, the magnitudes of the electric fields and magnetic fields are related by $E = cB$, where $c = 1/\sqrt{\epsilon_0\mu_0}$ is the speed of light. Show that in this case the electric energy and the magnetic energy densities are equal.

62 •• A solenoid of 2000 turns, area 4 cm^2, and length 30 cm carries a current of 4 A. (*a*) Calculate the magnetic energy stored in the solenoid from $\frac{1}{2}LI^2$. (*b*) Divide your answer in Part (*a*) by the volume of the solenoid to find the magnetic energy per unit volume in the solenoid. (*c*) Find B in the solenoid. (*d*) Compute the magnetic energy density from $u_m = B^2/2\mu_0$, and compare your answer with your result for Part (*b*).

63 •• [ISOLVE] A long cylindrical wire of radius $a = 2$ cm carries a current $I = 80$ A uniformly distributed over its cross-sectional area. Find the magnetic energy per unit length within the wire.

64 •• [SSM] You are given a length d of wire that has radius a and are told to wind it into an inductor in the shape of a cylinder with a circular cross section of radius r. The windings are to be as close together as possible without overlapping. Show that the self-inductance of this inductor is

$$L = \mu_0\left(\frac{rd}{4a}\right)$$

65 • Using the result of Problem 64, calculate the self-inductance of an inductor wound from 10 cm of wire with a diameter of 1 mm into a coil with radius $R = 0.25$ cm.

66 •• A toroid of mean radius 25 cm and circular cross section of radius 2 cm is wound with a superconducting wire of length 1000 m that carries a current of 400 A. (*a*) What is the number of turns on the coil? (*b*) What is the magnetic field at the mean radius? (*c*) Assuming that B is constant over the area of the coil, calculate the magnetic energy density and the total energy stored in the toroid.

*RL Circuits

67 • A coil of resistance 8 Ω and self-inductance 4 H is suddenly connected across a constant potential difference of 100 V. Let $t = 0$ be the time of connection, at which the current is zero. Find the current I and its rate of change dI/dt at times (*a*) $t = 0$, (*b*) $t = 0.1$ s, (*c*) $t = 0.5$ s, and (*d*) $t = 1.0$ s.

68 • The current in a coil with a self-inductance of 1 mH is 2 A at $t = 0$, when the coil is shorted through a resistor. The total resistance of the coil plus the resistor is 10 Ω. Find the current after (*a*) 0.5 ms and (*b*) 10 ms.

69 •• [SSM] In the circuit shown Figure 28-29, let $\mathcal{E}_0 = 12$ V, $R = 3$ Ω, and $L = 0.6$ H. The switch is closed at time $t = 0$. At time $t = 0.5$ s, find (*a*) the rate at which the battery supplies power, (*b*) the rate of joule heating, and (*c*) the rate at which energy is being stored in the inductor.

70 •• Rework Problem 69 for the times $t = 1$ s and $t = 100$ s.

71 •• The current in an RL circuit is zero at time $t = 0$ and increases to half its final value in 4 s. (*a*) What is the time constant of this circuit? (*b*) If the total resistance is 5 Ω, what is the self-inductance?

72 •• How many time constants must elapse before the current in an *RL* circuit that is initially zero reaches (*a*) 90 percent, (*b*) 99 percent, and (*c*) 99.9 percent of its final value?

73 •• A coil with inductance 4 mH and resistance 150 Ω is connected across a battery of emf 12 V and negligible internal resistance. (*a*) What is the initial rate of increase of the current? (*b*) What is the rate of increase when the current is half its final value? (*c*) What is the final current? (*d*) How long does it take for the current to reach 99 percent of its final value?

74 •• **ISOLVE** A large electromagnet has an inductance of 50 H and a resistance of 8 Ω. It is connected to a dc power source of 250 V. Find the time for the current to reach (*a*) 10 A and (*b*) 30 A.

75 ••• **SSM** Given the circuit shown in Figure 28-56, assume that the switch S has been closed for a long time so that steady currents exist in the inductor, and that the inductor *L* has negligible resistance. (*a*) Find the battery current, the current in the 100 Ω resistor, and the current through the inductor. (*b*) Find the initial voltage across the inductor when the switch S is opened. (*c*) Using a spreadsheet program, make graphs of the current and voltage across the inductor as a function of time.

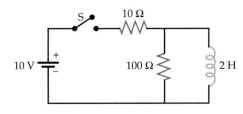

FIGURE 28-56 Problem 75

76 •• Compute the initial slope *dI/dt* at *t* = 0 from Equation 28-26, and show that if the current decreased steadily at this rate the current would be zero after one time constant.

77 •• An inductance *L* and resistance *R* are connected in series with a battery, as shown in Figure 28-31. A long time after switch S₁ is closed, the current is 2.5 A. When the battery is switched out of the circuit by opening switch S₁ and closing S₂, the current drops to 1.5 A in 45 ms. (*a*) What is the time constant for this circuit? (*b*) If *R* = 0.4 Ω, what is *L*?

78 • **ISOLVE** When the current in a certain coil is 5 A and the current is increasing at the rate of 10 A/s, the potential difference across the coil is 140 V. When the current is 5 A and the current is decreasing at the rate of 10 A/s, the potential difference is 60 V. Find the resistance and self-inductance of the coil.

79 •• For the circuit shown in Figure 28-57, (*a*) find the rate of change of the current in each inductor and in the resistor just after the switch is closed. (*b*) What is the final current? (Use the result from Problem 88.)

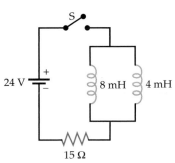

FIGURE 28-57 Problem 79

80 •• **SSM** For the circuit of Example 28-11, find the time at which the power dissipation in the resistor equals the rate at which magnetic energy is stored in the inductor.

81 ••• In the circuit shown in Figure 28-29, let $\mathcal{E}_0$ = 12 V, *R* = 3 Ω, and *L* = 0.6 H. The switch is closed at time *t* = 0. From time *t* = 0 to *t* = τ, find (*a*) the total energy that has been supplied by the battery, (*b*) the total energy that has been dissipated in the resistor, and (*c*) the energy that has been stored in the inductor. (*Hint:* Find the rates as functions of time and integrate from *t* = 0 to *t* = τ = *L/R*.)

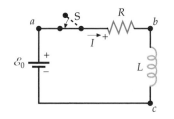

FIGURE 28-29 Problem 81

General Problems

82 • A circular coil of radius 3 cm has 6 turns. A magnetic field *B* = 5000 G is perpendicular to the coil. (*a*) Find the magnetic flux through the coil. (*b*) Find the magnetic flux through the coil if the coil makes an angle of 20° with the magnetic field.

83 • The magnetic field in Problem 82 is steadily reduced to zero in 1.2 s. Find the emf induced in the coil when (*a*) the magnetic field is perpendicular to the coil and (*b*) the magnetic field makes an angle of 20° with the normal to the coil.

84 • **ISOLVE** A 100-turn coil has a radius of 4 cm and a resistance of 25 Ω. At what rate must a perpendicular magnetic field change to produce a current of 4 A in the coil?

85 •• **SSM** Figure 28-58 shows an ac generator. The generator consists of a rectangular loop of dimensions *a* and *b* with *N* turns connected to slip rings. The loop rotates with an angular velocity ω in a uniform magnetic field $\vec{B}$. (*a*) Show that the potential difference between the two slip rings is $\mathcal{E}$ = *NBabω* sin ω*t*. (*b*) If *a* = 1 cm, *b* = 2 cm, *N* = 1000, and *B* = 2 T, at what angular frequency ω must the coil rotate to generate an emf whose maximum value is 110 V?

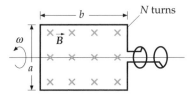

FIGURE 28-58 Problems 85 and 86

86 •• **ISOLVE** Prior to 1960, magnetic field strength was measured by means of a rotating coil gaussmeter. This device used a small loop of many turns rotating on an axis perpendicular to the magnetic field at fairly high speed, which was connected to an ac voltmeter by means of slip rings, like those shown in Figure 28-58. The sensing coil for a rotating coil gaussmeter has 400 turns and an area of 1.4 cm².

The coil rotates at 180 rpm. If the magnetic field strength is 0.45 T, find the maximum induced emf in the coil and the orientation of the coil relative to the field for which this maximum induced emf occurs.

87 •• Show that the effective inductance for two inductors L_1 and L_2 connected in series, so that none of the flux from either passes through the other, is given by $L_{eff} = L_1 + L_2$.

88 •• [SSM] Show that the effective inductance for two inductors L_1 and L_2 connected in parallel, so that none of the flux from either passes through the other, is given by

$$\frac{1}{L_{eff}} = \frac{1}{L_1} + \frac{1}{L_2}$$

89 •• [SSM] Figure 28-59(a) shows an experiment designed to measure the acceleration of gravity. A large plastic tube is encircled by a wire, which is arranged in single loops separated by a distance of 10 cm. A strong magnet is dropped through the top of the loop. As the magnet falls through each loop the voltage rises and then the voltage rapidly falls through 0 to a large negative value as the magnet passes through the loop and then returns to 0. The shape of the voltage signal is shown in Figure 28-59(b). (a) Explain how this experiment works. (b) Explain why the tube cannot be made of a conductive material. (c) Qualitatively explain the shape of the voltage signal in Figure 28-59(b). (d) The times at which the voltage crosses 0 as the magnet falls through each loop in succession are given in the table in the next column. Use these data to calculate a value for g.

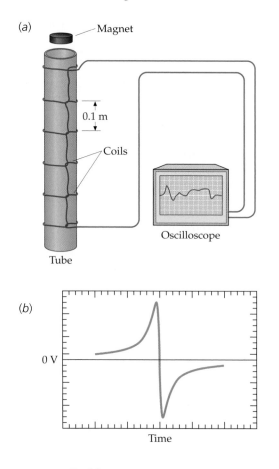

(a)

Magnet

0.1 m

Coils

Oscilloscope

Tube

(b)

0 V

Time

FIGURE 28-59 Problem 89

Loop Number	Zero Crossing Time(s)
1	0.011189
2	0.063133
3	0.10874
4	0.14703
5	0.18052
6	0.21025
7	0.23851
8	0.26363
9	0.28853
10	0.31144
11	0.33494
12	0.35476
13	0.37592
14	0.39107

90 •• The rectangular coil shown in Figure 28-60 has 80 turns, is 25 cm wide, is 30 cm long, and is located in a magnetic field $B = 1.4$ T directed out of the page, as shown, with only half of the coil in the region of the magnetic field. The resistance of the coil is 24 Ω. Find the magnitude and the direction of the induced current if the coil is moved with a speed of 2 m/s (a) to the right, (b) up, (c) to the left, and (d) down.

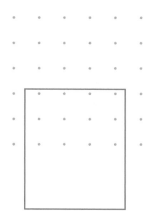

FIGURE 28-60 Problem 90

91 •• [SSM] Suppose the coil of Problem 90 is rotated about its vertical centerline at constant angular velocity of 2 rad/s. Find the induced current as a function of time.

92 •• Show that if the flux through each turn of an N-turn coil of resistance R changes from ϕ_{m1} to ϕ_{m2}, the total charge passing through the coil is given by $Q = N(\phi_{m1} - \phi_{m2})/R$.

93 •• A long solenoid has n turns per unit length and carries a current given by $I = I_0 \sin \omega t$. The solenoid has a circular cross section of radius R. Find the induced electric field at a radius r from the axis of the solenoid for (a) $r < R$ and (b) $r > R$.

94 ••• A coaxial cable consists of two very thin-walled conducting cylinders of radii r_1 and r_2 (Figure 28-61). Current I goes in one direction down the inner cylinder and in the opposite direction in the outer cylinder. (a) Use Ampère's law to find B. Show that $B = 0$, except in the region between the conductors. (b) Show that the magnetic energy density in the region between the cylinders is

$$u_m = \frac{\mu_0 I^2}{8\pi^2 r^2}$$

(c) Find the magnetic energy in a cylindrical shell volume element of length ℓ and volume $dV = \ell 2\pi r\, dr$, and integrate your result to show that the total magnetic energy in the volume of length ℓ is

$$U_m = \frac{\mu_0}{4\pi} I^2 \ell \ln\frac{r_2}{r_1}$$

(d) Use the result in Part (c) and $U_m = \frac{1}{2}LI^2$ to show that the self-inductance per unit length is

$$\frac{L}{\ell} = \frac{\mu_0}{2\pi}\ln\frac{r_2}{r_1}$$

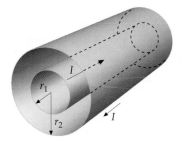

FIGURE 28-61 Problems 94 and 95

95 ••• Using Figure 28-61, compute the flux through a rectangular area of sides ℓ and $r_2 - r_1$ between the conductors. Show that the self-inductance per unit length can be found from $\phi_m = LI$ (see Part (d) of Problem 94).

96 ••• [SSM] Figure 28-62 shows a rectangular loop of wire, 0.30 m wide and 1.50 m long, in the vertical plane and perpendicular to a uniform magnetic field $B = 0.40$ T, directed inward as shown. The portion of the loop not in the magnetic field is 0.10 m long. The resistance of the loop is 0.20 Ω and its mass is 0.05 kg. The loop is released from rest at $t = 0$. (a) What is the magnitude and direction of the induced current when the loop has a downward velocity v? (b) What is the force that acts on the loop as a result of this current? (c) What is the net force acting on the loop? (d) Write the equation of motion of the loop. (e) Obtain an expression for the velocity of the loop as a function of time. (f) Integrate the

expression obtained in Part (e) to find the displacement y as a function of time. (g) Using a spreadsheet program, make a graph of the position y of the loop as a function of time for values of y between 0 m and 1.4 m (i.e., when the loop leaves the magnetic field). At what time t does $y = 1.4$ m? Compare this to the time it would have taken if $B = 0$.

FIGURE 28-62 Problems 96 and 97

97 ••• The loop of Problem 96 is attached to a plastic spring of spring constant κ (see Figure 28-62). (a) When $B = 0$, the period of small-amplitude vertical oscillations of the mass–spring system is 0.8 s. Find the spring constant κ. (b) When $B \neq 0$, a current is induced in the loop as a result of its up and down motion. Obtain an expression for the induced current as a function of time when $B = 0.40$ T, and the displacement of the center of the loop is $y = 0.05$ m downward. (c) Show that the force on the loop is of the form $-\beta v$, where $v = dy/dt$, and find an expression for β in terms of B, w, and R, where w is the width of the wire loop and R is its resistance. (d) Using a spreadsheet program, make graphs of the position y and the velocity v of the center of the loop as a function of time, use the parameters given.

98 ••• A coil of N turns and area A hangs from a wire that provides a linear restoring torque with torsion constant κ. The two ends of the coil are connected to each other, the coil has resistance R, and the moment of inertia of the coils is I. The plane of the coil is vertical, and parallel to a uniform horizontal magnetic field B when the wire is not twisted (i.e., $\theta = 0$). The coil is twisted and released from a small angle $\theta = \theta_0$. Show that the orientation of the coil will undergo damped harmonic oscillation according to $\theta(t) = \theta_0 e^{-\beta t}\cos\omega t$, where

$$\omega = \sqrt{\kappa/I} \quad \text{and} \quad \beta = \frac{N^2 B^2 A^2}{RI}.$$

Alternating Current Circuits

More than 99 percent of the electrical energy used today is produced by electrical generators in the form of alternating current, which has a great advantage over direct current, because electrical energy can be transported over long distances at very high voltages and low currents to reduce energy losses due to Joule heating. Electrical energy can then be transformed, with almost no energy loss, to lower and safer voltages and correspondingly higher currents for everyday use. The transformer that accomplishes these changes in potential difference and current works on the basis of magnetic induction. In North America, power is delivered by a sinusoidal current of frequency 60 Hz. Devices such as radios, television sets, and microwave ovens detect or generate alternating currents of much higher frequencies.

Alternating current is produced by motional emf or magnetic induction in an ac generator, which is designed to provide a sinusoidal emf.

➤ **In this chapter, we will see that when the generator output is sinusoidal, the current in an inductor, a capacitor, or a resistor is also sinusoidal, although it is generally not in phase with the generator's emf. When the emf and current are both sinusoidal, their maximum values are related. The study of sinusoidal** currents is particularly important because even currents that are not sinusoidal can be analyzed in terms of sinusoidal components using Fourier analysis.

29-1 Alternating Current Generators

Figure 29-1 shows a simple **ac generator** that consists of a coil of area A and N turns rotating in a uniform magnetic field. The ends of the coil are connected to rings, called slip rings, that rotate with the coil. They make electrical contact through stationary conducting brushes that are in contact with the rings.

When the normal to the plane of the coil makes an angle θ with a uniform magnetic field $\vec{B}$, as shown in the figure, the magnetic flux through the coil is

$$\phi_m = NBA \cos \theta \qquad 29\text{-}1$$

where A is the area of the flat surface bounded by a single turn of the coil and N is the number of turns. When the coil is mechanically rotated, the flux through the coil will change, and an emf will be induced. If ω is the angular velocity of rotation and the initial angle is δ, the angle at some later time t is given by

$$\theta = \omega t + \delta$$

Then

$$\phi_m = NBA \cos(\omega t + \delta) = NBA \cos(2\pi f t + \delta)$$

The emf in the coil will then be

$$\mathcal{E} = -\frac{d\phi_m}{dt} = -NBA\frac{d}{dt}\cos(\omega t + \delta) = +NBA\omega \sin(\omega t + \delta) \qquad 29\text{-}2$$

where $NBA\omega$ is the peak (maximum) emf. Thus,

$$\mathcal{E} = \mathcal{E}_{\text{peak}} \sin(\omega t + \delta) \qquad 29\text{-}3$$

where the emf amplitude is given by

$$\mathcal{E}_{\text{peak}} = NBA\omega \qquad 29\text{-}4$$

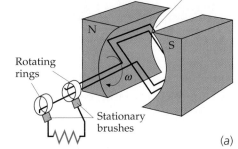

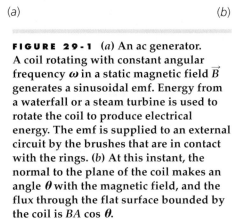

Rotating rings

Stationary brushes

N turns

(a)

(b)

FIGURE 29-1 (*a*) An ac generator. A coil rotating with constant angular frequency ω in a static magnetic field $\vec{B}$ generates a sinusoidal emf. Energy from a waterfall or a steam turbine is used to rotate the coil to produce electrical energy. The emf is supplied to an external circuit by the brushes that are in contact with the rings. (*b*) At this instant, the normal to the plane of the coil makes an angle θ with the magnetic field, and the flux through the flat surface bounded by the coil is $BA \cos \theta$.

We can thus produce a sinusoidal emf in a coil by rotating the coil with constant angular velocity in a magnetic field. Although practical generators are considerably more complicated, they produce a sinusoidal emf either via induction or via motional emf. In circuit diagrams, an ac generator is represented by the symbol ⊖.

The same coil in a static magnetic field that can be used to generate an alternating emf can also be used as an **ac motor.** Instead of mechanically rotating the coil to generate an emf, we apply an ac potential difference generated by another ac generator to the coil. This produces an ac current in the coil, and the magnetic field exerts forces on the wires producing a torque that rotates the coil. As the coil rotates in the magnetic field, a back emf is generated that tends to counter the

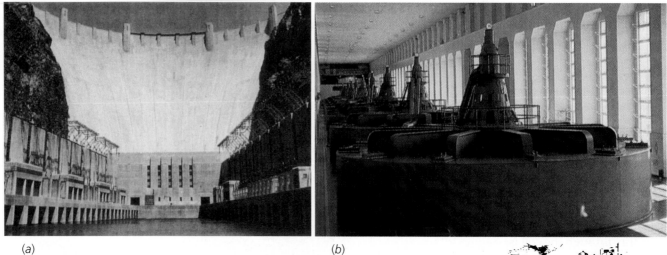

(a)

(b)

applied potential difference that produces the current. When the motor is first turned on, there is no back emf and the current is very large, being limited only by the resistance in the circuit. As the motor begins to rotate, the back emf increases and the current decreases.

EXERCISE A 250-turn coil has an area per turn of 3 cm². If it rotates in a magnetic field of 0.4 T at 60 Hz, what is $\mathcal{E}_{\text{peak}}$? (*Answer* $\mathcal{E}_{\text{peak}} = 11.3$ V)

29-2 Alternating Current in a Resistor

Figure 29-2 shows a simple ac circuit that consists of an ideal generator and a resistor. (A generator is ideal if its internal resistance, self-inductance, and capacitance are negligible.) The voltage drop across the resistor V_R is equal to the emf $\mathcal{E}$ of the generator. If the generator produces an emf given by Equation 29-3, we have

$$V_R = \mathcal{E} = \mathcal{E}_{\text{peak}} \sin(\omega t + \delta) = V_{R,\text{peak}} \sin(\omega t + \delta)$$

where $V_{R,\text{peak}} = \mathcal{E}_{\text{peak}}$. In this equation, the phase constant δ is arbitrary. It is convenient to choose $\delta = \pi/2$ so that

$$V_R = V_{R,\text{peak}} \sin\left(\omega t + \frac{\pi}{2}\right) = V_{R,\text{peak}} \cos \omega t$$

Applying Ohm's law, we have

$$V_R = IR \qquad\qquad 29\text{-}5$$

Thus,

$$V_{R,\text{peak}} \cos \omega t = IR \qquad\qquad 29\text{-}6$$

so the current in the resistor is

$$I = \frac{V_{R,\text{peak}}}{R} \cos \omega t = I_{\text{peak}} \cos \omega t \qquad\qquad 29\text{-}7$$

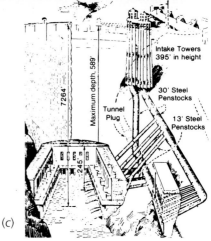

(c)

(*a*) The mechanical energy of falling water drives turbines (*b*) for the generation of electricity. (*c*) Schematic drawing of the Hoover Dam showing the intake towers and pipes (penstocks) that carry the water to the generators below.

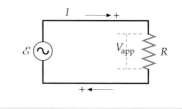

FIGURE 29-2 An ac generator in series with a resistor R.

where

$$I_{peak} = \frac{V_{R,peak}}{R}$$ 29-8

Note that the current through the resistor is in phase with the potential drop across the resistor, as shown in Figure 29-3.

The power dissipated in the resistor varies with time. Its instantaneous value is

$$P = I^2R = (I_{peak} \cos \omega t)^2 R = I_{peak}^2 R \cos^2 \omega t$$ 29-9

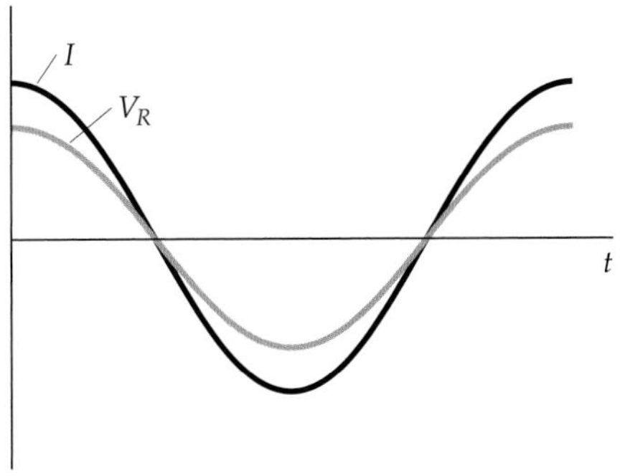

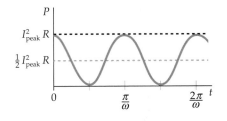

FIGURE 29-3 The voltage drop across a resistor is in phase with the current.

Figure 29-4 shows the power as a function of time. The power varies from zero to its peak value $I_{peak}^2 R$, as shown. We are usually interested in the average power over one or more complete cycles:

$$P_{av} = (I^2R)_{av} = I_{peak}^2 R(\cos^2 \omega t)_{av}$$

The average value of $\cos^2 \omega t$ over one or more periods is $\frac{1}{2}$. This can be seen from the identity $\cos^2 \omega t + \sin^2 \omega t = 1$. A plot of $\sin^2 \omega t$ looks the same as a plot of $\cos^2 \omega t$ except that the plot is shifted by 90°. Both have the same average value over one or more periods, and since their sum is 1, the average value of each must be $\frac{1}{2}$. The average power dissipated in the resistor is thus

$$P_{av} = (I^2R)_{av} = \frac{1}{2} I_{peak}^2 R$$ 29-10

FIGURE 29-4 Plot of the power dissipated in the resistor shown in Figure 29-2 versus time. The power varies from zero to a peak value $I_{peak}^2 R$. The average power is half the peak power.

Root-Mean-Square Values

Most ac ammeters and voltmeters are designed to measure the **root-mean-square (rms) values** of current and potential difference rather than the peak values. The rms value of a current I_{rms} is defined by

$$I_{rms} = \sqrt{(I^2)_{av}}$$ 29-11

DEFINITION—RMS CURRENT

For a sinusoidal current, the average value of I^2 is

$$(I^2)_{av} = \left[(I_{peak} \cos \omega t)^2\right]_{av} = \frac{1}{2} I_{peak}^2$$

Substituting $\frac{1}{2} I_{peak}^2$ for $(I^2)_{av}$ in Equation 29-11, we obtain

$$I_{rms} = \frac{1}{\sqrt{2}} I_{peak} \approx 0.707 I_{peak}$$ 29-12

RMS VALUE RELATED TO PEAK VALUE

The rms value of any quantity that varies sinusoidally equals the peak value of that quantity divided by $\sqrt{2}$.

Substituting I_{rms}^2 for $\frac{1}{2} I_{peak}^2$ in Equation 29-10, we obtain for the average power dissipated in the resistor

$$P_{av} = I_{rms}^2 R$$ 29-13

The rms current equals the steady dc current that would produce the same Joule heating as the actual ac current.

For the simple circuit in Figure 29-2, the average power delivered by the generator is:

$$P_{av} = (\mathcal{E}I)_{av} = [(\mathcal{E}_{peak} \cos \omega t)(I_{peak} \cos \omega t)]_{av} = \mathcal{E}_{peak} I_{peak} (\cos^2 \omega t)_{av}$$

or

$$P_{av} = \tfrac{1}{2} \mathcal{E}_{peak} I_{peak}$$

Using $I_{rms} = I_{peak}/\sqrt{2}$ and $\mathcal{E}_{rms} = \mathcal{E}_{peak}/\sqrt{2}$, this can be written

$$P_{av} = \mathcal{E}_{rms} I_{rms} \qquad\qquad 29\text{-}14$$

AVERAGE POWER DELIVERED BY A GENERATOR

The rms current is related to the rms potential drop in the same way that the peak current is related to the peak potential drop. We can see this by dividing each side of Equation 29-8 by $\sqrt{2}$ and using $I_{rms} = I_{peak}/\sqrt{2}$ and $V_{R,rms} = V_{R,peak}/\sqrt{2}$.

$$I_{rms} = \frac{V_{R,rms}}{R} \qquad\qquad 29\text{-}15$$

Equations 29-13, 29-14, and 29-15 are of the same form as the corresponding equations for direct-current circuits with I replaced by I_{rms} and V_R replaced by $V_{R,rms}$. We can therefore calculate the power input and the heat generated using the same equations that we used for direct current, if we use rms values for the current and potential drop.

EXERCISE The sinusoidal potential drop across a 12-Ω resistor has a peak value of 48 V. Find (*a*) the rms current, (*b*) the average power, and (*c*) the maximum power. (*Answer*   (*a*) 2.83 A, (*b*) 96 W, (*c*) 192 W)

The ac power supplied to domestic wall outlets and light fixtures in the United States has an rms potential difference of 120 V at a frequency of 60 Hz. This potential difference is maintained, independent of the current. If you plug a 1600-W space heater into a wall outlet it will draw a current of

$$I_{rms} = \frac{P_{av}}{V_{rms}} = \frac{1600 \text{ W}}{120 \text{ V}} = 13.3 \text{ A}$$

All appliances plugged into the outlets of a single 120-V circuit are connected in parallel. If you plug a 500-W toaster into another outlet of the same circuit, it will draw a current of 500 W/120 V = 4.17 A, and the total current through the parallel combination will be 17.5 A. Typical household wall outlets are rated at 15 A and are part of a circuit using wires rated at either 15 A or 20 A, with each circuit having several outlets. The wire in each circuit is rated at 15 A or 20 A, correspondingly. A total current greater than the rated current for the wiring is likely to overheat the wiring and is a fire hazard. Each circuit is therefore equipped with a circuit breaker (or a fuse in older houses) that trips (or blows) when the total current exceeds the 15-A or 20-A rating.

High-power domestic appliances, such as electric clothes dryers, kitchen ranges, and hot water heaters, typically require power delivered at 240-V rms. For a given power requirement, only half as much current is required at 240 V as at 120 V, but 240 V is more likely to deliver a fatal shock or to start a fire than 120 V.

SAWTOOTH WAVEFORM   **EXAMPLE 29-1**

Find (a) the average current and (b) the rms current for the sawtooth waveform shown in Figure 29-5. In the region $0 < t < T$, the current is given by $I = (I_0/T)t$.

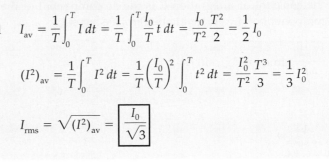

FIGURE 29-5

PICTURE THE PROBLEM The average of any quantity over a time interval T is the integral of the quantity over the interval divided by T. We use this to find both the average current, I_{av}, and the average of the current squared, $(I^2)_{av}$.

(a) Calculate I_{av} by integrating I from $t = 0$ to $t = T$ and dividing by T:

$$I_{av} = \frac{1}{T}\int_0^T I\,dt = \frac{1}{T}\int_0^T \frac{I_0}{T}t\,dt = \frac{I_0}{T^2}\frac{T^2}{2} = \frac{1}{2}I_0$$

(b) 1. Find $(I^2)_{av}$ by integrating I^2:

$$(I^2)_{av} = \frac{1}{T}\int_0^T I^2\,dt = \frac{1}{T}\left(\frac{I_0}{T}\right)^2\int_0^T t^2\,dt = \frac{I_0^2}{T^2}\frac{T^3}{3} = \frac{1}{3}I_0^2$$

2. The rms current is the square root of $(I^2)_{av}$:

$$I_{rms} = \sqrt{(I^2)_{av}} = \boxed{\frac{I_0}{\sqrt{3}}}$$

29-3 Alternating Current Circuits

Alternating current behaves differently than direct current in inductors and capacitors. When a capacitor becomes fully charged in a dc circuit, the capacitor blocks the current; that is, the capacitor acts like an open circuit. However, if the current alternates, charge continually flows onto the plates or off the plates of the capacitor. We will see that at high frequencies, a capacitor hardly impedes the current at all. That is, the capacitor acts like a short circuit. Conversely, an induction coil usually has a low resistance and is essentially a short circuit for direct current; however, when the current is changing, a back emf is generated in an inductor that is proportional to dI/dt. At high frequencies, the back emf is large and the inductor acts like an open circuit.

Inductors in Alternating Current Circuits

Figure 29-6 shows an inductor coil in series with an ac generator. When the current changes in the inductor, a back emf of magnitude $L\,dI/dt$ is generated due to the changing flux. Usually this back emf is much greater than the IR drop due to the resistance of the coil, so we normally neglect the resistance of the coil. The potential drop across the inductor V_L is then given by

$$V_L = L\frac{dI}{dt}$$

29-16

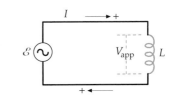

In this circuit, the potential drop V_L across the inductor equals the emf $\mathcal{E}$ of the generator. That is,

$$V_L = \mathcal{E} = \mathcal{E}_{max}\cos\omega t = V_{L,peak}\cos\omega t$$

where $V_{L,peak} = \mathcal{E}_{peak}$. Substituting for V_L in Equation 29-16 gives

$$V_{L,peak}\cos\omega t = L\frac{dI}{dt}$$

29-17

FIGURE 29-6 An ac generator in series with an inductor L. The arrow indicates the positive direction along the wire. Note that for a positive value of dI/dt, the voltage drop V_L across the inductor is positive. That is, if you traverse the inductor in the direction of the direction arrow you go in the direction of decreasing potential.

Rearranging, we obtain

$$dI = \frac{V_{L,\text{peak}}}{L} \cos \omega t \, dt \tag{29-18}$$

We solve for the current I by integrating both sides of the equation:

$$I = \frac{V_{L,\text{peak}}}{L} \int \cos \omega t \, dt = \frac{V_{L,\text{peak}}}{\omega L} \sin \omega t + C \tag{29-19}$$

where the constant of integration C is the dc component of the current. Setting the dc component of the current to be zero, we have

$$I = \frac{V_{L,\text{peak}}}{\omega L} \sin \omega t = I_{\text{peak}} \sin \omega t \tag{29-20}$$

where

$$I_{\text{peak}} = \frac{V_{L,\text{peak}}}{\omega L} \tag{29-21}$$

The potential drop $V_L = V_{L,\text{peak}} \cos \omega t$ across the inductor is 90° out of phase with the current $I = I_{\text{peak}} \sin \omega t$. From Figure 29-7, which shows I and V_L as functions of time, we can see that the peak value of the potential drop occurs 90° or one-fourth period prior to the corresponding peak value of the current. The potential drop across an inductor is said to *lead the current by 90°*. We can understand this physically. When I is zero but increasing, dI/dt is maximum, so the back emf induced in the inductor is at its maximum. One-quarter cycle later, I is maximum. At this time, dI/dt is zero, so V_L is zero. Using the trigonometric identity $\sin \theta = \cos\left(\theta - \frac{\pi}{2}\right)$, where $\theta = \omega t$, Equation 29-20 for the current can be written

$$I = I_{\text{peak}} \cos\left(\omega t - \frac{\pi}{2}\right) \tag{29-22}$$

The relation between the peak current and the peak potential drop (or between the rms current and rms potential drop) for an inductor can be written in a form similar to Equation 29-15 for a resistor. From Equation 29-21, we have

$$I_{\text{peak}} = \frac{V_{L,\text{peak}}}{\omega L} = \frac{V_{L,\text{peak}}}{X_L} \tag{29-23}$$

where

$$X_L = \omega L \tag{29-24}$$

DEFINITION—INDUCTIVE REACTANCE

is called the **inductive reactance.** Since $I_{\text{rms}} = I_{\text{peak}}/\sqrt{2}$ and $V_{L,\text{rms}} = V_{L,\text{peak}}/\sqrt{2}$ the rms current is given by

$$I_{\text{rms}} = \frac{V_{L,\text{rms}}}{X_L} \tag{29-25}$$

Like resistance, inductive reactance has units of ohms. As we can see from Equation 29-25, the larger the reactance for a given potential drop, the smaller

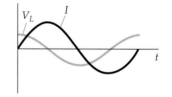

FIGURE 29-7 Current and potential drop across the inductor shown in Figure 29-6 as functions of time. The maximum potential drop occurs one-fourth period before the maximum current. Thus, the potential drop is said to lead the current by one-fourth period or 90°.

the peak current. Unlike resistance, the inductive reactance depends on the frequency of the current—the greater the frequency, the greater the reactance.

The *instantaneous* power delivered to the inductor from the generator is

$$P = V_L I = (V_{L,\text{peak}} \cos \omega t)(I_{\text{peak}} \sin \omega t) = V_{L,\text{peak}} I_{\text{peak}} \cos \omega t \sin \omega t$$

The *average* power delivered to the inductor is zero. We can see this by using the trigonometric identity

$$2 \cos \omega t \sin \omega t = \sin 2\omega t$$

The value of $\sin 2\omega t$ oscillates twice during each cycle and is negative as often as it is positive. Thus, on the average, no energy is dissipated in an inductor. (This is the case only if the resistance of the inductor is negligible.)

INDUCTIVE REACTANCE   **EXAMPLE 29-2**

The potential drop across a 40-mH inductor is sinusoidal with a peak potential drop of 120 V. Find the inductive reactance and the peak current when the frequency is (*a*) 60 Hz and (*b*) 2000 Hz.

PICTURE THE PROBLEM We calculate the inductive reactance at each frequency and use Equation 29-23 to find the peak current.

(*a*) 1. The peak current equals the peak potential drop divided by the inductive reactance. The peak potential drop equals the emf:

$$I_{\text{peak}} = \frac{V_{L,\text{peak}}}{X_L}$$

2. Compute the inductive reactance at 60 Hz:

$$X_{L1} = \omega_1 L = 2\pi f_1 L = (2\pi)(60 \text{ Hz})(40 \times 10^{-3} \text{ H})$$

$$= \boxed{15.1 \ \Omega}$$

3. Use this value of X_L to compute the peak current at 60 Hz:

$$I_{1,\text{peak}} = \frac{120 \text{ V}}{15.1 \ \Omega} = \boxed{7.95 \text{ A}}$$

(*b*) 1. Compute the inductive reactance at 2000 Hz:

$$X_{L2} = \omega_2 L = 2\pi f_2 L$$

$$= (2\pi)(2000 \text{ Hz})(40 \times 10^{-3} \text{ H}) = \boxed{503 \ \Omega}$$

2. Use this value of X_L to compute the peak current at 2000 Hz:

$$I_{2,\text{peak}} = \frac{120 \text{ V}}{503 \ \Omega} = \boxed{0.239 \text{ A}}$$

Capacitors in Alternating Current Circuits

When a capacitor is connected across the terminals of an ac generator (Figure 29-8), the voltage drop across the capacitor is

$$V_C = \frac{Q}{C} \qquad\qquad 29\text{-}26$$

where Q is the charge on the upper plate of the capacitor.

In this circuit, the potential drop V_C across the capacitor equals the emf $\mathcal{E}$ of the generator. That is,

$$V_C = \mathcal{E} = \mathcal{E}_{\text{peak}} \cos \omega t = V_{C,\text{peak}} \cos \omega t$$

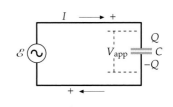

FIGURE 29-8 An ac generator in series with a capacitor *C*. The positive direction along the circuit is such that when the current is positive the charge *Q* on the upper capacitor plate is increasing, so the current is related to the charge by $I = +dQ/dt$.

where $V_{C,peak} = \mathcal{E}_{peak}$. Substituting for V_C in Equation 29-26 and solving for Q gives

$$Q = V_C C = V_{C,peak} C \cos \omega t$$

The current is

$$I = \frac{dQ}{dt} = -\omega V_{C,peak} C \sin \omega t = -I_{peak} \sin \omega t$$

where

$$I_{peak} = \omega V_{C,peak} C \qquad\qquad 29\text{-}27$$

Using the trigonometric identity $\sin \theta = -\cos\left(\theta + \dfrac{\pi}{2}\right)$, where $\theta = \omega t$, we obtain

$$I = -\omega C V_{C,peak} \sin \omega t = I_{peak} \cos\left(\omega t + \frac{\pi}{2}\right) \qquad\qquad 29\text{-}28$$

As with the inductor, the voltage drop $V_C = V_{C,peak} \cos \omega t$ across the capacitor is out of phase with the current

$$I = I_{peak} \cos\left(\omega t + \frac{\pi}{2}\right)$$

in the circuit. From Figure 29-9, we see that the maximum value of the potential drop occurs 90° or one-fourth period *after* the maximum value of the current. Thus, *the potential drop across a capacitor lags the current by 90°*. Again, we can understand this physically. The charge Q is proportional to the potential drop V_C. The maximum value of $dQ/dt = I$ occurs when the charge Q, and therefore when V_C, is zero. As the charge on the capacitor plate increases the current decreases until, one-fourth period later, the charge Q, and therefore V_C, is a maximum and the current is zero. The current then becomes negative as the charge Q decreases.

Again, we can relate the current to the potential drop in a form similar to Equation 29-8 for a resistor. From Equation 29-27, we have

$$I_{peak} = \omega C V_{C,peak} = \frac{V_{C,peak}}{1/(\omega C)} = \frac{V_{C,peak}}{X_C}$$

and, similarly,

$$I_{rms} = \frac{V_{C,rms}}{X_C} \qquad\qquad 29\text{-}29$$

where

$$X_C = \frac{1}{\omega C} \qquad\qquad 29\text{-}30$$

DEFINITION—CAPACITIVE REACTANCE

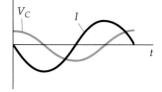

FIGURE 29-9 Current and potential drop across the capacitor shown in Figure 29-8 versus time. The maximum potential drop occurs one-fourth period after the maximum current. Thus, the potential drop is said to lag the current by 90°.

is called the **capacitive reactance** of the circuit. Like resistance and inductive reactance, capacitive reactance has units of ohms and, like inductive reactance, capacitive reactance depends on the frequency of the current. In this case, the greater the frequency, the smaller the reactance. The average power delivered to a capacitor from an ac generator is zero, as it is for an inductor. This occurs because the potential drop is proportional to cos ωt and the current is proportional to sin ωt and $(\cos \omega t \sin \omega t)_{av} = 0$. Thus, like inductors with no resistance, capacitors dissipate no energy.

Since charge cannot pass across the space between the plates of a capacitor, it may seem strange that there is a continuing alternating current in the circuit shown in Figure 29-8. Suppose we choose the time to be zero at the instant that the voltage drop V_C across the capacitor is both zero and increasing. (At this same instant, the charge Q on the upper plate of the capacitor is also both zero and increasing.) As V_C then increases, positive charge flows off the lower plate and onto the upper plate, and Q reaches its maximum value Q_{peak} a quarter period later. After Q reaches its maximum value Q continues to change, reaching zero at the half-period point, $-Q_{peak}$ at the three-quarter-period point, and zero (again) at the completion of the cycle at the full-period point. The charge Q_{peak} flows through the generator each quarter period. If we double the frequency, we halve the period. Thus, if we double the frequency we halve the time for the charge Q_{peak} to flow through the generator, so we have doubled the current amplitude I_{peak}. Hence, the greater the frequency, the less the capacitor impedes the flow of charge.

CAPACITIVE REACTANCE **E X A M P L E   2 9 - 3**

A 20-μF capacitor is placed across an ac generator that applies a potential drop with an amplitude (peak value) of 100 V. Find the capacitive reactance and the current amplitude when the frequency is 60 Hz and when the frequency is 6000 Hz.

PICTURE THE PROBLEM The capacitive reactance is $X_C = 1/(\omega C)$ and the peak current is $I_{peak} = V_{C,peak}/X_C$.

1. Calculate the capacitive reactance at 60 Hz and at 6000 Hz:

$$X_{C1} = \frac{1}{\omega_1 C} = \frac{1}{2\pi f_1 C} = \frac{1}{2\pi(60 \text{ Hz})(20 \times 10^{-6} \text{ F})} = \boxed{133 \ \Omega}$$

$$X_{C2} = \frac{1}{\omega_2 C} = \frac{1}{2\pi f_2 C} = \frac{1}{2\pi(6000 \text{ Hz})(20 \times 10^{-6} \text{ F})} = \boxed{1.33 \ \Omega}$$

2. Use these values of X_C to find the peak currents:

$$I_{1,peak} = \frac{V_{C,peak}}{X_{C1}} = \frac{100 \text{ V}}{133 \ \Omega} = \boxed{0.752 \text{ A}}$$

$$I_{2,peak} = \frac{V_{C,peak}}{X_{C2}} = \frac{100 \text{ V}}{1.33 \ \Omega} = \boxed{75.2 \text{ A}}$$

REMARKS Note that the capacitive reactance is inversely proportional to the frequency, so increasing the frequency by two orders of magnitude decreases the reactance by two orders of magnitude. The current is directly proportional to the frequency, as expected.

---

*29-4 Phasors

Until this point, the circuits considered contained an ideal ac generator and only a single passive element (i.e., resistor, inductor, or capacitor). In these circuits, the potential drop across the passive element equaled the emf of the generator. In circuits that contain an ideal ac generator and two or more additional elements connected in series, the sum of the potential drops across the elements is equal to the generator emf; which is the same as with dc circuits. However, in ac circuits these potential drops typically are not in phase, so the sum of their rms values does not equal the rms value of the generator emf.

Two-dimensional vectors, which are called phasors, can represent the phase relations between the current and the potential drops across resistors, capacitors, or inductors. In Figure 29-10, the potential drop across a resistor V_R is represented by a vector $\vec{V}_R$ that has magnitude $I_{peak}R$ and makes an angle θ with the x axis. This potential drop is in phase with the current. In general, the current in a steady-state ac circuit varies with time, as

$$I = I_{peak} \cos \theta = I_{peak} \cos (\omega t - \delta) \qquad 29\text{-}31$$

where ω is the angular frequency and δ is some phase constant. The potential drop across a resistor is then given by

$$V_R = IR = I_{peak}R \cos (\omega t - \delta) \qquad 29\text{-}32$$

The potential drop across a resistor is thus equal to the x component of the phasor vector $\vec{V}_R$, which rotates counterclockwise with an angular frequency ω. The current I may be written as the x component of a phasor $\vec{I}$ having the same direction as $\vec{V}_R$.

When several components are connected together in a series combination, their potential drops add. When several components are connected in parallel, their currents add. Unfortunately, adding sines or cosines of different amplitudes and phases algebraically is awkward. It is much easier to do this by vector addition.[†]

Let us look at how phasors are used. Any ac current or potential drop is written in the form $A \cos(\omega t - \delta)$, which in turn is treated as A_x, the x component of a phasor that makes an angle $(\omega t - \delta)$ with the positive x direction. Instead of adding two potential drops or currents algebraically, as $A \cos(\omega t - \delta_1) + B \cos(\omega t - \delta_2)$, we represent these quantities as phasors $\vec{A}$ and $\vec{B}$ and find the phasor sum $\vec{C} = \vec{A} + \vec{B}$ geometrically. The resultant potential drop or current is then the x component of the resultant phasor, $C_x = A_x + B_x$. The geometric representation conveniently shows the relative amplitudes and phases of the phasors.

Consider a circuit that contains an inductor L, a capacitor C, and a resistor R connected in series. They all carry the same current, which is represented as the x component of the current phasor $\vec{I}$. The potential drop across the inductor V_L is represented by a phasor $\vec{V}_L$ that has magnitude $I_{peak}X_L$ and leads the current phasor $\vec{I}$ by 90°. Similarly, the potential drop across the capacitor V_C is represented by a phasor $\vec{V}_C$ that has magnitude $I_{peak}X_C$ and lags the current by 90°. Figure 29-11 shows the phasors $\vec{V}_R$, $\vec{V}_L$, and $\vec{V}_C$. As time passes, the three phasors rotate counterclockwise with an angular frequency ω, so the relative positions of the vectors do not change. At any time, the instantaneous value of the potential drop across any of these elements equals the x component of the corresponding phasor.

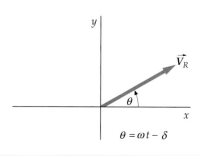

FIGURE 29-10 The potential drop across a resistor can be represented by a vector $\vec{V}_R$, which is called a phasor, that has magnitude $I_{peak}R$ and makes an angle $\theta = \omega t - \delta$ with the x axis. The phasor rotates with an angular frequency ω. The potential drop $V_R = IR$ is the x component of $\vec{V}_R$.

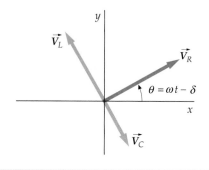

FIGURE 29-11 Phasor representations of the potential drops V_R, V_L, and V_C. Each vector rotates in the counterclockwise direction with an angular frequency ω. At any instant, the potential drop across an element equals the x component of the corresponding phasor, and the potential drop across the RLC-series combination, which equals the sum of the potential drops, equals the x component of the vector sum $\vec{V}_R + \vec{V}_L + \vec{V}_C$.

*29-5 LC and RLC Circuits Without a Generator

Figure 29-12 shows a simple circuit with inductance and capacitance but with no resistance. Such a circuit is called an **LC circuit**. We assume that the upper capacitor plate carries an initial positive charge Q_0 and that the switch is initially open.

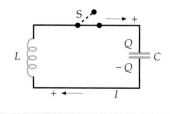

FIGURE 29-12 An *LC* circuit. When the switch is closed, the initially charged capacitor discharges through the inductor, producing a back emf.

---

† It is also easier to do using complex numbers.

After the switch is closed at $t = 0$, the charge begins to flow through the inductor. Let Q be the charge on the upper plate of the capacitor and let the positive direction around the circuit be counterclockwise, as shown. Then,

$$I = +\frac{dQ}{dt}$$

Applying Kirchhoff's loop rule to the circuit, we have

$$L\frac{dI}{dt} + \frac{Q}{C} = 0 \qquad\qquad 29\text{-}33$$

Substituting dQ/dt for I gives

$$L\frac{d^2Q}{dt^2} + \frac{Q}{C} = 0 \qquad\qquad 29\text{-}34$$

This equation is of the same form as Equation 14-2 for the acceleration of a mass on a spring:

$$m\frac{d^2x}{dt^2} + kx = 0$$

The behavior of an *LC* circuit is thus analogous to that of a mass on a spring, with L analogous to the mass m, Q analogous to the position x, and $1/C$ analogous to the spring constant k. Also, the current I is analogous to the velocity v, since $v = dx/dt$ and $I = dQ/dt$. In mechanics, the mass of an object describes the inertia of the object. The greater the mass, the more difficult it is to change the velocity of the object. Similarly, the inductance L can be thought of as the inertia of an ac circuit. The greater the inductance, the more opposition there is to changes in the current I.

If we divide each term in Equation 29-34 by L and rearrange, we obtain

$$\frac{d^2Q}{dt^2} = -\frac{1}{LC}Q \qquad\qquad 29\text{-}35$$

which is analogous to

$$\frac{d^2x}{dt^2} = -\frac{k}{m}x \qquad\qquad 29\text{-}36$$

In Chapter 14, we found that we could write the solution of Equation 29-36 for simple harmonic motion in the form

$$x = A\cos(\omega t - \delta)$$

where $\omega = \sqrt{k/m}$ is the angular frequency, A is the displacement amplitude, and δ is the phase constant, which depends on the initial conditions. The solution to Equation 29-35 is thus

$$Q = A\cos(\omega t - \delta)$$

with

$$\omega = \frac{1}{\sqrt{LC}} \qquad\qquad 29\text{-}37$$

The current I is found by differentiating:

$$I = \frac{dQ}{dt} = -\omega A \sin(\omega t - \delta)$$

If we choose our initial conditions to be $Q = Q_{peak}$ and $I = 0$ at $t = 0$, the phase constant δ is zero and $A = Q_{peak}$. Our solutions are then

$$Q = Q_{peak} \cos \omega t \qquad\qquad 29\text{-}38$$

and

$$I = -\omega Q_{peak} \sin \omega t = -I_{peak} \sin \omega t \qquad\qquad 29\text{-}39$$

where $I_{peak} = \omega Q_{peak}$.

Figure 29-13 shows graphs of Q and I versus time. The charge oscillates between the values $+Q_{peak}$ and $-Q_{peak}$ with angular frequency $\omega = 1/\sqrt{(LC)}$. The current oscillates between $+\omega Q_{peak}$ and $-\omega Q_{peak}$ with the same frequency. Also, the current leads the charge by 90° (see Problem 29-37). The current is maximum when the charge is zero and the current is zero when the charge is maximum.

In our study of the oscillations of a mass on a spring, we found that the total energy is constant, and that the total energy oscillates between potential energy and kinetic energy. We also have two kinds of energy in the LC circuit, electric energy and magnetic energy. The electric energy stored in the capacitor is

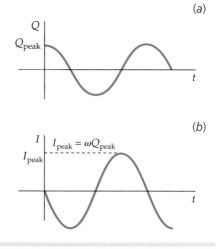

(a)

(b)

FIGURE 29-13 Graphs of (a) Q versus t and (b) I versus t for the LC circuit shown in Figure 29-12.

$$U_e = \frac{1}{2} Q V_C = \frac{1}{2} \frac{Q^2}{C}$$

Substituting $Q_{peak} \cos \omega t$ for Q, we have for the electric energy

$$U_e = \frac{1}{2} \frac{Q_{peak}^2}{C} \cos^2 \omega t \qquad\qquad 29\text{-}40$$

The electric energy oscillates between its maximum value $Q_0^2/(2C)$ and zero at an angular frequency of 2ω (see Problem 29-37). The magnetic energy stored in the inductor is

$$U_m = \frac{1}{2} L I^2 \qquad\qquad 29\text{-}41$$

Substituting $I = -\omega Q_{peak} \sin \omega t$ (Equation 29-39), we get

$$U_m = \frac{1}{2} L\omega^2 Q_{peak}^2 \sin^2 \omega t = \frac{1}{2} \frac{Q_{peak}^2}{C} \sin^2 \omega t \qquad\qquad 29\text{-}42$$

where we have used $\omega^2 = 1/LC$. The magnetic energy also oscillates between its maximum value of $Q_{peak}^2/2C$ and zero at an angular frequency of 2ω. The sum of the electrostatic energy and the magnetic energy is the total energy, which is constant in time:

$$U_{total} = U_e + U_m = \frac{1}{2} \frac{Q_{peak}^2}{C} \cos^2 \omega t + \frac{1}{2} \frac{Q_{peak}^2}{C} \sin^2 \omega t = \frac{1}{2} \frac{Q_{peak}^2}{C}$$

This sum equals the energy initially stored on the capacitor.

LC OSCILLATOR

EXAMPLE 29-4

A 2-μF capacitor is charged to 20 V and the capacitor is then connected across a 6-μH inductor. (*a*) What is the frequency of oscillation? (*b*) What is the peak value of the current?

PICTURE THE PROBLEM In (b), the current is maximum when dQ/dt is maximum, so the current amplitude is ωQ_{peak}. $Q = Q_{peak}$ when $V = V_{peak}$, where V is the voltage across the capacitor.

(*a*) The frequency of oscillation depends only on the values of the capacitance and the inductance:

$$f = \frac{\omega}{2\pi} = \frac{1}{2\pi\sqrt{LC}} = \frac{1}{2\pi\sqrt{(6 \times 10^{-6}\,\text{H})(2 \times 10^{-6}\,\text{F})}}$$

$$= \boxed{4.59 \times 10^4\,\text{Hz}}$$

(*b*) 1. The peak value of the current is related to the peak value of the charge:

$$I_{peak} = \omega Q_{peak} = \frac{Q_{peak}}{\sqrt{LC}}$$

2. The peak charge on the capacitor is related to the peak potential drop across the capacitor:

$$Q_{peak} = CV_{peak}$$

3. Substitute CV_{peak} for Q_{peak} and calculate I_{peak}:

$$I_{peak} = \frac{CV_{peak}}{\sqrt{LC}} = \frac{(2\,\mu\text{F})(20\,\text{V})}{\sqrt{(6\,\mu\text{H})(2\,\mu\text{F})}} = \boxed{11.5\,\text{A}}$$

EXERCISE A 5-μF capacitor is charged and is then discharged through an inductor. What should the value of the inductance be so that the current oscillates with frequency 8 kHz? (*Answer* 79.2 μH)

If we include a resistor in series with the capacitor and the inductor, as in Figure 29-14, we have an **RLC circuit.** Kirchhoff's loop rule gives

$$L\frac{dI}{dt} + IR + \frac{Q}{C} = 0 \qquad\qquad 29\text{-}43a$$

or

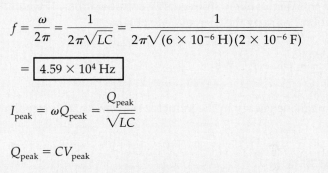

FIGURE 29-14 An *RLC* circuit.

$$L\frac{d^2Q}{dt^2} + R\frac{dQ}{dt} + \frac{1}{C}Q = 0 \qquad\qquad 29\text{-}43b$$

where we have used $I = dQ/dt$ as before. Equations 29-43a and 29-43b are analogous to the equation for a damped harmonic oscillator (see Equation 14-35):

$$m\frac{d^2x}{dt^2} + b\frac{dx}{dt} + kx = 0$$

The first term, $L\,dI/dt = L\,d^2Q/dt^2$, is analogous to the mass times the acceleration, $m\,dv/dt = m\,d^2x/dt^2$; the second term, $IR = R\,dQ/dt$, is analogous to the damping term, $bv = b\,dx/dt$; and the third term, Q/C, is analogous to the restoring force kx. In the oscillation of a mass on a spring, the damping constant b leads to a dissipation of mechanical energy. In an *RLC* circuit, the resistance R is analogous to the damping constant b and leads to a dissipation of electrical energy.

If the resistance is small, the charge and the current oscillate with (angular) frequency[†] that is very nearly equal to $\omega_0 = 1/\sqrt{LC}$, which is called the

---

† As in Chapter 14 when we discussed mechanical oscillations, we usually omit the word *angular* when the omission will not cause confusion.

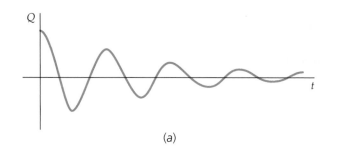

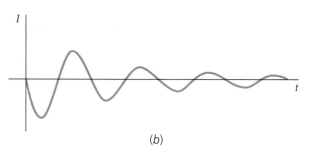

(a)

(b)

natural frequency of the circuit, but the oscillations are damped. We can understand this qualitatively from energy considerations. If we multiply each term in Equation 29-43a by the current I, we obtain

$$LI\frac{dI}{dt} + I^2R + I\frac{Q}{C} = 0 \qquad\qquad 29\text{-}44$$

The magnetic energy in the inductor is given by $\frac{1}{2}LI^2$ (see Equation 28-20). Note that

$$\frac{d(\frac{1}{2}LI^2)}{dt} = LI\frac{dI}{dt}$$

where $LI\,dI/dt$ is the first term in Equation 29-44. If $LI\,dI/dt$ is positive, it equals the rate at which electrical potential energy is transformed into magnetic energy. If $LI\,dI/dt$ is negative, it equals the rate at which magnetic energy is transformed back into electrical potential energy. Note that $LI\,dI/dt$ is positive or negative depending on whether I and dI/dt have the same sign or different signs. The second term in Equation 29-44 is I^2R, the rate at which electrical potential energy is dissipated in the resistor. I^2R is never negative. Note that

$$\frac{d(\frac{1}{2}Q^2/C)}{dt} = \frac{Q}{C}\frac{dQ}{dt} = I\frac{Q}{C}$$

where IQ/C is the third term in Equation 29-44. This is the rate of change of the electric potential energy of the capacitor, which may be positive or negative. The sum of the electric and magnetic energies is not constant for this circuit because energy is continually dissipated in the resistor. Figure 29-15 shows graphs of Q versus t and I versus t for a small resistance R in an RLC circuit. If we increase R, the oscillations become more heavily damped until a critical value of R is reached for which there is not even one oscillation. Figure 29-16 shows a graph of Q versus t in an RLC circuit when the value of R is greater than the critical damping value.

FIGURE 29-15 Graphs of (a) Q versus t and (b) I versus t for the RLC circuit shown in Figure 29-14 when the value of R is small enough so that the oscillations are underdamped.

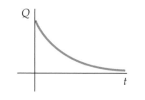

FIGURE 29-16 A graph of Q versus t for the RLC circuit shown in Figure 29-15 when the value of R is so large that the oscillations are overdamped.

*29-6 Driven *RLC* Circuits

Series *RLC* Circuit

Figure 29-17 shows a series RLC circuit being sinusoidally driven by an ac generator. If the potential drop applied by the generator to the series RLC combination is $V_{app} = V_{app,peak} \cos \omega t$, applying Kirchhoff's loop rule gives

$$V_{app,peak} \cos \omega t - L\frac{dI}{dt} - IR - \frac{Q}{C} = 0$$

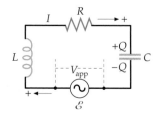

FIGURE 29-17 A series RLC circuit with an ac generator.

Using $I = dQ/dt$ and rearranging, we obtain

$$L \frac{d^2Q}{dt^2} + R \frac{dQ}{dt} + \frac{1}{C} Q = V_{\text{app,peak}} \cos \omega t \qquad \text{29-45}$$

This equation is analogous to Equation 14-51 for the forced oscillation of a mass on a spring:

$$m \frac{d^2x}{dt^2} + b \frac{dx}{dt} + m\omega_0^2 x = F_0 \cos \omega t$$

(In Equation 14-51, the force constant k was written in terms of the mass m and the natural angular frequency ω_0 using $k = m\omega_0^2$. The capacitance in Equation 29-45 could be similarly written in terms of L and the natural angular frequency using $1/C = L\omega_0^2$.)

We will discuss the solution of Equation 29-45 qualitatively as we did with Equation 14-51 for the forced oscillator. The current in the circuit consists of a transient current that depends on the initial conditions (e.g., the initial phase of the generator and the initial charge on the capacitor) and a steady-state current that does not depend on the initial conditions. We will ignore the transient current, which decreases exponentially with time and is eventually negligible, and concentrate on the steady-state current. The steady-state current obtained by solving Equation 29-45 is

$$I = I_{\text{peak}} \cos (\omega t - \delta) \qquad \text{29-46}$$

where the phase angle δ is given by

$$\tan \delta = \frac{X_L - X_C}{R} \qquad \text{29-47}$$

PHASE CONSTANT FOR A SERIES *RLC* CIRCUIT

The peak current is

$$I_{\text{peak}} = \frac{V_{\text{app,peak}}}{\sqrt{R^2 + (X_L - X_C)^2}} = \frac{V_{\text{app,peak}}}{Z} \qquad \text{29-48}$$

PEAK CURRENT IN A SERIES *RLC* CIRCUIT

where

$$Z = \sqrt{R^2 + (X_L - X_C)^2} \qquad \text{29-49}$$

IMPEDANCE OF A SERIES *RLC* CIRCUIT

The quantity $X_L - X_C$ is called the **total reactance,** and Z is called the **impedance.** Combining these results, we have

$$I = \frac{V_{\text{app,peak}}}{Z} \cos (\omega t - \delta) \qquad \text{29-50}$$

Equation 29-50 can also be obtained from a simple diagram using the phasor representations. Figure 29-18 shows the phasors representing the potential drops across the resistance, the inductance, and the capacitance. The x component of each of these vectors equals the instantaneous potential drop across the corresponding element. Since the sum of the x components equals the x component of the sum, the sum of the x components equals the sum of the potential drops across these elements, which by Kirchhoff's loop rule equals the instantaneous applied potential drop.

If we represent the potential drop applied across the series combination $V_{app} = V_{app,peak} \cos \omega t$ as a phasor $\vec{V}_{app}$ that has the magnitude $V_{app,peak}$, we have

$$\vec{V}_{app} = \vec{V}_R + \vec{V}_L + \vec{V}_C \qquad\qquad 29\text{-}51$$

In terms of the magnitudes,

$$V_{app,peak} = |\vec{V}_R + \vec{V}_L + \vec{V}_C| = \sqrt{V_{R,peak}^2 + (V_{L,peak} - V_{C,peak})^2}$$

But $V_R = I_{peak}R$, $V_L = I_{peak}X_L$, and $V_C = I_{peak}X_C$. Thus,

$$V_{app,peak} = I_{peak}\sqrt{R^2 + (X_L - X_C)^2} = I_{peak}Z$$

The phasor $\vec{V}_{app}$ makes an angle δ with $\vec{V}_R$, as shown in Figure 29-18. From the figure, we can see that

$$\tan \delta = \frac{|\vec{V}_L + \vec{V}_C|}{|\vec{V}_R|} = \frac{I_{peak}X_L - I_{peak}X_C}{I_{peak}R} = \frac{X_L - X_C}{R}$$

in agreement with Equation 29-47. Since $\vec{V}_{app}$ makes an angle ωt with the x axis, $\vec{V}_R$ makes an angle $\omega t - \delta$ with the x axis. This applied potential drop is in phase with the current, which is therefore given by

$$I = I_{peak}\cos(\omega t - \delta) = \frac{V_{app,peak}}{Z}\cos(\omega t - \delta)$$

This is Equation 29-50. The relation between the impedance Z, the resistance R, and the total reactance $X_L - X_C$ is best remembered by using the right triangle shown in Figure 29-19.

Resonance

When X_L and X_C are equal, the total reactance is zero, and the impedance Z has its smallest value R. Then I_{peak} has its greatest value and the phase angle δ is zero, which means that the current is in phase with the applied potential drop. Let ω_{res} be the value of ω for which X_L and X_C are equal. It is obtained from

$$X_L = X_C$$

$$\omega_{res}L = \frac{1}{\omega_{res}C}$$

or

$$\omega_{res} = \frac{1}{\sqrt{LC}}$$

which equals the natural frequency ω_0. When the frequency of the applied potential drop equals the natural frequency ω_0, the impedance is smallest, I_{peak} is

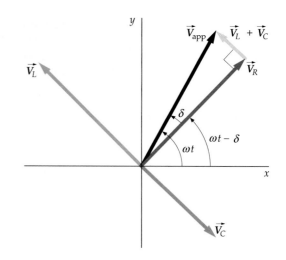

FIGURE 29-18 Phase relations among potential drops in a series *RLC* circuit. The potential drop across the resistor is in phase with the current. The potential drop across the inductor V_L leads the current by 90°. The potential drop across the capacitor lags the current by 90°. The sum of the vectors representing these potential drops gives a vector at an angle δ with the current representing the applied emf. For the case shown here, V_L is greater than V_C, and the current lags the applied potential drop by δ.

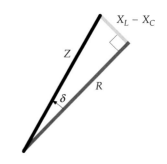

FIGURE 29-19 A right triangle relating capacitive and inductive reactance, resistance, impedance, and the phase angle in an *RLC* circuit.

greatest, and the circuit is said to be at **resonance.** The natural frequency ω_0 is therefore also called the **resonance frequency.** This resonance condition in a driven *RLC* circuit is similar to that in a driven simple harmonic oscillator.

Since neither an inductor nor a capacitor dissipates energy, the average power delivered to a series *RLC* circuit is the average power supplied to the resistor. The instantaneous power supplied to the resistor is

$$P = I^2 R = [I_{peak} \cos(\omega t - \delta)]^2 R$$

Averaging over one or more cycles and using $(\cos^2 \theta)_{av} = \frac{1}{2}$, we obtain for the average power

$$P_{av} = \frac{1}{2} I^2_{peak} R = I^2_{rms} R \qquad\qquad 29\text{-}52$$

Using $R/Z = \cos \delta$ from Figure 29-19 and $I_{peak} = V_{app,peak}/Z$, this can be written

$$P_{av} = \frac{1}{2} V_{app,peak} I_{peak} \cos \delta = V_{app,rms} I_{rms} \cos \delta \qquad\qquad 29\text{-}53$$

The quantity $\cos \delta$ is called the **power factor** of the *RLC* circuit. At resonance, δ is zero, and the power factor is 1.

The power can also be expressed as a function of the angular frequency ω. Using $I_{rms} = V_{app,rms}/Z$ Equation 29-52 becomes

$$P_{av} = I^2_{rms} R = V^2_{app,rms} \frac{R}{Z^2}$$

From the definition of impedance Z, we have

$$Z^2 = (X_L - X_C)^2 + R^2 = \left(\omega L - \frac{1}{\omega C}\right)^2 + R^2$$

$$= \frac{L^2}{\omega^2}\left(\omega^2 - \frac{1}{LC}\right)^2 + R^2$$

$$= \frac{L^2}{\omega^2}(\omega^2 - \omega_0^2)^2 + R^2$$

where we have used $\omega_0 = 1/\sqrt{LC}$. Using this expression for Z^2, we obtain the average power as a function of ω:

$$P_{av} = \frac{V^2_{app,rms} R \omega^2}{L^2(\omega^2 - \omega_0^2)^2 + \omega^2 R^2} \qquad\qquad 29\text{-}54$$

Figure 29-20 shows the average power supplied by the generator to the series combination as a function of generator frequency for two different values of the resistance R. These curves, called **resonance curves,** are the same as the power-versus-frequency curves for a driven damped oscillator (see Section 14-5). The average power is greatest when the generator frequency equals the resonance frequency. When the resistance is small, the resonance curve is narrow; when the resistance is large, the resonance curve is broad. A resonance curve can be characterized by the **resonance width** $\Delta\omega$. As shown in Figure 29-20, the resonance width is the frequency difference between the two points on the curve where the power is half its maximum value. When the width is small compared with the resonance frequency, the resonance is sharp; that is, the resonance curve is narrow.

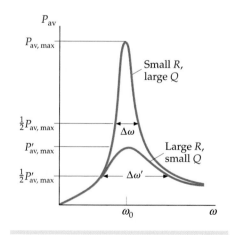

FIGURE 29-20 Plot of average power versus frequency for a series *RLC* circuit. The power is maximum when the frequency of the generator ω equals the natural frequency of the circuit $\omega_0 = 1/\sqrt{LC}$. If the resistance is small, the *Q* factor is large and the resonance is sharp. The resonance width $\Delta\omega$ of the curves is measured between points where the power is half its maximum value.

In Chapter 14, the Q factor for a mechanical oscillator is defined as $Q = \omega_0 m/b$ where m is the mass and b is the damping constant. We then saw that for an underdamped oscillator $Q = 2\pi E/|\Delta E|$, where E is the total energy of the system at the beginning of a cycle and ΔE is the energy dissipated during the cycle. The **Q factor** for an RLC circuit can be defined in a similar way. Since L is analogous to the mass m and R is analogous to the damping constant b, the Q factor for an RLC circuit is given by

$$Q = 2\pi \frac{E}{|\Delta E|} = \frac{\omega_0 L}{R} \qquad\qquad 29\text{-}55$$

When the resonance curve is reasonably narrow (that is, when Q is greater than about 2 or 3), the Q factor can be approximated by

$$Q = \frac{\omega_0}{\Delta\omega} = \frac{f_0}{\Delta f} \qquad\qquad 29\text{-}56$$

Q FACTOR FOR AN RLC CIRCUIT

Resonance circuits are used in radio receivers, where the resonance frequency of the circuit is varied either by varying the capacitance or the inductance. Resonance occurs when the natural frequency of the circuit equals one of the frequencies of the radio waves picked up at the antenna. At resonance, there is a relatively large current in the antenna circuit. If the Q factor of the circuit is sufficiently high, currents due to other station frequencies off resonance will be negligible compared with those currents due to the station frequency to which the circuit is tuned.

DRIVEN SERIES RLC CIRCUIT   **EXAMPLE 29-5**

A series RLC combination with $L = 2$ H, $C = 2$ μF, and $R = 20$ Ω is driven by an ideal generator with a peak emf of 100 V and a frequency that can be varied. Find (*a*) the resonance frequency f_0, (*b*) the Q value, (*c*) the width of the resonance Δf, and (*d*) the current amplitude at resonance.

PICTURE THE PROBLEM The resonance frequency is found from $\omega_0 = 1/\sqrt{LC}$ and the Q value is found from $Q = \omega_0 L/R$.

1. The resonance frequency is $f_0 = \omega_0/2\pi$:

$$f_0 = \frac{\omega_0}{2\pi} = \frac{1}{2\pi\sqrt{LC}}$$

$$= \frac{1}{2\pi\sqrt{(2\text{ H})(2 \times 10^{-6}\text{ F})}} = \boxed{79.6\text{ Hz}}$$

2. Use this result to calculate Q:

$$Q = \frac{\omega_0 L}{R} = \frac{2\pi(79.6\text{ Hz})(2\text{ H})}{20\ \Omega} = \boxed{50}$$

3. Use the value of Q to find the width of the resonance Δf:

$$\Delta f = \frac{f_0}{Q} = \frac{79.6\text{ Hz}}{50} = \boxed{1.59\text{ Hz}}$$

4. At resonance, the impedance is R and I_{peak} is $V_{app,peak}/R$:

$$I_{max} = \frac{V_{app,peak}}{R} = \frac{\mathcal{E}_{peak}}{R} = \frac{100\text{ V}}{20\ \Omega} = \boxed{5\text{ A}}$$

REMARKS The width of 1.59 Hz is less than 2 percent of the resonance frequency of 79.6 Hz, so the resonance peak is quite sharp.

DRIVEN SERIES RLC CIRCUIT CURRENT, PHASE, AND POWER   **EXAMPLE 29-6**   Try It Yourself

If the generator in Example 29-5 has a frequency of 60 Hz, find (*a*) the current amplitude, (*b*) the phase constant δ, (*c*) the power factor, and (*d*) the average power delivered.

PICTURE THE PROBLEM The current amplitude is the amplitude of the applied potential drop divided by the total impedance of the series combination. The phase angle δ is found from $\tan \delta = (X_L - X_C)/R$. You can use either Equation 29-52 or Equation 29-53 to find the average power delivered.

Cover the column to the right and try these on your own before looking at the answers.

Steps

Answers

(*a*) 1. Write the peak current in terms of $V_{app,peak}$ and the impedance.

$$I_{peak} = \frac{V_{app,peak}}{Z} = \frac{\mathcal{E}_{peak}}{Z}$$

2. Calculate the capacitive and inductive reactances and the total reactance.

$X_C = 1326\ \Omega$, $X_L = 754\ \Omega$

so

$X_L - X_C = -572\ \Omega$

3. Calculate the total impedance Z.

$Z = 573\ \Omega$

4. Use the results of steps 2 and 3 to calculate I_{peak}.

$I_{peak} = \boxed{0.175\ \text{A}}$

(*b*) Use the results of Part (*a*) steps 2 and 3 to calculate δ.

$\delta = \tan^{-1} \dfrac{X_L - X_C}{R} = \boxed{-88.0°}$

(*c*) Use your value of δ to compute the power factor.

$\cos \delta = 0.0349$

(*d*) Calculate the average power delivered from Equation 29-52.

$P_{av} = \frac{1}{2} I_{peak}^2 R = \boxed{0.305\ \text{W}}$

PLAUSIBILITY CHECK To check our result for the average power using the power factor found in Part (*c*), we have $P_{av} = \frac{1}{2} V_{app,peak} I_{peak} \cos \delta = \frac{1}{2} \mathcal{E}_{peak} I_{peak} \cos \delta = 0.305$ W. This is in agreement with our result for Part (*d*).

REMARKS The frequency of 60 Hz is well below the resonance frequency of 79.6 Hz. (Recall that the width as calculated in Example 29-5 is only 1.59 Hz.) As a result, the total reactance is much greater in magnitude than the resistance. This is always the case far from resonance. Similarly, an I_{peak} of 0.175 A is much less than an I_{peak} at resonance, which was found to be 5 A. Finally, we see from Figure 29-18 that a negative phase angle δ means that the current leads the applied potential drop.

DRIVEN SERIES RLC CIRCUIT AT RESONANCE   **EXAMPLE 29-7**   Try It Yourself

Find the peak potential drop across the resistor, the inductor, and the capacitor at resonance for the circuit in Example 29-5.

PICTURE THE PROBLEM The peak potential drop across the resistor is I_{peak} times R. Similarly, the peak potential drop across the inductor or capacitor is I_{peak} times the appropriate reactance. We found that at resonance $I_{peak} = 5$ A and $f_0 = 79.6$ Hz in Example 29-5.

Cover the column to the right and try these on your own before looking at the answers.

Steps   **Answers**

1. Calculate $V_{R,peak} = I_{peak}R$.

$$V_{R,peak} = I_{peak}R = \boxed{100 \text{ V}}$$

2. Express $V_{L,peak}$ in terms of I_{peak} and X_L.

$$V_{L,peak} = I_{peak}X_L = I_{peak}\omega_0 L = \boxed{5000 \text{ V}}$$

3. Express $V_{C,peak}$ in terms of I_{peak} and X_C.

$$V_{C,peak} = I_{peak}X_C = \frac{I_{peak}}{\omega_0 C} = \boxed{5000 \text{ V}}$$

REMARKS The inductive and capacitive reactances are equal, as we would expect, since we found the resonance frequency by setting them equal. The phasor diagram for the potential drops across the resistor, capacitor, and inductor is shown in Figure 29-21. The peak potential drop across the resistor is a relatively safe 100 V, equal to the peak emf of the generator. However, the peak potential drops across the inductor and the capacitor are a dangerously high 5000 V. These potential drops are 180° out of phase. At resonance, the potential drop across the inductor at any instant is the negative of that across the capacitor, so they always sum to zero, leaving the potential drop across the resistor equal to the emf in the circuit.

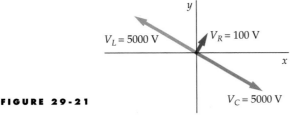

$V_L = 5000 \text{ V}$ $V_R = 100 \text{ V}$

$V_C = 5000 \text{ V}$

FIGURE 29-21

RC LOW-PASS FILTER   **EXAMPLE 29-8**

A resistor R and capacitor C are in series with a generator, as shown in Figure 29-22. The generator applies a potential drop across the RC combination given by $V_{app} = \sqrt{2}V_{app,rms} \cos \omega t$. Find the rms potential drop across the capacitor $V_{out,rms}$ as a function of frequency ω.

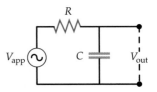

FIGURE 29-22 The peak output voltage decreases as frequency increases.

PICTURE THE PROBLEM The rms potential drop across the capacitor is the product of the rms current and the capacitive reactance. The rms current is found from the potential drop applied by the generator and the impedance of the series RC combination.

1. The potential drop across the capacitor is I_{rms} times X_C:

$$V_{out,rms} = I_{rms}X_C$$

2. The rms current depends on the applied rms potential drop and the impedance:

$$I_{rms} = \frac{V_{app,rms}}{Z}$$

3. In this circuit, only R and X_C contribute to the total impedance:

$$Z = \sqrt{R^2 + X_C^2}$$

4. Substitute these values and $X_C = 1/(\omega C)$ to find the output rms potential drop:

$$V_{out,rms} = I_{rms}X_C = \frac{V_{app,rms}}{Z}X_C = \frac{V_{app,rms}X_C}{\sqrt{R^2 + X_C^2}}$$

$$= \frac{V_{app,rms}\left(\dfrac{1}{\omega C}\right)}{\sqrt{R^2 + \left(\dfrac{1}{\omega C}\right)^2}} = \boxed{\frac{V_{app,rms}}{\sqrt{1 + \omega^2(RC)^2}}}$$

REMARKS This circuit is called an *RC low-pass filter,* since it transmits low frequencies with greater amplitude than high frequencies. In fact, the output potential drop equals the potential drop applied by the generator in the limit that $\omega \rightarrow 0$, but approaches zero for $\omega \rightarrow \infty$, as shown in the graph of the ratio of output potential drop to applied potential drop in Figure 29-23.

EXERCISE Find the output potential drop for this circuit if the capacitor is replaced by an inductor *L.* (*Answer* $V_{\text{out,rms}} = V_{\text{in,rms}}/\sqrt{1 + (R/L)^2/\omega^2}$. This circuit is a *high-pass filter.*)

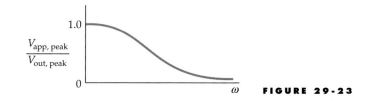

FIGURE 29-23

AN FM TUNER   **EXAMPLE 29-9**   **Put It in Context**

You have been tinkering with building a radio tuner using your new knowledge of physics. You know that the FM dial gives its frequencies in megahertz, and you would like to determine what percentage of change in an inductor would allow you to tune for the whole FM range. You decide to start at midrange and determine a percent increase and decrease needed for inductance. A variable inductor is usually an iron-core solenoid, and the inductance is increased by further inserting the core. The FM dial goes from 88 MHz to 108 MHz.

PICTURE THE PROBLEM We can relate inductance to the resonant frequency with $\omega = 2\pi f$ and $\omega = 1/\sqrt{LC}$. Then, if we find the percent change in frequency, we can determine the percent change in inductance. The capacitance C does not vary.

1. The resonant angular frequency ω is related to the inductance L:

$$\omega = 1/\sqrt{LC}$$

and

$$\omega = 2\pi f$$

so

$$f = \frac{1}{2\pi\sqrt{LC}}$$

2. L is inversely proportional to f^2:

$$L = af^{-2}$$

where

$$a = (4\pi^2 C)^{-1}$$

3. Express the fractional change in L in terms of the frequencies: When L is maximum, f is minimum and vice versa. The middle frequency f_{mid} is halfway between the maximum and minimum frequency, and L_{mid} is the inductance when $f = f_{\text{mid}}$:

$$\frac{\Delta L}{L} = \frac{L_{\text{max}} - L_{\text{min}}}{L_{\text{mid}}} = \frac{af_{\text{max}}^{-2} - af_{\text{min}}^{-2}}{af_{\text{mid}}^{-2}}$$

$$= f_{\text{mid}}^2 \left(\frac{1}{f_{\text{max}}^2} - \frac{1}{f_{\text{min}}^2} \right) = 98^2 \left(\frac{1}{108^2} - \frac{1}{88^2} \right)$$

$$= -0.417$$

4. The negative sign is not relevant, except as an indication that when the inductance increases the resonant frequency dcreases. Express the step 3 result as a percentage:

The inductance varies by about $\boxed{42 \text{ percent}}$

Parallel *RLC* Circuit

Figure 29-24 shows a resistor R, a capacitor C, and an inductor L connected in parallel across an ac generator. The total current I from the generator divides into three currents. The current I_R in the resistor, the current I_C in the capacitor, and the current I_L in the inductor. The instantaneous potential drop V_{app} is the same across each element. The current in the resistor is in phase with the potential drop and the phasor $\vec{I}_R$ has magnitude V_{peak}/R. Since the potential drop across an inductor *leads* the current in the inductor by 90°, I_L *lags* the potential drop by 90°, and the phasor $\vec{I}_L$ has magnitude V_{peak}/X_L. Similarly, the I_C leads the potential drop by 90° and the phasor $\vec{I}_C$ has magnitude V_{peak}/X_C. These currents are represented by phasors in Figure 29-25. The total current I is the x component of the vector sum of the individual currents as shown in the figure. The magnitude of the total current is

$$I = \sqrt{I_R^2 + (I_L - I_C)^2} = \sqrt{\left(\frac{V_{peak}}{R}\right)^2 + \left(\frac{V_{peak}}{X_L} - \frac{V_{peak}}{X_C}\right)^2} = \frac{V_{peak}}{Z} \qquad 29\text{-}57$$

where the total impedance Z is related to the resistance and the capacitive and inductive reactances by

$$\frac{1}{Z} = \sqrt{\left(\frac{1}{R}\right)^2 + \left(\frac{1}{X_L} - \frac{1}{X_C}\right)^2} \qquad 29\text{-}58$$

At resonance, the currents in the inductor and capacitor are 180° out of phase, so the total current is a minimum and is just the current in the resistor. We see from Equation 29-57 that this occurs if Z is maximum, so $1/Z$ is minimum. Then, we see from Equation 29-58 that if $X_L = X_C$, $1/Z$ has its minimum value $1/R$. Equating X_L with X_C and solving for ω obtains the resonant frequency, which equals the natural frequency $\omega_0 = 1/\sqrt{LC}$.

*29-7 The Transformer

A transformer is a device used to raise or lower the voltage in a circuit without an appreciable loss of power. Figure 29-26 shows a simple transformer consisting of two wire coils around a common iron core. The coil carrying the input power is

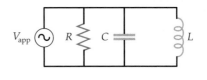

FIGURE 29-24 A parallel *RLC* circuit.

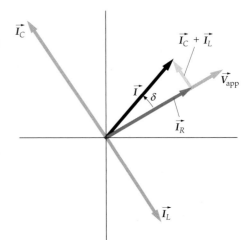

FIGURE 29-25 A phasor diagram for the currents in the parallel *RLC* circuit shown in Figure 29-24. The potential drop is the same across each element. The current in the resistor is in phase with the potential drop. The current in the capacitor leads the potential drop by 90° and the current in the inductor lags the potential drop by 90°. The phase difference δ between the total current and the potential drop depends on the relative magnitudes of the currents, which depend on the values of the resistance and of the capacitive and inductive reactances.

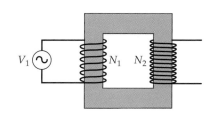

called the primary, and the other coil is called the secondary. Either coil of a transformer can be used for the primary or secondary. The transformer operates on the principle that an alternating current in one circuit induces an alternating emf in a nearby circuit due to the mutual inductance of the two circuits. The iron core increases the magnetic field for a given current and guides it so that nearly all the magnetic flux through one coil goes through the other coil. If no power were lost, the product of the potential difference across and the current in the secondary windings would equal the product of the potential drop across and the current in the primary windings. Thus, if the potential difference across the secondary coil is higher than the potential drop across the primary circuit, the current in the secondary coil is lower than the current in the primary coil, and vice versa. Power losses arise because of the Joule heating in the small resistances in both coils, or in current loops within the core,[†] and from hysteresis in the iron cores. We will neglect these losses and consider an ideal transformer of 100 percent efficiency, for which all of the power supplied to the primary coil appears in the secondary coil. Actual transformers are often 90 percent to 95 percent efficient.

FIGURE 29-26 A transformer with N_1 turns in the primary and N_2 turns in the secondary.

Consider a transformer with a potential drop V_1 across the primary coil of N_1 turns; the secondary coil of N_2 turns is an open circuit. Because of the iron core, there is a large flux through each coil even when the magnetizing current I_m in the primary circuit is very small. We can ignore the resistances of the coils, which are negligible in comparison with their inductive reactances. The primary circuit is then a simple circuit consisting of an ac generator and a pure inductance, like that discussed in Section 29-3. The current magnetizing in the primary coil and the voltage drop across the primary coil are out of phase by 90°, and the average power dissipated in the primary coil is zero. If ϕ_{turn} is the magnetic flux through a single turn of the primary coil, the potential drop across the primary coil is equal to the back emf, so

$$V_1 = N_1 \frac{d\phi_{turn}}{dt} \qquad 29\text{-}59$$

If there is no flux leakage out of the iron core, the flux through each turn is the same for both coils. Thus, the total flux through the secondary coil is $N_2 \phi_{turn}$, and the potential difference across the secondary coil is

$$V_2 = N_2 \frac{d\phi_{turn}}{dt} \qquad 29\text{-}60$$

Comparing Equations 29-59 and 29-60, we can see that

$$V_2 = \frac{N_2}{N_1} V_1 \qquad 29\text{-}61$$

(a)

If N_2 is greater than N_1, the potential difference across the secondary coil is greater than the potential drop across the primary coil, and the transformer is called a step-up transformer. If N_2 is less than N_1, the potential difference across the secondary coil is less than the potential drop across the primary coil, and the transformer is called a step-down transformer.

When we put a resistance R, called a load resistance, across the secondary coil, there will then be a current I_2 in the secondary circuit that is in phase with the potential drop V_2 across the resistance. This current sets up an additional

(b)

(a) A power box with a transformer for stepping down voltage for distribution to homes. (b) A suburban power substation where transformers step down voltage from high-voltage transmission lines.

---

† The induced currents, called eddy currents, can be greatly reduced by using a core of laminated metal to break up current paths.

flux ϕ'_{turn} through each turn that is proportional to $N_2 I_2$. This flux opposes the original flux set up by the original magnetizing current I_m in the primary. However, the potential drop across the primary coil is determined by the generator emf, which is unaffected by the secondary circuit. According to Equation 29-60, the flux in the iron core must change at the original rate; that is, the total flux in the iron core must be the same as when there is no load across the secondary. The primary coil thus draws an additional current I_1 to maintain the original flux ϕ_{turn}. The flux through each turn produced by this additional current is proportional to $N_1 I_1$. Since this flux equals $-\phi'_{turn}$, the additional current I_1 in the primary is related to the current I_2 in the secondary by

$$N_1 I_1 = -N_2 I_2 \qquad\qquad 29\text{-}62$$

These currents are 180° out of phase and produce counteracting fluxes. Since I_2 is in phase with V_2, the additional current I_1 is in phase with the potential drop across the primary circuit. The power input from the generator is $V_{1,rms} I_{1,rms}$, and the power output is $V_{2,rms} I_{2,rms}$. (The magnetizing current does not contribute to the power input because it is 90° out of phase with the generator voltage.) If there are no losses,

$$V_{1,rms} I_{1,rms} = V_{2,rms} I_{2,rms} \qquad\qquad 29\text{-}63$$

In most cases, the additional current in the primary I_1 is much greater than the original magnetizing current I_m that is drawn from the generator when there is no load. This can be demonstrated by putting a lightbulb in series with the primary coil. The lightbulb is much brighter when there is a load across the secondary circuit than when the secondary circuit is open. If I_m can be neglected, Equation 29-63 relates the total currents in the primary and secondary circuits.

DOORBELL TRANSFORMER | **EXAMPLE 29-10**

A doorbell requires 0.4 A at 6 V. It is connected to a transformer whose primary, containing 2000 turns, is connected to a 120-V ac line. (a) How many turns should there be in the secondary? (b) What is the current in the primary?

PICTURE THE PROBLEM We can find the number of turns from the turns ratio, which equals the voltage ratio. The primary current can be found by equating the power out to the power in.

1. The turns ratio can be obtained from Equation 29-61. Solve for the number of turns in the secondary, N_2:

$$\frac{N_2}{N_1} = \frac{V_2}{V_1}$$

so

$$N_2 = \frac{V_{2,rms}}{V_{1,rms}} N_1 = \frac{6\text{ V}}{120\text{ V}} 2000 \text{ turns} = \boxed{100 \text{ turns}}$$

2. Since we are assuming 100 percent efficiency in power transmission, the input and output currents are related by Equation 29-62. Solve for the current in the primary, I_1:

$$V_2 I_2 = V_1 I_1$$

so

$$I_1 = \frac{V_2}{V_1} I_2 = \frac{6\text{ V}}{120\text{ V}} (0.4\text{ A}) = \boxed{0.02 \text{ A}}$$

An important use of transformers is in the transport of electrical power. To minimize the $I^2 R$ heat loss (Joule heating) in transmission lines, it is economical to use a high voltage and a low current. On the other hand, safety and other considerations require that power be delivered to consumers at lower voltages and therefore with higher currents. Suppose, for example, that each person in a city with a population of 50,000 uses 1.2 kW of electric power. (The per capita consumption of power in the United States is actually somewhat higher than this.) At 120 V, the current required for each person would be

$$I = \frac{1200\ \text{W}}{120\ \text{V}} = 10\ \text{A}$$

The total current for 50,000 people would then be 500,000 A. The transport of such a current from a power-plant generator to a city many kilometers away would require conductors of enormous thickness, and the I^2R power loss would be substantial. Rather than transmit the power at 120 V, step-up transformers are used at the power plant to step up the voltage to some very large value, such as 600,000 V. For this voltage, the current needed is only

$$I = \frac{120\ \text{V}}{600,000\ \text{V}}\ (500,000\ \text{A}) = 100\ \text{A}$$

To reduce the voltage to a safer level for transport within a city, power substations are located just outside the city to step down the voltage to a safer value, such as 10,000 V. Transformers in boxes attached to the power poles outside each house again step down the voltage to 120 V (or 240 V) for distribution to the house. Because of the ease of stepping the voltage up or down with transformers, alternating current rather than direct current is in common use.

TRANSMISSION LOSSES   **E X A M P L E   2 9 - 1 1**

A transmission line has a resistance of 0.02 Ω/km. Calculate the I^2R power loss if 200 kW of power is transmitted from a power generator to a city 10 km away at (*a*) 240 V and (*b*) 4.4 kV.

PICTURE THE PROBLEM First, note that the total resistance of 10 km of wire is $R = (0.02\ \Omega/\text{km})(10\ \text{km}) = 0.2\ \Omega$. In each case, begin by finding the current needed to transmit 200 kW using $P = IV$, then find the power loss using I^2R.

(*a*) 1. Find the current needed to transmit 200 kW of power at 240 V:

$$I = \frac{P}{V} = \frac{200\ \text{kW}}{240\ \text{V}} = 833\ \text{A}$$

2. Calculate the power loss:

$$I^2R = (833\ \text{A})^2(0.2\ \Omega) = \boxed{139,000\ \text{W}}$$

(*b*) 1. Now, find the current needed to transmit 200 kW of power at 4.4 kV:

$$I = \frac{P}{V} = \frac{200\ \text{kW}}{4.4\ \text{kV}} = 45.5\ \text{A}$$

2. Calculate the power loss:

$$I^2R = (45.5\ \text{A})^2(0.2\ \Omega) = \boxed{414\ \text{W}}$$

REMARKS Note that with a transmission voltage of 240 V almost 70 percent of the power is wasted through heat loss, and there is an IR (voltage) drop across the transmission line of 167 V, so the power is delivered at only 73 V. However, with transmission at 4.4 kV only about 0.2 percent of the power is lost in transmission, and there is an IR drop across the transmission line of only 9 V, so the power is delivered with only a 0.2 percent voltage drop. This illustrates the advantages of high-voltage power transmission.

SUMMARY

1. Reactance is a frequency-dependent property of capacitors and inductors that is analogous to the resistance of a resistor.
2. Impedance is a frequency-dependent property of an ac circuit or circuit loop that is analogous to the resistance in a dc circuit.
3. Phasors are two-dimensional vectors that allow us to picture the phase relations in a circuit.
4. Resonance occurs when the frequency of the generator equals the natural frequency of the oscillating circuit.

Topic	Relevant Equations and Remarks	
1. Alternating Current Generators	An ac generator is a device that transforms mechanical energy into electrical energy. This transformation can be accomplished by using the mechanical energy to either rotate a conducting coil in a magnetic field or rotating a magnet in a conducting coil.	
EMF generated	$\mathcal{E} = \mathcal{E}_{\text{peak}} \sin(\omega t + \delta) = NBA\omega \sin(\omega t + \delta)$	**29-3, 29-4**
2. Current		
RMS current	$I_{\text{rms}} = \sqrt{(I^2)_{\text{av}}}$	**29-11**
RMS current and peak current	$I_{\text{rms}} = \dfrac{1}{\sqrt{2}} I_{\text{peak}}$	**29-12**
For a resistor	$I_{\text{rms}} = \dfrac{V_{R,\text{rms}}}{R}$	**29-15**
	potential drop and current in phase	
For an inductor	$I_{\text{rms}} = \dfrac{V_{L,\text{rms}}}{\omega L} = \dfrac{V_{L,\text{rms}}}{X_L}$	**29-25**
	potential drop leads current by 90°	
For a capacitor	$I_{\text{rms}} = \dfrac{V_{C,\text{rms}}}{1/\omega L} = \dfrac{V_{C,\text{rms}}}{X_C}$	**29-29**
	potential drop lags current by 90°	
3. Reactance		
Inductive reactance	$X_L = \omega L$	**29-24**
Capacitive reactance	$X_C = \dfrac{1}{\omega C}$	**29-30**
4. Average Power Dissipation		
By a resistor	$P_{\text{av}} = V_{R,\text{rms}} I_{\text{rms}} = I_{\text{rms}}^2 R$	**29-13, 29-15**
By an inductor or by a capacitor	$P_{\text{av}} = 0$	

5. ***Phasors**	Phasors are two-dimensional vectors that represent the current $\vec{I}$, the potential drop across a resistor $\vec{V}_R$, the potential drop across a capacitor $\vec{V}_C$, and the potential drop across an inductor $\vec{V}_L$ in an ac circuit. These phasors rotate in the counterclockwise direction with an angular velocity that is equal to the angular frequency ω of the current. $\vec{V}_R$ is in phase with the current, $\vec{V}_L$ leads the current by 90°, and $\vec{V}_C$ lags the current by 90°. The x component of each phasor equals the magnitude of the current or the corresponding potential drop at any instant.

6. *LC and RLC Series Circuits

If a capacitor is discharged through an inductor, the charge and the voltage on the capacitor oscillate with angular frequency

$$\omega = \frac{1}{\sqrt{LC}} \qquad \text{29-37}$$

The current in the inductor oscillates with the same frequency, but it is out of phase with the charge by 90°. The energy oscillates between electric energy in the capacitor and magnetic energy in the inductor. If the circuit also has resistance, the oscillations are damped because energy is dissipated in the resistor.

7. Series RLC Circuit Driven by an Applied Potential Drop of Frequency ω

Applied potential drop

$$V_{app} = V_{app,peak} \cos \omega t$$

Current

$$I = \frac{V_{app,peak}}{Z} \cos(\omega t - \delta) \qquad \text{29-50}$$

Impedance Z

$$Z = \sqrt{R^2 + (X_L - X_C)^2} \qquad \text{29-49}$$

Phase angle δ

$$\tan \delta = \frac{X_L - X_C}{R} \qquad \text{29-47}$$

Average power

$$P_{av} = I_{rms}^2 R = V_{app,rms} I_{rms} \cos \delta = \frac{V_{app,rms}^2 R \omega^2}{L^2(\omega^2 - \omega_0^2)^2 + \omega^2 R^2} \qquad \text{29-52, 29-53, 29-54}$$

Power factor

The quantity $\cos \delta$ in Equation 29-53 is called the power factor of the RLC circuit. At resonance, δ is zero, the power factor is 1, and

$$P_{av} = V_{app,rms} I_{rms}$$

Resonance

When the rms current is maximum, the circuit is said to be at resonance. The conditions for resonance are

$$X_L = X_C, \quad \text{so} \quad Z = \sqrt{R^2 + (X_L - X_C)^2} = R$$

$$\omega = \omega_0 = \frac{1}{\sqrt{LC}} \quad \text{and} \quad \delta = 0$$

8. Q Factor

The sharpness of the resonance curve is described by the Q factor

$$Q = \frac{\omega_0 L}{R} \qquad \text{29-55}$$

When the resonance curve is reasonably narrow, the Q factor can be approximated by

$$Q = \frac{\omega_0}{\Delta \omega} = \frac{f_0}{\Delta f} \qquad \text{29-56}$$

9. Transformers

A transformer is a device used to raise or lower the voltage in a circuit without an appreciable loss in power. For a transformer with N_1 turns in the primary and N_2 turns in the secondary, the potential difference across the secondary coil is related to the potential drop across the primary coil by

$$V_2 = \frac{N_2}{N_1}V_1 \qquad\qquad 29\text{-}61$$

If there are no power losses,

$$V_{1,\mathrm{rms}}I_{1,\mathrm{rms}} = V_{2,\mathrm{rms}}I_{2,\mathrm{rms}} \qquad\qquad 29\text{-}63$$

PROBLEMS

- Single-concept, single-step, relatively easy
- Intermediate-level, may require synthesis of concepts
- Challenging
- **SSM** Solution is in the *Student Solutions Manual*
- **iSOLVE** Problems available on iSOLVE online homework service
- **iSOLVE✓** These "Checkpoint" online homework service problems ask students additional questions about their confidence level, and how they arrived at their answer.

In a few problems, you are given more data than you actually need; in a few other problems, you are required to supply data from your general knowledge, outside sources, or informed estimates.

Conceptual Problems

1 • **SSM** As the frequency in the simple ac circuit in Figure 29-27 increases, the rms current through the resistor (a) increases. (b) does not change. (c) may increase or decrease depending on the magnitude of the original frequency. (d) may increase or decrease depending on the magnitude of the resistance. (e) decreases.

FIGURE 29-27 Problem 1

2 • If the rms voltage in an ac circuit is doubled, the peak voltage is (a) increased by a factor of 2. (b) decreased by a factor of 2. (c) increased by a factor of $\sqrt{2}$. (d) decreased by a factor of $\sqrt{2}$. (e) not changed.

3 • If the frequency in the circuit shown in Figure 29-28 is doubled, the inductance of the inductor will (a) increase by a factor of 2. (b) not change. (c) decrease by a factor of 2. (d) increase by a factor of 4. (e) decrease by a factor of 4.

FIGURE 29-28 Problems 3 and 4

4 • If the frequency in the circuit shown in Figure 29-28 is doubled, the inductive reactance of the inductor will (a) increase by a factor of 2. (b) not change. (c) decrease by a factor of 2. (d) increase by a factor of 4. (e) decrease by a factor of 4.

5 • **SSM** If the frequency in the circuit in Figure 29-29 is doubled, the capacitive reactance of the circuit will (a) increase by a factor of 2. (b) not change. (c) decrease by a factor of 2. (d) increase by a factor of 4. (e) decrease by a factor of 4.

FIGURE 29-29 Problem 5

6 • In a circuit consisting of a generator and an inductor, are there any times when the inductor absorbs power from the generator? Are there any times when the inductor supplies power to the generator?

7 • In a circuit consisting of a generator and a capacitor, are there any times when the capacitor absorbs power from the generator? Are there any times when the capacitor supplies power to the generator?

8 • The SI units of inductance times capacitance are (a) seconds squared. (b) hertz. (c) volts. (d) amperes. (e) ohms.

9 •• **SSM** Making LC circuits with oscillation frequencies of thousands of hertz or more is easy, but making LC circuits that have small frequencies is difficult. Why?

10 • True or false:

(a) An *RLC* circuit with a high *Q* factor has a narrow resonance curve.

(b) At resonance, the impedance of an *RLC* circuit equals the resistance *R*.

(c) At resonance, the current and generator voltage are in phase.

11 • Does the power factor depend on the frequency?

12 • **SSM** Are there any disadvantages to having a radio tuning circuit with an extremely large *Q* factor?

13 • What is the power factor for a circuit that has inductance and capacitance but no resistance?

14 • A transformer is used to change (a) capacitance, (b) frequency, (c) voltage, (d) power, (e) none of these.

15 • True or false: If a transformer increases the current, it must decrease the voltage.

16 •• An ideal transformer has N_1 turns on its primary and N_2 turns on its secondary. The power dissipated in a load resistance *R* connected across the secondary is P_2 when the primary voltage is V_1. The current in the primary windings is then (a) P_2/V_1. (b) $(N_1/N_2)(P_2/V_1)$. (c) $(N_2/N_1)(P_2/V_1)$. (d) $(N_2/N_1)^2(P_2/V_1)$.

17 • True or false:

(a) Alternating current in a resistance dissipates no power because the current is negative as often as the current is positive.

(b) At very high frequencies, a capacitor acts like a short circuit.

Estimation and Approximation

18 •• **SSM** The impedances of motors, transformers, and electromagnets have inductive reactance. Suppose that the phase angle of the total impedance of a large industrial plant is 25° when the plant is under full operation and using 2.3 MW of power. The power is supplied to the plant from a substation 4.5 km from the plant; the 60 Hz rms line voltage at the plant is 40,000 V. The resistance of the transmission line from the substation to the plant is 5.2 Ω. The cost per kilowatt-hour is 0.07 dollars. The plant pays only for the actual energy used. (a) What are the resistance and inductive reactance of the plant's total load? (b) What is the current in the power lines and what must be the rms voltage at the substation to maintain the voltage at the plant at 40,000 V? (c) How much power is lost in transmission? (d) Suppose that the phase angle of the plant's impedance were reduced to 18° by adding a bank of capacitors in series with the load. How much money would be saved by the electric utility during one month of operation, assuming the plant operates at full capacity for 16 h each day? (e) What must be the capacitance of this bank of capacitors?

Alternating Current Generators

19 • A 200-turn coil has an area of 4 cm² and rotates in a magnetic field of 0.5 T. (a) What frequency will generate a maximum emf of 10 V? (b) If the coil rotates at 60 Hz, what is the maximum emf?

20 • In what magnetic field must the coil of Problem 19 be rotating to generate a maximum emf of 10 V at 60 Hz?

21 • **SSM** A 2-cm by 1.5-cm rectangular coil has 300 turns and rotates in a magnetic field of 4000 G. (a) What is the maximum emf generated when the coil rotates at 60 Hz? (b) What must its frequency be to generate a maximum emf of 110 V?

22 • The coil of Problem 21 rotates at 60 Hz in a magnetic field *B*. What value of *B* will generate a maximum emf of 24 V?

Alternating Current in a Resistor

23 • **SSM** A 100-W lightbulb is plugged into a standard 120-V (rms) outlet. Find (a) I_{rms}, (b) I_{max}, and (c) the maximum power.

24 • **ISOLVE** A circuit breaker is rated for a current of 15 A rms at a voltage of 120 V rms. (a) What is the largest value of I_{max} that the breaker can carry? (b) What average power can be supplied by this circuit?

Alternating Current in Inductors and Capacitors

25 • What is the reactance of a 1-mH inductor at (a) 60 Hz, (b) 600 Hz, and (c) 6 kHz?

26 • **ISOLVE** An inductor has a reactance of 100 Ω at 80 Hz. (a) What is its inductance? (b) What is its reactance at 160 Hz?

27 • **ISOLVE** At what frequency would the reactance of a 10-μF capacitor equal that of a 1-mH inductor?

28 • What is the reactance of a 1-nF capacitor at (a) 60 Hz, (b) 6 kHz, and (c) 6 MHz?

29 • **SSM** An emf of 10 V maximum and frequency 20 Hz is applied to a 20-μF capacitor. Find (a) I_{max} and (b) I_{rms}.

30 • **ISOLVE** At what frequency is the reactance of a 10-μF capacitor (a) 1 Ω, (b) 100 Ω, and (c) 0.01 Ω?

31 •• **ISOLVE** Two ac voltage sources are connected in series with a resistor *R* = 25 Ω. One source is given by

$$V_1 = (5 \text{ V}) \cos(\omega t - \alpha),$$

and the other source is

$$V_2 = (5 \text{ V}) \cos(\omega t + \alpha),$$

with $\alpha = \pi/6$. (a) Find the current in *R* using a trigonometric identity for the sum of two cosines. (b) Use phasor diagrams to find the current in *R*. (c) Find the current in *R* if $\alpha = \pi/4$ and the amplitude of V_2 is increased from 5 V to 7 V.

LC and *RLC* Circuits Without a Generator

32 • **SSM** Show from the definitions of the henry and the farad that $1/\sqrt{LC}$ has the unit s^{-1}.

33 • (a) What is the period of oscillation of an *LC* circuit consisting of a 2-mH coil and a 20-μF capacitor? (b) What inductance is needed with an 80-μF capacitor to construct an *LC* circuit that oscillates with a frequency of 60 Hz?

34 •• An LC circuit has capacitance C_1 and inductance L_1. A second circuit has capacitance $C_2 = \frac{1}{2}C_1$ and $L_2 = 2L_1$, and a third circuit has capacitance $C_3 = 2C_1$ and $L_3 = \frac{1}{2}L_1$. (a) Show that each circuit oscillates with the same frequency. (b) In which circuit would the maximum current be greatest if the capacitor in each were charged to the same potential V?

35 •• ISOLVE A 5-μF capacitor is charged to 30 V and is then connected across a 10-mH inductor. (a) How much energy is stored in the system? (b) What is the frequency of oscillation of the circuit? (c) What is the maximum current in the circuit?

36 • ISOLVE A coil can be considered to be a resistance and an inductance in series. Assume that $R = 100\ \Omega$ and $L = 0.4$ H. The coil is connected across a 120-V rms, 60-Hz line. Find (a) the power factor, (b) the rms current, and (c) the average power supplied.

37 •• SSM An inductor and a capacitor are connected, as shown in Figure 29-30. With the switch open, the left plate of the capacitor has charge Q_0. The switch is closed and the charge and current vary sinusoidally with time. (a) Plot both Q versus t and I versus t and explain how to interpret these two plots to illustrate that the current leads the charge by 90°. (b) Using a trig identity, show the expression for the current (Equation 29-38) leads the expression for the charge (Equation 29-39) by 90°. That is, show

$$I = -I_{peak} \sin \omega t = I_{peak} \cos\left(\omega t + \frac{\pi}{2}\right).$$

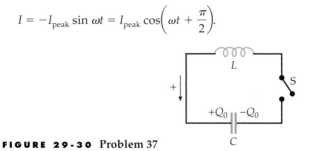

FIGURE 29-30 Problem 37

RL Circuits With a Generator

38 •• ISOLVE A resistance R and a 1.4-H inductance are in series across a 60-Hz ac voltage. The voltage across the resistor is 30 V and the voltage across the inductor is 40 V. (a) What is the resistance R? (b) What is the ac input voltage?

39 •• ISOLVE A coil has a dc resistance of 80 Ω and an impedance of 200 Ω at a frequency of 1 kHz. Neglect the wiring capacitance of the coil at this frequency. What is the inductance of the coil?

40 •• A single transmission line carries two voltage signals given by $V_1 = 10$ V cos $100t$ and $V_2 = 10$ V cos $10,000\ t$, where t is in seconds. A series inductor of 1 H and a shunting resistor of 1 kΩ are inserted into the transmission line, as indicated in Figure 29-31. (a) What is the voltage signal observed at the output side of the transmission line? (b) What is the ratio of the low-frequency amplitude to the high-frequency amplitude?

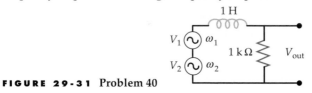

FIGURE 29-31 Problem 40

41 •• ISOLVE A coil with resistance and inductance is connected to a 120-V rms, 60-Hz line. The average power supplied to the coil is 60 W, and the rms current is 1.5 A. Find (a) the power factor, (b) the resistance of the coil, and (c) the inductance of the coil. (d) Does the current lag or lead the voltage? What is the phase angle δ?

42 •• ISOLVE A 36-mH inductor with a resistance of 40 Ω is connected to a source whose voltage is $\mathcal{E} = 345$ V cos $150\pi t$, where t is in seconds. Determine the maximum current in the circuit, the maximum and rms voltages across the inductor, the average power dissipation, and the maximum and average energy stored in the magnetic field of the inductor.

43 •• A coil of resistance R, inductance L, and negligible capacitance has a power factor of 0.866 at a frequency of 60 Hz. What is the power factor for a frequency of 240 Hz?

44 •• SSM A resistor and an inductor are connected in parallel across an emf $\mathcal{E} = \mathcal{E}_{max}$ as shown in Figure 29-32. Show that (a) the current in the resistor is $I_R = \mathcal{E}_{max}/R$ cos ωt, (b) the current in the inductor is $I_L = \mathcal{E}_{max}/X_L$ cos($\omega t - 90°$), and (c) $I = I_R + I_L = I_{max}$ cos($\omega t - \delta$), where tan $\delta = R/X_L$ and $I_{max} = \mathcal{E}_{max}/Z$ with $Z^{-2} = R^{-2} + X_L^{-2}$.

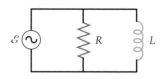

FIGURE 29-32 Problem 44

45 •• Figure 29-33 shows a load resistor $R_L = 20\ \Omega$ connected to a high-pass filter consisting of an inductor $L = 3.2$ mH and a resistor $R = 4\ \Omega$. The input voltage is $\mathcal{E} = 100$ V cos $2\pi ft$. Find the rms currents in R, L, and R_L if (a) $f = 500$ Hz and (b) $f = 2000$ Hz. (c) What fraction of the total power delivered by the voltage source is dissipated in the load resistor if the frequency is 500 Hz and if the frequency is 2000 Hz?

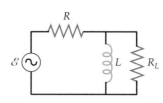

FIGURE 29-33 Problem 45

46 •• An ac source $\mathcal{E}_1 = 20$ V cos $2\pi ft$ in series with a battery whose emf is $\mathcal{E}_2 = 16$ V is connected to a circuit consisting of resistors $R_1 = 10\ \Omega$ and $R_2 = 8\ \Omega$ and an inductor $L = 6$ mH (Figure 29-34). Find the power dissipated in R_1 and R_2 if (a) $f = 100$ Hz, (b) $f = 200$ Hz, and (c) $f = 800$ Hz.

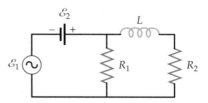

FIGURE 29-34 Problem 46

47 •• A 100-V rms voltage is applied to a series RC circuit. The rms voltage across the capacitor is 80 V. What is the voltage across the resistor?

Filters and Rectifiers

48 •• SSM The circuit shown in Figure 29-35 is called an RC high-pass filter because it transmits signals with a high-input frequency with greater amplitude than low-frequency signals. If the input voltage is $V_{in} = V_{peak} \cos \omega t$, show that the output voltage is $V_{out} = V_H \cos(\omega t - \delta)$ where

$$V_H = \frac{V_{peak}}{\sqrt{1 + \left(\frac{1}{\omega RC}\right)^2}}$$

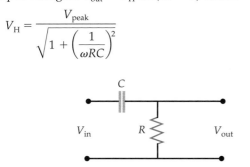

FIGURE 29-35 Problem 48

49 •• (a) Show that the phase constant δ in Problem 48 is given by

$$\tan \delta = -\left(\frac{1}{\omega RC}\right)$$

(b) What is the value of δ in the limit as $\omega \to 0$? (c) What is the value of δ in the limit $\omega \to \infty$?

50 •• Assume that the resistor of Problem 48 has value $R = 20 \text{ k}\Omega$ and the capacitor has value $C = 15 \text{ nF}$. (a) At what frequency f is $V_{out} = V_{in}/\sqrt{2}$? (This is known as the 3 dB frequency, or f_{3dB} for the circuit.) (b) Using a spreadsheet program, make a graph of V_{out} versus f. Use a logarithmic scale for each variable. Make sure that the scale extends from at least 0.1 f_{3dB} to 10 f_{3dB} (c) Make a graph of δ versus f and graph f on a logarithmic scale. What value does δ have at $f = f_{3dB}$?

51 ••• Show that if an arbitrary voltage signal is fed into the high-pass filter of Problem 48, in which the time variance of the signal is much slower than $1/(RC)$, the output of the circuit will be proportional to the time derivative of the input.

52 •• We define the output from the high-pass filter from Problem 48 in the decibel scale as

$$\beta = 20 \log_{10} \frac{V_H}{V_{peak}}$$

Show that for $f \ll f_{3dB}$, where f_{3dB} is defined in Problem 50, the output drops at a rate of 6 dB per octave. That is, every time the frequency is halved, the output drops by 6 dB.

53 •• SSM Show that the average power dissipated in the resistor of the high-pass filter of Problem 48 is given by

$$P_{ave} = \frac{V^2_{peak}}{2R} \left(\frac{(\omega RC)^2}{1 + (\omega RC)^2}\right)$$

54 •• One application of the high-pass filter of Problem 48 is that of a noise filter for electronic circuits (i.e., one that blocks out low-frequency noise). Using $R = 20 \text{ k}\Omega$, pick a value for C for a high-pass filter that attenuates an input voltage signal at $f = 60 \text{ Hz}$ by a factor of 10.

55 •• The circuit shown in Figure 29-36 is a low-pass filter. If the input voltage is

$V_{in} = V_{peak} \cos \omega t$ show that the output voltage is

$V_{out} = V_L \cos(\omega t - \delta)$ where

$$V_L = \frac{V_{peak}}{\sqrt{1 + (\omega RC)^2}}$$

Discuss the behavior of the output voltage in the limiting cases $\omega \to 0$ and $\omega \to \infty$.

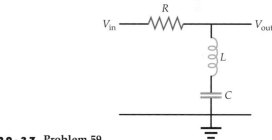

FIGURE 29-36 Problem 55

56 •• Show that δ for the low-pass filter of Problem 55 is given by the expression $\tan \delta = \omega RC$. Find the value of δ in the limit $\omega \to 0$ and $\omega \to \infty$.

57 •• SSM Using a spreadsheet program, make a graph of V_L versus $f = \omega/2\pi$ and δ versus f for the low-pass filter of Problem 55. Use $R = 10 \text{ k}\Omega$ and $C = 5 \text{ nF}$.

58 ••• Show that if an arbitrary voltage signal is fed into the low-pass filter of Problem 55, in which the time variance of the signal is much faster than $1/(RC)$, the output of the circuit will be proportional to the integral of the input.

59 ••• SSM Show the *trap* filter, shown in Figure 29-37, acts to reject signals at a frequency $\omega = 1/\sqrt{LC}$. How does the width of the frequency band rejected depend on the resistance R?

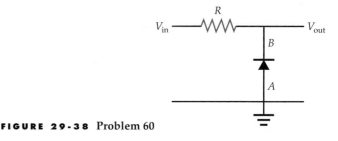

FIGURE 29-37 Problem 59

60 •• A half-wave rectifier for transforming an ac voltage into a dc voltage is shown in Figure 29-38. The diode in the figure can be thought of as a one-way valve for current, allowing current to pass in the forward (upward) direction when the voltage between points A and B is greater than +0.6 V. The resistance of the diode is effectively infinite when the voltage is less than +0.6 V. Using the same axes plot two cycles of both V_{in} and V_{out} versus t when $V_{in} = V_{peak} \cos \omega t$.

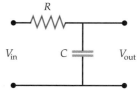

FIGURE 29-38 Problem 60

61 •• (*a*) The output of rectifier of Problem 60, Figure 29-39*a*, can be smoothed by putting its output through a low-pass filter. The resulting output is a dc voltage with a small amount of ripple on it, as shown in Figure 29-39*b*. If the input frequency $f = \omega/2\pi = 60$ Hz and the resistance is $R = 1$ kΩ, find an approximate value for C, so that the output voltage varies by less than 50 percent of the mean value over one cycle.

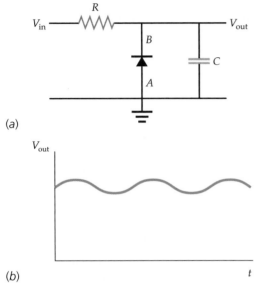

(a)

(b)

FIGURE 29-39 Problem 61

LC Circuits With a Generator

62 •• The generator voltage in Figure 29-40 is given by $\mathscr{E} = (100 \text{ V}) \cos 2\pi ft$. (*a*) For each branch, what is the amplitude of the current and what is its phase relative to the applied voltage? (*b*) What is the angular frequency ω so that the current in the generator vanishes? (*c*) At this resonance, what is the current in the inductor? What is the current in the capacitor? (*d*) Draw a phasor diagram showing the general relationships between the applied voltage, the generator current, the capacitor current, and the inductor current for the case where the inductive reactance is larger than the capacitive reactance.

FIGURE 29-40 Problem 62

63 •• 🟦SOLVE The charge on the capacitor of a series LC circuit is given by $Q = (15 \ \mu\text{C}) \cos(1250t + \frac{\pi}{4})$, where t is in seconds. (*a*) Find the current as a function of time. (*b*) Find C if $L = 28$ mH. (*c*) Write expressions for the electrical energy U_e, the magnetic energy U_m, and the total energy U.

64 ••• SSM One method for measuring the compressibility of a dielectric material uses an LC circuit with a parallel-plate capacitor. The dielectric is inserted between the plates and the change in resonance frequency is determined as the capacitor plates are subjected to a compressive stress. In such an arrangement, the resonance frequency is 120 MHz when a dielectric of thickness 0.1 cm and dielectric constant

$\kappa = 6.8$ is placed between the capacitor plates. Under a compressive stress of 800 atm, the resonance frequency decreases to 116 MHz. Find Young's modulus of the dielectric material.

65 ••• Figure 29-41 shows an inductance L and a parallel plate capacitor of width $w = 20$ cm and thickness 0.2 cm. A dielectric with dielectric constant $\kappa = 4.8$ that can completely fill the space between the capacitor plates can be slid between the plates. The inductor has an inductance $L = 2$ mH. When half the dielectric is between the capacitor plates (i.e., when $x = \frac{1}{2} w$), the resonant frequency of this LC combination is 90 MHz. (*a*) What is the capacitance of the capacitor without the dielectric? (*b*) Find the resonance frequency as a function of x.

FIGURE 29-41 Problem 65

RLC Circuits With a Generator

66 • A series RLC circuit in a radio receiver is tuned by a variable capacitor, so that it can resonate at frequencies from 500 to 1600 kHz. If $L = 1 \ \mu$H, find the range of capacitances necessary to cover this range of frequencies.

67 • (*a*) Find the power factor for the circuit in Example 29-5 when $\omega = 400$ rad/s. (*b*) At what angular frequency is the power factor 0.5?

68 • 🟦SOLVE An ac generator with a maximum emf of 20 V is connected in series with a 20-μF capacitor and an 80-Ω resistor. There is no inductance in the circuit. Find (*a*) the power factor, (*b*) the rms current, and (*c*) the average power if the angular frequency of the generator is 400 rad/s.

69 •• SSM Show that the formula $P_{av} = R\mathscr{E}^2_{rms}/Z^2$ gives the correct result for a circuit containing only a generator and (*a*) a resistor, (*b*) a capacitor, and (*c*) an inductor.

70 •• 🟦SOLVE A series RLC circuit with $L = 10$ mH, $C = 2 \ \mu$F, and $R = 5 \ \Omega$ is driven by a generator with a maximum emf of 100 V and a variable angular frequency ω. Find (*a*) the resonant frequency ω_0 and (*b*) I_{rms} at resonance. When $\omega = 8000$ rad/s, find (*c*) X_C and X_L, (*d*) Z and I_{rms}, and (*e*) the phase angle δ.

71 •• For the circuit in problem 70, let the generator frequency be $f = \omega/2\pi = 1$ kHz. Find (*a*) the resonance frequency $f_0 = \omega_0/2\pi$, (*b*) X_C and X_L, (*c*) the total impedance Z and I_{rms}, and (*d*) the phase angle δ.

72 •• Find the power factor and the phase angle δ for the circuit in Problem 70 when the generator frequency is (*a*) 900 Hz, (*b*) 1.1 kHz, and (*c*) 1.3 kHz.

73 •• Find (*a*) the Q factor and (*b*) the resonance width for the circuit in Problem 70. (*c*) What is the power factor when $\omega = 8000$ rad/s?

74 •• SSM 🟦SOLVE FM radio stations have carrier frequencies that are separated by 0.20 MHz. When the radio is tuned to a station, such as 100.1 MHz, the resonance width of the receiver circuit should be much smaller than 0.2 MHz, so that adjacent stations are not received. If $f_0 = 100.1$ MHz and $\Delta f = 0.05$ MHz, what is the Q factor for the circuit?

75 •• **ISOLVE** A coil is connected to a 60-Hz, 100-V ac generator. At this frequency, the coil has an impedance of 10 Ω and a reactance of 8 Ω. (*a*) What is the current in the coil? (*b*) What is the phase angle between the current and the applied voltage? (*c*) What series capacitance is required so that the current and voltage are in phase? (*d*) What is the voltage measured across the capacitor?

76 •• An 0.25-H inductor and a capacitor *C* are connected in series with a 60-Hz ac generator. An ac voltmeter is used to measure the rms voltages across the inductor and capacitor separately. The rms voltage across the capacitor is 75 V and that across the inductor is 50 V. (*a*) Find the capacitance *C* and the rms current in the circuit. (*b*) What would be the measured rms voltage across both the capacitor and inductor together?

77 •• (*a*) Show that Equation 29-47 can be written as

$$\tan \delta = \frac{L(\omega^2 - \omega_0^2)}{\omega R}$$

Find δ approximately at (*b*) very low frequencies and (*c*) very high frequencies.

78 •• (*a*) Show that in a series *RC* circuit with no inductance, the power factor is given by

$$\cos \delta = \frac{RC\omega}{\sqrt{1 + (RC\omega)^2}}$$

(*b*) Using a 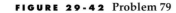 spreadsheet program, graph the power factor versus ω.

79 •• **SSM** In the circuit shown in Figure 29-42, the ac generator produces an rms voltage of 115 V when operated at 60 Hz. What is the rms voltage across points (*a*) *AB*, (*b*) *BC*, (*c*) *CD*, (*d*) *AC*, and (*e*) *BD*?

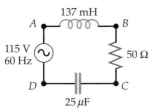

FIGURE 29-42 Problem 79

80 •• When an *RLC* series circuit is connected to a 120-V rms, 60-Hz line, the current is $I_{rms} = 11$ A and the current leads the voltage by 45°. (*a*) Find the power supplied to the circuit. (*b*) What is the resistance? (*c*) If the inductance $L = 0.05$ H, find the capacitance *C*. (*d*) What capacitance or inductance should you add to make the power factor 1?

81 •• **ISOLVE** A series *RLC* circuit is driven at a frequency of 500 Hz. The phase angle between the applied voltage and current is determined from an oscilloscope measurement to be δ = 75°. If the total resistance is known to be 35 Ω and the inductance is 0.15 H, what is the capacitance of the circuit?

82 •• A series *RLC* circuit with *R* = 400 Ω, *L* = 0.35 H, and *C* = 5 μF is driven by a generator of variable frequency *f*. (*a*) What is the resonance frequency f_0? Find *f* and f/f_0 when the phase angle δ is (*b*) 60° and (*c*) −60°.

83 •• Sketch the impedance *Z* versus ω for (*a*) a series *LR* circuit, (*b*) a series *RC* circuit, and (*c*) a series *RLC* circuit.

84 •• **SSM** Show that Equation 29-48 can be written as

$$I_{max} = \frac{\omega \mathcal{E}_{max}}{\sqrt{L^2(\omega^2 - \omega_0^2)^2 + \omega^2 R^2}}$$

85 •• In a series *RLC* circuit, $X_C = 16$ Ω and $X_L = 4$ Ω at some frequency. The resonance frequency is $\omega_0 = 10^4$ rad/s. (*a*) Find *L* and *C*. If *R* = 5 Ω and $\mathcal{E}_{max} = 26$ V, find (*b*) the *Q* factor, and (*c*) the maximum current.

86 •• In a series *RLC* circuit connected to an ac generator whose maximum emf is 200 V, the resistance is 60 Ω and the capacitance is 8 μF. The inductance can be varied from 8 mH to 40 mH, by the insertion of an iron core in the solenoid. The angular frequency of the generator is 2500 rad/s. If the capacitor voltage is not to exceed 150 V, find (*a*) the maximum current and (*b*) the range of inductance that is safe to use.

87 •• A certain electrical device draws 10 A rms and has an average power of 720 W when connected to a 120-V rms, 60-Hz power line. (*a*) What is the impedance of the device? (*b*) What series combination of resistance and reactance is this device equivalent to? (*c*) If the current leads the emf, is the reactance inductive or capacitive?

88 •• **SSM** A method for measuring inductance is to connect the inductor in series with a known capacitance, a known resistance, an ac ammeter, and a variable-frequency signal generator. The frequency of the signal generator is varied and the emf is kept constant until the current is maximum. (*a*) If *C* = 10 μF, $\mathcal{E}_{max} = 10$ V, *R* = 100 Ω, and *I* is maximum at ω = 5000 rad/s, what is *L*? (*b*) What is I_{max}?

89 •• A resistor and a capacitor are connected in parallel across a sinusoidal emf $\mathcal{E} = \mathcal{E}_{max} \cos \omega t$, as shown in Figure 29-43. (*a*) Show that the current in the resistor is $I_R = (\mathcal{E}_{max}/R) \cos \omega t$. (*b*) Show that the current in the capacitor branch is $I_C = (\mathcal{E}_{max}/X_C) \cos(\omega t + 90°)$. (*c*) Show that the total current is given by $I = I_R + I_C = I_{max} \cos(\omega t + \delta)$, where $\tan \delta = R/X_C$ and $I_{max} = \mathcal{E}_{max}/Z$ with $Z^{-2} = R^{-2} + X_C^{-2}$.

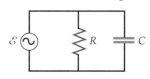

FIGURE 29-43 Problem 89

90 •• **SSM** In the circuit shown in Figure 29-44, *R* = 10 Ω, $R_L = 30$ Ω, *L* = 150 mH, and *C* = 8 μF; the frequency of the ac source is 10 Hz and its amplitude is 100 V. (*a*) Using phasor diagrams, determine the impedance of the circuit when switch S is closed. (*b*) Determine the impedance of the circuit when switch S is open. (*c*) What are the voltages across the load resistor R_L when switch S is closed and when it is open? (*d*) Repeat Parts (*a*), (*b*), and (*c*) with the frequency of the source changed to 1000 Hz. (*e*) Which arrangement is a better low-pass filter, S open or S closed?

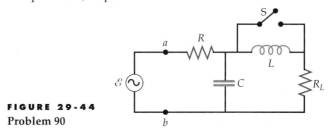

FIGURE 29-44
Problem 90

91 •• In the circuit shown in Figure 29-45, $R_1 = 2\,\Omega$, $R_2 = 4\,\Omega$, $L = 12$ mH, $C = 30\,\mu$F, and $\mathcal{E} = (40\text{ V})\cos\omega t$. (a) Find the resonance frequency. (b) At the resonance frequency, what are the rms currents in each resistor and the rms current supplied by the source emf?

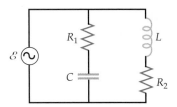

FIGURE 29-45 Problems 91, 103, and 104

92 •• For the circuit in Figure 29-24, derive an expression for the Q of the circuit, assuming the resonance is sharp.

93 •• ‖SOLVE‖ For the circuit in Figure 29-24, $L = 4$ mH. (a) What capacitance C will result in a resonance frequency of 4 kHz? (b) When C has the value found in Part (a), what should be the resistance R, so that the Q of the circuit is 8?

94 •• If the capacitance of C in Problem 93 is reduced to half the value found in Problem 93, what then are the resonance frequency and the Q of the circuit? What should be the resistance R to give $Q = 8$?

95 •• ‖SOLVE‖ A series circuit consists of a 4.0-nF capacitor, a 36-mH inductor, and a 100-Ω resistor. The circuit is connected to a 20-V ac source whose frequency can be varied over a wide range. (a) Find the resonance frequency f_0 of the circuit. (b) At resonance, what is the rms current in the circuit and what are the rms voltages across the inductor and capacitor? (c) What is the rms current and what are the rms voltages across the inductor and capacitor at $f = f_0 + \frac{1}{2}\Delta f$, where Δf is the width of the resonance?

96 ••• In the parallel circuit shown in Figure 29-46, $V_{\max} = 110$ V. (a) What is the impedance of each branch? (b) For each branch, what is the current amplitude and its phase relative to the applied voltage? (c) Give the current phasor diagram, and use it to find the total current and its phase relative to the applied voltage.

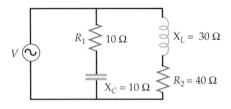

FIGURE 29-46 Problem 96

97 ••• ‖SSM‖ (a) Show that Equation 29-47 can be written as

$$\tan\delta = \frac{Q(\omega^2 - \omega_0^2)}{\omega\omega_0}$$

(b) Show that near resonance

$$\tan\delta \approx \frac{2Q(\omega_0 - \omega)}{\omega}$$

(c) Sketch a plot of δ versus x, where $x = \omega/\omega_0$, for a circuit with high Q and for one with low Q.

98 ••• Show by direct substitution that the current given by Equation 29-46 with δ and $I_{\max}$ given by Equations 29-47 and 29-48, respectively, satisfies Equation 29-45. (Hint: Use trigonometric identities for the sine and cosine of the sum of two angles, and write the equation in the form $A\sin\omega t + B\cos\omega t = 0$. Because this equation must hold for all times, $A = 0$ and $B = 0$.)

99 ••• An ac generator is in series with a capacitor and an inductor in a circuit with negligible resistance. (a) Show that the charge on the capacitor obeys the equation

$$L\frac{d^2Q}{dt^2} + \frac{Q}{C} = \mathcal{E}_{\max}\cos\omega t$$

(b) Show by direct substitution that this equation is satisfied by $Q = Q_{\max}\cos\omega t$ if,

$$Q_{\max} = -\frac{\mathcal{E}_{\max}}{L(\omega^2 - \omega_0^2)}$$

(c) Show that the current can be written as $I = I_{\max}\cos(\omega t - \delta)$, where

$$I_{\max} = \frac{\omega\mathcal{E}_{\max}}{L|\omega^2 - \omega_0^2|} = \frac{\mathcal{E}_{\max}}{|X_L - X_C|}$$

and $\delta = -90°$ for $\omega < \omega_0$ and $\delta = 90°$ for $\omega > \omega_0$.

100 ••• Figure 29-20 shows a plot of average power P_{av} versus generator frequency ω for an RLC circuit with a generator. The average power P_{av} is given by Equation 29-54. The full width at half-maximum, $\Delta\omega$, is the width of the resonance curve between the two points, where P_{av} is one-half its maximum value. Show that for a sharply peaked resonance, $\Delta\omega \approx R/L$ and, hence, that $Q \approx \omega_0/\Delta\omega$ in this case (Equation 29-56). (Hint: At resonance, the denominator of the expression on the right of Equation 29-54 is $\omega^2 R^2$. The half-power points will occur when the denominator is twice the value near resonance; that is, when $L^2(\omega^2 - \omega_0^2)^2 = \omega^2 R^2 \approx \omega_0^2 R^2$. Let ω_1 and ω_2 be the solutions of this equation. For a sharply peaked resonance, $\omega_1 \approx \omega_0$ and $\omega_2 \approx \omega_0$. Then, using the fact that $\omega + \omega_0 \approx 2\omega_0$, one finds that $\Delta\omega = \omega_2 - \omega_1 \approx R/L$.)

101 • Show by direct substitution that

$$L\frac{d^2Q}{dt^2} + R\frac{dQ}{dt} + \frac{1}{C}Q = 0$$

(Equation 29-43b) is satisfied by

$$Q = Q_0 e^{-Rt/2L}\cos\omega't$$

where

$$\omega' = \sqrt{(1/LC - (R/2L)^2}$$

and Q_0 is the charge on the capacitor at $t = 0$.

102 ••• ‖SSM‖ One method for measuring the magnetic susceptibility of a sample uses an LC circuit consisting of an air-core solenoid and a capacitor. The resonant frequency of the circuit without the sample is determined and then measured again with the sample inserted in the solenoid. Suppose the solenoid is 4 cm long, 0.3 cm in diameter, and has 400 turns of fine wire. Assume that the sample that is inserted in the solenoid is also 4 cm long and fills the air space. Neglect end effects. (In practice, a test sample of known susceptibility of the same shape as the unknown is used to calibrate the instrument.) (*a*) What is the inductance of the empty solenoid? (*b*) What should be the capacitance of the capacitor so that the resonance frequency of the circuit without a sample is 6.0000 MHz? (*c*) When a sample is inserted in the solenoid, the resonance frequency drops to 5.9989 MHz. Determine the sample's susceptibility.

103 ••• (*a*) Find the angular frequency ω for the circuit in Problem 91 so that the magnitude of the reactance of the two parallel branches are equal. (*b*) At that frequency, what is the power dissipation in each of the two resistors?

104 ••• (*a*) For the circuit of Problem 91, find the angular frequency ω for which the power dissipation in the two resistors is the same. (*b*) At that angular frequency, what is the reactance of each of the two parallel branches? (*c*) Draw a phasor diagram showing the current through each of the two parallel branches. (*d*) What is the impedance of the circuit?

*The Transformer

105 • SSM An ac voltage of 24 V is required for a device whose impedance is 12 Ω. (*a*) What should the turn ratio of a transformer be, so that the device can be operated from a 120-V line? (*b*) Suppose the transformer is accidentally connected reversed (i.e., with the secondary winding across the 120-V line and the 12-Ω load across the primary). How much current will then flow in the primary winding?

106 • A transformer has 400 turns in the primary and 8 turns in the secondary. (*a*) Is this a step-up or a step-down transformer? (*b*) If the primary is connected across 120 V rms, what is the open-circuit voltage across the secondary? (*c*) If the primary current is 0.1 A, what is the secondary current, assuming negligible magnetization current and no power loss?

107 • ISOLVE The primary of a step-down transformer has 250 turns and is connected to a 120-V rms line. The secondary is to supply 20 A at 9 V. Find (*a*) the current in the primary and (*b*) the number of turns in the secondary, assuming 100 percent efficiency.

108 • A transformer has 500 turns in its primary, which is connected to 120 V rms. Its secondary coil is tapped at three places to give outputs of 2.5 V, 7.5 V, and 9 V. How many turns are needed for each part of the secondary coil?

109 • The distribution circuit of a residential power line is operated at 2000 V rms. This voltage must be reduced to 240 V rms for use within residences. If the secondary side of the transformer has 400 turns, how many turns are in the primary?

110 •• SSM An audio oscillator (ac source) with an internal resistance of 2000 Ω and an open-circuit rms output voltage of 12 V is to be used to drive a loudspeaker with a resistance of 8 Ω. What should be the ratio of primary to secondary turns of a transformer, so that maximum power is transferred to the speaker? Suppose a second identical speaker is connected in parallel with the first speaker. How much power is then supplied to the two speakers combined?

111 •• One use of a transformer is for *impedance matching*. For example, the output impedance of a stereo amplifier is matched to the impedance of a speaker by a transformer. In Equation 29-63, the currents I_1 and I_2 can be related to the impedance Z in the secondary because $I_2 = V_2/Z$. Using Equations 29-61 and 29-62, show that

$$I_1 = \mathcal{E}/[(N_1/N_2)^2 Z]$$

and, therefore, $Z_{\text{eff}} = (N_1/N_2)^2 Z$.

General Problems

112 • A 5-kW electric clothes dryer runs on 240 V rms. Find (*a*) I_{rms} and (*b*) I_{max}. (*c*) Find the same quantities for a dryer of the same power that operates at 120 V rms.

113 • Find the reactance of a 10.0-μF capacitor at (*a*) 60 Hz, (*b*) 6 kHz, and (*c*) 6 MHz.

114 •• ISOLVE A resistance R carries a current $I = 5$ A sin $120\pi t + 7$ A sin $240\pi t$. (*a*) What is the rms current? (*b*) If the resistance R is 12 Ω, what is the power dissipated in the resistor? (*c*) What is the rms voltage across the resistor?

115 •• SSM Figure 29-47 shows the voltage V versus time t for a *square-wave* voltage. If $V_0 = 12$ V, (*a*) what is the rms voltage of this waveform? (*b*) If this alternating waveform is rectified by eliminating the negative voltages, so that only the positive voltages remain, what now is the rms voltage of the rectified waveform?

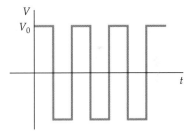

FIGURE 29-47 Problem 115

116 •• ISOLVE A pulsed current has a constant value of 15 A for the first 0.1 s of each second and is then 0 for the next 0.9 s of each second. (*a*) What is the rms value for this current waveform? (*b*) Each current pulse is generated by a voltage pulse of maximum value 100 V. What is the average power delivered by the pulse generator?

117 •• A circuit consists of two capacitors, a 24-V battery, and an ac voltage connected, as shown in Figure 29-48. The ac voltage is given by $\mathcal{E} = 20$ V cos $120\pi t$, where t is in seconds. (*a*) Find the charge on each capacitor as a function of time. Assume transient effects have had sufficient time to decay. (*b*) What is the steady-state current? (*c*) What is the maximum energy stored in the capacitors? (*d*) What is the minimum energy stored in the capacitors?

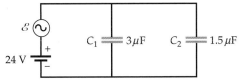

FIGURE 29-48 Problem 117

118 •• What are the average values and rms values of current for the two current waveforms shown in Figure 29-49?

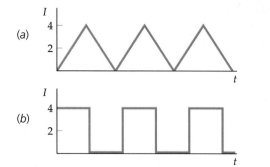

FIGURE 29-49 Problem 118

119 •• [iSOLVE] In the circuit shown in Figure 29-50, $\mathcal{E}_1 = (20\ V)\cos 2\pi ft$, $f = 180$ Hz; $\mathcal{E}_2 = 18$ V, and $R = 36\ \Omega$. Find the maximum, minimum, average, and rms values of the current through the resistor.

FIGURE 29-50
Problems 119 through 121

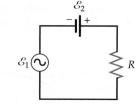

120 •• [SSM] Repeat Problem 119 if the resistor R is replaced by a 2-μF capacitor.

121 •• Repeat Problem 119 if the resistor R is replaced by a 12-mH inductor.

Maxwell's Equations and Electromagnetic Waves

THE 70-M ANTENNA AT GOLDSTONE, CALIFORNIA. THE GOLDSTONE DEEP SPACE COMMUNICATIONS COMPLEX, LOCATED IN THE MOJAVE DESERT IN CALIFORNIA, IS ONE OF THREE COMPLEXES THAT COMPRISE NASA'S DEEP SPACE NETWORK. THIS NETWORK PROVIDES RADIO COMMUNICATIONS FOR ALL OF NASA'S INTERPLANETARY SPACECRAFT AND IS ALSO UTILIZED FOR RADIO ASTRONOMY AND RADAR OBSERVATIONS OF THE SOLAR SYSTEM AND THE UNIVERSE.

? **Did you ever wonder** **whether a radio antenna generates a wave equally in all directions? This topic is discussed in Section 30-3.**

Maxwell's equations, first proposed by the great Scottish physicist James Clerk Maxwell, relate the electric and magnetic field vectors $\vec{E}$ and $\vec{B}$ and their sources, which are electric charges and currents. These equations summarize the experimental laws of electricity and magnetism—the laws of Coulomb, Gauss, Biot–Savart, Ampère, and Faraday. These experimental laws hold in general except for Ampère's law, which applies only to steady continuous currents.

➤ **In this chapter, we will see how Maxwell was able to generalize Ampère's law with the invention of the displacement current (Section 30-1). Maxwell was then able to show that the generalized laws of electricity and magnetism imply the existence of electromagnetic waves.**

Maxwell's equations play a role in classical electromagnetism analogous to that of Newton's laws in classical mechanics. In principle, all problems in classical electricity and magnetism can be solved using Maxwell's equations, just as all problems in classical mechanics can be solved using Newton's laws. Maxwell's equations are considerably more complicated than Newton's laws, however, and their application to most problems involves mathematics beyond the scope of this book. Nevertheless, Maxwell's equations are of great theoretical importance. For example, Maxwell showed that these equations can be combined to yield a wave equation for the electric and magnetic field vectors $\vec{E}$ and $\vec{B}$. Such **electromagnetic waves** are caused by accelerating charges, (e.g., the charges in an alternating current in an antenna). These electromagnetic waves were first produced in the laboratory by Heinrich Hertz in 1887. Maxwell showed that his equations predicted the speed of electromagnetic waves in free space to be

$$c = \frac{1}{\sqrt{\mu_0 \epsilon_0}}$$

30-1

THE SPEED OF ELECTROMAGNETIC WAVES

where ϵ_0, the permittivity of free space, is the constant appearing in Coulomb's and Gauss's laws and μ_0, the permeability of free space, is the constant appearing in the Biot–Savart law and Ampère's law. Maxwell noticed with great excitement the coincidence that the measure for the speed of light equaled $1/\sqrt{\mu_0 \epsilon_0}$, and Maxwell correctly surmised that light itself is an electromagnetic wave. Today, the value of c is defined as 2.99792458×10^8 m/s, the value of μ_0 is defined as $4\pi \times 10^7$ N/A², and the value of ϵ_0 is defined by Equation 30-1.

30-1 Maxwell's Displacement Current

Ampère's law (Equation 27-15) relates the line integral of the magnetic field around some closed curve C to the current that passes through any surface bounded by that curve:

$$\oint_C \vec{B} \cdot d\vec{\ell} = \mu_0 I_s, \text{ for any closed curve } C$$

30-2

Maxwell recognized a flaw in Ampère's law. Figure 30-1 shows two different surfaces, S_1 and S_2, bounded by the same curve C, which encircles a wire carrying current to a capacitor plate. The current through surface S_1 is I, but there is no current through surface S_2 because the charge stops on the capacitor plate. Thus, there is ambiguity in the phrase "the current through any surface bounded by the curve." Such a problem arises when the current is not continuous.

Maxwell showed that the law can be generalized to include all situations if the current I in the equation is replaced by the sum of the conduction current I and another term I_d, called **Maxwell's displacement current**, defined as

$$I_d = \epsilon_0 \frac{d\phi_e}{dt}$$

30-3

DEFINITION—DISPLACEMENT CURRENT

where ϕ_e is the flux of the electric field through the same surface bounded by the curve C. The generalized form of Ampère's law is then

$$\oint_C \vec{B} \cdot d\vec{\ell} = \mu_0(I + I_d) = \mu_0 I + \mu_0 \epsilon_0 \frac{d\phi_e}{dt}$$

30-4

GENERALIZED FORM OF AMPÈRE'S LAW

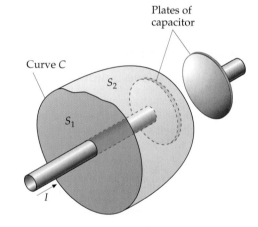

FIGURE 30-1 Two surfaces S_1 and S_2 bounded by the same curve C. The current I passes through surface S_1 but not through surface S_2. Ampère's law, which relates the line integral of the magnetic field around the curve C to the total current passing through any surface bounded by C, is not valid when the current is not continuous, as when it stops at the capacitor plate here.

We can understand this generalization by considering Figure 30-1 again. Let us call the sum $I + I_d$ the generalized current. According to the argument just stated, the same generalized current must cross any surface bounded by the curve C. Thus, there can be no net generalized current into or out of the volume bounded by the two surfaces S_1 and S_2, which together form a closed surface. If there is a net conduction current I into the volume, there must be an equal net displacement current I_d out of the volume. In the volume in the figure, there is a net conduction current I into the volume that increases the charge Q_{inside} within the volume:

$$I = \frac{dQ_{inside}}{dt}$$

The flux of the electric field out of the volume is related to the charge by Gauss's law:

$$\phi_{e, net} = \oint_S E_n \, dA = \frac{1}{\epsilon_0} Q_{inside}$$

Solving for the charge gives

$$Q_{inside} = \epsilon_0 \, \phi_{e, net}$$

and taking the derivative of each side gives

$$\frac{dQ_{inside}}{dt} = \epsilon_0 \frac{d\phi_{e, net}}{dt}$$

The rate of increase of the charge is thus proportional to the rate of increase of the net flux out of the volume:

$$\frac{dQ_{inside}}{dt} = \epsilon_0 \frac{d\phi_{e, net}}{dt} = I_d$$

Thus, the net conduction current into the volume equals the net displacement current out of the volume. The generalized current is thus continuous, and this is *always* the case.

It is interesting to compare Equation 30-4 to Equation 28-5:

$$\mathcal{E} = \oint_C \vec{E} \cdot d\vec{\ell} = -\frac{d\phi_m}{dt} = -\int_S \frac{\partial B_n}{\partial t} \, dA \qquad\qquad 30\text{-}5$$

which in this chapter will be referred to as Faraday's law. (Equation 30-5 is a restricted form of Faraday's law, a form that does not include motional emfs. Equation 30-5 does include emfs associated with a time varying magnetic field.) According to Faraday's law, a changing magnetic flux produces an electric field whose line integral around a closed curve is proportional to the rate of change of magnetic flux through any surface bounded by the curve. Maxwell's modification of Ampère's law shows that a changing electric flux produces a magnetic field whose line integral around a curve is proportional to the rate of change of the electric flux. We thus have the interesting reciprocal result that a changing magnetic field produces an electric field (Faraday's law) and a changing electric field produces a magnetic field (generalized form of Ampère's law). Note, there is no magnetic analog of a conduction current I. This is because the magnetic monopole, the magnetic analog of an electric charge, does not exist.[†]

[†] The question of the existence of magnetic monopoles has theoretical importance. There have been numerous attempts to observe magnetic monopoles but to date no one has been successful at doing so.

CALCULATING DISPLACEMENT CURRENT **EXAMPLE 30-1**

A parallel-plate capacitor has closely spaced circular plates of radius R. Charge is flowing onto the positive plate and off the negative plate at the rate $I = dQ/dt = 2.5$ A. Compute the displacement current through surface S passing between the plates (Figure 30-2) by directly computing the rate of change of the flux of $\vec{E}$ through surface S.

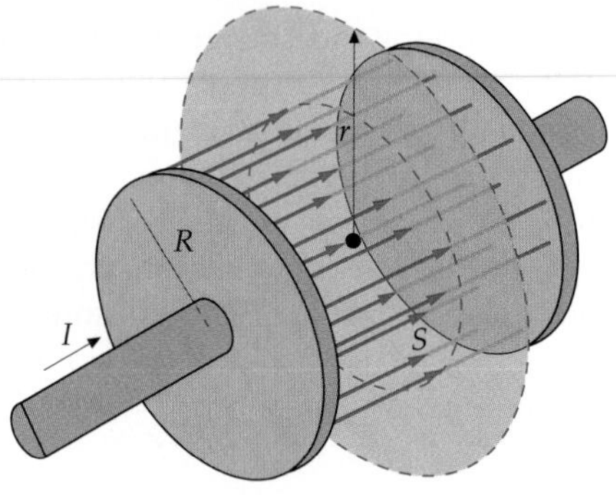

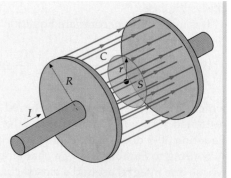

FIGURE 30-2 The surface S passes between the capacitor plates. The charge Q is increasing at 2.5 C/s = 2.5 A. The distance between the plates is not drawn to scale. The plates are much closer together than the plates shown in the figure.

PICTURE THE PROBLEM The displacement current is $I_d = \epsilon_0\, d\phi_e/dt$, where ϕ_e is the electric flux through the surface between the plates. Since the parallel plates are closely spaced, in the region between the plates the electric field is uniform and perpendicular to the plates. Outside the capacitor the electric field is negligible. Thus, the electric flux is simply $\phi_e = EA$, where E is the electric field between the plates and A is the plate area.

1. The displacement current is found by taking the time derivative of the electric flux:

$$I_d = \epsilon_0 \frac{d\phi_e}{dt}$$

2. The flux equals the electric field magnitude times the plate area:

$$\phi_e = EA$$

3. The electric field is proportional to the charge density on the plates, which we treat as uniformly distributed:

$$E = \frac{\sigma}{\epsilon_0} = \frac{Q/A}{\epsilon_0}$$

4. Substitute these results to calculate I_d:

$$I_d = \epsilon_0 \frac{d(EA)}{dt} = \epsilon_0 A \frac{dE}{dt} = \epsilon_0 A \frac{d}{dt}\left(\frac{Q}{A\epsilon_0}\right)$$

$$= \frac{dQ}{dt} = \boxed{2.5\ \text{A}}$$

REMARKS Note that the displacement current through the surface passing between the plates of the capacitor is equal to the conduction current in the wires carrying current to and from the capacitor.

CALCULATING $\vec{B}$ FROM DISPLACEMENT CURRENT **EXAMPLE 30-2**

The circular plates in Example 30-1 have a radius of $R = 3.0$ cm. Find the magnetic field strength B at a point between the plates a distance $r = 2.0$ cm from the axis of the plates when the current into the positive plate is 2.5 A.

PICTURE THE PROBLEM We find B from the generalized form of Ampère's law (Equation 30-4). We chose a circular path C of radius $r = 2.0$ cm about the centerline joining the plates, as shown in Figure 30-3. We then calculate the displacement current through the surface S bounded by C. By symmetry, $\vec{B}$ is tangent to C and has the same magnitude everywhere on C.

1. We find B from the generalized form of Ampère's law:

$$\oint_C \vec{B} \cdot d\vec{\ell} = \mu_0(I + I_d)$$

where

$$I_d = \epsilon_0 \frac{d\phi_e}{dt}$$

FIGURE 30-3 The space distance between the plates is not drawn to scale. The plates are much closer together than they appear.

2. The line integral is B times the circumference of the circle:

$$\oint_C \vec{B} \cdot d\vec{\ell} = B(2\pi r)$$

3. Since there is no conduction current between the plates of the capacitor, $I = 0$. The generalized current is just the displacement current:

$$\oint_C \vec{B} \cdot d\vec{\ell} = \mu_0 I + \mu_0 \epsilon_0 \frac{d\phi_e}{dt}$$

$$B(2\pi r) = 0 + \mu_0 \epsilon_0 \frac{d\phi_e}{dt}$$

4. The electric flux equals the product of the uniform field E and the area of the flat surface bounded by the curve:

$$\phi_e = \pi r^2 E = \pi r^2 \frac{\sigma}{\epsilon_0} = \pi r^2 \frac{Q}{\epsilon_0 \pi R^2}$$

$$= \frac{Q r^2}{\epsilon_0 R^2}$$

5. Substitute these results into step 3 and solve for B:

$$B(2\pi r) = \mu_0 \epsilon_0 \frac{d}{dt}\left(\frac{Q r^2}{\epsilon_0 R^2}\right) = \mu_0 \frac{r^2}{R^2}\frac{dQ}{dt}$$

$$B = \frac{\mu_0}{2\pi}\frac{r}{R^2}\frac{dQ}{dt} = \frac{\mu_0}{2\pi}\frac{r}{R^2}I$$

$$= (2 \times 10^{-7}\ \text{T·m/A})\,\frac{0.02\ \text{m}}{(0.03\ \text{m})^2}\,(2.5\ \text{A})$$

$$= \boxed{1.11 \times 10^{-5}\ \text{T}}$$

30-2 Maxwell's Equations

Maxwell's equations are

$$\oint_S E_n\, dA = \frac{1}{\epsilon_0} Q_{\text{inside}} \qquad\qquad 30\text{-}6a$$

$$\oint_S B_n\, dA = 0 \qquad\qquad 30\text{-}6b$$

$$\oint_C \vec{E} \cdot d\vec{\ell} = -\frac{d}{dt}\int_S B_n\, dA = -\int_S \frac{\partial B_n}{\partial t}\, dA \qquad\qquad 30\text{-}6c$$

$$\oint_C \vec{B} \cdot d\vec{\ell} = \mu_0(I + I_d)$$

$$= \mu_0 I + \mu_0 \epsilon_0 \frac{d}{dt}\int_S E_n\, dA = \mu_0 I + \mu_0 \epsilon_0 \int_S \frac{\partial E_n}{\partial t}\, dA \qquad\qquad 30\text{-}6d$$

MAXWELL'S EQUATIONS[†]

Equation 30-6a is Gauss's law; it states that the flux of the electric field through any closed surface equals $1/\epsilon_0$ times the net charge inside the surface. As discussed in Chapter 22, Gauss's law implies that the electric field due to a point charge varies inversely as the square of the distance from the charge. This law describes how electric field lines diverge from a positive charge and converge on a negative charge. Its experimental basis is Coulomb's law.

[†] In all four equations, the integration paths C and the integration surfaces S are at rest and the integrations take place at an instant in time.

Equation 30-6b, sometimes called Gauss's law for magnetism, states that the flux of the magnetic field vector $\vec{B}$ is zero through *any* closed surface. This equation describes the experimental observation that magnetic field lines do not diverge from any point in space or converge on any point; that is, it implies that isolated magnetic poles do not exist.

Equation 30-6c is Faraday's law; it states that the integral of the electric field around any closed curve C, which is the emf, equals the (negative) rate of change of the magnetic flux through any surface S bounded by the curve. (S is not a closed surface, so the magnetic flux through S is not necessarily zero.) Faraday's law describes how electric field lines encircle any area through which the magnetic flux is changing, and it relates the electric field vector $\vec{E}$ to the rate of change of the magnetic field vector $\vec{B}$.

Equation 30-6d, which is Ampère's law modified to include Maxwell's displacement current, states that the line integral of the magnetic field $\vec{B}$ around any closed curve C equals μ_0 times the current through any surface S bounded by the curve plus $\mu_0 \epsilon_0$ times the rate of change of the electric flux through the same surface S. This law describes how the magnetic field lines encircle an area through which a current is passing or through which the electric flux is changing.

In Section 30-4, we show how wave equations for both the electric field $\vec{E}$ and the magnetic field $\vec{B}$ can be derived from Maxwell's equations.

30-3 Electromagnetic Waves

Figure 30-4 shows the electric and magnetic field vectors of an electromagnetic wave. The electric and magnetic fields are perpendicular to each other and perpendicular to the direction of propagation of the wave. Electromagnetic waves are thus transverse waves. The electric and magnetic fields are in phase and, at each point in space and at each instant in time, their magnitudes are related by

$$E = cB \qquad\qquad\qquad 30\text{-}7$$

where $c = 1/\sqrt{\mu_0 \epsilon_0}$ is the speed of the wave. The direction of propagation of an electromagnetic wave is the direction of the cross product $\vec{E} \times \vec{B}$.

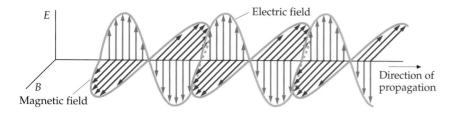

FIGURE 30-4 The electric and magnetic field vectors in an electromagnetic wave. The fields are in phase, perpendicular to each other, and perpendicular to the direction of propagation of the wave.

The Electromagnetic Spectrum

The various types of electromagnetic waves—light, radio waves, X rays, gamma rays, microwaves, and others—differ only in wavelength and frequency, which are related to the speed c in the usual way, $f\lambda = c$. Table 30-1 gives the **electromagnetic spectrum** and the names usually associated with the various frequency and wavelength ranges. These ranges are often not well defined and sometimes overlap. For example, electromagnetic waves with wavelengths of approximately 0.1 nm are usually called X rays, but if the electromagnetic waves originate from nuclear radioactivity, they are called gamma rays.

The human eye is sensitive to electromagnetic radiation with wavelengths from approximately 400 nm to 700 nm, which is the range called **visible light.** The shortest wavelengths in the visible spectrum correspond to violet light and

TABLE 30-1

The Electromagnetic Spectrum

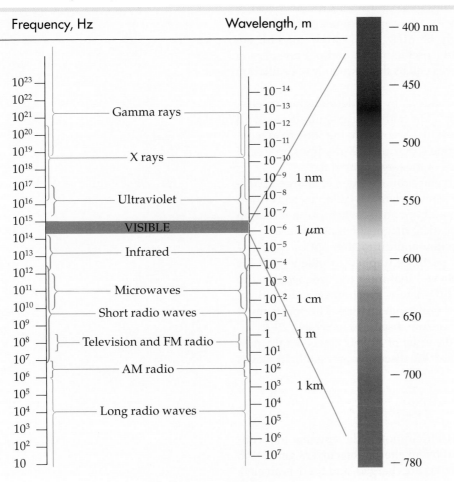

the longest wavelengths to red light, with all the colors of the rainbow falling between these extremes. Electromagnetic waves with wavelengths just beyond the visible spectrum on the short-wavelength side are called **ultraviolet rays,** and those with wavelengths just beyond the visible spectrum on the long-wavelength side are called **infrared waves.** Heat radiation given off by bodies at ordinary temperatures is in the infrared region of the electromagnetic spectrum. There are no limits on the wavelengths of electromagnetic radiation; that is, all wavelengths (or frequencies) are theoretically possible.

The differences in wavelengths of the various kinds of electromagnetic waves have important physical consequences. As we know, the behavior of waves depends strongly on the relative sizes of the wavelengths and the physical objects or apertures the waves encounter. Since the wavelengths of light are in the rather narrow range from approximately 400 nm to 700 nm, they are much smaller than most obstacles, so the ray approximation (introduced in Section 15-4) is often valid. The wavelength and frequency are also important in determining the kinds of interactions between electromagnetic waves and matter. X rays, for example, have very short wavelengths and high frequencies. They easily penetrate many materials that are opaque to lower-frequency light waves, which are absorbed by the materials. Microwaves have wavelengths of the order of a few centimeters and frequencies that are close to the natural resonance frequencies of water molecules in solids and liquids. Microwaves are therefore readily absorbed by the water molecules in foods, which is the mechanism by which food is heated in microwave ovens.

Production of Electromagnetic Waves

Electromagnetic waves are produced when free electric charges accelerate or when electrons bound to atoms and molecules make transitions to lower energy states. Radio waves, which have frequencies from approximately 550 kHz to 1600 kHz for AM and from approximately 88 MHz to 108 MHz for FM, are produced by macroscopic electric currents oscillating in radio transmission antennas. The frequency of the emitted waves equals the frequency of oscillation of the charges.

A continuous spectrum of X rays is produced by the deceleration of electrons when they crash into a metal target. The radiation produced is called **bremsstrahlung** (German for braking radiation). Accompanying the broad, continuous bremsstrahlung spectrum is a discrete spectrum of X-ray lines produced by transitions of inner electrons in the atoms of the target material.

Synchrotron radiation arises from the circular orbital motion of charged particles (usually electrons or positrons) in nuclear accelerators called synchrotrons. Originally considered a nuisance by accelerator scientists, synchrotron radiation X rays are now produced and used as a medical diagnostic tool because of the ease of manipulating the beams with reflection and diffraction optics. Synchrotron radiation is also emitted by charged particles trapped in magnetic fields associated with stars and galaxies. It is believed that most low-frequency radio waves reaching the earth from outer space originate as synchrotron radiation.

Heat is radiated by the thermally excited molecular charges. The spectrum of heat radiation is the blackbody radiation spectrum discussed in Section 20-4.

Light waves, which have frequencies of the order of 10^{14} Hz, are generally produced by transitions of bound atomic charges. We discuss sources of light waves in Chapter 31.

Electric Dipole Radiation

Figure 30-5 is a schematic drawing of an electric-dipole radio antenna that consists of two conducting rods along a line fed by an alternating current generator. At time $t = 0$ (Figure 30-5a), the ends of the rods are charged, and there is an electric field near the rod parallel to the rod. There is also a magnetic field, which is not shown, encircling the rods due to the current in the rods. The fluctuations in these fields move out away from the rods with the speed of light. After one-fourth period, at $t = T/4$ (Figure 30-5b), the rods are uncharged, and the electric field near the rod is zero. At $t = T/2$ (Figure 30-5c), the rods are again charged, but the charges are opposite those at $t = 0$. The electric and magnetic fields at a great distance from the antenna are quite different from the fields near the antenna. Far from the antenna, the electric and magnetic fields oscillate in phase with simple harmonic motion, perpendicular to each other and to the direction of propagation of the wave. Figure 30-6 shows the electric and magnetic fields far from an electric dipole antenna.

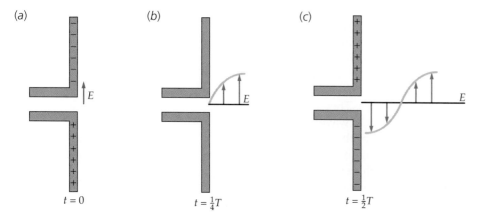

(a) $t = 0$ (b) $t = \frac{1}{4}T$ (c) $t = \frac{1}{2}T$

FIGURE 30-5 An electric dipole radio antenna for radiating electromagnetic waves. Alternating current is supplied to the antenna by a generator (not shown). The fluctuations in the electric field due to the fluctuations in the charges in the antenna propagates outward at the speed of light. There is also a fluctuating magnetic field (not shown) perpendicular to the paper due to the current in the antenna.

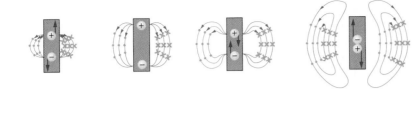

FIGURE 30-6 Electric field lines (in red) and magnetic field lines (in blue) produced by an oscillating electric dipole. Each magnetic field line is a circle with the dipole along its axis. The cross product $\vec{E} \times \vec{B}$ is directed away from the dipole at all points.

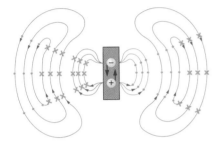

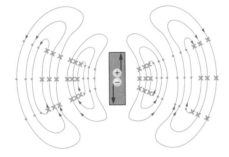

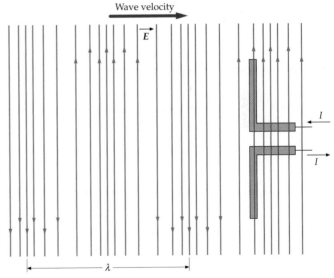

Electromagnetic waves of radio or television frequencies can be detected by an electric dipole antenna placed parallel to the electric field of the incoming wave, so that it induces an alternating current in the antenna (Figure 30-7). These electromagnetic waves can also be detected by a loop antenna placed perpendicular to the magnetic field, so that the changing magnetic flux through the loop induces a current in the loop (Figure 30-8). Electromagnetic waves of frequency in the visible light range are detected by the eye or by photographic film, both of which are mainly sensitive to the electric field.

FIGURE 30-7 An electric dipole antenna for detecting electromagnetic waves. The alternating electric field of the incoming wave produces an alternating current in the antenna.

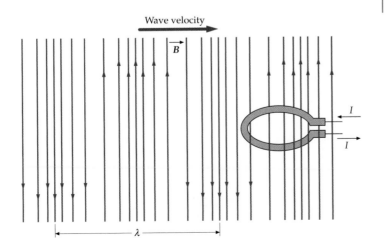

FIGURE 30-8 Loop antenna for detecting electromagnetic radiation. The alternating magnetic flux through the loop due to the magnetic field of the radiation induces an alternating current in the loop.

The radiation from a dipole antenna, such as that shown in Figure 30-5, is called electric dipole radiation. Many electromagnetic waves exhibit the characteristics of electric dipole radiation. An important feature of this type of radiation is that the intensity of the electromagnetic waves radiated by a dipole antenna is zero along the axis of the antenna and maximum in the radial direction (away from the axis). If the dipole is in the y direction with its center at the origin, as in Figure 30-9, the intensity is zero along the y axis and maximum in the xz plane. In the direction of a line making an angle θ with the y axis, the intensity is proportional to $\sin^2 \theta$.

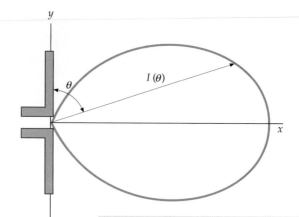

FIGURE 30-9 Polar plot of the intensity of electromagnetic radiation from an electric dipole antenna versus angle. The intensity $I(\theta)$ is proportional to the length of the arrow. The intensity is maximum perpendicular to the antenna at $\theta = 90°$ and minimum along the antenna at $\theta = 0°$ or $\theta = 180°$.

EMF Induced in a Loop Antenna   **EXAMPLE 30-3**

A loop antenna consisting of a single 10-cm radius loop of wire is used to detect electromagnetic waves for which E_{rms} = 0.15 V/m. Find the rms emf induced in the loop if the wave frequency is (*a*) 600 kHz and (*b*) 60 MHz.

PICTURE THE PROBLEM The induced emf in the wire is related to the rate of change of the magnetic flux through the loop by Faraday's law (Equation 30-5). Using Equation 30-7, we can obtain the rms value of the magnetic field from the given rms value of the electric field.

(*a*) 1. Faraday's law relates the magnitude of the emf to the rate of change of the magnetic flux through the flat stationary surface bounded by the loop:

$$|\mathcal{E}| = \frac{d\phi_m}{dt}$$

2. The wavelength of a 600 kHz wave traveling at speed c is $\lambda = c/f = 500$ m. Over the flat surface bounded by the 10-cm radius loop, $\vec{B}$ is quite uniform.

$$\phi_m = BA = \pi r^2 B, \text{ so } |\mathcal{E}| = \frac{d\phi_m}{dt} = \pi r^2 \frac{\partial B}{\partial t}$$

and

$$\mathcal{E}_{rms} = \pi r^2 \left(\frac{\partial B}{\partial t}\right)_{rms}$$

3. Compute dB_{rms}/dt from a sinusoidal B:

$$B = B_0 \sin(kx - \omega t)$$

$$\frac{\partial B}{\partial t} = -\omega B_0 \cos(kx - \omega t)$$

4. Calculate the rms value of $\partial B/\partial t$. The rms value of any sinusoidal function of time equals $1/\sqrt{2}$, and the peak value divided by $\sqrt{2}$ equals the rms value:

$$\left(\frac{\partial B}{\partial t}\right)_{rms} = \omega B_0 \left[-\cos(kx - \omega t)\right]_{rms} = \omega B_0/\sqrt{2} = \omega B_{rms}$$

5. Using Equation 30-7 ($E = cB$), relate the rms value of $\partial B/\partial t$ to E_{rms}:

$$E = cB$$

so

$$B_{rms} = \frac{E_{rms}}{c}$$

6. Substituting into the step 3 result gives:

$$\left(\frac{\partial B}{\partial t}\right)_{rms} = \omega B_{rms} = \omega \frac{E_{rms}}{c} = \frac{2\pi f}{c} E_{rms}$$

7. Substituting the step 6 result into the step 2 result, calculate $\mathcal{E}_{rms}$ at $f = 600$ kHz:

$$\mathcal{E}_{rms} = \pi r^2 \left(\frac{\partial B}{\partial t}\right)_{rms} = \pi r^2 \frac{2\pi f}{c} E_{rms}$$

$$= \pi (0.1\text{ m})^2 \frac{2\pi (6 \times 10^5\text{ Hz})}{3 \times 10^8\text{ m/s}} (0.15\text{ V/m})$$

$$= \boxed{5.92 \times 10^{-5}\text{ V} = 59.2\ \mu\text{V}}$$

(b) The induced emf is proportional to the frequency (step 4), so at 60 MHz it will be 100 times greater than at 600 kHz:

$$\mathcal{E}_{rms} = (100)(5.92 \times 10^{-5}\text{ V}) = 0.00592\text{ V}$$

$$= \boxed{5.92\text{ mV}}$$

REMARKS For part (b) the frequency is 60 MHz, so $\lambda = c/f = 5$ m. $\vec{B}$ is not as uniform over the surface bounded by the 10-cm radius loop when $\lambda = 5$ m as it is when $\lambda = 500$ m, as in part (a). Thus, $\vec{B}$ on the surface when $\lambda = 5$ m is uniform enough that the part (b) result is sufficiently accurate for most purposes.

Energy and Momentum in an Electromagnetic Wave

Like other waves, electromagnetic waves carry energy and momentum. The energy carried is described by the intensity, which is the average power per unit area incident on a surface perpendicular to the direction of propagation. The momentum per unit time per unit area carried by an electromagnetic wave is called the **radiation pressure.**

Intensity Consider an electromagnetic wave traveling toward the right and a cylindrical region of length L and cross-sectional area A with its axis oriented from left to right. The average amount of electromagnetic energy U_{av} within this region equals $u_{av}V$, where u_{av} is the average energy density and $V = LA$ is the volume of the region. In the time it takes the electromagnetic wave to travel the distance L, all of this energy passes through the right end of the region. The time Δt for the wave to travel the distance L is L/c, so the power P_{av} (the energy per unit time) passing out the right end of the region is

$$P_{av} = U_{av}/\Delta t = u_{av}LA/(L/c) = u_{av}Ac$$

and the intensity I (the average power per unit area) is

$$I = P_{av}/A = u_{av}c$$

The total energy density in the wave u is the sum of the electric and magnetic energy densities. The electric energy density u_e (Equation 24-13) and magnetic energy density u_m (Equation 28-20) are given by

$$u_e = \frac{1}{2}\epsilon_0 E^2 \quad \text{and} \quad u_m = \frac{B^2}{2\mu_0}$$

In an electromagnetic wave in free space, E equals cB, so we can express the magnetic energy density in terms of the electric field:

$$u_m = \frac{B^2}{2\mu_0} = \frac{(E/c)^2}{2\mu_0} = \frac{E^2}{2\mu_0 c^2} = \frac{1}{2}\epsilon_0 E^2$$

where we have used $c^2 = 1/(\epsilon_0\mu_0)$. Thus, the electric and magnetic energy densities are equal. Using $E = cB$, we may express the total energy density in several useful ways:

$$u = u_e + u_m = \epsilon_0 E^2 = \frac{B^2}{\mu_0} = \frac{EB}{\mu_0 c}$$ 30-8

ENERGY DENSITY IN AN ELECTROMAGNETIC WAVE

To compute the average energy density, we replace the instantaneous fields E and B by their rms values $E_{rms} = E_0/\sqrt{2}$ and $B_{rms} = B_0/\sqrt{2}$, where E_0 and B_0 are the maximum values of the fields. The intensity is then

$$I = u_{av}c = \frac{E_{rms}B_{rms}}{\mu_0} = \frac{1}{2}\frac{E_0 B_0}{\mu_0} = |\vec{S}|_{av}$$ 30-9

INTENSITY OF AN ELECTROMAGNETIC WAVE

where the vector

$$\vec{S} = \frac{\vec{E} \times \vec{B}}{\mu_0}$$ 30-10

DEFINITION—POYNTING VECTOR

is called the **Poynting vector** after its discoverer, John Poynting. The average magnitude of $\vec{S}$ is the intensity of the wave, and the direction of $\vec{S}$ is the direction of propagation of the wave.

Radiation Pressure   We now show by a simple example that an electromagnetic wave carries momentum. Consider a wave moving along the x axis that is incident on a stationary charge, as shown in Figure 30-10. For simplicity, we assume that $\vec{E}$ is in the y direction and $\vec{B}$ is in the z direction, and we neglect the time dependence of the fields. The particle experiences a force $q\vec{E}$ in the y direction and is thus accelerated by the electric field. At any time t, the velocity in the y direction is

$$v_y = at = \frac{qE}{m}t$$

After a short time t_1, the charge has acquired kinetic energy equal to

$$K = \frac{1}{2}mv_y^2 = \frac{1}{2}\frac{mq^2E^2t_1^2}{m^2} = \frac{1}{2}\frac{q^2E^2}{m}t_1^2$$ 30-11

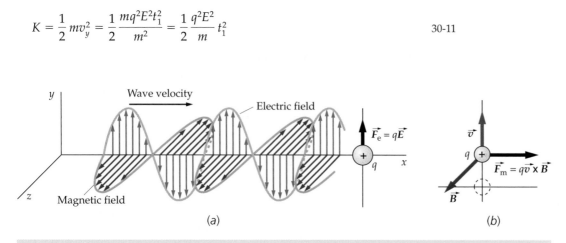

(a)   (b)

FIGURE 30-10 An electromagnetic wave incident on a point charge that is initially at rest on the x axis. (a) The electric force $q\vec{E}$ accelerates the charge in the upward direction. (b) When the velocity $\vec{v}$ of the charge is upward, the magnetic force $q\vec{v} \times \vec{B}$ accelerates the charge in the direction of the wave.

When the charge is moving in the y direction, it experiences a magnetic force

$$\vec{F}_{m} = q\vec{v} \times \vec{B} = qv_{y}\hat{j} \times B\hat{k} = qv_{y}B\hat{i} = \frac{q^2EB}{m}\,t\hat{i}$$

Note that this force is in the direction of propagation of the wave. Using $dp_x = F_x\,dt$, we find for the momentum p_x transferred by the wave to the particle in time t_1:

$$p_x = \int_0^{t_1} F_x\,dt = \int_0^{t_1} \frac{q^2EB}{m}\,t\,dt = \frac{1}{2}\frac{q^2EB}{m}\,t_1^2$$

If we use $B = E/c$, this becomes

$$p_x = \frac{1}{c}\left(\frac{1}{2}\frac{q^2E^2}{m}\,t_1^2\right) \tag{30-12}$$

Comparing Equations 30-11 and 30-12, we see that the momentum acquired by the charge in the direction of the wave is $1/c$ times the energy. Although our simple calculation was not rigorous, the results are correct. The magnitude of the momentum carried by an electromagnetic wave is $1/c$ times the energy carried by the wave:

$$p = \frac{U}{c} \tag{30-13}$$

MOMENTUM AND ENERGY IN AN ELECTROMAGNETIC WAVE

Since the intensity is the energy per unit area per unit time, the intensity divided by c is the momentum carried by the wave per unit area per unit time. The momentum carried per unit time is a force. The intensity divided by c is thus a force per unit area, which is a pressure. This pressure is the radiation pressure P_r:

$$P_r = \frac{I}{c} \tag{30-14}$$

RADIATION PRESSURE AND INTENSITY

We can relate the radiation pressure to the electric or magnetic fields by using Equation 30-9 to relate I to E and B, and Equation 30-7 to eliminate either E or B:

$$P_r = \frac{I}{c} = \frac{E_0 B_0}{2\mu_0 c} = \frac{E_{\text{rms}} B_{\text{rms}}}{\mu_0 c} = \frac{E_0^2}{2\mu_0 c^2} = \frac{B_0^2}{2\mu_0} \tag{30-15}$$

RADIATION PRESSURE IN TERMS OF E AND B

Consider an electromagnetic wave incident normally on some surface. If the surface absorbs energy U from the electromagnetic wave, it also absorbs momentum p given by Equation 30-13, and the pressure exerted on the surface equals the radiation pressure. If the wave is reflected, the momentum transferred is $2p$ because the wave now carries momentum in the opposite direction. The pressure exerted on the surface by the wave is then twice that given by Equation 30-15.

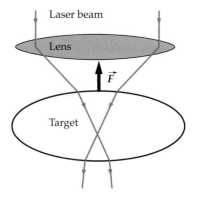

"Laser tweezers" make use of the momentum carried by electromagnetic waves to manipulate targets on a molecular scale. The two rays shown are refracted as they pass through a transparent target, such as a biological cell, or on an even smaller scale, as a tiny transparent bead attached to a large molecule within a cell. At each refraction, the rays are bent downward, which increases the downward component of momentum of the rays. The target thus exerts a downward force on the laser beams, and the laser beams exert an upward force on the target, which pulls the target toward the laser source. The force is typically of the order of piconewtons. Laser tweezers have been used to accomplish such astonishing feats as stretching out coiled DNA.

RADIATION PRESSURE 3 M FROM A LIGHTBULB   **EXAMPLE 30-4**

A lightbulb emits spherical electromagnetic waves uniformly in all directions. Find (a) the intensity, (b) the radiation pressure, and (c) the electric and magnetic field magnitudes at a distance of 3 m from the lightbulb, assuming that 50 W of electromagnetic radiation is emitted.

PICTURE THE PROBLEM At a distance r from the lightbulb, the energy is spread uniformly over an area $4\pi r^2$. The intensity is the power divided by the area. The radiation pressure can then be found from $P_r = I/c$.

(a) 1. Divide the power output by the area to find the intensity:

$$I = \frac{50 \text{ W}}{4\pi r^2}$$

2. Substitute $r = 3$ m:

$$I = \frac{50 \text{ W}}{4\pi(3 \text{ m})^2} = \boxed{0.442 \text{ W/m}^2}$$

(b) The radiation pressure is the intensity divided by the speed of light:

$$P_r = \frac{I}{c} = \frac{0.442 \text{ W/m}^2}{3 \times 10^8 \text{ m/s}} = \boxed{1.47 \times 10^{-9} \text{ Pa}}$$

(c) 1. B_0 is related to P_r by Equation 30-15:

$$B_0 = \sqrt{2\mu_0 P_r}$$
$$= [2(4\pi \times 10^{-7} \text{ T·m/A})(1.47 \times 10^{-9} \text{ Pa})]^{1/2}$$
$$= 6.08 \times 10^{-8} \text{ T}$$

2. The maximum value of the electric field E_0 is c times B_0:

$$E_0 = cB_0 = (3 \times 10^8 \text{ m/s})(6.08 \times 10^{-8} \text{ T})$$
$$= 18.2 \text{ V/m}$$

3. The electric and magnetic field magnitudes at that point are of the form:

$$\boxed{\begin{array}{l} E = E_0 \sin \omega t \quad \text{and} \quad B = B_0 \sin \omega t \\ \text{with } E_0 = 18.2 \text{ V/m} \quad \text{and} \quad B_0 = 6.08 \times 10^{-8} \text{ T} \end{array}}$$

REMARKS Only about 2 percent of the power consumed by incandescent bulbs is transformed into visible light. Note that the radiation pressure calculated in Part (b) is very small compared with the atmospheric pressure, which is of the order of 10^5 Pa.

A LASER ROCKET   **EXAMPLE 30-5**

You are stranded in space a distance of 20 m from your spaceship. You carry a 1-kW laser. If your total mass, including your space suit and laser, is 95 kg, how long will it take you to reach the spaceship if you point the laser directly away from it?

PICTURE THE PROBLEM The laser emits light, which carries with it momentum. By momentum conservation, you are given an equal and opposite momentum toward the spaceship. The momentum carried by light is $p = U/c$, where U is the energy of the light. If the power of the laser is $P = dU/dt$, then the rate of change of momentum produced by the laser is $dp/dt = (dU/dt)/c = P/c$. This is the force exerted on you, which is constant.

1. The time taken is related to the distance and the acceleration. We assume that you are initially at rest relative to the spaceship:

$$x = \frac{1}{2}at^2; \quad t = \sqrt{\frac{2x}{a}}$$

2. Your acceleration is the force divided by your mass, and the force is the power divided by c:

$$a = \frac{F}{m} = \frac{P/c}{m} = \frac{P}{mc}$$

3. Use this acceleration to calculate the time t:

$$t = \sqrt{\frac{2x}{a}} = \sqrt{\frac{2xmc}{P}}$$

$$= \sqrt{\frac{2(20 \text{ m})(95 \text{ kg})(3 \times 10^8 \text{ m/s})}{1000 \text{ W}}}$$

$$= 3.38 \times 10^4 \text{ s} = \boxed{9.38 \text{ h}}$$

REMARKS Note that the acceleration is extremely small—only about 10^{-9} g. Your speed when you reach the spaceship would be $v = at = 1.19$ mm/s, which is practically imperceptible.

EXERCISE How long would it take you to reach the spaceship if you took off one of your shoelaces and threw it as fast as you could in the direction opposite the ship? (To answer this, you must first estimate the mass of the shoelace and the maximum speed that you can throw the shoelace.) (*Answer* About 5 h for a 10-g shoelace thrown at 10 m/s)

*30-4 The Wave Equation for Electromagnetic Waves

In Section 15-1, we saw that waves on a string obey a partial differential equation called the **wave equation:**

$$\frac{\partial^2 y(x, t)}{\partial x^2} = \frac{1}{v^2} \frac{\partial^2 y(x, t)}{\partial t^2} \qquad \text{30-16}$$

where $y(x, t)$ is the wave function, which for string waves is the displacement of the string. The velocity of the wave is given by $v = \sqrt{F/\mu}$, where F is the tension and μ is the linear mass density. The general solution to this equation is

$$y(x, t) = f_1(x - vt) + f_2(x + vt)$$

The general solution functions can be expressed as a superposition of harmonic wave functions of the form

$$y(x, t) = y_0 \sin(kx - \omega t) \quad \text{and} \quad y(x, t) = y_0 \sin(kx + \omega t)$$

where $k = 2\pi/\lambda$ is the wave number and $\omega = 2\pi f$ is the angular frequency.

Maxwell's equations imply that $\vec{E}$ and $\vec{B}$ obey wave equations similar to Equation 30-16. We consider only free space, in which there are no charges or currents, and we assume that the electric and magnetic fields $\vec{E}$ and $\vec{B}$ are functions of time and one space coordinate only, which we will take to be the x coordinate. Such a wave is called a **plane wave,** because $\vec{E}$ and $\vec{B}$ are uniform throughout any plane perpendicular to the x axis. For a plane electromagnetic wave traveling parallel to the x axis, the x components of the fields are zero, so the vectors $\vec{E}$ and $\vec{B}$ are perpendicular to the x axis and each obeys the wave equation:

$$\frac{\partial^2 \vec{E}}{\partial x^2} = \frac{1}{c^2} \frac{\partial^2 \vec{E}}{\partial t^2} \qquad \text{30-17a}$$

WAVE EQUATION FOR $\vec{E}$

$$\frac{\partial^2 \vec{B}}{\partial x^2} = \frac{1}{c^2}\frac{\partial^2 \vec{B}}{\partial t^2}$$

30-17b

WAVE EQUATION FOR $\vec{B}$

where $c = 1/\sqrt{\mu_0 \epsilon_0}$ is the speed of the waves. (*Note:* Dimensional reasoning helps in remembering these equations. For each equation, the numerators on both sides are the same and the denominators on both sides have the dimension of length squared.)

*Derivation of the Wave Equation

We can relate the space derivative of one of the field vectors to the time derivative of the other field vector by applying Faraday's law (Equation 30-6c) and the modified version of Ampère's law (Equation 30-6d) to appropriately chosen curves in space. We first relate the space derivative of E_y to the time derivative of B_z by applying Equation 30-6c (Faraday's law) to the rectangular curve of sides Δx and Δy lying in the xy plane (Figure 30-11). The circulation of $\vec{E}$ around C (the line integral of $\vec{E}$ around curve C) is

$$\oint_C \vec{E} \cdot d\vec{\ell} = E_y(x_2)\Delta y - E_y(x_1)\Delta y$$

where $E_y(x_1)$ is the value of E_y at the point x_1 and $E_y(x_2)$ is the value of E_y at the point x_2. The contributions of the type $E_x\Delta x$ from the top and bottom of this curve are zero because $E_x = 0$. Since Δx is very small (compared to the wavelength), we can approximate the difference in E_y on the left and right sides of this curve (at x_1 and at x_2) by

$$E_y(x_2) - E_y(x_1) = \Delta E_y \approx \frac{\partial E_y}{\partial x}\Delta x$$

Then

$$\oint_C \vec{E} \cdot d\vec{\ell} \approx \frac{\partial E_y}{\partial x}\Delta x\,\Delta y$$

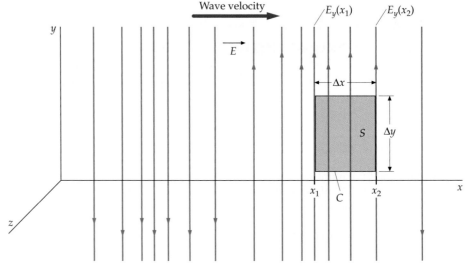

FIGURE 30-11 A rectangular curve in the xy plane for the derivation of Equation 30-18.

Faraday's law is

$$\oint_C \vec{E} \cdot d\vec{\ell} = -\int_S \frac{\partial B_n}{\partial t}\,dA$$

The flux of $\partial B_n/\partial t$ through the rectangular surface bounded by this curve is approximately

$$\int_S B_n\,dA \approx \frac{\partial B_z}{\partial t}\Delta x\,\Delta y$$

Faraday's law then gives

$$\frac{\partial E_y}{\partial x}\Delta x\,\Delta y = -\frac{\partial B_z}{\partial t}\Delta x\,\Delta y$$

or

$$\frac{\partial E_y}{\partial x} = -\frac{\partial B_z}{\partial t}$$ 30-18

Equation 30-18 implies that if there is a component of the electric field E_y that depends on x, there must be a component of the magnetic field B_z that depends on time or, conversely, if there is a component of the magnetic field B_z that depends on time, there must be a component of the electric field E_y that depends on x. We can get a similar equation relating the space derivative of the magnetic field B_z to the time derivative of the electric field E_y by applying Ampère's law (Equation 30-6d) to the curve of sides Δx and Δz in the xz plane shown in Figure 30-12.

For the case of no conduction currents ($I = 0$), Equation 30-6d is

$$\oint_C \vec{B} \cdot d\vec{\ell} = \mu_0\epsilon_0 \int_S \frac{\partial E_n}{\partial t}\, dA$$

The details of this calculation are similar to those for Equation 30-18. The result is

$$\frac{\partial B_z}{\partial x} = -\mu_0\epsilon_0 \frac{\partial E_y}{\partial t}$$ 30-19

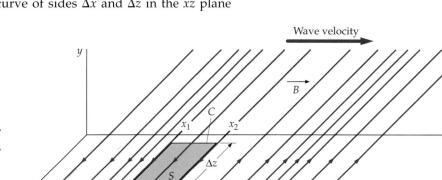

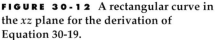

We can eliminate either B_z or E_y from Equations 30-18 and 30-19 by differentiating both sides of either equation with respect to either x or t. If we differentiate both sides of Equation 30-18 with respect to x, we obtain

FIGURE 30-12 A rectangular curve in the xz plane for the derivation of Equation 30-19.

$$\frac{\partial}{\partial x}\left(\frac{\partial E_y}{\partial x}\right) = -\frac{\partial}{\partial x}\left(\frac{\partial B_z}{\partial t}\right)$$

Interchanging the order of the time and space derivatives on the term to the right of the equal sign gives

$$\frac{\partial^2 E_y}{\partial x^2} = -\frac{\partial}{\partial t}\left(\frac{\partial B_z}{\partial x}\right)$$

Using Equation 30-19, we substitute for $\partial B_z/\partial x$ to obtain

$$\frac{\partial^2 E_y}{\partial x^2} = -\frac{\partial}{\partial t}\left(-\mu_0\epsilon_0 \frac{\partial E_y}{\partial t}\right)$$

which yields the wave equation

$$\frac{\partial^2 E_y}{\partial x^2} = \mu_0\epsilon_0 \frac{\partial^2 E_y}{\partial t^2}$$ 30-20

Comparing Equation 30-20 with Equation 30-16, we see that E_y obeys a wave equation for waves with speed $c = 1/\sqrt{\mu_0\epsilon_0}$, which is Equation 30-1.

If we had instead chosen to eliminate E_y from Equations 30-18 and 30-19 (by differentiating Equation 30-18 with respect to t, for example), we would have obtained an equation identical to Equation 30-20 except with B_z replacing E_y. We can thus see that both the electric field E_y and the magnetic field B_z obey a wave equation for waves traveling with the velocity $1/\sqrt{\mu_0\epsilon_0}$, which is the velocity of light.

By following the same line of reasoning as used above, and applying Equation 30-6c (Faraday's law) to the curve in the xz plane (Figure 30-12), we would obtain

$$\frac{\partial E_z}{\partial x} = \frac{\partial B_y}{\partial t} \qquad\qquad 30\text{-}21$$

Similarly, the application of Equation 30-6d to the curve in the xy plane (Figure 30-11) gives

$$\frac{\partial B_y}{\partial x} = \mu_0 \epsilon_0 \frac{\partial E_z}{\partial t} \qquad\qquad 30\text{-}22$$

We can use these results to show that, for a wave propagating in the x direction, the components E_z and B_y also obey the wave equation.

To show that the magnetic field B_z is in phase with the electric field E_y, consider the harmonic wave function of the form

$$E_y = E_{y0} \sin(kx - \omega t) \qquad\qquad 30\text{-}23$$

If we substitute this solution into Equation 30-18, we have

$$\frac{\partial B_z}{\partial t} = -\frac{\partial E_y}{\partial x} = -kE_{y0} \cos(kx - \omega t)$$

To solve for B_z, we take the integral of $\partial B_z / \partial t$ with respect to time. Doing so yields

$$B_z = \int \frac{\partial B_z}{\partial t}\, dt = \frac{k}{\omega} E_{y0} \sin(kx - \omega t) + f(x) \qquad\qquad 30\text{-}24$$

where $f(x)$ is an arbitrary function of x.

EXERCISE Verify Equation 30-24 by taking $\partial B_z / \partial t$, where $B_z = (k/\omega)E_{y0} \sin(kx - \omega t) + f(x)$.

The result should be $-kE_{y0} \cos(kx - \omega t)$, which is the right hand side of the previous equation.

We next substitute the solution (Equation 30-23) into Equation 30-19 and obtain

$$\frac{\partial B_z}{\partial x} = -\mu_0 \epsilon_0 \frac{\partial E_y}{\partial t} = \omega \mu_0 \epsilon_0 E_{y0} \cos(kx - \omega t)$$

Solving for B_z gives

$$B_z = \int \frac{\partial B_z}{\partial x}\, dx = \frac{\omega \mu_0 \epsilon_0}{k} E_{y0} \sin(kx - \omega t) + g(t) \qquad\qquad 30\text{-}25$$

where $g(t)$ is an arbitrary function of time. Equating the right sides of Equations 30-24 and 30-25 gives

$$\frac{k}{\omega} E_{y0} \sin(kx - \omega t) + f(x) = \frac{\omega \mu_0 \epsilon_0}{k} E_{y0} \sin(kx - \omega t) + g(t)$$

Substituting c for ω/k and $1/c^2$ for $\mu_0\epsilon_0$ gives

$$\frac{1}{c} E_{y0} \sin(kx - \omega t) + f(x) = \frac{1}{c} E_{y0} \sin(kx - \omega t) + g(t)$$

which implies $f(x) = g(t)$ for all values of x and t. These remain equal only if $f(x) = g(t) = $ constant (independent of both x and t). Thus, Equation 30-24 becomes

$$B_z = \frac{k}{\omega} E_{y0} \sin(kx - \omega t) + \text{constant} = B_{z0} \sin(kx - \omega t) \qquad \text{30-26}$$

where $B_{z0} = (k/\omega)E_{y0} = (1/c)E_{y0}$. The integration constant was dropped because it plays no part in the wave. It merely allows for the presence of a static uniform magnetic field. Since the electric and magnetic fields oscillate in phase with the same frequency, we have the general result that the magnitude of the electric field is c times the magnitude of the magnetic field for an electromagnetic wave:

$$E = cB$$

which is Equation 30-7.

We see that Maxwell's equations imply wave equations 30-17a and 30-17b for the electric and magnetic fields; and that if E_y varies harmonically, as in Equation 30-23, the magnetic field B_z is in phase with E_y and has an amplitude related to the amplitude of E_y by $B_z = E_y/c$. The electric and magnetic fields are perpendicular to each other and to the direction of the wave propagation, as shown in Figure 30-4.

$\vec{B}(x, t)$ *FOR A* *LINEARLY POLARIZED PLANE WAVE*   **EXAMPLE   30-6**

The electric field of an electromagnetic wave is given by $\vec{E}(x, t) = E_0 \cos(kx - \omega t)\hat{k}$. (*a*) What is the direction of propagation of the wave? (*b*) What is the direction of the magnetic field in the $x = 0$ plane at time $t = 0$? (*c*) Find the magnetic field of the same wave. (*d*) Compute $\vec{E} \times \vec{B}$.

PICTURE THE PROBLEM   The argument of the cosine gives the direction of propagation. $\vec{B}$ is perpendicular to both $\vec{E}$ and to the direction of propagation. $\vec{B}$ and $\vec{E}$ are in phase.

(*a*) The argument of the cosine function $(kx - \omega t)$ tells us the direction of propagation:

| The direction of propagation is the direction of increasing x, which is the direction of $\hat{i}$. |

(*b*) 1. $\vec{B}$ is in phase with $\vec{E}$ and is perpendicular to both $\vec{E}$ and the direction of propagation $\hat{k}$. (That is, $\vec{B}$ is perpendicular to both $\hat{i}$ and $\hat{k}$.) That means:

$$\vec{B}(x, t) = \pm B_0 \cos(kx - \omega t)\hat{j}$$

2. $\vec{E} \times \vec{B}$ is in the direction of propagation $\hat{i}$. Use the expressions for $\vec{E}$ and $\vec{B}$ and take the cross product:

$$\vec{E} \times \vec{B} = E_0 \cos(kx - \omega t)\hat{k} \times (\pm B_0 \cos(kx - \omega t)\hat{j})$$
$$= E_0(\pm B_0) \cos^2(kx - \omega t)(\hat{k} \times \hat{j})$$
$$= E_0(\pm B_0) \cos^2(kx - \omega t)(-\hat{i})$$

3. Choose the sign so that $\vec{E} \times \vec{B}$ is in the $\hat{i}$ direction:

$$\vec{E} \times \vec{B} = E_0(-B_0) \cos^2(kx - \omega t)(-\hat{i})$$

so

$$\vec{B}(x, t) = -B_0 \cos(kx - \omega t)\hat{j}$$

4. Evaluate $\vec{B}$ when both x and t equal zero.

$$\vec{B}(0, 0) = -B_0 \cos[k(0) - \omega(0)]\hat{j} = -B_0\hat{j}$$

$\therefore$ | $\vec{B}(0, 0)$ is in the negative y direction. |

(c) In an electromagnetic wave, $E_0 = cB_0$ and $\vec{B}$ and $\vec{E}$ are in phase. Thus:

$$\vec{B}(x, t) = \boxed{-B_0 \cos(kx - \omega t)\,\hat{j}, \text{ where } B_0 = E_0/c}$$

(d) Calculate $\vec{E} \times \vec{B}$. Let $\theta = kx - \omega t$ and do the calculation:

$$\vec{E} \times \vec{B} = (E_0 \cos\theta\,\hat{k}) \times (-B_0 \cos\theta\,\hat{j})$$

$$= -E_0 B_0 \cos^2\theta\,(\hat{k} \times \hat{j})$$

$$= \boxed{E_0 B_0 \cos^2\theta\,\hat{i}, \text{ where } \theta = kx - \omega t}$$

REMARKS The Part (d) result confirms the Part (a) result, because for an electromagnetic wave $\vec{E} \times \vec{B}$ is always in the direction of propagation.

$\vec{B}(x, t)$ FOR A CIRCULAR POLARIZED PLANE WAVE   **EXAMPLE 30-7**

The electric field of an electromagnetic wave is given by $\vec{E}(x, t) = E_0 \sin(kx - \omega t)\,\hat{j} + E_0 \cos(kx - \omega t)\,\hat{k}$. **(a) Find the magnetic field of the same wave. (b) Compute $\vec{E} \cdot \vec{B}$ and $\vec{E} \times \vec{B}$.**

PICTURE THE PROBLEM We can solve this using the principle of superposition. The given electric field is the superposition of two fields, the one given in Equation 30-23 and the one given in the problem statement of Example 30-6.

(a) 1. From the phase (the argument of the trig functions) we can see that the direction of propagation is the positive x direction:

The phase is for a wave traveling in the positive x direction.

2. The given electric field can be considered as the superposition of $\vec{E}_1 = E_0 \sin(kx - \omega t)\,\hat{j}$ and $\vec{E}_2 = E_0 \cos(kx - \omega t)\,\hat{k}$. Find the magnetic fields $\vec{B}_1$ and $\vec{B}_2$ associated with these electric fields, respectively. Use the procedure followed in Example 30-6:

For $\vec{E}_1 = E_0 \sin(kx - \omega t)\,\hat{j}$, $\vec{B}_1 = B_0 \sin(kx - \omega t)\,\hat{k}$

and

For $\vec{E}_2 = E_0 \cos(kx - \omega t)\,\hat{k}$, $\vec{B}_2 = -B_0 \cos(kx - \omega t)\,\hat{j}$

where

$E_0 = cB_0$

3. The superposition of magnetic fields gives the resultant magnetic fields:

$$\boxed{\begin{aligned} \vec{B}(x, t) &= \vec{B}_1 + \vec{B}_2 \\ &= B_0 \sin(kx - \omega t)\,\hat{k} - B_0 \cos(kx - \omega t)\,\hat{j} \\ \text{where} \\ B_0 &= E_0/c \end{aligned}}$$

(b) 1. Let $\theta = kx - \omega t$ to simplify the notation and calculate $\vec{E} \cdot \vec{B}$:

$$\vec{E} \cdot \vec{B} = (E_0 \sin\theta\,\hat{j} + E_0 \cos\theta\,\hat{k}) \cdot (B_0 \sin\theta\,\hat{k} - B_0 \cos\theta\,\hat{j})$$

$$= E_0 B_0 \sin^2\theta\,\hat{j} \cdot \hat{k} - E_0 B_0 \sin\theta\cos\theta\,\hat{j} \cdot \hat{j}$$

$$+ E_0 B_0 \cos\theta\sin\theta\,\hat{k} \cdot \hat{k} - E_0 B_0 \cos^2\theta\,\hat{k} \cdot \hat{j}$$

$$= 0 - E_0 B_0 \sin\theta\cos\theta + E_0 B_0 \cos\theta\sin\theta - 0 = \boxed{0}$$

2. Calculate $\vec{E} \times \vec{B}$:

$$\vec{E} \times \vec{B} = (E_0 \sin\theta\,\hat{j} + E_0 \cos\theta\,\hat{k}) \times (-B_0 \cos\theta\,\hat{j} + B_0 \sin\theta\,\hat{k})$$

$$= -E_0 B_0 \sin\theta\cos\theta\,(\hat{j} \times \hat{j}) + E_0 B_0 \sin^2\theta\,(\hat{j} \times \hat{k})$$

$$- E_0 B_0 \cos^2\theta\,(\hat{k} \times \hat{j}) + E_0 B_0 \cos\theta\sin\theta\,(\hat{k} \times \hat{k})$$

$$= 0 + E_0 B_0 \sin^2\theta\,\hat{i} + E_0 B_0 \cos^2\theta\,\hat{i} + 0 = \boxed{E_0 B_0\,\hat{i}}$$

REMARKS We see that $\vec{E}$ and $\vec{B}$ are perpendicular to one another, and that $\vec{E} \times \vec{B}$ is in the direction of propagation of the wave. This type of electromagnetic wave is said to be *circularly polarized*. At a fixed value of x, both $\vec{E}$ and $\vec{B}$ rotate in a circle in a plane perpendicular to $\hat{i}$ with angular frequency ω.

EXERCISE Calculate $\vec{E} \cdot \vec{E}$ and $\vec{B} \cdot \vec{B}$. [*Answer* $\vec{E} \cdot \vec{E} = E_y^2 + E_z^2 = E_0^2 \sin^2(kx - \omega t) + E_0^2 \cos^2(kx - \omega t) = E_0^2$ and $\vec{B} \cdot \vec{B} = B_y^2 + B_z^2 = B_0^2 \cos^2(kx - \omega t) + B_0^2 \sin^2(kx - \omega t) = B_0^2$]

REMARKS The fields $\vec{E}$ and $\vec{B}$ are constant in magnitude.

SUMMARY

1. Maxwell's equations summarize the fundamental laws of physics that govern electricity and magnetism.

2. Electromagnetic waves include light, radio and television waves, X rays, gamma rays, microwaves, and others.

Topic	Relevant Equations and Remarks	
1. Maxwell's Displacement Current	Ampère's law can be generalized to apply to currents that are not steady (and not continuous) if the conduction current I is replaced by $I + I_d$, where I_d is Maxwell's displacement current:	
	$$I_d = \epsilon_0 \frac{d\phi_e}{dt}$$	30-3
Generalized form of Ampère's law	$$\oint_C \vec{B} \cdot d\vec{\ell} = \mu_0(I + I_d) = \mu_0 I + \mu_0 \epsilon_0 \frac{d\phi_e}{dt}$$	30-4
2. Maxwell's Equations	The laws of electricity and magnetism are summarized by Maxwell's equations.	
Gauss's law	$$\oint_S E_n \, dA = \frac{1}{\epsilon_0} Q_{\text{inside}}$$	30-6a
Gauss's law for magnetism (isolated magnetic poles do not exist)	$$\oint_S B_n \, dA = 0$$	30-6b
Faraday's law	$$\oint_C \vec{E} \cdot d\vec{\ell} = -\frac{d}{dt}\int_S B_n \, dA = -\int_S \frac{\partial B_n}{\partial t} \, dA$$	30-6c
Ampère's law modified	$$\oint_C \vec{B} \cdot d\vec{\ell} \, \mu_0(I + I_d)$$ $$= \mu_0 I + \mu_0 \epsilon_0 \frac{d}{dt}\int_S E_n \, dA = \mu_0 I + \mu_0 \epsilon_0 \int_S \frac{\partial E_n}{\partial t} \, dA$$	30-6d
3. Electromagnetic Waves	In an electromagnetic wave, the electric and magnetic field vectors are perpendicular to each other and to the direction of propagation. Their magnitudes are related by	
	$$E = cB$$	30-7
Wave speed	$$c = \frac{1}{\sqrt{\mu_0 \epsilon_0}} \approx 3 \times 10^8 \text{ m/s}$$	30-1

Electromagnetic spectrum	The various types of electromagnetic waves—light, radio waves, X rays, gamma rays, microwaves, and others—differ only in wavelength and frequency. The human eye is sensitive to the range from about 400 nm to 700 nm.	
Electric dipole radiation	Electromagnetic waves are produced when free electric charges accelerate. Oscillating charges in an electric dipole antenna radiate electromagnetic waves with an intensity that is greatest in directions perpendicular to the antenna. There is no radiated intensity along the axis of the antenna. Perpendicular to the antenna and far away from it, the electric field of the electromagnetic wave is parallel to the antenna.	
Energy density in an electromagnetic wave	$u = u_e + u_m = \epsilon_0 E^2 = \dfrac{B^2}{\mu_0} = \dfrac{EB}{\mu_0 c}$	30-8
Intensity of an electromagnetic wave	$I = u_{av} c = \dfrac{E_{rms} B_{rms}}{\mu_0} = \dfrac{1}{2} \dfrac{E_0 B_0}{\mu_0} = \left| \vec{S} \right|_{av}$	30-9
Poynting vector	$\vec{S} = \dfrac{\vec{E} \times \vec{B}}{\mu_0}$	30-10
Momentum and energy in an electromagnetic wave	$p = \dfrac{U}{c}$	30-13
Radiation pressure and intensity	$P_r = \dfrac{I}{c}$	30-14

4. ***The Wave Equation for Electromagnetic Waves**

Maxwell's equations imply that the electric and magnetic field vectors in free space obey a wave equation.

$$\frac{\partial^2 \vec{E}}{\partial x^2} = \frac{1}{c^2} \frac{\partial^2 \vec{E}}{\partial t^2} \qquad \text{30-17}a$$

$$\frac{\partial^2 \vec{B}}{\partial x^2} = \frac{1}{c^2} \frac{\partial^2 \vec{B}}{\partial t^2} \qquad \text{30-17}b$$

PROBLEMS

- Single-concept, single-step, relatively easy
- •• Intermediate-level, may require synthesis of concepts
- ••• Challenging
- SSM Solution is in the *Student Solutions Manual*
- iSOLVE Problems available on iSOLVE online homework service
- iSOLVE✓ These "Checkpoint" online homework service problems ask students additional questions about their confidence level, and how they arrived at their answer.

In a few problems, you are given more data than you actually need; in a few other problems, you are required to supply data from your general knowledge, outside sources, or informed estimates.

Conceptual Problems

1 • SSM True or false:

(a) Maxwell's equations apply only to fields that are constant over time.
(b) The wave equation can be derived from Maxwell's equations.

(c) Electromagnetic waves are transverse waves.
(d) In an electromagnetic wave in free space, the electric and magnetic fields are in phase.
(e) In an electromagnetic wave in free space, the electric and magnetic field vectors $\vec{E}$ and $\vec{B}$ are equal in magnitude.
(f) In an electromagnetic wave in free space, the electric and magnetic energy densities are equal.

2 •• Theorists have speculated about the possible existence of magnetic monopoles, and there have been several, as yet unsuccessful, experimental searches for such monopoles. Suppose magnetic monopoles were found and that the magnetic field at a distance r from a monopole of strength q_m is given by $B = (\mu_0/4\pi)q_m/r^2$. How would Maxwell's equations have to be modified to be consistent with such a discovery?

3 • Which waves have greater frequencies, light waves or X rays?

4 • SSM Are the frequencies of ultraviolet radiation greater or less than those of infrared radiation?

5 • What kind of waves have wavelengths of the order of a few meters?

6 • The detection of radio waves can be accomplished with either a dipole antenna or a loop antenna. The dipole antenna detects the (pick one) *electric/magnetic* field of the wave, and the loop antenna detects the *electric/magnetic* field of the wave.

7 • A transmitter uses a loop antenna with the loop in the horizontal plane. What should be the orientation of a dipole antenna at the receiver for optimum signal reception?

8 • SSM A helium-neon laser has a red beam. It is shone in turn on a red plastic filter (of the kind used for theater lighting) and a green plastic filter. (A red theater-lighting filter transmitts only red light.) On which filter will the laser exert a larger force?

Estimation and Approximation

9 •• Estimate the intensity and total power needed in a laser beam to lift a 15-μm diameter plastic bead against the force of gravity. Make any assumptions you think reasonable.

10 ••• Some science fiction writers have used solar sails to propel interstellar spaceships. Imagine a giant sail erected on a spacecraft subjected to the solar radiation pressure. (*a*) Show that the spacecraft's acceleration is given by

$$a = \frac{P_S A}{4\pi r^2 \, cm}$$

where P_S is the power output of the sun and is equal to 3.8×10^{26} W, A is the surface area of the sail, m is the total mass of the spacecraft, r is the distance from the sun, and c is the speed of light. (*b*) Show that the velocity of the spacecraft at a distance r from the sun is found from

$$v^2 = v_0^2 + \left(\frac{P_S A}{2\pi mc}\right)\left(\frac{1}{r_0} - \frac{1}{r}\right)$$

where v_0 is the initial velocity at r_0. (*c*) Compare the relative accelerations due to the radiation pressure and the gravitational force. Use reasonable values for A and m. Will such a system work?

11 •• The intensity of sunlight striking the earth's upper atmosphere (called the solar constant) is 1.37 kW/m². (*a*) Find E_{rms} and B_{rms} due to the sun at the upper atmosphere of the earth. (*b*) Find the average power output of the sun. (*c*) Find the intensity and the radiation pressure at the surface of the sun.

12 •• SSM Estimate the radiation pressure force exerted on the earth by the sun, and compare the radiation pressure force to the gravitational attraction of the sun. At the earth's orbit the intensity of sunlight is 1.37 kW/m²

13 •• SSM Repeat Problem 12 for the planet Mars. Which planet has the larger ratio of radiation pressure to gravitational attraction. Why?

14 •• SSM In the new field of laser cooling and trapping, the forces associated with radiation pressure are used to slow down atoms from thermal speeds of hundreds of meters per second at room temperature to speeds of just a few meters per second or slower. An isolated atom will absorb radiation only at specific resonant frequencies. If the frequency of the laser-beam radiation is one of the resonant frequencies of the target atom, then the radiation is absorbed via a process called resonant absorption. The effective cross-sectional area of the atom for resonant absorption is approximately equal to λ^2, where λ is the wavelength of the laser beam. (*a*) Estimate the acceleration of a rubidium atom (atomic mass 85 g/mol) in a laser beam whose wavelength is 780 nm and intensity is 10 W/m². (*b*) About how long would it take such a light beam to slow a rubidium atom in a gas at room temperature (300 K) down to near-zero velocity?

Maxwell's Displacement Current

15 • iSOLVE ✓ A parallel-plate capacitor in air has circular plates of radius 2.3 cm separated by 1.1 mm. Charge is flowing onto the upper plate and off the lower plate at a rate of 5 A. (*a*) Find the time rate of change of the electric field between the plates. (*b*) Compute the displacement current between the plates and show that the displacement current equals 5 A.

16 • iSOLVE In a region of space, the electric field varies according to $E = (0.05$ N/C$) \sin 2000t$, where t is in seconds. Find the maximum displacement current through a 1-m² area perpendicular to $\vec{E}$.

17 •• For Problem 15, show that at a distance r from the axis of the plates the magnetic field between the plates is given by $B = (1.89 \times 10^{-3}$ T/m$)r$, if r is less than the radius of the plates.

18 •• (*a*) Show that for a parallel-plate capacitor the displacement current is given by $I_d = C\, dV/dt$, where C is the capacitance and V is the voltage across the capacitor. (*b*) A 5-nF parallel-plate capacitor is connected to an emf $\mathcal{E} = \mathcal{E}_0 \cos \omega t$, where $\mathcal{E}_0 = 3$ V and $\omega = 500\pi$. Find the displacement current between the plates as a function of time. Neglect any resistance in the circuit.

19 •• SSM iSOLVE ✓ Current of 10 A flows into a capacitor having plates with areas of 0.5 m². (*a*) What is the displacement current between the plates? (*b*) What is dE/dt between the plates for this current? (*c*) What is the line integral of $\vec{B} \cdot d\vec{\ell}$ around a circle of radius 10 cm that lies within the plates and parallel to the plates?

20 •• A parallel-plate capacitor with circular plates is given a charge Q_0. Between the plates is a leaky dielectric having a dielectric constant of κ and a resistivity ρ. (a) Find the conduction current between the plates as a function of time. (b) Find the displacement current between the plates as a function of time. What is the total (conduction plus displacement) current? (c) Find the magnetic field produced between the plates by the leakage discharge current as a function of time. (d) Find the magnetic field between the plates produced by the displacement current as a function of time. (e) What is the total magnetic field between the plates during discharge of the capacitor?

21 •• The leaky capacitor of Problem 20 has plate separation d. It is being charged such that the voltage across the capacitor is given by $V(t) = (0.01 \text{ V/s})t$. (a) Find the conduction current as a function of time. (b) Find the displacement current. (c) Find the time for which the displacement current is equal to the conduction current.

22 •• The space between the plates of a capacitor is filled with a material of resistivity $\rho = 10^4 \ \Omega\cdot\text{m}$ and dielectric constant $\kappa = 2.5$. The parallel plates are circular with a radius of 20 cm and are separated by 1 mm. The voltage across the plates is given by $V_0 \cos \omega t$, with $V_0 = 40$ V and $\omega = 120\pi$ rad/s. (a) What is the displacement current density? (b) What is the conduction current between the plates? (c) At what angular frequency is the total current 45° out of phase with the applied voltage?

23 ••• [SSM] Show that the generalized form of Ampère's law (Equation 30-4) and the Biot–Savart law give the same result in a situation in which they both can be used. Figure 30-13 shows two charges $+Q$ and $-Q$ on the x axis at $x = -a$ and $x = +a$, with a current $I = -dQ/dt$ along the line between them. Point P is on the y axis at $y = R$. (a) Use the Biot–Savart law to show that the magnitude of B at point P is

$$B = \frac{\mu_0 Ia}{2\pi R} \frac{1}{\sqrt{R^2 + a^2}}$$

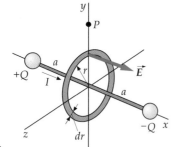

FIGURE 30-13 Problem 23

(b) Consider a circular strip of radius r and width dr in the yz plane with its center at the origin. Show that the flux of the electric field through this strip is

$$E_x \, dA = \frac{Q}{\epsilon_0} a(r^2 + a^2)^{-3/2} r \, dr$$

(c) Use your result from Part (b) to find the total flux ϕ_e through a circular area of radius R. Show that

$$\epsilon_0 \, \phi_e = Q\left(1 - \frac{a}{\sqrt{a^2 + R^2}}\right)$$

(d) Find the displacement current I_d, and show that

$$I + I_d = I \frac{a}{\sqrt{a^2 + R^2}}$$

(e) Finally, show that Equation 30-4 gives the same result for B as the result found in Part (a).

Maxwell's Equations and the Electromagnetic Spectrum

24 •• Show that the normal component of the magnetic field $\vec{B}$ is continuous across a surface, by applying Gauss's law for $\vec{B}$ ($\int B_n \, dA = 0$) to a pillbox Gaussian surface that has a face on each side of the surface.

25 • [SSM] [ISOLVE ✓] Find the wavelength for (a) a typical AM radio wave with a frequency of 1000 kHz and (b) a typical FM radio wave with a frequency of 100 MHz.

26 • [SSM] [ISOLVE] What is the frequency of a 3-cm microwave?

27 • [ISOLVE] What is the frequency of an X ray with a wavelength of 0.1 nm?

Electric Dipole Radiation

28 •• [ISOLVE] The intensity of radiation from an electric dipole is proportional to $\sin^2 \theta/r^2$, where θ is the angle between the electric dipole moment and the position vector $\vec{r}$. A radiating electric dipole lies along the z axis (its dipole moment is in the z direction). Let I_1 be the intensity of the radiation at a distance $r = 10$ m and at angle $\theta = 90°$. Find the intensity (in terms of I_1) at (a) $r = 30$ m, $\theta = 90°$; (b) $r = 10$ m, $\theta = 45°$; and (c) $r = 20$ m, $\theta = 30°$.

29 •• (a) For the situation described in Problem 28, at what angle is the intensity at $r = 5$ m equal to I_1? (b) At what distance is the intensity equal to I_1 at $\theta = 45°$?

30 •• [ISOLVE] The transmitting antenna of a station is a dipole located atop a mountain 2000 m above sea level. The intensity of the signal on a nearby mountain 4 km distant and also 2000 m above sea level is 4×10^{-12} W/m². What is the intensity of the signal at sea level and 1.5 km from the transmitter? (See Problem 28.)

31 ••• A radio station that uses a vertical dipole antenna broadcasts at a frequency of 1.20 MHz with total power output of 500 kW. The radiation pattern is as shown in Figure 30-8 (i.e., the intensity of the signal varies as $\sin^2 \theta$, where θ is the angle between the direction of propagation and the vertical and is independent of azimuthal angle). Calculate the intensity of the signal at a horizontal distance of 120 km from the station. What is the intensity at that point as measured in photons per square centimeter per second?

32 ••• [SSM] At a distance of 30 km from a radio station broadcasting at a frequency of 0.8 MHz, the intensity of the electromagnetic wave is 2×10^{-13} W/m². The transmitting antenna is a vertical dipole. What is the total power radiated by the station?

33 ••• [ISOLVE ✓] A small private plane approaching an airport is flying at an altitude of 2500 m above ground. The airport's flight control system transmits 100 W at 24 MHz, using a vertical dipole antenna. What is the intensity of the signal at the plane's receiving antenna when the plane's position on a map is 4 km from the airport?

Energy and Momentum in an Electromagnetic Wave

34 • ☑**SOLVE**✓ An electromagnetic wave has an intensity of 100 W/m². Find (a) the radiation pressure P_r, (b) E_{rms}, and (c) B_{rms}.

35 • ☑**SOLVE**✓ The amplitude of an electromagnetic wave is $E_0 = 400$ V/m. Find (a) E_{rms}, (b) B_{rms}, (c) the intensity I, and (d) the radiation pressure P_r.

36 • ☑**SOLVE** The rms value of the electric field in an electromagnetic wave is $E_{rms} = 400$ V/m. (a) Find B_{rms}, (b) the average energy density, and (c) the intensity.

37 • Show that the units of $E = cB$ are consistent; that is, show that when B is in teslas and c is in meters per second, the units of cB are volts per meter or newtons per coulomb.

38 • **SSM** ☑**SOLVE** The rms value of the magnitude of the magnetic field in an electromagnetic wave is $B_{rms} = 0.245 \ \mu T$. Find (a) E_{rms}, (b) the average energy density, and (c) the intensity.

39 •• ☑**SOLVE** (a) An electromagnetic wave of intensity 200 W/m² is incident normally on a rectangular black card with sides of 20 cm and 30 cm that absorbs all the radiation. Find the force exerted on the card by the radiation. (b) Find the force exerted by the same wave if the card reflects all the radiation incident on it.

40 •• Find the force exerted by the electromagnetic wave on the reflecting card in Part (b) of Problem 39 if the radiation is incident at an angle of 30° to the normal.

41 •• **SSM** An AM radio station radiates an isotropic sinusoidal wave with an average power of 50 kW. What are the amplitudes of E_{max} and B_{max} at a distance of (a) 500 m, (b) 5 km, and (c) 50 km?

42 •• ☑**SOLVE**✓ A laser beam has a diameter of 1.0 mm and average power of 1.5 mW. Find (a) the intensity of the beam, (b) E_{rms}, (c) B_{rms}, and (d) the radiation pressure.

43 •• **SSM** ☑**SOLVE** Instead of sending power by a 750-kV, 1000-A transmission line, one desires to beam this energy via an electromagnetic wave. The beam has a uniform intensity within a cross-sectional area of 50 m². What are the rms values of the electric and the magnetic fields?

44 •• ☑**SOLVE**✓ A laser pulse has an energy of 20 J and a beam radius of 2 mm. The pulse duration is 10 ns and the energy density is constant within the pulse. (a) What is the spatial length of the pulse? (b) What is the energy density within the pulse? (c) Find the electric and magnetic amplitudes of the laser pulse.

45 •• **SSM** The electric field of an electromagnetic wave oscillates in the y direction and the Poynting vector is given by

$$\vec{S}(x, t) = (100 \text{ W/m}^2) \cos^2[10x - (3 \times 10^9)t]\hat{i}$$

where x is in meters and t is in seconds. (a) What is the direction of propagation of the wave? (b) Find the wavelength and the frequency. (c) Find the electric and magnetic fields.

46 •• A parallel-plate capacitor is being charged. The capacitor consists of two circular parallel plates of area A and separation d. (a) Show that the displacement current in the capacitor gap has the same value as the conduction current in the capacitor leads. (b) What is the direction of the Poynting vector in the region of space between the capacitor plates? (c) Calculate the Poynting vector S in this region and show that the flux of S into this region is equal to the rate of change of the energy stored in the capacitor.

47 •• ☑**SOLVE**✓ A pulsed laser fires a 1000-MW pulse of 200-ns duration at a small object of mass 10 mg suspended by a fine fiber 4 cm long. If the radiation is completely absorbed without other effects, what is the maximum angle of deflection of this pendulum?

48 •• The mirrors used in a particular type of laser are 99.99% reflecting. (a) If the laser has an average output power of 15 W, what is the average power of the radiation incident on one of the mirrors? (b) What is the force due to radiation pressure on one of the mirrors?

49 •• A 10-cm by 15-cm card has a mass of 2 g and is perfectly reflecting. The card hangs in a vertical plane and is free to rotate about a horizontal axis through the top edge. The card is illuminated uniformly by an intense light that causes the card to make an angle of 1° with the vertical. Find the intensity of the light.

*The Wave Equation for Electromagnetic Waves

50 • Show by direct substitution that Equation 30-17a is satisfied by the wave function

$$E_y = E_0 \sin(kx - \omega t) = E_0 \sin k(x - ct)$$

where $c = \omega/k$.

51 • Use the known values of μ_0 and ϵ_0 in SI units to compute $c = 1/\sqrt{\epsilon_0\mu_0}$, and show that it is approximately 3×10^8 m/s.

52 ••• **SSM** (a) Using arguments similar to those given in the text, show that for a plane wave, in which E and B are independent of y and z,

$$\frac{\partial E_z}{\partial x} = \frac{\partial B_y}{\partial t} \quad \text{and} \quad \frac{\partial B_y}{\partial x} = \mu_0\epsilon_0 \frac{\partial E_z}{\partial t}$$

(b) Show that E_z and B_y also satisfy the wave equation.

53 ••• Show that any function of the form $y(x, t) = f(x - vt)$ or $y(x, t) = g(x + vt)$ satisfies the wave Equation 30-16.

General Problems

54 • (a) Show that if E is in volts per meter and B is in teslas, the units of the Poynting vector $\vec{S} = (\vec{E} \times \vec{B})/\mu_0$ are watts per square meter. (b) Show that if the intensity I is in watts per square meter, the units of radiation pressure $P_r = I/c$ are newtons per square meter.

55 •• A loop antenna that may be rotated about a vertical axis is used to locate an unlicensed amateur radio transmitter. If the output of the receiver is proportional to the intensity of the received signal, how does the output of the receiver vary with the orientation of the loop antenna?

56 •• An electromagnetic wave has a frequency of 100 MHz and is traveling in a vacuum. The magnetic field is given by $\vec{B}\,(z, t) = (10^{-8}\ \text{T}) \cos(kz - \omega t)\hat{\imath}$. (a) Find the wavelength and the direction of propagation of this wave. (b) Find the electric vector $\vec{E}\,(z, t)$. (c) Give Poynting's vector, and find the intensity of this wave.

57 •• **SSM** A circular loop of wire can be used to detect electromagnetic waves. Suppose a 100-MHz FM radio station radiates 50 kW uniformly in all directions. What is the maximum rms voltage induced in a loop of radius 30 cm at a distance of 10^5 m from the station?

58 •• **SOLVE** The electric field from a radio station some distance from the transmitter is given by $E = (10^{-4}\ \text{N/C}) \cos 10^6 t$, where t is in seconds. (a) What voltage is picked up on a 50-cm wire oriented along the electric field direction? (b) What voltage can be induced in a loop of radius 20 cm?

59 •• A circular capacitor of radius a has a thin wire of resistance R connecting the centers of the two plates. A voltage $V_0 \sin \omega t$ is applied between the plates. (a) What is the current drawn by this capacitor? (b) What is the magnetic field as a function of radial distance r from the centerline within the plates of this capacitor? (c) What is the phase angle between the current and the applied voltage?

60 •• **SOLVE** A 20-kW beam of radiation is incident normally on a surface that reflects half of the radiation. What is the force on this surface?

61 •• **SSM** The electric fields of two harmonic waves of angular frequency ω_1 and ω_2 are given by $\vec{E}_1 = E_{1,0} \cos(k_1 x - \omega_1 t)\hat{\jmath}$ and by $\vec{E}_2 = E_{2,0} \cos(k_2 x - \omega_2 t + \delta)\hat{\jmath}$. Find (a) the instantaneous Poynting vector for the resultant wave motion and (b) the time-average Poynting vector. If the direction of propagation of the second wave is reversed so $\vec{E}_2 = E_{2,0} \cos(k_2 x + \omega_2 t + \delta)\hat{\jmath}$, find (c) the instantaneous Poynting vector for the resultant wave motion and (d) the time-average Poynting vector.

62 •• **SSM** **SOLVE** At the surface of the earth, there is an approximate average solar flux of 0.75 kW/m². A family wishes to construct a solar energy conversion system to power their home. If the conversion system is 30 percent efficient and the family needs a maximum of 25 kW, what effective surface area is needed for perfectly absorbing collectors?

63 •• **SOLVE** ✔ Suppose one has an excellent radio capable of detecting a signal as weak as 10^{-14} W/m². This radio has a 2000-turn coil antenna that has a radius of 1 cm wound on an iron core that increases the magnetic field by a factor of 200. The radio frequency is 140 KHz. (a) What is the amplitude of the magnetic field in this wave? (b) What is the emf induced in the antenna? (c) What would be the emf induced in a 2-m wire oriented in the direction of the electric field?

64 •• Show that

$$\frac{\partial B_z}{\partial x} = -\mu_0 \epsilon_0 \frac{\partial E_y}{\partial t}$$

(Equation 30-19) follows from

$$\oint_C \vec{B} \cdot d\vec{\ell} = \mu_0 \epsilon_0 \int_S \frac{\partial E_n}{\partial t}\, dA$$

(Equation 30-6d with $I = 0$) by integrating along a suitable curve C and over a suitable surface S in a manner that parallels the derivation of Equation 30-18.

65 ••• **SSM** A long cylindrical conductor of length L, radius a, and resistivity ρ carries a steady current I that is uniformly distributed over its cross-sectional area. (a) Use Ohm's law to relate the electric field E in the conductor to I, ρ, and a. (b) Find the magnetic field B just outside the conductor. (c) Use the results from Part (a) and Part (b) to compute the Poynting vector $\vec{S} = (\vec{E} \times \vec{B})/\mu_0$ at $r = a$ (the edge of the conductor). In what direction is $\vec{S}$? (d) Find the flux $\oint S_n\, dA$ through the surface of the conductor into the conductor, and show that the rate of energy flow into the conductor equals $I^2 R$, where R is the resistance of the cylinder. (Here, S_n is the *inward* component of $\vec{S}$ perpendicular to the surface of the conductor.)

66 ••• A long solenoid of n turns per unit length has a current that slowly increases with time. The solenoid has radius R, and the current in the windings has the form $I(t) = at$. (a) Find the induced electric field at a distance $r < R$ from the solenoid axis. (b) Find the magnitude and direction of the Poynting vector $\vec{S}$ at the cylindrical surface $r = R$ just inside the solenoid windings. (c) Calculate the flux $\oint S_n\, dA$ into the solenoid, and show that the flux equals the rate of increase of the magnetic energy inside the solenoid. (Here, S_n is the *inward* component of $\vec{S}$ perpendicular to the surface of the solenoid.)

67 ••• **SSM** **SOLVE** Small particles might be blown out of solar systems by the radiation pressure of sunlight. Assume that the particles are spherical with a radius r and a density of 1 g/cm³ and that they absorb all the radiation in a cross-sectional area of πr^2. The particles are a distance R from the sun, which has a power output of 3.83×10^{26} W. What is the radius r for which the radiation force of repulsion just balances the gravitational force of attraction to the sun?

68 ••• When an electromagnetic wave is reflected at normal incidence on a perfectly conducting surface, the electric field vector of the reflected wave at the reflecting surface is the negative of that of the incident wave. (a) Explain why this should be. (b) Show that the superposition of incident and reflected waves results in a standing wave. (c) What is the relationship between the magnetic field vector of the incident waves and reflected waves at the reflecting surface?

69 ••• **SSM** An intense point source of light radiates 1 MW isotropically. The source is located 1 m above an infinite, perfectly reflecting plane. Determine the force that acts on the plane.

Properties of Light

LIGHT IS TRANSMITTED BY TOTAL INTERNAL REFLECTION THROUGH TINY GLASS FIBERS.

? **How large must the angle of incidence of the light on the wall of the tube be so that no light escapes? (See Example 31-5.)**

The human eye is sensitive to electromagnetic radiation with wavelengths from approximately 400 nm to 700 nm. The shortest wavelengths in the visible spectrum correspond to violet light and the longest to red light. The perceived colors of light are the result of the physiological and psychological response of the eye–brain sensing system to the different frequencies of visible light. Although the correspondence between perceived color and frequency is quite good, there are many interesting deviations. For example, a mixture of red light and green light is perceived by the eye–brain sensing system as yellow even in the absence of light in the yellow region of the spectrum.

➤ In this chapter, we study how light is produced; how its speed is measured; and how light is scattered, reflected, refracted, and polarized.

31-1 Wave-Particle Duality

The wave nature of light was first demonstrated by Thomas Young, who observed the interference pattern of two coherent light sources produced by illuminating a pair of narrow, parallel slits with a single source. The wave theory of light culminated in 1860 with Maxwell's prediction of electromagnetic waves. The particle nature of light was first proposed by Albert Einstein in 1905 in his explanation of the photoelectric effect.[†] A particle of light called a **photon** has energy E that is related to the frequency f and wavelength λ of the light wave by the Einstein equation

$$E = hf = \frac{hc}{\lambda} \qquad\qquad 31\text{-}1$$

EINSTEIN'S EQUATION FOR PHOTON ENERGY

where c is the speed of light and h is Planck's constant:

$$h = 6.626 \times 10^{-34}\,\text{J·s} = 4.136 \times 10^{-15}\,\text{eV·s}$$

Since energies are often given in electron volts and wavelengths are given in nanometers, it is convenient to express the combination hc in eV·nm. We have

$$hc = (4.136 \times 10^{-15}\,\text{eV·s})(2.998 \times 10^{8}\,\text{m/s}) = 1.240 \times 10^{-6}\,\text{eV·m}$$

or

$$hc = 1240\,\text{eV·nm} \qquad\qquad 31\text{-}2$$

The propagation of light is governed by its wave properties, whereas the exchange of energy between light and matter is governed by its particle properties. This wave–particle duality is a general property of nature. For example, the propagation of electrons (and other so-called particles) is also governed by wave properties, whereas the exchange of energy between the electrons and other particles is governed by particle properties.

31-2 Light Spectra

Newton was the first to recognize that white light is a mixture of light of all colors of approximately equal intensity. He demonstrated this by letting sunlight fall on a glass prism and observing the spectrum of the refracted light (Figure 31-1). Because the angle of refraction produced by a glass prism depends slightly on wavelength, the refracted beam is spread out in space into its component colors or wavelengths, like a rainbow.

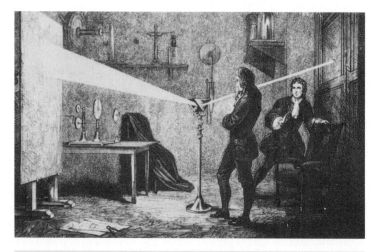

FIGURE 31-1 Newton demonstrating the spectrum of sunlight with a glass prism.

---

† The photoelectric effect is discussed in Chapter 34.

Figure 31-2 shows a spectroscope, which is a device for analyzing the spectra of a light source. Light from the source passes through a narrow slit, traverses a lens to make the beam parallel, and falls on a glass prism. The refracted beam is viewed with a telescope, which is mounted on a rotating platform so that the angle of the refracted beam, which depends on the wavelength, can be measured. The spectrum of the light source can thus be analyzed in terms of its component wavelengths. The spectrum of sunlight contains a continuous range of wavelengths and is therefore called a **continuous spectrum.** The light emitted by the atoms in low-pressure gases, such as mercury atoms in a fluorescent lamp, contains only a discrete set of wavelengths. Each wavelength emitted by the source produces a separate image of the collimating slit in the spectroscope. Such a spectrum is called a **line spectrum.** The continuous visible spectrum and the line spectra from several elements are shown in the photograph.

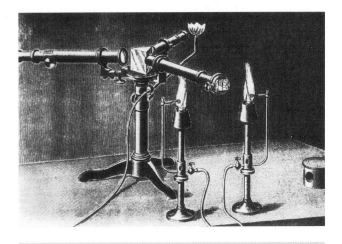

FIGURE 31-2 A late nineteenth-century spectroscope belonging to Gustav Kirchhoff. Modern student spectroscopes usually share the same general design.

31-3 Sources of Light

Line Spectra

The most common sources of visible light are transitions of the outer electrons in atoms. Normally an atom is in its ground state with its electrons at their lowest allowed energy levels consistent with the exclusion principle. (The exclusion principle, which was first enunciated by Wolfgang Pauli in 1925 to explain the electronic structure of atoms, states that no two electrons in an atom can be in the same quantum state.) The lowest energy electrons are closest to the nucleus and are tightly bound, forming a stable inner core. The one or two electrons in the highest energy states are much farther from the nucleus and are relatively easily excited to vacant higher energy states. These outer electrons are responsible for the energy changes in the atom that result in the emission or absorption of visible light.

The continuous visible spectrum (top) and the line spectra of (from top to bottom) hydrogen, helium, barium, and mercury.

When an atom collides with another atom or with a free electron, or when the atom absorbs electromagnetic energy, the outer electrons can be excited to higher energy states. After a time of approximately 10 ns (1 ns = 10^{-9} s), these outer electrons spontaneously make transitions to lower energy states with the emission of a photon. This process, called **spontaneous emission,** is random; the photons emitted from two different atoms are not correlated. The emitted light is thus incoherent. By conservation of energy, the energy of an emitted photon is the energy difference $|\Delta E|$ between the initial state and the final state. The frequency of the light wave is related to the energy by the Einstein equation, $|\Delta E| = hf$. The wavelength of the emitted light is then

$$\lambda = \frac{c}{f} = \frac{hc}{hf} = \frac{hc}{|\Delta E|} \qquad\qquad 31\text{-}3$$

The photon energies corresponding to shortest wavelengths (400 nm) and longest (700 nm) wavelengths in the visible spectrum are

$$E_{400\,\text{nm}} = \frac{hc}{\lambda} = \frac{1240\ \text{eV·nm}}{400\ \text{nm}} = 3.10\ \text{eV} \qquad\qquad 31\text{-}4a$$

and

$$E_{700\,\text{nm}} = \frac{hc}{\lambda} = \frac{1240\ \text{eV·nm}}{700\ \text{nm}} = 1.77\ \text{eV} \qquad\qquad 31\text{-}4b$$

Because the energy levels in atoms form a discrete set, the emission spectrum of light from single atoms or from atoms in low-pressure gases consists of a set of sharp discrete lines that are characteristic of the element. These narrow lines are broadened somewhat by Doppler shifts, due to the motion of the atom relative to the observer and by collisions with other atoms; but, generally, if the gas density is low enough, the lines are narrow and well separated from one another. The study of the line spectra of hydrogen and other atoms led to the first understanding of the energy levels of atoms.

Continuous Spectra When atoms are close together and interact strongly, as in liquids and solids, the energy levels of the individual atoms are spread out into energy bands, resulting in essentially continuous bands of energy levels. When the bands overlap, as they often do, the result is a continuous spectrum of possible energies and a continuous emission spectrum. In an incandescent material such as a hot metal filament, electrons are randomly accelerated by frequent collisions, resulting in a broad spectrum of thermal radiation. The rate at which an object radiates thermal energy is proportional to the fourth power of its absolute temperature.[†] The radiation emitted by an object at temperatures below approximately 600°C is concentrated in the infrared and is not visible. As an object is heated, the energy radiated extends to shorter and shorter wavelengths. Between approximately 600°C and 700°C, enough of the radiated energy is in the visible spectrum for the object to glow a dull red. At higher and higher temperatures, the object becomes bright red and then white. For a given temperature, the wavelength λ_{peak} at which the emitted power is a maximum varies inversely with the temperature, a result known as Wien's displacement law. The surface of the sun at $T = 6000$ K emits a continuous spectrum of approximately constant intensity over the visible range of wavelengths.

[†] This is known as the Stefan–Boltzmann law. This and other properties of thermal radiation, such as Wien's displacement law, are discussed more fully in Section 20-4.

Absorption, Scattering, Spontaneous Emission, and Stimulated Emission

When radiation is emitted, an atom makes a transition from an excited state to a state of lower energy; when radiation is absorbed, an atom makes a transition from a lower state to a higher state. When atoms are irradiated with a continuous spectrum of radiation, the transmitted spectrum shows dark lines corresponding to the absorption of light at discrete wavelengths. The absorption spectra of atoms were the first line spectra observed. Since atoms and molecules at normal temperatures are in either their ground states or low-lying excited states, only transitions from a ground state (or a near ground state) to a more highly excited state are observed. Thus, absorption spectra usually have far fewer lines than emission spectra have.

Figure 31-3 illustrates several interesting phenomena that can occur when a photon is incident on an atom. In Figure 31-3a, the energy of the incoming photon is too small to excite the atom to an excited state, so the atom remains in its ground state and the photon is said to be scattered. Since the incoming and outgoing or scattered photons have the same energy, the scattering is said to be elastic. If the wavelength of the incident light is large compared with the size of the atom, the scattering can be described in terms of classical electromagnetic theory and is called **Rayleigh scattering** after Lord Rayleigh, who worked out the theory in 1871. The probability of Rayleigh scattering varies as $1/\lambda^4$. This means that blue light is scattered much more readily than red light, which accounts for the bluish color of the sky. The removal of blue light by Rayleigh scattering also accounts for some of the reddish color of the transmitted light seen in sunsets.

Inelastic scattering, also called **Raman scattering,** occurs when an incident photon with just the right amount of energy is absorbed and the molecule undergoes a transition to a more energetic state. Then the molecule emits a photon as it undergoes a transition to a less energetic state, whose energy differs from that of the initial state. If the energy of the scattered photon hf' is less than that of the incident photon hf (Figure 31-3b), it is called **Stokes Raman scattering.** If the energy of the scattered photon is greater than that of the incident photon (Figure 31-3c), it is called **anti-Stokes Raman scattering.**

In Figure 31-3d, the energy of the incident photon is just equal to the difference in energy between the initial state and a more energetic state. The atom absorbs the photon and makes a transition to the more excited state in a process called **resonance absorption.**

In Figure 31-3e, an atom in an excited state spontaneously undergoes a transition to a less energetic state, in a process called **spontaneous emission.** Often an atom in an excited state undergoes transitions to one or more intermediate states as it returns to the ground state. A common example occurs when an atom is excited by ultraviolet light and emits visible light as it returns via multiple transitions to its ground state. This process, often called **fluorescence,** occurs in a thin film lining the inside of the glass tubes of fluorescent light bulbs. Since the lifetime of a typical excited atomic energy state is of the order of 10 ns, this process appears to occur instantaneously. However, some excited states have much longer lifetimes—of the order of milliseconds or occasionally seconds or even minutes. Such a state is called a **metastable state. Phosphorescent materials** have very long-lived metastable states and emit light long after the original excitation.

Figure 31-3f illustrates the photoelectric effect, in which the absorption of the photon ionizes the atom by causing the emission of an electron. Figure 31-3g illustrates **stimulated emission.** This process occurs if the atom or molecule is initially in an excited state of energy E_H, and the energy of the incident photon is equal to $E_H - E_L$, where E_L is the energy of a lower state. In this case, the oscillating electromagnetic field associated with the incident photon can stimulate the excited atom or molecule, which then emits a photon in the same direction as the incident photon and in phase with it. The photons from the stimulated atoms or molecules can stimulate the emission of additional photons propagating in the same direction

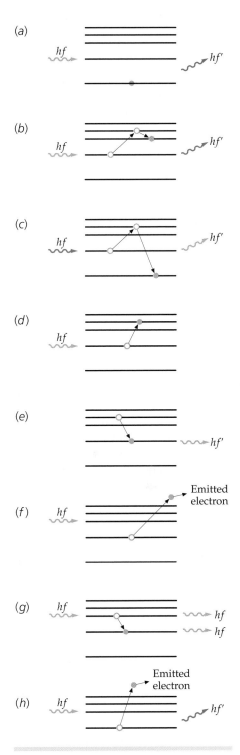

FIGURE 31-3 Photon-atom and photon-molecule interactions. (*a*) Elastic scattering (*b*) Stokes Raman scattering (*c*) Anti-Stokes Raman scattering (*d*) Resonance absorption (*e*) Spontaneous emission (*f*) Photoelectric effect (*g*) Stimulated emission (*h*) Compton scattering.

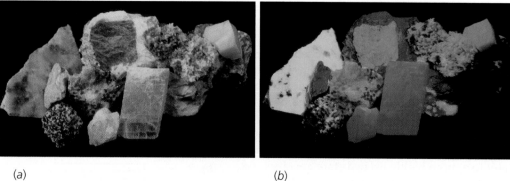

(a) (b)

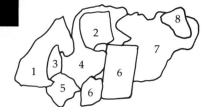

(c)

with the same phase. This process amplifies the initially emitted photon, yielding a beam of light originating from different atoms that is coherent. As a result, interference of the light from a large number of atoms can easily be observed.

Figure 31-3*h* illustrates **Compton scattering,** which occurs if the energy of the incident photon is much greater than the ionization energy. Note that in Compton scattering, a photon is absorbed and a photon is emitted, whereas in the photoelectric effect, a photon is absorbed with none emitted.

A collection of minerals in (*a*) daylight and in (*b*) ultraviolet light (sometimes called *black light*). Identified by number in the schematic (*c*), they are 1, powerllite; 2, willemite; 3, scheelite; 4, calcite; 5, calcite and willemite composite; 6, optical calcite; 7, willemite; and 8, opal. The change in color is due to the minerals fluorescing under the ultraviolet light. In optical calcite, both fluorescence and phosphorescence occur.

RESONANT ABSORPTION AND EMISSION **EXAMPLE 3 1 - 1**

The first excited state of potassium is $E_1 = 1.62$ eV above the ground state E_0, which we take to be zero. The second and third excited states of potassium have energy levels at $E_2 = 2.61$ eV and $E_3 = 3.07$ eV above the ground state. (*a*) What is the maximum wavelength of radiation that can be absorbed by potassium in its ground state? Calculate the wavelength of the emitted photon when the atom makes a transition from (*b*) the second excited state (E_2) to the ground state and from (*c*) the third excited state (E_3) to the second excited state (E_2).

PICTURE THE PROBLEM The ground state and the first three excited energy levels are shown in Figure 31-4. (*a*) Since the wavelength is related to the energy of a photon by $\lambda = hc/\Delta E$, longer wavelengths correspond to smaller energy differences. The smallest energy difference for a transition originating at the ground state is from the ground state to the first excited state. (*b*) The wavelengths of the photons given off when the atom de-excites are related to the energy differences by $\lambda = hc/|\Delta E|$.

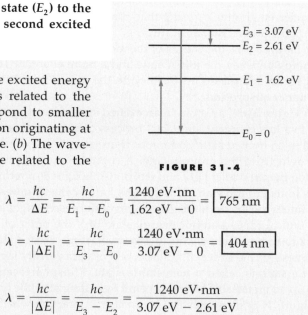

FIGURE 31-4

(*a*) Calculate the wavelength of radiation absorbed in a transition from the ground state to the first excited state:

$$\lambda = \frac{hc}{\Delta E} = \frac{hc}{E_1 - E_0} = \frac{1240 \text{ eV·nm}}{1.62 \text{ eV} - 0} = \boxed{765 \text{ nm}}$$

(*b*) For the transition from E_3 to the ground state, the photon energy is $E_3 - E_0 = E_3$. Calculate the wavelength of radiation emitted in this transition:

$$\lambda = \frac{hc}{|\Delta E|} = \frac{hc}{E_3 - E_0} = \frac{1240 \text{ eV·nm}}{3.07 \text{ eV} - 0} = \boxed{404 \text{ nm}}$$

(*c*) For the transition from E_3 to E_2, the photon energy is $E_3 - E_2$. Calculate the wavelength of radiation emitted in this transition:

$$\lambda = \frac{hc}{|\Delta E|} = \frac{hc}{E_3 - E_2} = \frac{1240 \text{ eV·nm}}{3.07 \text{ eV} - 2.61 \text{ eV}}$$

$$= \boxed{2700 \text{ nm}}$$

REMARKS The wavelength of radiation emitted in the transition from E_1 to the ground state E_0 is 765 nm, the same as that for radiation absorbed in the transition from the ground state to E_1. This transition and the transmission from E_3 to the ground state both result in photons in the visible spectrum.

Lasers

The *laser* (*l*ight *a*mplification by *s*timulated *e*mission of *r*adiation) is a device that produces a strong beam of coherent photons by stimulated emission. Consider a system consisting of atoms that have a ground state of energy E_1 and an excited metastable state of energy E_2. If these atoms are irradiated by photons of energy $E_2 - E_1$, those atoms in the ground state can absorb a photon and make the transition to state E_2, whereas those atoms already in the excited state may be stimulated to decay back to the ground state. The relative probabilities of absorption and stimulated emission, first worked out by Einstein, are equal. Ordinarily, nearly all the atoms of the system at normal temperature will initially be in the ground state, so absorption will be the main effect. To produce more stimulated-emission transitions than absorption transitions, we must arrange to have more atoms in the excited state than in the ground state. This condition, called population inversion, can be achieved by a method called optical pumping in which atoms are *pumped* up to levels of energy greater than E_2 by the absorption of an intense auxiliary radiation. The atoms then decay down to state E_2 either by spontaneous emission or by nonradiative transitions, such as those due to collisions.

Figure 31-5 shows a schematic diagram of the first laser, a ruby laser built by Theodore Maiman in 1960. The laser consists of a ruby rod a few centimeters long surrounded by a helical gaseous flashtube that emits a broad spectrum of light. The ends of the ruby rod are flat and perpendicular to the axis of the rod. Ruby is a transparent crystal of Al_2O_3 with a small amount (about 0.05 percent) of chromium. It appears red because the chromium ions (Cr^{3+}) have strong absorption bands in the blue and green regions of the visible spectrum, as shown in Figure 31-6. The energy levels of chromium—important for the operation of a ruby laser—are shown in Figure 31-7. When the flashtube is fired, there is an intense burst of light that lasts several milliseconds. Photon absorption excites many of the chromium ions to the bands of energy levels indicated by the shading in Figure 31-7. The excited chromium ions then rapidly drop down to a closely spaced pair of metastable states labeled E_2 in the figure. These metastable states are approximately 1.79 eV above the ground state. The expected lifetime for a chromium ion to remain in one of these metastable states is about 5 ms, after which the chromium ion spontaneously emits a photon and decays to the ground state. A millisecond is a long time for an atomic process. Consequently, if the flash is intense enough, the number of chromium ions populating the two metastable states will exceed the population of chromium ions in the ground state. It follows that during the time that the flashtube is firing, the populations of ions in the ground state and the metastable states are inverted. When the chromium ions in the state E_2 decay to the ground state by spontaneous emission, they emit photons of energy 1.79 eV and wavelength 694.3 nm. These photons have just the right energy to stimulate chromium ions in the metastable states to emit photons of the same energy (and wavelength) as they undergo the transition to the ground state. The photons also have just the right energy to stimulate chromium ions in the ground state to absorb a photon as they undergo the transition to one of the metastable states. These are competing processes, and the stimulated emission process dominates as long as the population of chromium ions in the metastable states exceeds the population in the ground state.

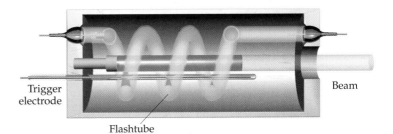

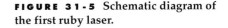

FIGURE 31-5 Schematic diagram of the first ruby laser.

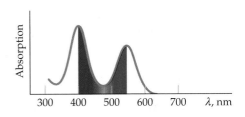

FIGURE 31-6 Absorption versus wavelength for Cr^{3+} in ruby. Ruby appears red because of the strong absorption of green and blue light by the chromium ions.

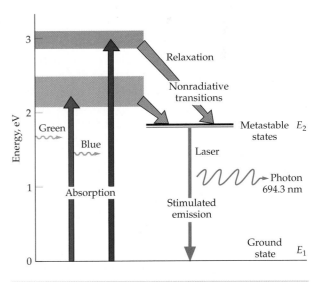

FIGURE 31-7 Energy levels in a ruby laser. To make the population of the metastable states greater than that of the ground state, the ruby crystal is subjected to intense radiation that contains energy in the green and blue wavelengths. This excites atoms from the ground state to the bands of energy levels indicated by the shading, from which the atoms decay to the metastable states by nonradiative transitions. Then, by stimulated emission, the atoms undergo the transition from the metastable states to the ground state.

Silvered end

Partially
silvered end

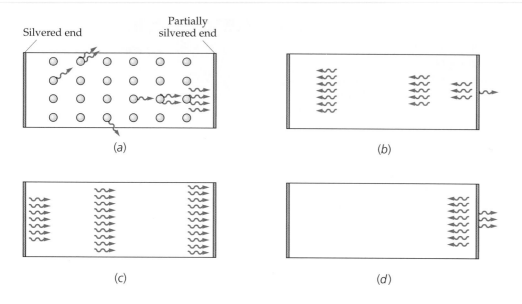

(a)

(b)

(c)

(d)

FIGURE 31-8 Buildup of photon beam in a laser. (*a*) When irradiated, some atoms spontaneously emit photons, some of which travel to the right and stimulate other atoms to emit photons parallel to the axis of the crystal. (*b*) Of the four photons that strike the right face, one is transmitted and three are reflected. As the reflected photons traverse the laser crystal, they stimulate other atoms to emit photons, and the beam builds up. By the time the beam reaches the right face again (*c*), it comprises many photons. (*d*) Some of these photons are transmitted, the rest of the photons are reflected.

In the ruby laser, one end of the crystal is fully silvered, so it is 100 percent reflecting; the other end of the crystal, called the output coupler, is partially silvered, leaving it about 85 percent reflecting. When photons traveling parallel to the axis of the crystal strike the silvered ends, all are reflected from the back face and 85 percent are reflected from the front face, with 15 percent of the photons escaping through the partially silvered front face. During each pass through the crystal, the photons stimulate more and more atoms so that an intense beam is emitted from the partially silvered end (Figure 31-8). Because the duration of each flash of the flashtube is between two and three seconds, the laser beam is produced in pulses lasting a few milliseconds. Modern ruby lasers generate intense light beams with energies ranging from 50 J to 100 J. The beam can have a diameter as small as 1 mm and an angular divergence as small as 0.25 milliradian to about 7 milliradians.

Population inversion is achieved somewhat differently in the continuous helium–neon laser. The energy levels of helium and neon that are important for the operation of the laser are shown in Figure 31-9. Helium has an excited energy state $E_{2,He}$ that is 20.61 eV above its ground state. Helium atoms are excited to state $E_{2,He}$ by an electric discharge. Neon has an excited state $E_{3,Ne}$ that is 20.66 eV above its ground state. This is just 0.05 eV above the first excited state of helium. The neon atoms are excited to state $E_{3,Ne}$ by collisions with excited helium atoms. The kinetic energy of the helium atoms provides the extra 0.05 eV of energy needed to excite the neon atoms. There is another excited state of neon $E_{2,Ne}$ that is 18.70 eV above its ground state and 1.96 eV below state $E_{3,Ne}$. Since state $E_{2,Ne}$ is normally unoccupied, population inversion between states $E_{3,Ne}$ and $E_{2,Ne}$ is obtained immediately. The stimulated emission that occurs between these states results in photons of energy 1.96 eV and wavelength 632.8 nm, which produces a bright red light. After stimulated emission, the atoms in state $E_{2,Ne}$ decay to the ground state by spontaneous emission.

Note that there are four energy levels involved in the helium–neon laser, whereas the ruby laser involved only three levels. In a three-level laser, population inversion is

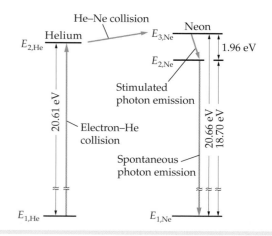

FIGURE 31-9 Energy levels of helium and neon that are important for the helium–neon laser. The helium atoms are excited by electrical discharge to an energy state 20.61 eV above the ground state. They collide with neon atoms, exciting some neon atoms to an energy state 20.66 eV above the ground state. Population inversion is thus achieved between this level and one 1.96 eV below it. The spontaneous emission of photons of energy 1.96 eV stimulates other atoms in the upper state to emit photons of energy 1.96 eV.

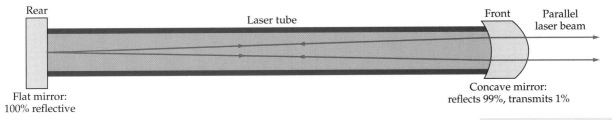

Rear

Laser tube

Front Parallel laser beam

Flat mirror: 100% reflective

Concave mirror: reflects 99%, transmits 1%

difficult to achieve because more than half the atoms in the ground state must be excited. In a four-level laser, population inversion is easily achieved because the state after stimulated emission is not the ground state but an excited state that is normally unpopulated.

Figure 31-10 shows a schematic diagram of a helium–neon laser commonly used for physics demonstrations. The helium–neon laser consists of a gas tube that contains 15 percent helium gas and 85 percent neon gas. A totally reflecting flat mirror is mounted at one end of the gas tube and a 99 percent reflecting concave mirror is placed at the other end of the gas tube. The concave mirror focuses parallel light at the flat mirror and also acts as a lens that transmits part of the light, so that the light emerges as a parallel beam.

A laser beam is coherent, very narrow, and intense. Its coherence makes the laser beam useful in the production of holograms, which we discuss in Chapter 33. The precise direction and small angular spread of the laser beam make it useful as a surgical tool for destroying cancer cells or reattaching a detached retina. Lasers are also used by surveyors for precise alignment over large distances. Distances can be accurately measured by reflecting a laser pulse from a mirror and measuring the time the pulse takes to travel to the mirror and back. The distance to the moon has been measured to within a few centimeters using a mirror placed on the moon for that purpose. Laser beams are also used in fusion research. An intense laser pulse is focused on tiny pellets of deuterium–tritium in a combustion chamber. The beam heats the pellets to temperatures of the order of 10^8 K in a very short time, causing the deuterium and tritium to fuse and release energy.

Laser technology is advancing so quickly that it is possible to mention only a few of the recent developments. In addition to the ruby laser, there are many other solid-state lasers with output wavelengths that range from approximately 170 nm to 3900 nm. Lasers that generate more than 1 kW of continuous power have been constructed. Pulsed lasers can now deliver nanosecond pulses of power exceeding 10^{14} W. Various gas lasers can now produce beams of wavelengths that range from the far infrared to the ultraviolet. Semiconductor lasers (also known as diode lasers or junction lasers) have shrunk in just 10 years from the size of a pinhead to mere billionths of a meter. Liquid lasers that use chemical dyes can be tuned over a range of wavelengths (approximately 70 nm for continuous lasers and more than 170 nm for pulsed lasers). A relatively new laser, the free-electron laser, extracts light energy from a beam of free electrons moving through a spatially varying magnetic field. The free-electron laser has the potential for very high power and high efficiency and can be tuned over a large range of wavelengths. There appears to be no limit to the variety and uses of modern lasers.

FIGURE 31-10 Schematic drawing of a helium–neon laser. The use of a concave mirror rather than a second plane mirror makes the alignment of the mirrors less critical than it is for the ruby laser. The concave mirror on the right also serves as a lens that focuses the emitted light into a parallel beam.

31-4 The Speed of Light

Prior to the seventeenth century the speed of light was thought by many to be infinite, and an effort to measure the speed of light was made by Galileo. He and a partner stood on hilltops about three kilometers apart, each with a lantern and a shutter to cover it. Galileo proposed to measure the time it took for light to travel back and forth between the experimenters. First, one would uncover his lantern, and when the other saw the light, he would uncover his. The time between the first partner's uncovering his lantern and his seeing the light from the other lantern would be the time it took for light to travel back and forth between the experimenters. Though this method is sound in principle, the speed of light is so great that the time interval to be measured is much smaller than fluctuations in human response time, so Galileo was unable to obtain a value for the speed of light.

The first indication of the true magnitude of the speed of light came from astronomical observations of the period of Io, one of the moons of Jupiter. This period is determined by measuring the time between eclipses of Io behind Jupiter. The eclipse period is about 42.5 h, but measurements made when the earth is moving away from Jupiter along path *ABC* in Figure 31-11 give a greater time for this period than do measurements made when the earth is moving toward Jupiter along path *CDA* in the figure. Since these measurements differ from the average value by only about 15 s, the discrepancies were difficult to measure accurately. In 1675, the astronomer Ole Römer attributed these discrepancies to the fact that the speed of light is finite, and that during the 42.5 h between eclipses of Jupiter's moon, the distance between the earth and Jupiter changes, making the path for the light longer or shorter. Römer devised the following method for measuring the cumulative effect of these discrepancies. Jupiter is moving much more slowly than the earth, so we can neglect its motion. When the earth is at point *A*, nearest to Jupiter, the distance between the earth and Jupiter is changing negligibly. The period of Io's eclipse is measured, providing the time between the beginnings of successive eclipses. Based on this measurement, the number of occultations during 6 months is computed, and the time when an eclipse should begin a half-year later when the earth is at point *C* is predicted. When the earth is actually at point *C*, the observed beginning of the eclipse is about 16.6 min later than predicted. This is the time it takes light to travel a distance equal to the diameter of the earth's orbit. This calculation neglects the distance traveled by Jupiter toward the earth. However, because the orbital speed of Jupiter is so much slower than that of the earth, the distance Jupiter moves toward (or away from) the earth during the 6 months is much less than the diameter of the earth's orbit.

EXERCISE Calculate (*a*) the distance traveled by the earth between successive eclipses of Io and (*b*) the speed of light, given that the time between successive eclipses is 15 s longer than average when the earth is moving directly away from Jupiter. (*Answer* (*a*) 4.59×10^6 km (*b*) 3.06×10^8 m/s)

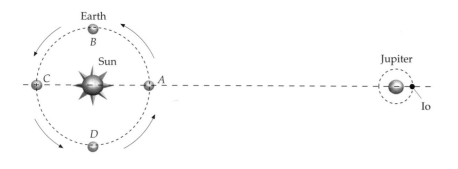

FIGURE 31-11 Römer's method of measuring the speed of light. The time between eclipses of Jupiter's moon Io appears to be greater when the earth is moving along path *ABC* than when the earth is moving along path *CDA*. The difference is due to the time it takes light to travel the distance traveled by the earth along the line of sight during one period of Io. (The distance traveled by Jupiter in one earth year is negligible.)

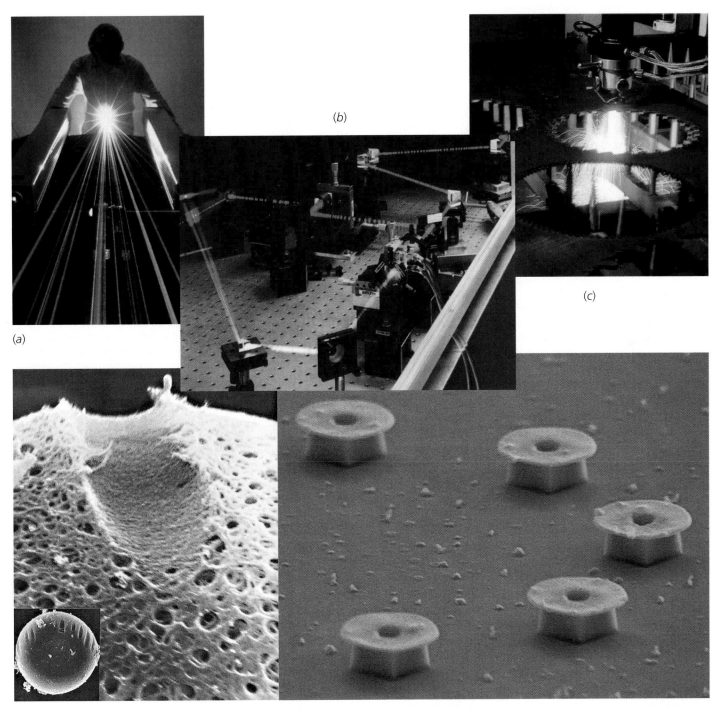

(*a*) Beams from a krypton laser and an argon laser, split into their component wavelengths. In these gas lasers, krypton and argon atoms have been stripped of multiple electrons, forming positive ions. The light-emitting energy transitions occur when excited electrons in the ions decay from one upper energy level to another. Here, several energy transitions are occurring at once, each corresponding to emitted light of a different wavelength. (*b*) A femtosecond pulsed laser. By a technique known as *modelocking*, different excited modes within a laser's cavity can be made to interfere with one another and create a series of ultrashort pulses, which are picoseconds long, that correspond to the time it takes light to bounce back and forth once within the cavity. Ultrashort pulses have been used as probes to study the behavior of molecules during chemical reactions. (*c*) A carbon dioxide laser takes just 2 minutes to cut out a steel saw blade. (*d*) A groove etched in the zona pellucida (protective outer covering) of a mouse egg by a *laser scissor* facilitates implantation. This technique has already been applied in human fertility therapies. Several effects contribute to the ability of the finely focused laser to cut on such a delicate scale—photon absorption may heat the target, break molecular bonds, or drive chemical reactions. (*e*) The so-called nanolasers shown are semiconductor disks mere microns in diameter and fractions of a micron in width. These tiny lasers work like their larger counterparts. Exploiting quantum effects that prevail on this microscopic scale, nanolasers promise great efficiency and they are being explored as ultrafast, low-energy switching devices.

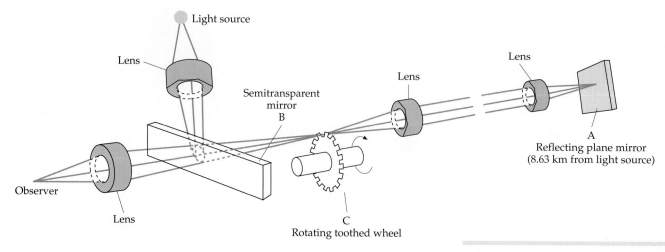

Light source

Lens

Lens

Lens

Semitransparent
mirror
B

A
Reflecting plane mirror
(8.63 km from light source)

Observer

Lens

C
Rotating toothed wheel

FIGURE 31-12 Fizeau's method of
measuring the speed of light. Light from
the source is reflected by mirror B and is
transmitted through a gap in the toothed
wheel to mirror A. The speed of light is
determined by measuring the angular
speed of the wheel that will permit the
reflected light to pass through the next
gap in the toothed wheel so that an
image of the source is observed.

The French physicist Armand Fizeau made the first nonastronomical measurement of the speed of light in 1849. On a hill in Paris, Fizeau placed a light source and a system of lenses arranged so that the light reflected from a semitransparent mirror was focused on a gap in a toothed wheel, as shown in Figure 31-12. On a distant hill (about 8.63 km away) Fizeau placed a mirror to reflect the light back, to be viewed by an observer as shown. The toothed wheel was rotated, and the speed of rotation was varied. At low speeds of rotation, no light was visible because the light that passed through a gap in the rotating wheel and was reflected back by the mirror was obstructed by the next tooth of the wheel. The speed of rotation was then increased. The light suddenly became visible when the rotation speed was such that the reflected light passed through the next gap in the wheel. The time for the wheel to rotate through the angle between successive gaps equals the time for the light to make the round trip to the distant mirror and back.

Fizeau's method was improved upon by Jean Foucault, who replaced the toothed wheel with a rotating mirror, as shown in Figure 31-13. Light strikes the rotating mirror, the light is reflected to a distant fixed mirror, the light is then reflected back to the rotating mirror, and then to an observing telescope. During the time taken for the light to travel from the rotating mirror to the distant fixed mirror and back, the mirror rotates through a small angle. By measuring the angle θ, the time for the light to travel to the distant mirror and back is determined. In approximately 1850, Foucault measured the speed of light in air and in water, and he showed that the speed of light is less in water than the speed of light in air. Using essentially the same method, the American physicist A. A. Michelson made more precise measurements of the speed of light in approximately 1880. A half-century later, Michelson made even more precise measurements of the speed of light, using an octagonal rotating mirror (Figure 31-14). In these measurements, the mirror rotates through one-eighth of a turn during the time it takes for the light to travel to the fixed mirror and back. The rotation rate is varied until another face of the mirror is in the right position for the reflected light to enter the telescope.

Another method of determining the speed of light involves the measurement of the electrical constants ϵ_0 and μ_0 to determine c from $c = 1/\sqrt{\epsilon_0 \mu_0}$.

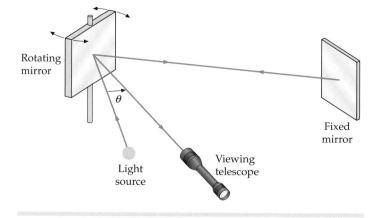

Rotating
mirror

θ

Fixed
mirror

Light
source

Viewing
telescope

FIGURE 31-13 Simplified drawing of Foucault's method of
measuring the speed of light.

The various methods we have discussed for measuring the speed of light are all in general agreement. Today, the speed of light is defined to be exactly

$$c = 299{,}792{,}458 \text{ m/s} \qquad\qquad 31\text{-}5$$

and the standard unit of length, the meter, is defined in terms of this speed and the standard unit of time. The meter is the distance light travels (in a vacuum) in $1/299{,}792{,}458$ s. The value 3×10^8 m/s for the speed of light is accurate enough for nearly all calculations. The speed of radio waves and all other electromagnetic waves (in a vacuum) is the same as the speed of light.

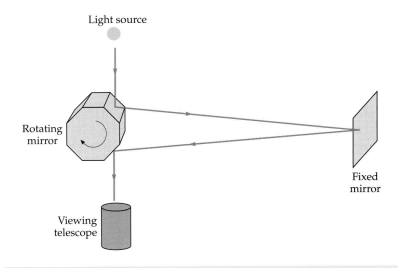

Light source

Rotating mirror

Fixed mirror

Viewing telescope

FIGURE 31-14 Simplified drawing of Michelson's method of measuring the speed of light at Mt. Wilson in the late 1920s.

THE SPEED OF LIGHT   **EXAMPLE 31-2**

What is the speed of light in feet per nanosecond?

PICTURE THE PROBLEM This is an exercise in unit conversions. There are $\sim$30 cm = 0.3 m in 1 ft.

1. Convert m/s to ft/ns:

$$c = 3 \times 10^8 \text{ m/s} \times \left(\frac{1 \text{ ft}}{0.3 \text{ m}}\right) \times \left(\frac{1 \text{ s}}{10^9 \text{ ns}}\right) = \boxed{1 \text{ ft/ns}}$$

FIZEAU'S DETERMINATION OF C   **EXAMPLE 31-3**

In Fizeau's experiment, his wheel had 720 teeth, and light was observed when the wheel rotated at 25.2 revolutions per second. If the distance from the wheel to the distant mirror was 8.63 km, what was Fizeau's value for the speed of light?

PICTURE THE PROBLEM The time taken for the light to travel from the wheel to the mirror and back is the time for the wheel to rotate one Nth of a revolution, where $N = 720$ is the total number of teeth.

1. The speed is the distance divided by the time. The distance from the wheel to the mirror is L:

$$c = \frac{2L}{\Delta t}$$

2. The angular displacement equals the angular speed times the time:

$$\Delta\theta = \omega\,\Delta t$$

3. Solve for the time:

$$\Delta t = \frac{\Delta\theta}{\omega}$$

4. Substitute for Δt and solve for c:

$$c = \frac{2L\omega}{\Delta\theta} = \frac{2(8.63 \times 10^3 \text{ m})(25.2 \text{ rev/s})}{\dfrac{1}{720} \text{ rev}}$$

$$= \boxed{3.14 \times 10^8 \text{ m/s}}$$

REMARKS This result is about 5 percent too high.

EXERCISE Space travelers on the moon use electromagnetic waves to communicate with the space control center on earth. Use $c = 3.00 \times 10^8$ m/s to calculate the time delay for their signal to reach the earth, which is 3.84×10^8 m away. (*Answer* 1.28 s each way)

Large distances are often given in terms of the distance traveled by light in a given time. For example, the distance to the sun is 8.33 light-minutes, written 8.33 c-min. A light-year is the distance light travels in one year. We can easily find a conversion factor between light-years and meters. The number of seconds in one year is

$$1 \text{ y} = 1 \text{ y} \times \frac{365.24 \text{ d}}{1 \text{ y}} \times \frac{24 \text{ h}}{1 \text{ d}} \times \frac{3600 \text{ s}}{1 \text{ h}} = 3.156 \times 10^7 \text{ s}$$

(*Note:* There are approximately π times 10^7 seconds per year, which is how some individuals remember the approximate value of the conversion.) The number of meters in one light-year is thus

$$1c\text{-year} = (2.998 \times 10^8 \text{ m/s})(3.156 \times 10^7 \text{ s}) = 9.46 \times 10^{15} \text{ m} \qquad \text{31-6}$$

31-5 The Propagation of Light

The propagation of light is governed by the wave equation discussed in Chapter 30. But long before Maxwell's theory of electromagnetic waves, the propagation of light and other waves was described empirically by two interesting and very different principles attributed to the Dutch physicist Christian Huygens (1629–1695) and the French mathematician Pierre de Fermat (1601–1665).

Huygens's Principle

Figure 31-15 shows a portion of a spherical wavefront emanating from a point source. The wavefront is the locus of points of constant phase. If the radius of the wavefront is r at time t, its radius at time $t + \Delta t$ is $r + c \Delta t$, where c is the speed of the wave. However, if a part of the wave is blocked by some obstacle or if the wave passes through a different medium, as in Figure 31-16, the determination of the new wavefront position at time $t + \Delta t$ is much more difficult. The propagation of any wavefront through space can be described using a geometric method discovered by Huygens in approximately 1678, which is now known as **Huygens's principle** or **Huygens's construction:**

> Each point on a primary wavefront serves as the source of spherical secondary wavelets that advance with a speed and frequency equal to those of the primary wave. The primary wavefront at some later time is the envelope of these wavelets.
>
> HUYGENS'S PRINCIPLE

Figure 31-17 shows the application of Huygens's principle to the propagation of a plane wave and the propagation of a spherical wave. Of course, if each point on a wavefront were really a point source, there would be waves in the backward direction as well. Huygens ignored these back waves.

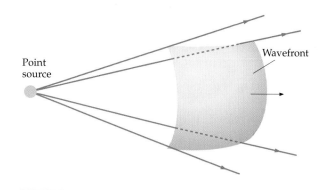

FIGURE 31-15 Spherical wavefront from a point source.

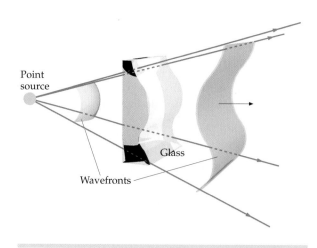
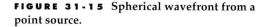

FIGURE 31-16 Wavefront from a point source before and after passing through a piece of glass of varied thickness.

Huygens's principle was later modified by Augustin Fresnel, so that the new wavefront was calculated from the old wavefront by superposition of the wavelets considering their relative amplitudes and phases. Kirchhoff later showed that the Huygens–Fresnel principle was a consequence of the wave equation (Equation 30-17), thus putting it on a firm mathematical basis. Kirchhoff showed that the intensity of each wavelet depends on the angle and is zero at 180° (the backward direction).

We will use Huygens's principle to derive the laws of reflection and refraction in Section 31-8. In Chapter 33, we apply Huygens's principle with Fresnel's modification to calculate the diffraction pattern of a single slit. Because the wavelength of light is so small, we can often use the ray approximation to describe its propagation.

Fermat's Principle

The propagation of light can also be described by Fermat's principle:

> The path taken by light traveling from one point to another is such that the time of travel is a minimum.[†]

FERMAT'S PRINCIPLE

In Section 31-8, we will use Fermat's principle to derive the laws of reflection and refraction.

31-6 Reflection and Refraction

The speed of light in a transparent medium such as air, water, or glass is less than the speed $c = 3 \times 10^8$ m/s in vacuum. A transparent medium is characterized by the **index of refraction**, n, which is defined as the ratio of the speed of light in a vacuum, c, to the speed in the medium, v:

$$n = \frac{c}{v} \qquad\qquad 31\text{-}7$$

DEFINITION—INDEX OF REFRACTION

For water, $n = 1.33$, whereas for glass n ranges from approximately 1.50 to 1.66, depending on the type of glass. Diamond has a very high index of refraction—approximately 2.4. The index of refraction of air is approximately 1.0003, so for most purposes we can assume that the speed of light in air is the same as the speed of light in vacuum.

When a beam of light strikes a boundary surface separating two different media, such as an air–glass interface, part of the light energy is reflected and part of the light energy enters the second medium. If the incident light is not perpendicular to the surface, then the transmitted beam is not parallel to the incident beam. The change in direction of the transmitted ray is called **refraction**. Figure 31-18 shows a light ray striking a smooth air–glass interface. The angle θ_1 between the incident ray and the normal (the line perpendicular to the surface) is called the **angle of incidence**, and the plane defined by these two lines is called the **plane of incidence**. The reflected ray lies in the plane of incidence and makes an angle θ_1' with the normal that is equal to the angle of incidence as shown in the figure:

$$\theta_1' = \theta_1 \qquad\qquad 31\text{-}8$$

LAW OF REFLECTION

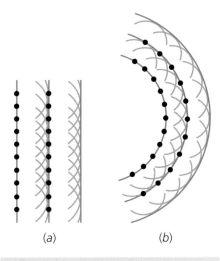

FIGURE 31-17 Huygens's construction for the propagation to the right of (*a*) a plane wave and (*b*) an outgoing spherical, or circular, wave.

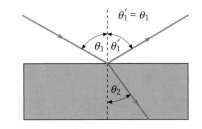

FIGURE 31-18 The angle of reflection θ_1' equals the angle of incidence θ_1. The angle of refraction θ_2 is less than the angle of incidence if the light speed in the second medium is less than that in the incident medium.

[†] A more complete and general statement is that the time of travel is stationary with respect to variations in path; that is, if t is expressed in terms of some parameter x, the path taken will be such that $dt/dx = 0$. The important characteristic of a stationary path is that the time taken along nearby paths will be approximately the same as that along the true path.

This result is known as the **law of reflection.** The law of reflection holds for any type of wave. Figure 31-19 illustrates the law of reflection for rays of light and for wavefronts of ultrasonic waves.

The ray that enters the glass in Figure 31-18 is called the refracted ray, and the angle θ_2 is called the angle of refraction. When a wave crosses a boundary where the wave speed is reduced, as in the case of light entering glass from air, the angle of refraction is less than the angle of incidence θ_1, as shown in Figure 31-18; that is, the refracted ray is bent toward the normal. If, on the other hand, the light beam originates in the glass and is refracted into the air, then the refracted ray is bent away from the normal.

The angle of refraction θ_2 depends on the angle of incidence and on the relative speed of light waves in the two mediums. If v_1 is the wave speed in the incident medium and v_2 is the wave speed in the transmission medium, the angles of incidence and refraction are related by

$$\frac{1}{v_1}\sin\theta_1 = \frac{1}{v_2}\sin\theta_2 \qquad 31\text{-}9a$$

Equation 31-9a holds for the refraction of any kind of wave incident on a boundary separating two media.

In terms of the indexes of refraction of the two media n_1 and n_2, Equation 31-9a is

$$n_1\sin\theta_1 = n_2\sin\theta_2 \qquad 31\text{-}9b$$

SNELL'S LAW OF REFRACTION

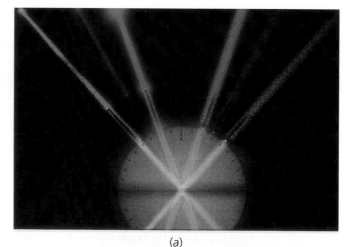

(a)

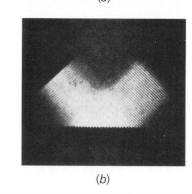

(b)

FIGURE 31-19 (*a*) Light rays reflecting from an air–glass interface showing equal angles of incidence and reflection. (*b*) Ultrasonic plane waves in water reflecting from a steel plate.

This result was discovered experimentally in 1621 by the Dutch scientist Willebrord Snell and is known as **Snell's law** or the **law of refraction.** It was independently discovered a few years later by the French mathematician and philosopher René Descartes.

Reflection and refraction of a beam of light incident on a glass slab.

Physical Mechanisms for Reflection and Refraction

The physical mechanism of the reflection and refraction of light can be understood in terms of the absorption and reradiation of the light by the atoms in the reflecting or refracting medium. When light traveling in air strikes a glass surface, the atoms in the glass absorb the light and reradiate it at the same frequency in all directions. The waves radiated backward by the glass atoms interfere constructively at an angle equal to the angle of incidence to produce the reflected wave.

The transmitted wave is the result of the interference of the incident wave and the wave produced by the absorption and reradiation of light energy by the atoms in the medium. For light entering glass from air, there is a phase lag between the reradiated wave and the incident wave. There is, therefore, also a phase lag between the resultant wave and the incident wave. This phase lag means that the position of a wave crest of the transmitted wave is retarded relative to the position of a wave crest of the incident wave in the medium. As a result, a transmitted wave crest does not travel as far in a given time as the original incident wave crest; that is, the velocity of the transmitted wave is less than that of the incident wave. The index of refraction is therefore greater than 1. The frequency of the light in the second medium is the same as the frequency of the incident light—the atoms absorb and reradiate the light at the same frequency—but the wave speed is different, so the wavelength of the transmitted light is different from that of the incident light. If λ is the wavelength of light in a vacuum, then $\lambda f = c$, and if λ' is the wavelength in a medium in which it has speed v, then $\lambda' f = v$. Combining these two relations gives $\lambda / \lambda' = c/v$, or

$$\lambda' = \frac{\lambda}{c/v} = \frac{\lambda}{n} \qquad\qquad 31\text{-}10$$

where $n = c/v$ is the index of refraction of the medium.

Specular Reflection and Diffuse Reflection

Figure 31-20a shows a bundle of light rays from a point source P that are reflected from a flat surface. After reflection, the rays diverge exactly as if they came from a point P' behind the surface. (This point is called the *image point*. We will study the formation of images by reflecting and refracting surfaces in the next chapter.) When these rays enter the eye, they cannot be distinguished from rays actually diverging from a source at P'.

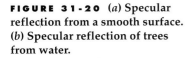

FIGURE 31-20 (a) Specular reflection from a smooth surface. (b) Specular reflection of trees from water.

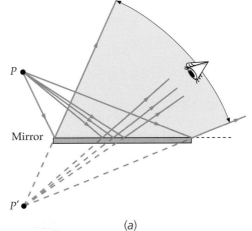

(a)

(b)

Reflection from a smooth surface is called **specular reflection.** It differs from **diffuse reflection,** which is illustrated in Figure 31-21. Here, because the surface is rough, the rays from a point reflect in random directions and do not diverge from any point, so there is no image. The reflection of light from the page of this book is diffuse reflection. The glass used in picture frames is sometimes ground slightly to give diffuse reflection and thereby cut down on glare from the light used to illuminate the picture. Diffuse reflection from the surface of the road allows you to see the road when you are driving at night because some of the light from your headlights reflects back toward you. In wet weather the reflection is mostly specular; therefore, little light is reflected back toward you, which makes the road difficult to see.

Relative Intensity of Reflected and Transmitted Light

The fraction of light energy reflected at a boundary, such as an air–glass interface, depends in a complicated way on the angle of incidence, the orientation of the electric field vector associated with the wave, and the indexes of refraction of the two media. For the special case of normal incidence ($\theta_1 = \theta_1' = 0$), the reflected intensity can be shown to be

$$I = \left(\frac{n_1 - n_2}{n_1 + n_2}\right)^2 I_0 \qquad \text{31-11}$$

where I_0 is the incident intensity and n_1 and n_2 are the indexes of refraction of the two media. For a typical case of reflection from an air–glass interface for which $n_1 = 1$ and $n_2 = 1.5$, Equation 31-11 gives $I = I_0/25$. Only about 4 percent of the energy is reflected; the remainder of the energy is transmitted.

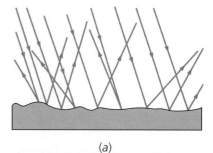

(a)

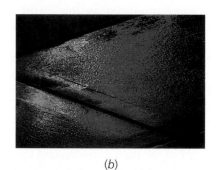

(b)

FIGURE 31-21 (a) Diffuse reflection from a rough surface. (b) Diffuse reflection of colored lights from a sidewalk.

REFRACTION FROM AIR TO WATER   **EXAMPLE 31-4**

Light traveling in air enters water with an angle of incidence of 45°. If the index of refraction of water is 1.33, what is the angle of refraction?

PICTURE THE PROBLEM The angle of refraction is found using Snell's law. Let subscripts 1 and 2 refer to the air and water, respectively. Then $n_1 = 1$, $\theta_1 = 45°$, $n_2 = 1.33$, and θ_2 is the angle of refraction (Figure 31-22).

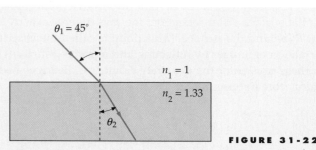

FIGURE 31-22

1. Use Snell's law to solve for $\sin \theta_2$, the sine of the angle of refraction:

$$n_1 \sin \theta_1 = n_2 \sin \theta_2$$

so

$$\sin \theta_2 = \frac{n_1}{n_2} \sin \theta_1$$

2. Find the angle whose sine is 0.532:

$$\theta_2 = \sin^{-1}\left(\frac{n_1}{n_2} \sin \theta_1\right) = \sin^{-1}\left(\frac{1.00}{1.33} \sin 45°\right)$$

$$= \sin^{-1}(0.532) = \boxed{32.1°}$$

REMARKS Note that the light is bent toward the normal as the light travels into the medium with the larger index of refraction.

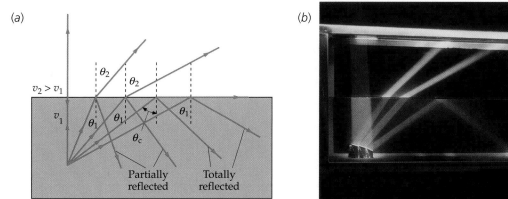

(a)

(b)

FIGURE 31-23 (*a*) Total internal reflection. As the angle of incidence is increased, the angle of refraction is increased until, at a critical angle of incidence θ_c, the angle of refraction is 90°. For angles of incidence greater than the critical angle, there is no refracted ray. (*b*) A photograph of refraction and total internal reflection from a water–air interface.

Total Internal Reflection

Figure 31-23 shows a point source in glass with rays striking the glass–air interface at various angles. All the rays not perpendicular to the interface are bent away from the normal. As the angle of incidence is increased, the angle of refraction increases until a critical angle of incidence θ_c is reached for which the angle of refraction is 90°. For incident angles greater than this critical angle, there is no refracted ray. All the energy is reflected. This phenomenon is called **total internal reflection.** The critical angle can be found in terms of the indexes of refraction of the two media by solving Equation 31-9*b* ($n_1 \sin \theta_1 = n_2 \sin \theta_2$) for $\sin \theta_1$ and setting θ_2 equal to 90°. That is,

$$\sin \theta_c = \frac{n_2}{n_1} \sin 90° = \frac{n_2}{n_1}$$ 31-12

CRITICAL ANGLE FOR TOTAL INTERNAL REFLECTION

Note that total internal reflection occurs only when the incident light is in the medium with the higher index of refraction. Mathematically, if n_2 is greater than n_1, Snell's law of refraction cannot be satisfied because there is no real-valued angle whose sine is greater than 1.

TOTAL INTERNAL REFLECTION   **EXAMPLE 31-5**   Try It Yourself

A particular glass has an index of refraction of $n = 1.50$. What is the critical angle for total internal reflection for light leaving this glass and entering air, for which $n = 1.00$ (Figure 31-24)?

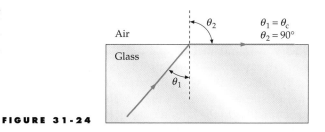

FIGURE 31-24

Cover the column to the right and try these on your own before looking at the answers.

Steps   **Answers**

1. Make a diagram showing the incident and refracted rays. For the critical angle the angle of refraction is 90°.

2. Apply the law of refraction Equation 31-9*b*. The critical $\theta_c = \boxed{41.8°}$
   angle is the angle of incidence.

You are enjoying a nice break at the pool. While under the water, you look up and notice that you see objects above water level in a circle of light of radius approximately 2.0 m, and the rest of your vision is the color of the sides of the pool. How deep are you in the pool?

PICTURE THE PROBLEM We can determine the depth of the pool from the radius of the light and the angle at which the light is entering our eye from the edge of the circle. At the edge of the circle the light is entering the water at 90°, so the angle of refraction at the air–water surface is the critical angle for total internal refraction at the water–air surface. From Figure 31-25, we see that the depth y is related to this angle and the radius of the circle R by $\tan \theta_c = R/y$. The critical angle is found from Equation 31-12 with $n_2 = 1$ and $n_1 = 1.33$.

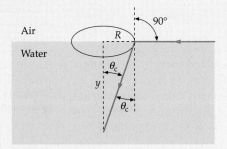

FIGURE 31-25

1. The depth y is related to the radius of the circle R and the critical angle θ_c:

$$\tan \theta_c = R/y$$

2. Solve for the depth y:

$$y = \frac{R}{\tan \theta_c}$$

3. Find the critical angle for total internal refraction at a water–air surface:

$$\sin \theta_c = \frac{n_2}{n_1} = \frac{1}{1.33} = 0.752$$

$$\theta_c = 48.8°$$

4. Solve for the depth y:

$$y = \frac{R}{\tan \theta_c} = \frac{2.0 \text{ m}}{\tan 48.8°} = \boxed{1.75 \text{ m}}$$

Figure 31-26a shows light incident normally on one of the short sides of a 45–45–90° glass prism. If the index of refraction of the prism is 1.5, the critical angle for total internal reflection is 41.8°, as you found in Example 31-5. Since the angle of incidence of the ray on the glass–air interface is 45°, the light will be totally reflected and will exit perpendicular to the other face of the prism, as shown. In Figure 31-26b, the light is incident perpendicular to the hypotenuse of the prism and is totally reflected twice so that it emerges at 180° to its original direction. Prisms are used to change the directions of light rays. In binoculars, two prisms are used on each side. These prisms reflect the light, thus shortening the required length, and reinvert the image (first inverted by a lens).[†] Diamonds have a very high index of refraction ($n \approx 2.4$), so nearly all the light that enters a diamond is eventually reflected back out, giving the diamond its sparkle.

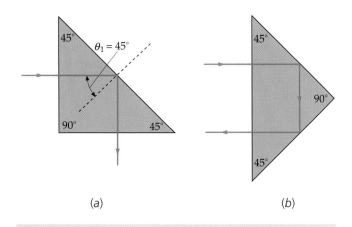

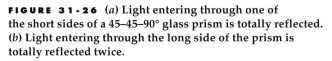

FIGURE 31-26 (a) Light entering through one of the short sides of a 45–45–90° glass prism is totally reflected. (b) Light entering through the long side of the prism is totally reflected twice.

<hr />

† The image produced by the objective lens of a telescope is discussed in Section 32-4.

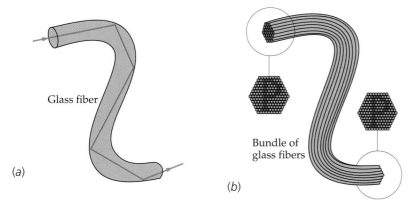

(a)

(b)

(c)

Fiber Optics

An interesting application of total internal reflection is the transmission of a beam of light down a long, narrow, transparent glass fiber (Figure 31-27a). If the beam begins approximately parallel to the axis of the fiber, it will strike the walls of the fiber at angles greater than the critical angle (if the bends in the fiber are not too sharp) and no light energy will be lost through the walls of the fiber. A bundle of such fibers can be used for imaging, as illustrated in Figure 31-27b. Fiber optics has many applications in medicine and in communications. In medicine, light is transmitted along tiny fibers to visually probe various internal organs without surgery. In communications, the rate at which information can be transmitted is related to the signal frequency. A transmission system using light of frequencies of the order of 10^{14} Hz can transmit information at a much greater rate than one using radio waves, which have frequencies of the order of 10^6 Hz. In telecommunication systems, a single glass fiber the thickness of a human hair can transmit audio or video information equivalent to 32,000 voices speaking simultaneously.

Mirages

When the index of refraction of a medium changes gradually, the refraction is continuous, leading to a gradual bending of the light. An interesting example of this is the formation of a mirage. On a hot and sunny day, the surface of exposed rocks, pavement, and sand often gets very hot. In this case there is often a layer of air near the ground that is warmer, and therefore less dense, than the air just above it. The speed

FIGURE 31-27 (*a*) A light pipe. Light inside the pipe is always incident at an angle greater than the critical angle, so no light escapes the pipe by refraction. (*b*) Light from the object is transported by a bundle of glass fibers to form an image of the object at the other end of the pipe. (*c*) Light emerging from a bundle of glass fibers.

(a)

(b)

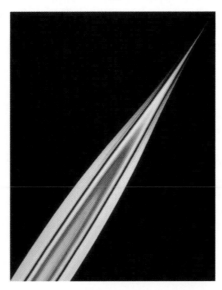

(*a*) In this demonstration at the Naval Research Laboratory, a combination of laser sources generates different colors that excite adjacent fiber sensor elements, leading to a separation of the information as indicated by the separation of the colors. (*b*) The tip of a light guide preform is softened by heat and drawn into a long, tiny fiber. The colors in the preform indicate a layered structure of differing compositions, which is retained in the fiber.

(a)

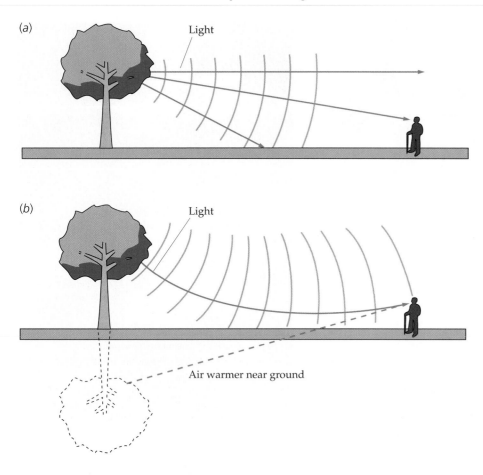

(c)

(b)

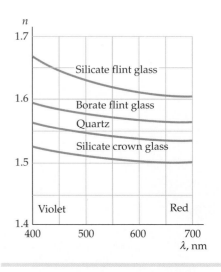

FIGURE 31-28 A mirage. (*a*) When the air is at a uniform temperature, the wavefronts of the light from the tree are spherical. (*b*) When the air near the ground is warmer, the wavefronts are not spherical and the light from the tree is continuously refracted into a curved path. (*c*) Apparent reflections of motorcycles on a hot road.

of any light waves is slightly greater in this less dense layer, so a light beam passing from the cooler layer into the warmer layer is bent. Figure 31-28*a* shows the light from a tree when all the surrounding air is at the same temperature. The wavefronts are spherical, and the rays are straight lines. In Figure 31-28*b*, the air near the ground is warmer, resulting in the wavefronts traveling faster there. The portions of the wavefronts near the hot ground get ahead of the higher portions, creating a nonspherical wavefront and causing a curving of the rays. Thus, the two rays shown initially heading for the ground are bent upward. As a result, the viewer sees an image of the tree looking as if it were reflected off a water surface on the ground. When driving on a hot sunny day, you may have noticed apparent wet spots on the highway ahead that disappear as you approach them. These mirages are due to the refraction of light from the sky by a layer of air that has been heated due to its proximity to the hot pavement.

Dispersion

The index of refraction of a material has a slight dependence on wavelength. For many materials, *n* decreases slightly as the wavelength increases, as shown in Figure 31-29. The dependence of the index of refraction on wavelength (and therefore on frequency) is called **dispersion.** When a beam of white light is incident at some angle on the surface of a glass prism, the angle of refraction (which is measured relative to the normal) for the shorter wavelengths is slightly smaller than the angle of refraction for the longer wavelengths. The light of shorter wavelength (toward the violet end of the spectrum) is therefore bent more toward the normal than that of longer wavelength. The beam of white light is thus spread out or dispersed into its component colors or wavelengths (Figure 31-30).

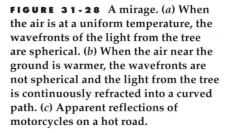

FIGURE 31-29 Index of refraction versus wavelength for various materials.

FIGURE 31-30 A beam of white light incident on a glass prism is dispersed into its component colors. The index of refraction decreases as the wavelength increases so that the longer wavelengths (red) are bent less than the shorter wavelengths (blue).

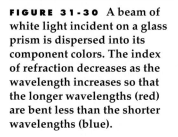

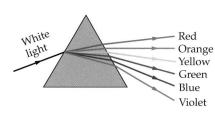

Rainbows   The rainbow is a familiar example of dispersion, in this case the dispersion of sunlight. Figure 31-31 is a diagram originally drawn by Descartes, showing parallel rays of light from the sun entering a spherical water drop. First, the rays are refracted as they enter the drop. The rays are then reflected from the water–air interface on the other side of the drop and finally are refracted again as they leave the drop.

From Figure 31-31, we can see that the angle made by the emerging rays and the diameter (along ray 1) reaches a maximum around ray 7 and then decreases. The concentration of rays emerging at approximately the maximum angle gives rise to the rainbow. By construction, using the law of refraction, Descartes showed that the maximum angle is about 42°. To observe a rainbow, we must therefore look at the water drops at an angle of 42° relative to the line back to the sun, as shown in Figure 31-32. The angular radius of the rainbow is therefore 42°.

The separation of the colors in the rainbow results from the fact that the index of refraction of water depends slightly on the wavelength of light. The angular radius of the bow will therefore depend slightly on the wavelength of the light. The observed rainbow is made up of light rays from many different droplets of water (Figure 31-33). The color seen at a particular angular radius corresponds to the wavelength of light that allows the light to reach the eye from the droplets at that angular radius. Because n_{water} is smaller for red light than for blue light, the red part of the rainbow is at a slightly greater angular radius than the blue part of the rainbow, so red is at the outer side of the rainbow.

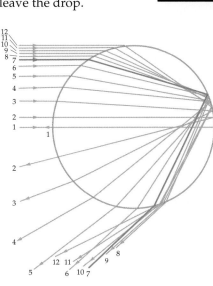

FIGURE 31-31 Descartes's construction of parallel rays of light entering a spherical water drop. Ray 1 enters the drop along a diameter and is reflected back along its incident path. Ray 2 enters slightly above the diameter and emerges below the diameter at a small angle with the diameter. The rays entering farther and farther away from the diameter emerge at greater and greater angles up to ray 7, shown as the heavy line. Rays entering above ray 7 emerge at smaller and smaller angles with the diameter.

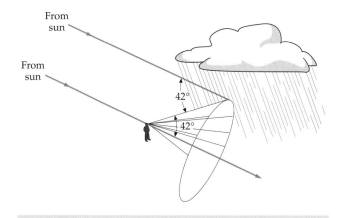

FIGURE 31-32 A rainbow is viewed at an angle of 42° from the line to the sun, as predicted by Descartes's construction, as shown in Figure 31-31.

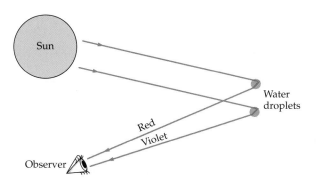

FIGURE 31-33 The rainbow results from light rays from many different water droplets.

(a)

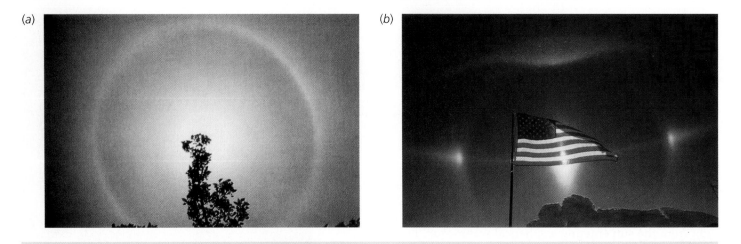

(b)

(*a*) **This 22° halo around the sun results from refraction by hexagonal ice crystals that are randomly oriented in the upper atmosphere.** (*b*) **When the ice crystals are not randomly oriented but are falling with their flat bases horizontal, only parts of the halo on each side of the sun, called** *sun dogs*, **are seen.**

When a light ray strikes a surface separating water and air, part of the light is reflected and part of the light is refracted. A secondary rainbow results from the light rays that are reflected twice within a droplet (Figure 31-34). The secondary bow has an angular radius of 51°, and its color sequence is the reverse of that of the primary bow; that is, the violet is on the outside in the secondary bow. Because of the small fraction of light reflected from a water–air interface, the secondary bow is considerably fainter than the primary bow.

*Calculating the Angular Radius of the Rainbow We can calculate the angular radius of the rainbow from the laws of reflection and refraction. Figure 31-35 shows a ray of light incident on a spherical water droplet at point *A*. The angle of refraction θ_2 is related to the angle of incidence θ_1 by Snell's law of refraction:

$$n_{air} \sin \theta_1 = n_{water} \sin \theta_2 \qquad\qquad 31\text{-}13$$

Point *P* in Figure 31-35 is the intersection of the line of the incident ray and the line of the emerging ray. The angle ϕ_d is called the angle of deviation of the ray, and ϕ_d and 2β form a straight angle. Thus,

$$\phi_d + 2\beta = \pi \qquad\qquad 31\text{-}14$$

We wish to relate the angle of deviation ϕ_d to the angle of incidence θ_1. From the triangle *AOB*, we have

$$2\theta_2 + \alpha = \pi \qquad\qquad 31\text{-}15$$

Similarly, from the triangle *AOP*, we have

$$\theta_1 + \beta + \alpha = \pi \qquad\qquad 31\text{-}16$$

Eliminating α from Equations 31-15 and 31-16 and solving for β gives

$$\beta = \pi - \theta_1 - \alpha = \pi - \theta_1 - (\pi - 2\theta_2) = 2\theta_2 - \theta_1$$

Substituting this value for β into Equation 31-14 gives the angle of deviation:

$$\phi_d = \pi - 2\beta = \pi - 4\theta_2 + 2\theta_1 \qquad\qquad 31\text{-}17$$

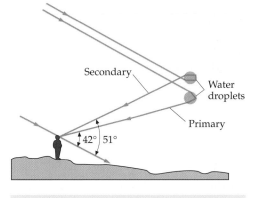

FIGURE 31-34 The secondary rainbow results from light rays that are reflected twice within a water droplet.

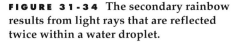

FIGURE 31-35 Light ray incident on a spherical water drop. The refracted ray strikes the back of the water droplet at point *B*. It makes an angle θ_2 with the radial line *OB* and is reflected at an equal angle. The ray is refracted again at point *C*, where it leaves the droplet.

Equation 31-17 can be combined with Equation 31-13 to eliminate θ_2 and give the angle of deviation ϕ_d in terms of the angle of incidence θ_1:

$$\phi_d = \pi + 2\theta_1 - 4 \sin^{-1}\left(\frac{n_{air}}{n_{water}} \sin \theta_1\right) \qquad 31\text{-}18$$

Figure 31-36 shows a plot of ϕ_d versus θ_1. The angle of deviation ϕ_d has its minimum value when $\theta_1 \approx 60°$. At an angle of incidence of 60°, the angle of deviation is $\phi_{d,min} = 138°$. This angle is called the **angle of minimum deviation.** At incident angles that are slightly greater or slightly smaller than 60°, the angle of deviation is approximately the same. Therefore, the intensity of the light reflected by the water droplet will be a maximum at the angle of minimum deviation. We can see from Figure 31-35 that the maximum value of β corresponds to the minimum value of ϕ_d. Thus, angular radius of the intensity maximum, given by $2\beta_{max}$, is

$$2\beta_{max} = \pi - \phi_{d,min} = 180° - 138° = 42° \qquad 31\text{-}19$$

The index of refraction of water varies slightly with wavelength. Thus, for each wavelength (color), the intensity maxima occurs at an angular radius slightly different than that of neighboring wavelengths.

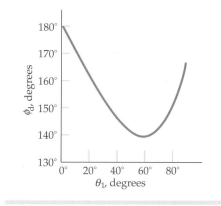

FIGURE 31-36 Plot of the angle of deviation ϕ_d as a function of incident angle θ_1. The angle of deviation has its minimum value of 138° when the angle of incidence is 60°. Since $d\phi_d/d\theta_1 = 0$ at minimum deviation, the deviation of rays with incident angles slightly less or slightly greater than 60° will be approximately the same.

31-7 Polarization

In a transverse mechanical wave, the vibration is perpendicular to the direction of propagation of the wave. If the vibration remains parallel to a plane, the wave is said to be **plane polarized** or **linearly polarized.** We can visualize polarization most easily by considering mechanical waves on a taut horizontal string. Let the x axis be along the string, let the z axis be vertical, and let the y axis be horizontal, perpendicular to both the x and z axes. If one end of the string is shaken up and down, the resulting waves on the string are linearly polarized with each element of the string vibrating up and down. Similarly, if one end is shaken back and forth along the y axis, the displacements of the string are linearly polarized with each element vibrating parallel with the y axis. If one end of the string is moved with constant speed in an ellipse in the $x = 0$ plane, the resulting wave is said to be **elliptically polarized.** In this case, each element of the string moves in an ellipse in a plane of constant x. Unpolarized waves can be produced by moving the end of the string in the $x = 0$ plane in a random way. Then the vibrations will have both y and z components that vary randomly. A linearly polarized electromagnetic wave is one in which the electric field remains parallel to a line. A wave produced by an electric dipole antenna is polarized with the electric field vector at any field point remaining parallel with the plane containing the field point and the antenna axis. Waves produced by numerous sources are usually unpolarized. A typical light source, for example, contains millions of atoms acting independently. The electric field for such a wave can be resolved into x and y components that vary randomly, because there is no correlation between the individual atoms producing the light.

The polarization of electromagnetic waves can be demonstrated with microwaves, which have wavelengths on the order of centimeters. In a typical microwave generator, polarized waves are radiated by an electric dipole antenna. In Figure 31-37, the electric dipole antenna is vertical, so the electric field vector $\vec{E}$ of the horizontally radiated waves is also vertical. An absorber can be made of a screen of parallel straight wires. When the wires are vertical, as in Figure 31-37a, the electric field parallel to the wires sets up currents in the wires and energy is absorbed. When the wires are horizontal and therefore perpendicular to $\vec{E}$, as in Figure 31-37b, no currents are set up and the waves are transmitted.

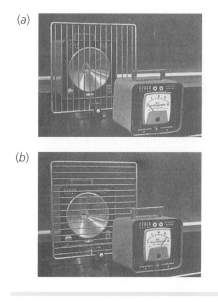

(a)

(b)

FIGURE 31-37 Demonstration showing the polarization of microwaves. The electric field of the microwaves is vertical, parallel to the vertical dipole antenna. (a) When the metal wires of the absorber are vertical, electric currents are set up in the wires and energy is absorbed, as indicated by the low reading on the microwave detector. (b) When the wires are horizontal, no currents are set up and the microwaves are transmitted, as indicated by the high reading on the detector.

There are four phenomena that produce polarized electromagnetic waves from unpolarized waves: (1) absorption, (2) reflection, (3) scattering, and (4) birefringence (also called double refraction), each of which is examined in the upcoming sections.

Polarization by Absorption

Several naturally occurring crystals, when cut into appropriate shapes, absorb and transmit light differently depending on the polarization of the light. These crystals can be used to produce linearly polarized light. In 1938, E. H. Land invented a simple commercial polarizing sheet called Polaroid. This material contains long-chain hydrocarbon molecules that are aligned when the sheet is stretched in one direction during the manufacturing process. These chains become conducting at optical frequencies when the sheet is dipped in a solution containing iodine. When light is incident with its electric field vector parallel to the chains, electric currents are set up along the chains, and the light energy is absorbed, just as the microwaves are absorbed by the wires in Figure 31-37. If the electric field is perpendicular to the chains, the light is transmitted. The direction perpendicular to the chains is called the **transmission axis.** We will make the simplifying assumption that all the light is transmitted when the electric field is parallel to the transmission axis and all the light is absorbed when it is perpendicular to the transmission axis. In reality, Polaroid absorbs some of the light, even when the electric field is parallel to the transmission axis.

Consider an unpolarized light beam incident on a polarizing sheet with its transmission axis along the x direction, as shown in Figure 31-38. The beam is incident on a second polarizing sheet, the analyzer, whose transmission axis makes an angle θ with the x axis. If E is the electric field amplitude of the incident beam, the component parallel with the transmission axis is $E_\parallel = E \cos \theta$, and the component perpendicular to the transmission axis is $E_\perp = E \sin \theta$. The sheet absorbs $E_\perp$ and transmits $E_\parallel$, so the transmitted beam has an electric field amplitude of $E_\parallel = E \cos \theta$ and is linearly polarized in the direction of the transmission axis. Because the intensity of light is proportional to the square of the magnitude of the electric field amplitude, the intensity I of light transmitted by the sheet is given by

$$I = I_0 \cos^2 \theta \qquad \text{31-20}$$

LAW OF MALUS

where I_0 is the intensity of the incident beam. If we have an incident beam of unpolarized light of intensity I_0 incident on a polarizing sheet, the direction of the incident electric field varies from location to location on the sheet, and at each location it fluctuates in time. At each location the angle between the electric field and the transmission axis is, on average, 45°, so applying Equation 31-20 gives $I = I_0 \cos^2 45° = \frac{1}{2} I_0$, where I is the intensity of the transmitted beam.

When two polarizing elements are placed in succession in a beam of unpolarized light, the first polarizing element is called the **polarizer** and the second polarizing element is called the **analyzer.** If the polarizer and the analyzer are crossed, that is, if their transmission axes are perpendicular to each other, no light gets through. Equation 31-20 is known as the **law of Malus** after its discoverer, E. L. Malus (1775–1812). It applies to any two polarizing elements whose transmission axes make an angle θ with each other.

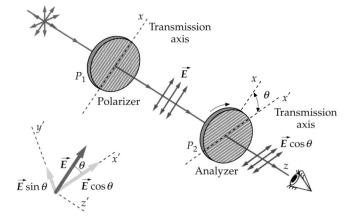

FIGURE 31-38 A vertically polarized beam is incident on a polarizing sheet with its transmission axes making an angle θ with the vertical. Only the component $E \cos \theta$ is transmitted through the second sheet, and the transmitted beam is linearly polarized in the direction of the transmission axis. If the intensity between the sheets is I_0, the intensity transmitted by both sheets is $I_0 \cos^2 \theta$.

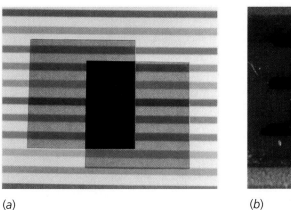

(a)

(b)

(*a*) Cross polarizers block out all of the light. (*b*) In a liquid crystal display, the crystal is between crossed polarizers. Light incident on the crystal is transmitted because the crystal rotates the direction of polarization of the light 90°. The light is reflected back out through the crystal by a mirror behind the crystal, and a uniform background is seen. When a voltage is applied across a small segment of the crystal, the polarization is not rotated, so no light is transmitted and the segment appears black.

INTENSITY TRANSMITTED

E X A M P L E   3 1 - 7

FIGURE 31-39

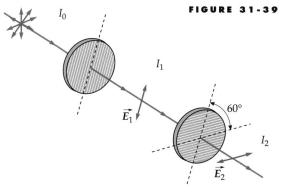

Unpolarized light of intensity 3.0 W/m² is incident on two polarizing sheets whose transmission axes make an angle of 60° (Figure 31-39). What is the intensity of light transmitted by the second sheet?

PICTURE THE PROBLEM The incident light is unpolarized, so the intensity transmitted by the first polarizing sheet is half the incident intensity. The second sheet further reduces the intensity by a factor of $\cos^2 \theta$, with $\theta = 60°$.

1. The intensity I_1 transmitted by the first sheet is half the intensity I_0 of unpolarized light incident on the first sheet:

$I_1 = \frac{1}{2} I_0$

2. The intensity I_2 transmitted by the second sheet is related to the intensity I_1 of the light incident on the second sheet by Equation 31-20:

$I_2 = I_1 \cos^2 \theta$

3. Combine these results and substitute the given data:

$I_2 = \frac{1}{2} I_0 \cos^2 60° = \frac{1}{2} (3.0 \text{ W/m}^2)(0.500)^2$

$= \boxed{0.375 \text{ W/m}^2}$

REMARKS Half the intensity passes through the first sheet no matter what the orientation of that sheet's transmission axis. Note that the second sheet rotates the plane of polarization by 60°.

Polarization by Reflection

When unpolarized light is reflected from a plane surface boundary between two transparent media, such as air and glass or air and water, the reflected light is partially polarized. The degree of polarization depends on the angle of incidence and on the ratio of the wave speeds in the two media. For a certain angle of incidence called the polarizing angle θ_p, the reflected light is completely polarized. At the polarizing angle, the reflected and refracted rays are perpendicular to each other. David Brewster (1781–1868), a Scottish scientist and an inventor of numerous instruments (including the kaleidoscope), discovered this experimentally in 1812. The polarizing angle is also referred to as Brewster's angle.

Figure 31-40 shows light incident at the polarizing angle θ_p for which the reflected light is completely polarized. The electric field of the incident light can be resolved into components parallel and perpendicular to the plane of incidence. The reflected light is linearly polarized with its electric field perpendicular to the plane of incidence. We can relate the polarizing angle to the indexes of refraction of the media using Snell's law (the law of refraction). If n_1 is the index of refraction of the first medium and n_2 is the index of refraction of the second medium, the law of refraction gives

$$n_1 \sin \theta_p = n_2 \sin \theta_2$$

where θ_2 is the angle of refraction. From Figure 31-40, we can see that the sum of the angle of reflection and the angle of refraction is 90°. Since the angle of reflection equals the angle of incidence, we have

$$\theta_2 = 90° - \theta_p$$

Then

$$n_1 \sin \theta_p = n_2 \sin(90° - \theta_p) = n_2 \cos \theta_p$$

or

$$\tan \theta_p = \frac{n_2}{n_1} \qquad \text{31-21}$$

POLARIZING ANGLE

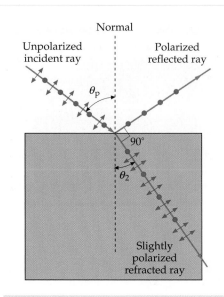

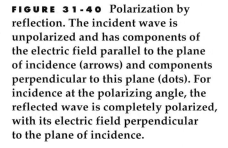

FIGURE 31-40 Polarization by reflection. The incident wave is unpolarized and has components of the electric field parallel to the plane of incidence (arrows) and components perpendicular to this plane (dots). For incidence at the polarizing angle, the reflected wave is completely polarized, with its electric field perpendicular to the plane of incidence.

Although the reflected light is completely polarized for this angle of incidence, the transmitted light is only partially polarized because only a small fraction of the incident light is reflected. If the incident light itself is polarized with the electric field in the plane of incidence, there is no reflected light when the angle of incidence is θ_p. We can understand this qualitatively from Figure 31-41. If we consider the molecules next to the surface of the second medium to be oscillating parallel to the electric field of the refracted ray, there can be no reflected ray because for an electric dipole antenna no energy is radiated along the line of oscillation. (Each of the oscillating molecules are a small electric dipole antenna.)

Because of the polarization of reflected light, sunglasses that contain a polarizing sheet can be very effective in cutting out glare. If light is reflected from a horizontal surface, such as a lake surface or snow on the ground, the electric field of the reflected light will be predominantly horizontal and the plane of incidence on the glasses will be predominantly vertical. Polarized sunglasses with a vertical transmission axis will then reduce glare by absorbing much of the reflected light. If you have polarized sunglasses, you can observe this effect by looking through the glasses at reflected light and then rotating the glasses 90°; much more of the light will be transmitted.

Polarization by Scattering

The phenomenon of absorption and reradiation is called **scattering**. Scattering can be demonstrated by passing a light beam through a container of water to which a small amount of powdered milk has been added. The milk particles absorb light and reradiate it, making the light beam visible. Similarly, laser beams can be made visible by introducing chalk or smoke particles into the air to scatter the light. A familiar example of light scattering is that from air molecules, which tend to scatter short wavelengths more than long wavelengths, thereby giving the sky its blue color.

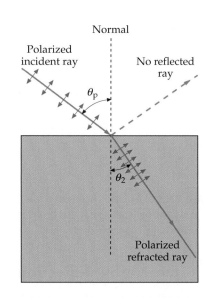

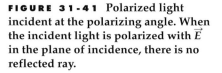

FIGURE 31-41 Polarized light incident at the polarizing angle. When the incident light is polarized with $\vec{E}$ in the plane of incidence, there is no reflected ray.

We can understand polarization by scattering if we think of a scattering molecule as an electric dipole antenna that radiates waves with a maximum intensity in directions perpendicular to the antenna axis and zero intensity in the direction along the antenna axis. The electric field vector of the scattered light perpendicular to the direction of propagation is in the plane of the antenna axis and the field point. Figure 31-42 shows a beam of unpolarized light that initially travels along the z axis, striking a molecule at the origin. The electric field in the light beam has components in both the x and y directions perpendicular to the direction of motion of the light beam. These fields set up oscillations of the charges within the molecule in the z = 0 plane, and there is no oscillation along the z direction. These oscillations can be thought of as a superposition of an oscillation along the x axis and another along the y axis, with each of these oscillations producing dipole radiation. Thus, the oscillation along the x axis produces no radiation along the x axis, which means the light radiated along the x axis is produced only by the oscillation along the y axis. It follows that the light radiated along the x axis is polarized with its electric field parallel with the y axis. There is nothing special about the choice of axes for this discussion, so the result can be generalized. That is, the light scattered in a direction perpendicular to the incident light beam is polarized with its electric field perpendicular to both the incident beam and the direction of propagation of the scattered light. This can be seen easily by examining the scattered light with a piece of polarizing sheet.

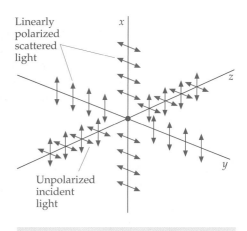

FIGURE 31-42 Polarization by scattering. Unpolarized light propagating in the z direction is incident on a scattering center at the origin. The light scattered in the z = 0 plane along the x direction is polarized parallel with the y axis (and the light scattered in the y direction is polarized parallel with the x axis).

Polarization by Birefringence

Birefringence is a complicated phenomenon that occurs in calcite and other noncubic crystals and in some stressed plastics, such as cellophane. Most materials are **isotropic,** that is, the speed of light passing through the material is the same in all directions. Because of their atomic structure, birefringent materials are **anisotropic.** The speed of light depends on the plane of polarization and on the direction of propagation of the light. When a light ray is incident on such materials, it may be separated into two rays called the *ordinary ray* and the *extraordinary ray.* These rays are polarized in mutually perpendicular directions, and they travel with different speeds. Depending on the relative orientation of the material and the incident light beam, the rays may also travel in different directions.

There is one particular direction in a birefringent material in which both rays propagate with the same speed. This direction is called the **optic axis** of the material. (The optic axis is actually a *direction* rather than a line in the material.) Nothing unusual happens when light travels in the direction of the optic axis. However, when light is incident at an angle to the optic axis, as shown in Figure 31-43, the rays travel in different directions and emerge separated in space. If the material is rotated, the extraordinary ray (the e ray in the figure) revolves in space around the ordinary ray (o ray).

If light is incident on a birefringent plate perpendicular to its crystal face and perpendicular to the optic axis, the two rays travel in the same direction but at different speeds. The number of wavelengths in the two rays in the plate is different because the wavelengths ($\lambda = v/f$) of the rays differ. The rays emerge with a phase difference that depends on the thickness of the plate and on the wavelength of the incident light. In a **quarter-wave plate,** the thickness is such that there is a 90° phase difference between the waves of a particular wavelength when they emerge. In a **half-wave plate,** the rays emerge with a phase difference of 180°.

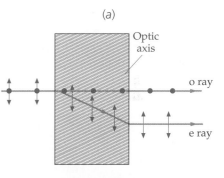

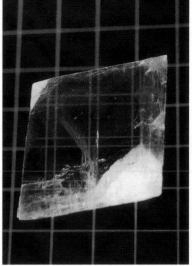

FIGURE 31-43 (*a*) A narrow beam of light incident on a birefringent crystal such as calcite is split into two beams, called the ordinary ray (o ray) and the extraordinary ray (e ray), that have mutually perpendicular polarizations. If the crystal is rotated, the extraordinary ray rotates in space. (*b*) A double image of the cross hatching is produced by this birefringent crystal of calcium carbonate.

(a)

(b)

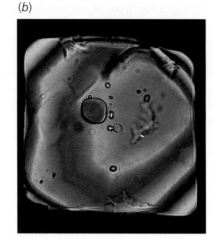

(c)

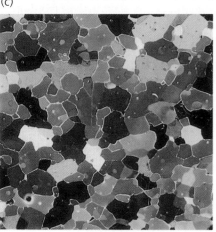

(d)

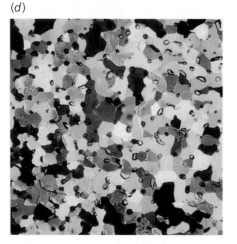

When the transmission axes of two polarizing sheets are perpendicular, the polarizers are said to be crossed and no light is transmitted. However, many materials are birefringent or become so under stress. Such materials rotate the direction of polarization of the light so that light of a particular wavelength is transmitted through both polarizers. When a birefringent material is viewed between crossed polarizers, information about its internal structure is revealed. (a) A shocked quartz grain from the site of a meteorite crater. The layered structure, evidenced by the parallel lines, arises from the shock of the impact of the meteor. (b) A grain of quartz typically found in silicic volcanic rocks. No shock lines are seen. (c) Thin sections of ice core from the antarctic ice sheet reveal bubbles of trapped CO_2, which appear amber-colored. This sample was taken from a depth of 194 m, corresponding to air trapped 1600 years ago, whereas the sample in (d) is from a depth of 56 m, corresponding to air trapped 450 years ago. Ice core measurements have replaced the less reliable technique of analyzing carbon in tree rings to compare current atmospheric CO_2 levels with those of the recent past. (e) Robert Mark of the Princeton School of Architecture examines the stress patterns in a plastic model of the nave structure of Chartres Cathedral.

(e)

Suppose that the incident light is linearly polarized so that the electric field vector is at 45° to the optic axis, as illustrated in Figure 31-44. The ordinary and extraordinary rays start out in phase and have equal amplitudes. With a quarter-wave plate, the waves emerge with a phase difference of 90°, so the resultant electric field has components $E_x = E_0 \sin \omega t$ and $E_y = E_0 \sin(\omega t + 90°) = E_0 \cos \omega t$. The electric field vector thus rotates in a circle and the wave is circularly polarized.

With a half-wave plate, the waves emerge with a phase difference of 180°, so the resultant electric field is linearly polarized with components $E_x = E_0 \sin \omega t$ and $E_y = E_0 \sin(\omega t + 180°) = -E_0 \sin \omega t$. The net effect is that the direction of polarization of the wave is rotated by 90° relative to that of the incident light, as shown in Figure 31-45.

Interesting and beautiful patterns can be observed by placing birefringent materials, such as cellophane or stressed plastic, between two polarizing sheets with their transmission axes perpendicular to each other. Ordinarily, no light is transmitted through crossed polarizing sheets. However, if we place a birefringent material between the crossed polarizing sheets, the material acts as a half-wave plate for light of a certain color depending on the material's thickness. The direction of polarization is rotated and some light gets through both sheets. Various glasses and plastics become birefringent when under stress. The stress patterns can be observed when the material is placed between crossed polarizing sheets.

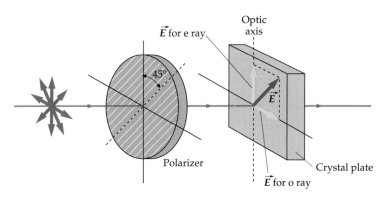

FIGURE 31-44 Polarized light emerging from the polarizer is incident on a birefringent crystal so that the electric field vector makes a 45° angle with the optic axis, which is perpendicular to the light beam. The ordinary and extraordinary rays travel in the same direction but at different speeds. The polarization of the emerging light depends on the thickness of the crystal and the wavelength of the light.

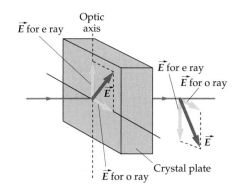

FIGURE 31-45 If the birefringent crystal in Figure 31-44 is a half-wave plate, and if the electric field vector of the incident light makes an angle of 45° with the optic axis, then the direction of polarization of the emerging light is rotated by 90°.

31-8 Derivation of the Laws of Reflection and Refraction

The laws of reflection and refraction can be derived from either Huygens's principle or Fermat's principle.

Huygens's Principle

Reflection Figure 31-46 shows a plane wavefront AA' striking a mirror at point A. As can be seen from the figure, the angle ϕ_1 between the wavefront and the mirror is the same as the angle of incidence θ_1, which is the angle between the normal to the mirror and the rays (which are perpendicular to the wavefronts). According to Huygens's principle, each point on a given wavefront can be considered a point source of secondary wavelets. The position of the wavefront after a time t is found by constructing wavelets of radius ct with their centers on the wavefront AA'. Wavelets that have not yet reached the mirror form the portion of the new wavefront BB'. Wavelets that have already reached the mirror are reflected and form the portion of the new wavefront $B''B$. By a similar construction, the wavefront $C''C$ is obtained from the Huygens's wavelets originating on the wavefront $B''B$. Figure 31-47 is an enlargement of a portion of Figure 31-46 showing AP, which is part of the initial position of the wavefront. During the time t, the wavelet from point P reaches the mirror at point B, and the wavelet from point A reaches point B''. The reflected wavefront $B''B$ makes an angle ϕ_1' with the mirror that is equal to the angle of reflection θ_1' between the reflected ray

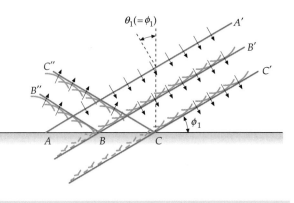

FIGURE 31-46 Plane wave reflected at a plane mirror. The angle θ_1 between the incident ray and the normal to the mirror is the angle of incidence. It is equal to the angle ϕ_1 between the incident wavefront and the mirror.

and the normal to the mirror. The triangles $AB''B$ and APB are both right triangles with a common side AB and equal sides $AB'' = BP = ct$. Hence, these triangles are congruent, and the angles ϕ_1 and ϕ_1' are equal, implying that the angle of reflection θ_1' equals the angle of incidence θ_1.

Refraction Figure 31-48 shows a plane wave incident on an air–glass interface. We apply Huygens's construction to find the wavefront in the transmitted wave. Line AP indicates a portion of the wavefront in medium 1 that strikes the glass surface at an angle ϕ_1. In time t, the wavelet from P travels the distance $v_1 t$ and reaches the point B on the line AB separating the two media, while the wavelet from point A travels a shorter distance $v_2 t$ into the second medium. The new wavefront BB' is not parallel to the original wavefront AP because the speeds v_1 and v_2 are different. From the triangle APB,

$$\sin \phi_1 = \frac{v_1 t}{AB}$$

or

$$AB = \frac{v_1 t}{\sin \phi_1} = \frac{v_1 t}{\sin \theta_1}$$

using the fact that the angle ϕ_1 equals the angle of incidence θ_1. Similarly, from triangle $AB'B$,

$$\sin \phi_2 = \frac{v_2 t}{AB}$$

or

$$AB = \frac{v_2 t}{\sin \phi_2} = \frac{v_2 t}{\sin \theta_2}$$

where $\theta_2 = \phi_2$ is the angle of refraction. Equating the two values for AB, we obtain

$$\frac{\sin \theta_1}{v_1} = \frac{\sin \theta_2}{v_2} \qquad\qquad \text{31-22}$$

Substituting $v_1 = c/n_1$ and $v_2 = c/n_2$ in this equation and multiplying by c, we obtain $n_1 \sin \theta_1 = n_2 \sin \theta_2$, which is Snell's law.

Fermat's Principle

Reflection Figure 31-49 shows two paths in which light leaves point A, strikes the plane surface, which we can consider to be a mirror, and travels to point B. The problem for the application of Fermat's principle to reflection can be stated as follows: At what point P in the figure must the light strike the mirror so that it will travel from point A to point B in the least time? Since the light is traveling in the same medium for this problem, the time will be minimum when the distance is minimum. In Figure 31-49 the distance APB is the same as the distance $A'PB$, where point A' lies along the perpendicular from A to the mirror and is equidistant behind the mirror. As we vary point P, the distance $A'PB$ is least when the points A', P, and B lie on a straight line. We can see from the figure that this occurs when the angle of incidence equals the angle of reflection.

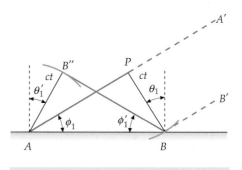

FIGURE 31-47 Geometry of Huygens's construction for the calculation of the law of reflection. The wavefront AP initially strikes the mirror at point A. After a time t, the Huygens wavelet from P strikes the mirror at point B, and the Huygens wavelet from point A reaches point B.

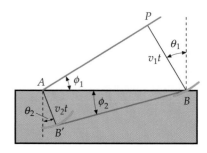

FIGURE 31-48 Application of Huygens's principle to the refraction of plane waves at the surface separating a medium in which the wave speed is v_1 from a medium in which the wave speed v_2 is less than v_1. The angle of refraction in this case is less than the angle of incidence.

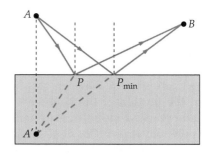

FIGURE 31-49 Geometry for deriving the law of reflection from Fermat's principle. The time it takes for the light to travel from point A to point B is a minimum for light striking the surface at point P.

Refraction   The derivation of Snell's law of refraction from Fermat's principle is slightly more complicated. Figure 31-50 shows the possible paths for light traveling from point A in air to point B in glass. Point P_1 is on the straight line between A and B, but this path is not the one for the shortest travel time because light travels with a smaller speed in the glass. If we move slightly to the right of P_1, the total path length is greater, but the distance traveled in the slower medium is less than for the path through P_1. It is not apparent from the figure which path is the path of least time, but it is not surprising that a path slightly to the right of the straight-line path takes less time because the time gained by traveling a shorter distance in the glass more than compensates for the time lost traveling a longer distance in the air. As we move the point of intersection of the possible path to the right of point P_1, the total time of travel from point A to point B decreases until we reach a minimum at point P_{min}. Beyond this point, the time saved by traveling a shorter distance in the glass does not compensate for the greater time required for the greater distance traveled in the air.

Figure 31-51 shows the geometry for finding the path of least time. If L_1 is the distance traveled in medium 1 with index of refraction n_1, and L_2 is the distance traveled in medium 2 with index of refraction n_2, the time for light to travel the total path AB is

$$ t = \frac{L_1}{v_1} + \frac{L_2}{v_2} = \frac{L_1}{c/n_1} + \frac{L_2}{c/n_2} = \frac{n_1 L_1}{c} + \frac{n_2 L_2}{c} \qquad \text{31-23} $$

We wish to find the point P_{min} for which this time is a minimum. We do this by expressing the time in terms of a single parameter x, as shown in the figure, indicating the position of point P_{min}. In terms of the distance x,

$$ L_1^2 = a^2 + x^2 \quad \text{and} \quad L_2^2 = b^2 + (d-x)^2 \qquad \text{31-24} $$

Figure 31-52 shows the time t as a function of x. At the value of x for which the time is a minimum, the slope of the graph of t versus x is zero:

$$ \frac{dt}{dx} = 0 $$

Differentiating each term in Equation 31-23 with respect to x and setting the result equal to zero, we obtain

$$ \frac{dt}{dx} = \frac{1}{c}\left(n_1 \frac{dL_1}{dx} + n_2 \frac{dL_2}{dx} \right) = 0 \qquad \text{31-25} $$

We can compute these derivatives from Equations 31-24. We have

$$ 2L_1 \frac{dL_1}{dx} = 2x \quad \text{or} \quad \frac{dL_1}{dx} = \frac{x}{L_1} $$

where x/L_1 is just $\sin \theta_1$ and θ_1 is the angle of incidence. Thus,

$$ \frac{dL_1}{dx} = \sin \theta_1 \qquad \text{31-26} $$

Similarly,

$$ 2L_2 \frac{dL_2}{dx} = 2(d-x)(-1) $$

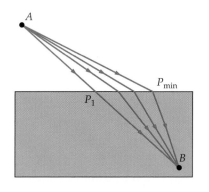

FIGURE 31-50 Geometry for deriving Snell's law from Fermat's principle. The point P_{min} is the point at which light must strike the glass in order that the travel time from point A to point B is a minimum.

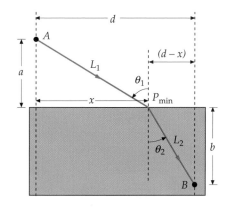

FIGURE 31-51 Geometry for calculating the minimum time in the derivation of Snell's law from Fermat's principle.

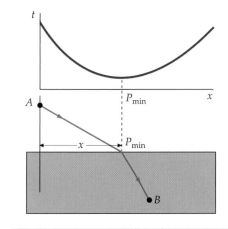

FIGURE 31-52 Graph of the time it takes for light to travel from point A to point B versus x, measured along the refracting surface. The time is a minimum at the point at which the angles of incidence and refraction obey Snell's law.

or

$$\frac{dL_2}{dx} = -\frac{d-x}{L_2} = -\sin\theta_2 \qquad\qquad\qquad 31\text{-}27$$

where θ_2 is the angle of refraction. From Equation 31-25,

$$n_1\frac{dL_1}{dx} + n_2\frac{dL_2}{dx} = 0 \qquad\qquad\qquad 31\text{-}28$$

Substituting the results of Equations 31-26 and 31-27 for dL_1/dx and dL_2/dx gives

$$n_1\sin\theta_1 + n_2(-\sin\theta_2) = 0$$

or

$$n_1\sin\theta_1 = n_2\sin\theta_2$$

which is Snell's law.

SUMMARY

Topic	Relevant Equations and Remarks
1. **Visible Light**	The human eye is sensitive to electromagnetic radiation with wavelengths from approximately 400 nm (violet) to 700 nm (red). The photon energies range from approximately 1.8 eV to 3.1 eV. A uniform mixture of wavelengths, such as the wavelengths emitted by the sun, appears white to our eyes.
2. **Wave–Particle Duality**	Light propagates like a wave, but interacts with matter like a particle.
Photon energy	$E = hf = \dfrac{hc}{\lambda}$ **31-1**
Planck's constant	$h = 6.626 \times 10^{-34}\,\text{J·s} = 4.136 \times 10^{-15}\,\text{eV·s}$
hc	$hc = 1240\,\text{eV·nm}$ **31-2**
3. **Emission of Light**	Light is emitted when an outer atomic electron makes a transition from an excited state to a state of lower energy.
Line spectra	Atoms in dilute gases emit a discrete set of wavelengths called a line spectra. The photon energy $E = hf = hc/\lambda$ equals the difference in energy of the initial and final states of the atom.
Continuous spectra	Atoms in high-density gases, liquids, or solids have continuous bands of energy levels, so they emit a continuous spectrum of light. Thermal radiation is visible if the temperature of the emitting object is above approximately 600°C.
Spontaneous emission	An atom in an excited state will spontaneously make a transition to a lower state with the emission of a photon. This process is random, with a characteristic lifetime of about 10^{-8} s. The photons from two or more atoms are not correlated, so the light is incoherent.

Stimulated emission	Stimulated emission occurs if an atom is initially in an excited state and a photon of energy equal to the energy difference between that state and a lower state is incident on the atom. The oscillating electromagnetic field of the incident photon stimulates the excited atom to emit another photon in the same direction and in phase with the incident photon. The emitted light is coherent.
4. Lasers	A laser produces an intense, coherent, and narrow beam of photons as the result of stimulated emission. The operation of a laser depends on population inversion, in which there are more atoms in an excited state than in the ground state or a lower state.
5. Speed of Light	The SI unit of length, the meter, is defined so that the speed of light in vacuum is exactly

$$c = 299{,}792{,}458 \text{ m/s} \qquad \text{31-5}$$

v in a transparent medium	$$v = \frac{c}{n} \qquad \text{31-7}$$

where n is the index of refraction.

6. Huygens's Principle	Each point on a primary wavefront serves as the source of spherical secondary wavelets that advance with a speed and frequency equal to that of the primary wave. The primary wavefront at some later time is the envelope of these wavelets.
7. Reflection and Refraction	When light is incident on a surface separating two media in which the speed of light differs, part of the light energy is transmitted and part of the light energy is reflected.
Law of reflection	The reflected ray lies in the plane of incidence and makes an angle θ_1' with the normal that is equal to the angle of incidence.

$$\theta_1' = \theta_1 \qquad \text{31-8}$$

Reflected intensity, normal incidence	$$I = \left(\frac{n_1 - n_2}{n_1 + n_2}\right)^2 I_0 \qquad \text{31-11}$$
Index of refraction	$$n = \frac{c}{v} \qquad \text{31-7}$$
Law of refraction (Snell's law)	$$n_1 \sin \theta_1 = n_2 \sin \theta_2 \qquad \text{31-9}b$$
Total internal reflection	When light is traveling in a medium with an index of refraction n_1 and is incident on the boundary of a second medium with a lower index of refraction $n_2 < n_1$, the light is totally reflected if the angle of incidence is greater than the critical angle θ_c given by
Critical angle	$$\sin \theta_c = \frac{n_2}{n_1} \qquad \text{31-12}$$
Dispersion	The speed of light in a medium, and therefore the index of refraction of that medium, depends on the wavelength of light. Because of dispersion, a beam of white light incident on a refracting prism is dispersed into its component colors. Similarly, the reflection and refraction of sunlight by raindrops produces a rainbow.
8. Polarization	Transverse waves can be polarized. The four phenomena that produce polarized electromagnetic waves from unpolarized waves are: (1) absorption, (2) scattering, (3) reflection, and (4) birefringence.
Malus's law	When two polarizers have their transmission axes at an angle θ, the intensity transmitted by the second polarizer is reduced by the factor $\cos^2 \theta$.

$$I = I_0 \cos^2 \theta \qquad \text{31-20}$$

PROBLEMS

- Single-concept, single-step, relatively easy
- •• Intermediate-level, may require synthesis of concepts
- ••• Challenging
- SSM Solution is in the *Student Solutions Manual*
- iSOLVE Problems available on iSOLVE online homework service
- iSOLVE✓ These "Checkpoint" online homework service problems ask students additional questions about their confidence level, and how they arrived at their answer.

In a few problems, you are given more data than you actually need; in a few other problems, you are required to supply data from your general knowledge, outside sources, or informed estimates.

Conceptual Problems

1 • Why is helium needed in a helium–neon laser? Why not just use neon?

2 •• When a beam of visible white light passes through a gas of atomic hydrogen and is viewed with a spectroscope, dark lines are observed at the wavelengths of the emission series. The atoms that participate in the resonance absorption then emit this same wavelength light as they return to the ground state. Explain why the observed spectrum nevertheless exhibits pronounced dark lines.

3 • How does a thin layer of water on the road affect the light you see reflected off the road from your own headlights? How does it affect the light you see reflected from the headlights of an oncoming car?

4 • A ray of light passes from air into water, striking the surface of the water with an angle of incidence of 45°. Which of the following four quantities change as the light enters the water: (1) wavelength, (2) frequency, (3) speed of propagation, (4) direction of propagation? (*a*) 1 and 2 only; (*b*) 2, 3, and 4 only; (*c*) 1, 3, and 4 only; (*d*) 3 and 4 only; or (*e*) 1, 2, 3, and 4.

5 •• SSM The density of the atmosphere decreases with height, as does the index of refraction. Explain how one can see the sun after it has set. Why does the setting sun appear flattened?

6 • A physics student playing pocket billiards wants to strike her cue ball so that it hits a cushion and then hits the eight ball squarely. She chooses several points on the cushion and for each point measures the distance from it to the cue ball and to the eight ball. She aims at the point for which the sum of these distances is least. (*a*) Will her cue ball hit the eight ball? (*b*) How is her method related to Fermat's principle?

7 • A swimmer at point *S* in Figure 31-53 develops a leg cramp while swimming near the shore of a calm lake and calls for help. A lifeguard at point *L* hears the call. The lifeguard can run 9 m/s and swim 3 m/s. She knows physics and chooses a path that will take the least time to reach the swimmer. Which of the paths shown in Figure 31-53 does the lifeguard take?

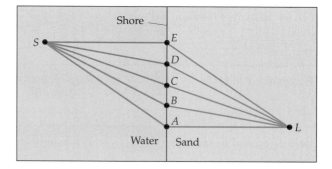

FIGURE 31-53 Problem 7

8 • Two polarizers have their transmission axes at an angle θ. Unpolarized light of intensity I is incident upon the first polarizer. What is the intensity of the light transmitted by the second polarizer? (*a*) $I \cos^2 \theta$, (*b*) $(I \cos^2 \theta)/2$, (*c*) $(I \cos^2 \theta)/4$, (*d*) $I \cos \theta$, (*e*) $(I \cos \theta)/4$, or (*f*) none of the answers are correct.

9 • Which of the following is *not* a phenomenon whereby polarized light can be produced from unpolarized light? (*a*) absorption (*b*) reflection (*c*) birefringence (*d*) diffraction (*e*) scattering

10 •• SSM We learned in Chapter 30, Section 30-3, that an oscillating electric dipole produces electromagnetic radiation (see Figure 30-8). Assuming that the light reflected off and refracted into the surface of a piece of transparent material is caused by such dipoles, show that the condition for Brewster's angle (Equation 31-21) is exactly the same as saying that the refracted ray is perpendicular to the axis of the radiating dipoles for light polarized in the plane of incidence.

11 •• Draw a diagram to explain how Polaroid sunglasses reduce glare from sunlight reflected from a smooth horizontal surface, such as the surface found on a pool of water. Your diagram should clearly indicate the direction of polarization of the light as it propagates from the sun to the reflecting surface and then through the sunglasses into the eye.

12 • iSOLVE True or false:

(*a*) Light and radio waves travel with the same speed through a vacuum.

(*b*) Most of the light incident normally on an air–glass interface is reflected.

(c) The angle of refraction of light is always less than the angle of incidence.

(d) The index of refraction of water is the same for all wavelengths in the visible spectrum.

(e) Longitudinal waves cannot be polarized.

13 •• **iSOLVE** ✓ Of the following statements about the speeds of the various colors of light in glass, which are true?

(a) All colors of light have the same speed in glass.

(b) Violet has the highest speed, red the lowest.

(c) Red has the highest speed, violet the lowest.

(d) Green has the highest speed, red and violet the lowest.

(e) Red and violet have the highest speed, green the lowest.

14 •• **SSM** It is a common experience that on a calm, sunny day one can hear voices of persons in a boat over great distances. Explain this phenomenon, keeping in mind that sound is reflected from the surface of the water and that the temperature of the air just above the water's surface is usually less than that at a height of 10 m or 20 m above the water.

15 • The human eye perceives color using a structure called a cone, located on the retina. The molecules of the cones come in three types that respond in a process similar to resonance absorption to red, green, and blue light, respectively. Use this fact to explain why the color of a blue object (450 nm in air) does not appear to change when immersed in clear colorless water, in spite of the fact that the wavelength of the light is shortened in accordance with Equation 31-10.

Estimation and Approximation

16 • Estimate the time required for light to make the round trip in Galileo's experiment to determine the speed of light.

17 • Ole Römer's method for measuring the speed of light requires the precise prediction of the time of occurrence for the eclipse of Jupiter's moon Io. Assuming an eclipse took place on June 1 at midnight when the earth was in location A, as shown in Figure 31-11, predict the expected time of the eclipse one-quarter year later at location B, assuming (a) the speed of light is infinite and (b) the speed of light is the presently defined value of 2.998×10^8 m/s.

18 •• If the angle of incidence is small enough, the approximation $\sin \theta \approx \theta$ may be used to simplify Snell's law. Calculate the angle of incidence that would make the error in calculating the angle of refraction using this small angle approximation no worse than 1 percent when compared to the exact formula. This approximation will be used in connection with image formation by spherical surfaces in Chapter 32.

Sources of Light

19 • **iSOLVE** A pulse from a ruby laser has an average power of 10 MW and lasts 1.5 ns. (a) What is the total energy of the pulse? (b) How many photons are emitted in this pulse?

20 • **iSOLVE** A helium–neon laser emits light of wavelength 632.8 nm and has a power output of 4 mW. How many photons are emitted per second by this laser?

21 • **iSOLVE** ✓ The first excited state of an atom of a gas is 2.85 eV above the ground state. (a) What is the wavelength of radiation for resonance absorption? (b) If the gas is irradiated with monochromatic light of 320 nm wavelength, what is the wavelength of the Raman scattered light?

22 •• A gas is irradiated with monochromatic ultraviolet light of 368 nm wavelength. Scattered light of the same wavelength and of 658 nm wavelength is observed. Assuming that the gas atoms were in their ground state prior to irradiation, find the energy difference between the ground state and the atomic state excited by the irradiation.

23 •• Sodium has excited states 2.11 eV, 3.2 eV, and 4.35 eV above the ground state. (a) What is the maximum wavelength of radiation that will result in resonance fluorescence? What is the wavelength of the fluorescent radiation? (b) What wavelength will result in excitation of the state 4.35 eV above the ground state? If that state is excited, what are the possible wavelengths of resonance fluorescence that might be observed?

24 •• **SSM** Singly ionized helium is a hydrogen-like atom with a nuclear charge of $2e$. Its energy levels are given by $E_n = -4E_0/n^2$, where $E_0 = 13.6$ eV. If a beam of visible white light is sent through a gas of singly ionized helium, at what wavelengths will dark lines be found in the spectrum of the transmitted radiation?

The Speed of Light

25 • Mission Control sends a brief wake-up call to astronauts in a far away spaceship. Five seconds after the call is sent, Mission Control can hear the groans of the astronauts. How far away (at most) from the earth is the spaceship? (a) 7.5×10^8 m. (b) 15×10^8 m. (c) 30×10^8 m. (d) 45×10^8 m. (e) The spaceship is on the moon.

26 • **iSOLVE** The spiral galaxy in the Andromeda constellation is about 2×10^{19} km away from us. How many light-years is this?

27 • **iSOLVE** On a spacecraft sent to Mars to take pictures, the camera is triggered by radio waves, which like all electromagnetic waves, travel with the speed of light. What is the time delay between sending the signal from the earth and receiving the signal on Mars? (Take the distance to Mars to be 9.7×10^{10} m.)

28 • The distance from a point on the surface of the earth to one on the surface of the moon is measured by aiming a laser light beam at a reflector on the surface of the moon and measuring the time required for the light to make a round trip. The uncertainty in the measured distance Δx is related to the uncertainty in the time Δt by $\Delta x = c \Delta t$. If the time intervals can be measured to ± 1 ns, find the uncertainty of the distance in meters.

29 •• **SSM** **iSOLVE** In Galileo's attempt to determine the speed of light, he and his assistant were located on hilltops about 3 km apart. Galileo flashed a light and received a return flash from his assistant. (a) If his assistant had an instant reaction, what time difference would Galileo need to be able to measure for this method to be successful? (b) How does this time compare with human reaction time, which is about 0.2 s?

Reflection and Refraction

30 • ISOLVE✓ Calculate the fraction of light energy reflected from an air–water interface at normal incidence.

31 •• SSM A ray of light is incident on one of a pair of mirrors set at right angles to each other. The plane of incidence is perpendicular to both mirrors. Show that after reflecting off each mirror the ray will emerge in the opposite direction, regardless of the angle of incidence.

32 •• (a) A beam of light in air is incident on an air–water interface. Using a spreadsheet or graphing program, plot the angle of refraction as a function of the angle of incidence from 0° to 90°. (b) Repeat Part (a), but for a beam of light initially in water, incident on a water–air interface. For Part (b), what is the meaning of your graph for angles of incidence that are greater than the critical angle?

33 • ISOLVE✓ Find the speed of light in water and in glass.

34 • ISOLVE The index of refraction for silicate flint glass is 1.66 for light with a wavelength of 400 nm and 1.61 for light with a wavelength of 700 nm. Find the angles of refraction for light of these wavelengths that is incident on this glass at an angle of 45°.

35 •• ISOLVE✓ A slab of glass with an index of refraction of 1.5 is submerged in water with an index of refraction of 1.33. Light in the water is incident on the glass. Find the angle of refraction if the angle of incidence is (a) 60°, (b) 45°, and (c) 30°.

36 •• Repeat Problem 35 for a beam of light initially in the glass that is incident on the glass–water interface at the same angles.

37 •• SSM ISOLVE✓ Light is incident normally on a slab of glass with an index of refraction $n = 1.5$. Reflection occurs at both surfaces of the slab. Approximately what percentage of the incident light energy is transmitted by the slab?

38 •• This problem is a refraction analogy. A band is marching down a football field with a constant speed v_1. About midfield, the band comes to a section of muddy ground that has a sharp boundary making an angle of 30° with the 50-yd line, as shown in Figure 31-54. In the mud, the marchers move with speed $v_2 = v_1/2$. Diagram how each line of marchers is bent as it encounters the muddy section of the field so that the band is eventually marching in a different direction. Indicate the original direction by a ray, the final direction by a second ray, and find the angles between the rays and the line perpendicular to the boundary. Is their direction of motion bent toward the perpendicular to the boundary or away from it?

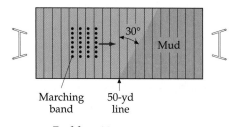

FIGURE 31-54 Problem 38

39 •• In Figure 31-55, light is initially in a medium (e.g., air) of index of refraction n_1. It is incident at angle θ_1 on the surface of a liquid (e.g., water) of index of refraction n_2. The light passes through the layer of water and enters glass of index of refraction n_3. If θ_3 is the angle of refraction in the glass, show that $n_1 \sin \theta_1 = n_3 \sin \theta_3$. That is, show that the second medium can be neglected when finding the angle of refraction in the third medium.

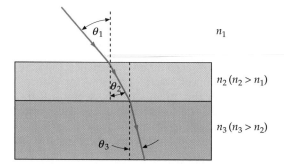

FIGURE 31-55 Problem 39

40 ••• SSM Figure 31-56 shows a beam of light incident on a glass plate of thickness d and index of refraction n. (a) Find the angle of incidence so that the perpendicular separation between the ray reflected from the top surface and the ray reflected from the bottom surface and exiting the top surface is a maximum. (b) What is this angle of incidence if the index of refraction of the glass is 1.60? What is the separation of the two beams if the thickness of the glass plate is 4.0 cm?

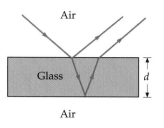

FIGURE 31-56 Problem 40

Total Internal Reflection

41 • What is the critical angle for total internal reflection for light traveling initially in water that is incident on a water–air interface?

42 •• ISOLVE A glass surface ($n = 1.50$) has a layer of water ($n = 1.33$) on it. Light in the glass is incident on the glass–water interface. Find the critical angle for total internal reflection.

43 •• ISOLVE✓ A point source of light is located 5 m below the surface of a large pool of water. Find the area of the largest circle on the pool's surface through which light coming directly from the source can emerge.

44 •• Light is incident normally on the largest face of an isosceles-right-triangle prism. What is the speed of light in this prism if the prism is just barely able to produce total internal reflection?

45 •• [ISOLVE]✓ A point source of light is located at the bottom of a steel tank, and an opaque circular card of radius 6 cm is placed over it. A transparent fluid is gently added to the tank so that the card floats on the surface with its center directly above the light source. No light is seen by an observer above the surface until the fluid is 5 cm deep. What is the index of refraction of the fluid?

46 •• [SSM] An optical fiber allows rays of light to propagate long distances through total internal reflection. As shown in Figure 31-57, the fiber consists of a core material with index of refraction n_2 and radius b, surrounded by a cladding material of index $n_3 < n_2$. The numerical aperture of the fiber is defined as $\sin \theta_1$, where θ_1 is the angle of incidence of a ray of light impinging the end of the fiber that reflects off the core-cladding interface at the critical angle. Using the figure as a guide, show that the numerical aperture is given by

$$\sqrt{n_2^2 - n_3^2}$$

assuming the ray is incident from air. (*Hint: Use of the Pythagorean theorem may be required.*)

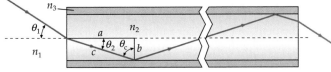

Incident ray

FIGURE 31-57 Problems 46, 47, and 48

47 • Find the maximum angle of incidence of a ray that would propagate through an optical fiber with a core index of refraction of 1.492, a core radius of 50 μm, and a cladding index of 1.489. See Problem 46 and Figure 31-57.

48 •• Calculate the difference in time needed for two pulses of light to travel down 15 km of the fiber of Problem 47. Assume that one pulse enters the fiber at normal incidence, and the second pulse enters the fiber at the maximum angle of incidence calculated in Problem 47 (see Figure 31-57). In fiber optics, this effect is known as modal dispersion.

49 ••• Investigate how a thin film of water on a glass surface affects the critical angle for total reflection. Take $n = 1.5$ for glass and $n = 1.33$ for water. (*a*) What is the critical angle for total internal reflection at the glass–water interface? (*b*) Is there any range of incident angles that are greater than θ_c for glass-to-air refraction and for which light rays will leave the glass and the water and pass into the air?

50 ••• A laser beam is incident on a plate of glass of thickness 3 cm. The glass has an index of refraction of 1.5 and the angle of incidence is 40°. The top and bottom surfaces of the glass are parallel and both produce reflected beams of nearly the same intensity. What is the perpendicular distance d between the two adjacent reflected beams?

Dispersion

51 •• [SSM] [ISOLVE] A beam of light strikes the plane surface of silicate flint glass at an angle of incidence of 45°. The index of refraction of the glass varies with wavelength, as shown in the graph in Figure 31-26. How much smaller is the

angle of refraction for violet light of wavelength 400 nm than that for red light of wavelength 700 nm?

52 •• Different colors (frequencies) of light travel at different speeds (a phenomena referred to as dispersion). This can cause problems in fiber-optic communications systems where pulses of light must travel very long distances in glass. Assuming a fiber is made of silicate crown glass, calculate the difference in time needed for two short pulses of light to travel 15 km of fiber if the first pulse has a wavelength of 700 nm and the second pulse has a wavelength of 500 nm.

Polarization

53 • [ISOLVE]✓ What is the polarizing angle for (*a*) water with $n = 1.33$ and (*b*) glass with $n = 1.5$?

54 • Light known to be polarized in the horizontal direction is incident on a polarizing sheet. It is observed that only 15 percent of the intensity of the incident light is transmitted through the sheet. What angle does the transmission axis of the sheet make with the horizontal? (*a*) 8.6°, (*b*) 21°, (*c*) 23°, (*d*) 67°, or (*e*) 81°.

55 • Two polarizing sheets have their transmission axes crossed so that no light gets through. A third sheet is inserted between the first two so that its transmission axis makes an angle θ with that of the first sheet. Unpolarized light of intensity I_0 is incident on the first sheet. Find the intensity of the light transmitted through all three sheets if (*a*) $\theta = 45°$ and (*b*) $\theta = 30°$.

56 •• A horizontal 5 mW laser beam polarized in the vertical direction is incident on a pair of polarizers. The first is oriented so that its transmission axis is vertical and the second is oriented with its transmission axis at an angle of 27° with respect to the first. What is the power of the transmitted beam?

57 •• The polarizing angle for a certain substance is 60°. (*a*) What is the angle of refraction of light incident at this angle? (*b*) What is the index of refraction of this substance?

58 •• Two polarizing sheets have their transmission axes crossed and a third sheet is inserted so that its transmission axis makes an angle θ with that of the first sheet, as in Problem 55. Find the intensity of the transmitted light as a function of θ. Show that the intensity transmitted through all three sheets is maximum when $\theta = 45°$.

59 •• If the middle polarizing sheet in Problem 58 is rotating at an angular velocity ω about an axis parallel to the light beam, find the intensity transmitted through all three sheets as a function of time. Assume that $\theta = 0$ at time $t = 0$.

60 •• [SSM] A stack of $N + 1$ ideal polarizing sheets is arranged with each sheet rotated by an angle of $\pi/(2N)$ rad with respect to the preceding sheet. A plane, linearly polarized light wave of intensity I_0 is incident normally on the stack. The incident light is polarized along the transmission axis of the first sheet and is therefore perpendicular to the transmission axis of the last sheet in the stack. (*a*) Show that the transmitted intensity through the stack is given by the expression

$$I_0 \cos^{2N}\left(\frac{\pi}{2N}\right)$$

(b) Using a spreadsheet or graphing program, plot the transmitted intensity as a function of N for values of N from 2 to 100. (c) What is the direction of polarization of the transmitted beam in each case?

61 •• The device described in Problem 60 could serve as a polarization *rotator,* one that takes the linear plane of polarization from one direction to another. The efficiency of such a device is measured by taking the ratio of the output intensity at the desired polarization to the input intensity. The result of the previous problem suggests that the best way to do this would be to use a large number N. But in a real polarizer, a small amount of intensity is lost regardless of the input polarization. For each polarizer, assume the transmitted intensity is 98 percent of the amount predicted by the law of Malus and use a spreadsheet or graphing program to determine the optimum number of sheets you should use to rotate the polarization 90°.

62 •• SSM Show that a linearly polarized wave can be thought of as a superposition of a right and a left circularly polarized wave.

63 •• Suppose that the middle sheet in Problem 55 is replaced by two polarizing sheets. If the angles between the directions of polarization of adjacent sheets is 30°, what is the intensity of the transmitted light? How does this compare with the intensity obtained in Problem 55, Part (a)?

64 •• SSM Show that the electric field of a circularly polarized wave propagating in the x direction can be expressed by

$$\vec{E} = E_0 \sin(kx - \omega t)\hat{j} + E_0 \cos(kx - \omega t)\hat{k}$$

65 •• A circularly polarized wave is said to be *right circularly polarized* if the electric and magnetic fields rotate clockwise when viewed along the direction of propagation and *left circularly polarized* if the fields rotate counterclockwise. What is the sense of the circular polarization for the wave described by the expression in Problem 64? What would be the corresponding expression for a circularly polarized wave of the opposite sense?

General Problems

66 • ISOLVE A beam of monochromatic red light with a wavelength of 700 nm in air travels in water. (a) What is the wavelength in water? (b) Does a swimmer underwater observe the same color or a different color for this light?

67 •• The critical angle for total internal reflection for a substance is 45°. What is the polarizing angle for this substance?

68 •• SSM Figure 31-58 shows two plane mirrors that make an angle θ with each other. Show that the angle between the incident and reflected rays is 2θ.

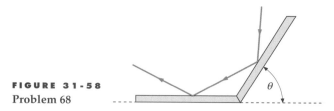

FIGURE 31-58
Problem 68

69 •• A silver coin sits on the bottom of a swimming pool that is 4 m deep. A beam of light reflected from the coin emerges from the pool making an angle of 20° with respect to the water's surface and enters the eye of an observer. Draw a ray from the coin to the eye of the observer. Extend this ray, which goes from the water–air interface to the eye, straight back until it intersects with the vertical line drawn through the coin. What is the apparent depth of the swimming pool to this observer?

70 •• Fishermen always insist on silence because noise on shore will scare fish away. Suppose a fisherman cast a baited hook 20 m from the shore of a calm lake to a point where the depth is 15 m. Show that noise on shore cannot possibly be sensed by fish at that point. (*Note:* The speed of sound in air is 330 m/s; the speed of sound in water is 1450 m/s.)

71 •• SSM ISOLVE✓ A swimmer at the bottom of a pool 3 m deep looks up and sees a circle of light. If the index of refraction of the water in the pool is 1.33, find the radius of the circle.

72 •• Show that when a mirror is rotated through an angle θ, the reflected beam of light is rotated through 2θ.

73 •• Use Figure 31-25 to calculate the critical angles for total internal reflection for light initially in silicate flint glass that is incident on a glass–air interface if the light is (a) violet light of wavelength 400 nm and (b) red light of wavelength 700 nm.

74 •• Show that for normally incident light, the intensity transmitted through a glass slab with an index of refraction of n is approximately given by

$$I_T = I_0 \left[\frac{4n}{(n + 1)^2} \right]^2$$

75 •• A ray of light begins at the point $x = -2$ m, $y = 2$ m, strikes a mirror in the xz plane at some point x, and reflects through the point $x = 2$ m, $y = 6$ m. (a) Find the value of x that makes the total distance traveled by the ray a minimum. (b) What is the angle of incidence on the reflecting plane? What is the angle of reflection?

76 •• SSM A Brewster window is used in lasers to preferentially transmit light of one polarization, as shown in Figure 31-59. Show that if θ_{P1} is the polarizing angle for the n_1/n_2 interface, then θ_{P2} is the polarizing angle for the n_2/n_1 interface.

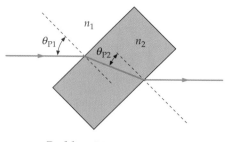

FIGURE 31-59 Problem 76

77 •• From the data provided in Figure 31-29, calculate the polarization angle for an air–glass interface, using light of wavelength 550 nm in each of the four types of glass shown in the figure.

78 ••• Light passes symmetrically through a prism with an apex angle of α, as shown in Figure 31-60. (*a*) Show that the angle of deviation δ is given by

$$\sin \frac{\alpha + \delta}{2} = n \sin \frac{\alpha}{2}$$

(*b*) If the refractive index for red light is 1.48 and the refractive index for violet light is 1.52, what is the angular separation of visible light for a prism with an apex angle of 60°?

FIGURE 31-60
Problems 78 and 89

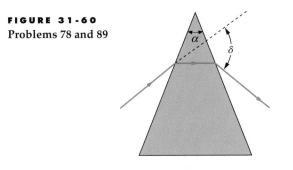

79 •• SSM (*a*) For a light ray inside a transparent medium that has a planar interface with a vacuum, show that the polarizing angle and the critical angle for internal reflection satisfy $\tan \theta_p = \sin \theta_c$. (*b*) Which angle is larger?

80 •• iSOLVE Light is incident from air on a transparent substance at an angle of 58° with the normal. The reflected and refracted rays are observed to be mutually perpendicular. (*a*) What is the index of refraction of the transparent substance? (*b*) What is the critical angle for total internal reflection in this substance?

81 •• iSOLVE A light ray in dense flint glass with an index of refraction of 1.655 is incident on the glass surface. An unknown liquid condenses on the surface of the glass. Total internal reflection on the glass–liquid interface occurs for an angle of incidence on the glass–liquid interface of 53.7°. (*a*) What is the refractive index of the unknown liquid? (*b*) If the liquid is removed, what is the angle of incidence for total internal reflection? (*c*) For the angle of incidence found in Part (*b*), what is the angle of refraction of the ray into the liquid film? Does a ray emerge from the liquid film into the air above? Assume the glass and liquid have perfect planar surfaces.

82 •• Given that the index of refraction for red light in water is 1.3318 and that the index of refraction for blue light in water is 1.3435, find the angular separation of these colors in the primary rainbow. (Use the equation given in Problem 86.)

83 •• (*a*) Use the result from Problem 74 to find the ratio of the transmitted intensity to the incident intensity through N parallel slabs of glass for light of normal incidence. (*b*) Find this ratio for three slabs of glass with $n = 1.5$. (*c*) How many slabs of glass with $n = 1.5$ will reduce the intensity to 10 percent of the incident intensity?

84 •• Light is incident on a slab of transparent material at an angle θ_1, as shown in Figure 31-61. The slab has a thickness t and an index of refraction n. Show that

$$n = \frac{\sin \theta_1}{\sin[\arctan(d/t)]}$$

where d is the distance shown in the figure and $\arctan(d/t)$ is the angle whose tangent is d/t.

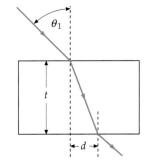

FIGURE 31-61 Problem 84

85 •• SSM Suppose rain falls vertically from a stationary cloud 10,000 m above a confused marathoner running in a circle with constant speed of 4 m/s. The rain has a terminal speed of 9 m/s. (*a*) What is the angle that the rain appears to make with the vertical to the marathoner? (*b*) What is the apparent motion of the cloud as observed by the marathoner? (*c*) A star on the axis of the earth's orbit appears to have a circular orbit of angular diameter of 41.2 seconds of arc. How is this angle related to the earth's speed in its orbit and the velocity of photons received from this distant star? (*d*) What is the speed of light as determined from the data in Part (*c*)?

86 ••• Equation 31-18 gives the relation between the angle of deviation ϕ_d of a light ray incident on a spherical drop of water in terms of the incident angle θ_1 and the index of refraction of water. (*a*) Assume that $n_{air} = 1$, and differentiate ϕ_d with respect to θ_1. [*Hint:* If $y = \arcsin x$, then $dy/dx = (1 - x^2)^{-1/2}$.] (*b*) Set $d\phi_d/d\theta_1 = 0$ and show that the angle of incidence $\theta_{1\,m}$ for minimum deviation is given by

$$\cos \theta_{1m} = \sqrt{\frac{n^2 - 1}{3}}$$

and find θ_{1m} for water, where the index of refraction for water is 1.33.

87 ••• iSOLVE Show that a light ray incident on a rectangular glass slab of thickness t at angle of incidence θ_1 emerges parallel to the incident ray but displaced from it by an amount s given by

$$s = \frac{t \sin(\theta_1 - \theta_2)}{\cos(\theta_2)}$$

where θ_2 is the angle of refraction.

88 •• Use the result of Problem 87 to find the lateral displacement of a laser beam incident at an angle of 30° on a 15 mm thick rectangular slab of glass with index of refraction 1.5.

89 ••• Show that the angle of deviation δ is a minimum if the angle of incidence is such that the ray passes through the prism symmetrically, as shown in Figure 31-60.

Optical Images

NOTE THAT THE PHOTOGRAPH SHOWS EVIDENCE THAT THE HANDS ON THE MIRROR IMAGE OF AN ORDINARY CLOCK ROTATE NOT CLOCKWISE, BUT COUNTERCLOCKWISE. IT IS ALSO TRUE THAT IF YOU COULD LOOK AT A CLOCK FACE FROM BEHIND THE CLOCK IT WOULD LOOK JUST LIKE THE MIRROR IMAGE OF THE CLOCK FACE.

? **When looked at from behind, do the hands of a clock rotate clockwise or counterclockwise? A number of observations concerning mirror images are discussed in Section 32-1.**

Because the wavelength of light is very small compared with most obstacles and openings, diffraction—the bending of waves around corners—is often negligible, and the ray approximation, in which waves are considered to propagate in straight lines, accurately describes observations. ➤ **In this chapter, we apply the laws of reflection and refraction to the formation of images by mirrors and lenses.**

32-1 Mirrors

Plane Mirrors

Figure 32-1 shows a bundle of light rays emanating from a point source P and reflected from a plane mirror. After reflection, the rays diverge exactly as if they came from a point P' behind the plane of the mirror. The point P' is called the **image** of the **object** P. When these reflected rays enter the eye, they cannot be distinguished from rays diverging from a source at P' with no mirror present. This image at P' is called a **virtual image** because the light does not actually emanate from it. The plane of the mirror is the perpendicular bisector of the line from the object point P to the image point P' as shown. The image can be seen by an eye located anywhere in the shaded region indicated, in which a straight line from

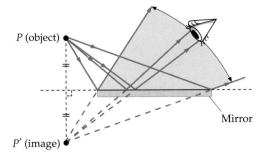

FIGURE 32-1 Image formed by a plane mirror. The rays from point P that strike the mirror and enter the eye appear to come from the image point P' behind the mirror. The image can be seen by an eye located anywhere in the shaded region.

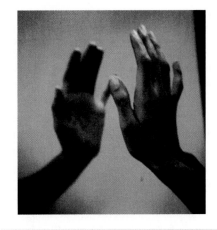

the image to the eye passes through the mirror. The object need not be directly in front of the mirror. As long as the object is not behind the plane of the mirror, there is some location at which the eye can be located to view the image.

If you hold up your right hand and look in the mirror, the image you see is neither magnified nor reduced, but it looks like a left hand (Figure 32-2). This right-to-left reversal is a result of **depth inversion**—the hand is transformed from a right hand to a left hand because the front and the back of the hand are reversed by the mirror. Depth inversion is also illustrated in Figure 32-3. Figure 32-4 shows the image of a simple rectangular coordinate system. The mirror transforms a right-handed coordinate system, for which $\hat{i} \times \hat{j} = \hat{k}$, into a left-handed coordinate system, for which $\hat{i} \times \hat{j} = -\hat{k}$.

Figure 32-5 shows an arrow of height y standing parallel to a plane mirror a distance s from the mirror. We can locate the image of the arrowhead (and of any other point on the arrow) by drawing two rays. One ray, drawn perpendicular to the mirror, hits the mirror at point A and is reflected back onto itself. The other ray, making an angle θ with the normal to the mirror, is reflected, making an equal angle θ with the x axis. The extension of these two rays back behind the mirror locates the image of the arrowhead, as shown by the dashed lines in the figure. We can see from this figure that the image is the same distance behind the mirror as the object is in front of the mirror, and that the image is upright (points in the same direction as the object) and the image is the same size as the object.

FIGURE 32-2 The image of a right hand in a plane mirror is transformed to a left hand. This right-to-left reversal is a result of depth inversion.

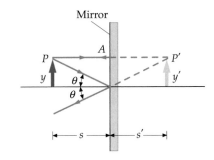

FIGURE 32-3 A person lying down with her feet against the mirror. The image is depth inverted.

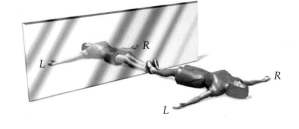
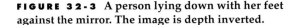

FIGURE 32-5 Ray diagram for locating the image of an arrow in a plane mirror.

FIGURE 32-4 Image of a rectangular coordinate system in a plane mirror. The arrow along the z axis is reversed in the image. The image of the original right-handed coordinate system, for which $\hat{i} \times \hat{j} = \hat{k}$, is a left-handed coordinate system, for which $\hat{i} \times \hat{j} = -\hat{k}$.

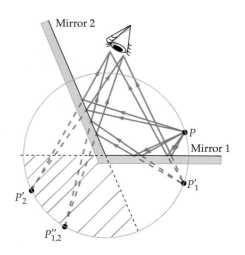

FIGURE 32-6 Images formed by two plane mirrors. P'_1 is the image of the object P in mirror 1, and P'_2 is the image of the object in mirror 2. Point $P''_{1,2}$ is the image of P'_1 in mirror 2, which is seen when light rays from the object reflect first from mirror 1 and then from mirror 2. The image P'_2 does not have an image in mirror 1 because it is behind that mirror.

The formation of multiple images by two plane mirrors that make an angle with each other is illustrated in Figure 32-6. We frequently see this phenomenon in clothing stores that provide adjacent mirrors. The light from source point P that is reflected from mirror 1 strikes mirror 2 just as if it came from the image point P'_1. The image P'_1 is the object for mirror 2. Its image is behind mirror 2 at point $P''_{1,2}$. This image will be formed whenever the image point P'_1 is in front of the plane of mirror 2. The image at point P'_2 is due to rays from P that reflect directly from mirror 2. Since P'_2 is behind the plane of mirror 1, it cannot serve as an object point for a further image in mirror 1. The image at point P'_2 cannot serve as an object for mirror 1 because the geometry dictates that none of the rays from P that reflect directly from mirror 2 can then strike mirror 1. An alternative way of stating this is that since P'_2 is behind the plane of mirror 1, the image at P'_2 cannot serve as an object for mirror one. The number of multiple images formed by two mirrors depends on the angle between the mirrors and the position of the object.

EXERCISE Show that a source point and all consequent image points formed by two plane mirrors are equidistant from the intersection of the two planes.

Suppose your friend Ben is standing at point P, is wearing a sweatshirt with BEN printed on it, and is waving his right hand. Also, suppose that you are standing at the location of the eye. You can see an image of Ben at all three image locations. For the images at P'_1 and P'_2, Ben is waving his left hand and the printing on his sweatshirt appears as **NƎB**. However, for the image at $P''_{1,2}$, Ben is waving his right hand and the printing appears as BEN. For the image at $P''_{1,2}$ depth inversion occurs twice, once for each reflection, so the result is as if no depth inversion occurs.

EXERCISE Which of the images of himself can Ben see? (*Answer* Ben can see only the image at P'_1.)

Figure 32-7 illustrates the fact that a horizontal ray reflected from two perpendicular vertical mirrors is reflected back along a parallel path no matter what angle the ray makes with the mirrors. If three mirrors are placed perpendicular to each other, like the sides of an inside corner of a box, any ray incident on any of the mirrors from any direction is reflected back on a path parallel to that of the incident

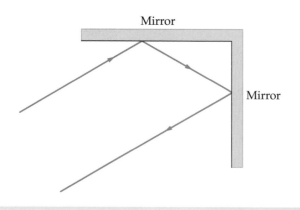

FIGURE 32-7 A horizontal ray striking one of two perpendicular plane mirrors is reflected from the second mirror in the direction opposite the original direction for any angle of incidence.

ray. A set of three mirrors arranged in this manner is called a corner-cube reflector. An array of corner-cube reflectors was placed on the moon in the Sea of Tranquility by the Apollo 11 astronauts in 1969. A laser beam from the earth that is directed at the mirrors is reflected back to the same place on the earth. Such a beam has been used to measure the distance to the mirrors to within a few centimeters by measuring the time it takes for the light to reach the mirrors and return.

Spherical Mirrors

Figure 32-8 shows a bundle of rays from a point source P on the axis of a concave spherical mirror reflecting from the mirror and converging at point P'. (A concave mirror is shaped like a cave when you look into it.) The rays then diverge from this point, just as if there were an object at the point. This image is called a **real image,** because light actually does emanate from the image point. The image can be seen by an eye at the left of the image looking into the mirror. It could also be observed on a small viewing screen[†] or on a small piece of photographic film placed at the image point. A virtual image, such as that formed by a plane mirror as discussed in the previous section, cannot be observed on a screen at the image point because there is no light at the image point. Despite this distinction between real and virtual images, the eye makes no distinction between them. The light rays diverging from a real image and those appearing to diverge from a virtual image are the same to the eye.

From Figure 32-9, we can see that only rays that strike the spherical mirror at points near the axis AV are reflected through the image point. Rays almost parallel with the axis and near to it are called **paraxial rays.** Rays that strike the mirror at points far from the axis upon reflection pass near the image point, but not through it. Such rays cause the image to appear blurred, an effect called **spherical aberration.** The image can be sharpened by blocking off all but the central part of the mirror, so that rays far from the axis do not strike it. The image is then sharper, but its brightness is reduced because less light is reflected to the image point.

We wish to obtain an equation relating the position of the image point to the position of the object point. To do this, we draw two rays (Figure 32-10a) from an arbitrarily positioned object point P. One ray passes through point C, the center of curvature of the mirror, and the other ray strikes point A, an arbitrarily positioned point on the mirror. The image point P' is where these two rays intersect after reflecting off the mirror. Using the law of reflection, we obtain the location of P'. The ray passing through point C strikes the mirror at normal incidence, so the ray reflects back upon itself. The ray striking the mirror at A makes angle θ with the normal, so, as shown, the reflected ray also makes angle θ with the normal. (Any line normal to a spherical surface passes through the center of curvature.) The image distance s' and object distance s are measured from the plane tangent to the mirror at its vertex V. The angle β is an exterior angle to the triangle PAC, therefore, $\beta = \alpha + \theta$. Similarly, from the triangle PAP', $\gamma = \alpha + 2\theta$. Eliminating θ from these equations gives $2\beta = \alpha + \gamma$. By assuming all rays are paraxial, we can substitute using the small-angle approximations: $\alpha \approx \ell/s$, $\beta \approx \ell/r$, and $\gamma \approx \ell/s'$. Equation 32-1 follows directly:

$$\frac{1}{s} + \frac{1}{s'} = \frac{2}{r}$$

32-1

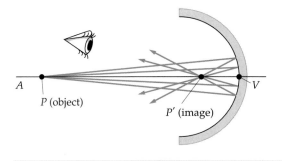

FIGURE 32-8 Rays from a point object P on the axis AV of a concave spherical mirror form an image at P'. The image is sharp if the rays strike the mirror near the axis and if the rays are almost parallel with the axis.

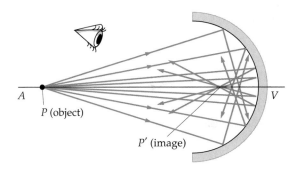

FIGURE 32-9 Spherical aberration of a mirror. Nonparaxial rays that strike the mirror at points far from the axis AV are not reflected through the image point P' formed by the paraxial rays. The nonparaxial rays blur the image.

---

† A viewing screen must produce either diffuse reflection or diffuse transmission of the light. Ground glass is commonly used for this purpose. The screen must be small, so that it does not block all of the light from the source from reaching the mirror.

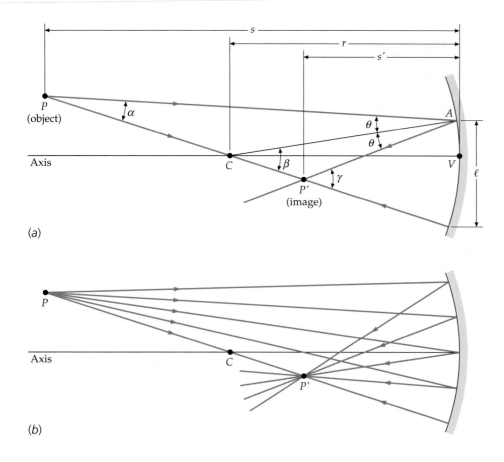

(a)

(b)

FIGURE 32-10 (*a*) Geometry for calculating the image distance *s′* from the object distance *s* and the radius of curvature *r*. The angle β is an exterior angle to the triangle *PAC*; therefore, $\beta = \alpha + \theta$. Similarly, from the triangle *PAP′*, $\gamma = \alpha + 2\theta$. Eliminating θ from these equations gives $2\beta = \alpha + \gamma$. Equation 32-1 follows directly, if we assume the following small-angle approximations: $\alpha \approx \ell/s$, $\beta \approx \ell/r$, and $\gamma \approx \ell/s′$. (*b*) All paraxial rays from object point *P* pass through image point *P′* after reflecting off the mirror.

This equation relates the object and image distances with the radius of curvature. The striking thing about this equation is that it contains absolutely nothing about the location of point *A*. Therefore, the equation is valid for *any* choice for the location of point *A*, as long as point *A* is on the surface of the mirror and all rays are paraxial. That is, as shown in Figure 32-10*b*, *all* paraxial rays emanating from an object point will, upon reflection, pass through a *single* image point.

Equation 32-1 specifies the image position in terms of its distance from the mirror. We now specify the image position in terms of its distance from the axis. We first draw a single ray (Figure 32-11) that reflects off the mirror at its vertex. The two right triangles formed are similar. Corresponding sides of similar triangles are equal, so

$$\frac{y'}{y} = -\frac{s'}{s}$$

32-2

FIGURE 32-11 Geometry for calculating the position *y′* of the image point with respect to its distance from the axis.

The negative sign takes into account that y'/y is negative as P and P' are on opposite sides of the axis. Thus, if y is positive, y' is negative and if y is negative, y' is positive.

EXERCISE For the image point and object point shown in Figure 32-11, show that

$$\frac{y'}{y} = -\frac{r/2}{s - (r/2)}$$

(*Hint:* Solve Equation 32-1 for s' and substitute your result into Equation 32-2.)

When the object distance is large compared with the radius of curvature of the mirror, the term $1/s$ in Equation 32-1 is much smaller than $2/r$ and can be neglected. That is, as $s \to \infty$, $s' \to \frac{1}{2}r$, where s' is the image distance. This distance is called the **focal length** f of the mirror, and the plane on which parallel rays incident on the mirror are focused is called the **focal plane**. The intersection of the axis with the focal plane is called the **focal point** F, as illustrated in Figure 32-12a. (Again, only paraxial rays are focused at a single point.)

$$f = \tfrac{1}{2}r \hspace{6cm} \text{32-3}$$

FOCAL LENGTH FOR A MIRROR

EXERCISE Show that solving Equation 32-1 for s' gives

$$s' = \frac{r}{2 - \dfrac{r}{s}}$$

Then show that as $s \to \infty$, $s' \to \frac{1}{2}r$.

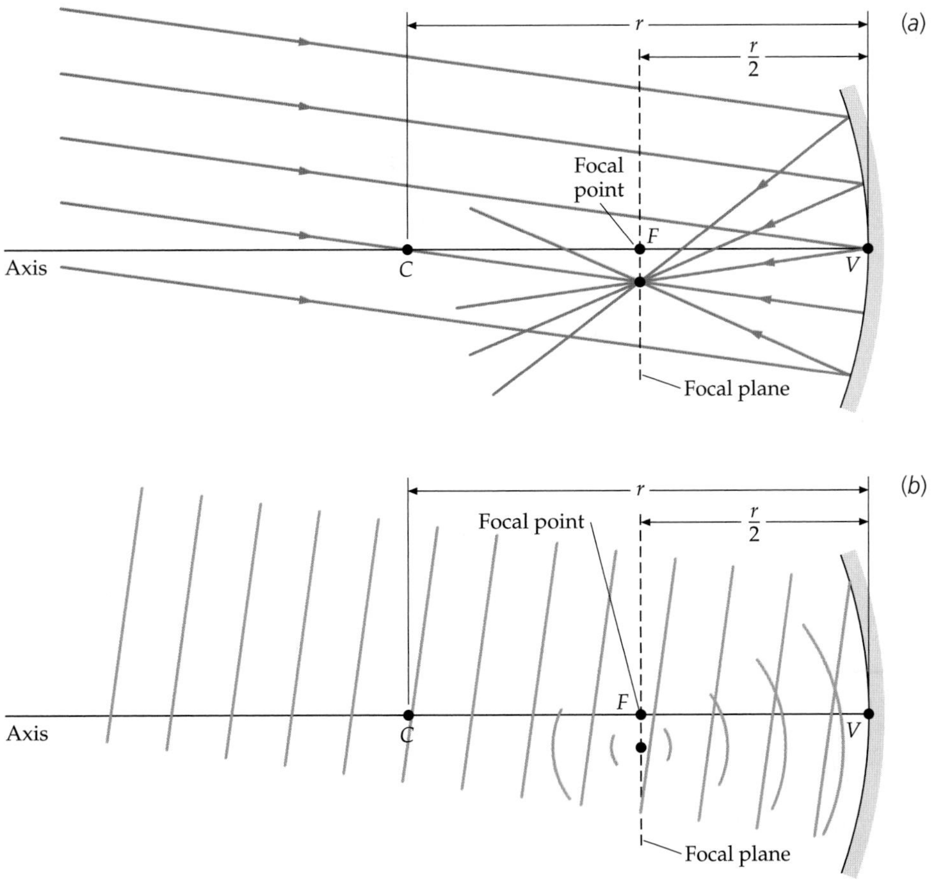

FIGURE 32-12 (*a*) Parallel rays strike a concave mirror and are reflected to a point on the focal plane a distance $r/2$ to the left of the mirror. (*b*) The incoming wavefronts are plane waves; upon reflection, they become spherical wavefronts that converge to, and then diverge from, the focal point.

The focal length of a spherical mirror is half the radius of curvature. In terms of the focal length f, Equation 32-1 is

$$\frac{1}{s} + \frac{1}{s'} = \frac{1}{f}$$

32-4

MIRROR EQUATION

Equation 32-4 is called the **mirror equation.**

When an object point is very far from the mirror, the rays are parallel, and the wavefronts are approximately planes (Figure 32-12*b*). In Figure 32-12*b*, note that the last part of each wavefront to reflect off the concave mirror surface is the part just below the vertex *V*. This results in a spherical wavefront upon reflection. Figure 32-13 shows both the wavefronts and the rays for plane waves striking a convex mirror. In this case, the central part of the wavefront strikes the mirror first, and the reflected waves appear to come from the focal point behind the mirror.

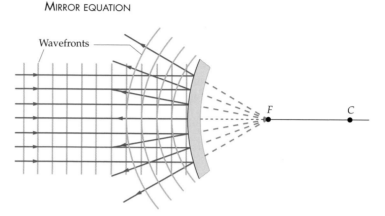

Figure 32-14 illustrates a property of waves called **reversibility.** If we reverse the direction of a reflected ray, the law of reflection assures us that the reflected ray will be along the original incoming ray, but in the opposite direction. (Reversibility holds also for refracted rays, which are discussed in later sections.) Thus, if we have a real image of an object formed by a reflecting (or refracting) surface, we can place an object at the image point and a new, real image will be formed at the position of the original object.

FIGURE 32-13 Reflection of plane waves from a convex mirror. The outgoing wavefronts are spherical, as if emanating from the focal point *F* behind the mirror. The rays are normal to the wavefronts and appear to diverge from *F*.

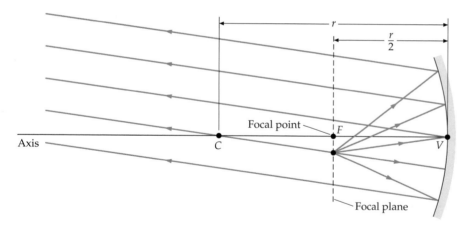

FIGURE 32-14 Reversibility. Rays diverging from a point source on the focal plane of a concave mirror are reflected from the mirror as parallel rays. The rays follow the same paths as in Figure 32-12*a* but in the reverse direction.

IMAGE IN A CONCAVE MIRROR   **EXAMPLE 32-1**

A point object is 12 cm from a concave mirror and 3 cm above the axis of the mirror. The radius of curvature of the mirror is 6 cm. Find (*a*) the focal length of the mirror and (*b*) the image distance. (*c*) Find the position of the image relative to the axis.

PICTURE THE PROBLEM The focal length of a spherical mirror is half the radius of curvature. Once the focal length is known, the image distance can be found using the mirror equation (Equation 32-4), and the distance of the image from the axis can be found using Equation 32-2. The image distance from the mirror is the distance from the plane tangent to the mirror at its vertex.

(*a*) The focal length is half the radius of curvature:

$$f = \tfrac{1}{2}r = \tfrac{1}{2}(6 \text{ cm}) = \boxed{3 \text{ cm}}$$

(*b*) 1. Use the mirror equation to find a relation for the image distance *s'*:

$$\frac{1}{s} + \frac{1}{s'} = \frac{1}{f}$$

or

$$\frac{1}{12 \text{ cm}} + \frac{1}{s'} = \frac{1}{3 \text{ cm}}$$

2. Solve for *s'*:

$$\frac{1}{s'} = \frac{4}{12 \text{ cm}} - \frac{1}{12 \text{ cm}} = \frac{3}{12 \text{ cm}}$$

$$s' = \boxed{4 \text{ cm}}$$

(*c*) 1. Use Equation 32-2 to find the distance *y'* of image from the axis:

$$\frac{y'}{y} = -\frac{s'}{s}$$

2. Solve for *y'*:

$$y' = -\frac{s'}{s}y = -\frac{4 \text{ cm}}{12 \text{ cm}}(3 \text{ cm}) = \boxed{-1 \text{ cm}}$$

EXERCISE A concave mirror has a focal length of 4 cm. (*a*) What is the mirror's radius of curvature? (*b*) Find the image distance for an object 2 cm from the mirror. (*Answer* (*a*) 8 cm (*b*) *s'* = −4 cm)

REMARKS A negative image distance means the image is on the opposite side of the mirror from the reflected light, as is the case with a plane mirror.

EXERCISE What is the radius of curvature of a plane mirror? (*Answer* Infinity)

Ray Diagrams for Mirrors

A useful method to locate images is by geometric construction of a **ray diagram,** as illustrated in Figure 32-15, where the object is a human figure perpendicular to the axis a distance *s* from the mirror. By the judicious choice of rays from the head of the figure, we can quickly locate the image. There are three **principal rays** that are convenient to use:

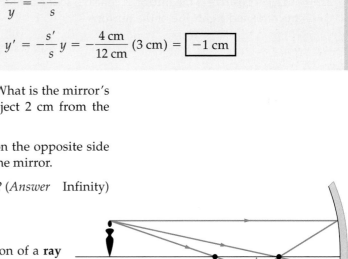

1. The **parallel ray,** drawn parallel to the axis. This ray is reflected through the focal point.

2. The **focal ray,** drawn through the focal point. This ray is reflected parallel to the axis.

3. The **radial ray,** drawn through the center of curvature. This ray strikes the mirror perpendicular to its surface and is thus reflected back on itself.

FIGURE 32-15 Ray diagram for the location of the image by geometric construction.

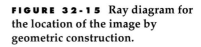

PRINCIPAL RAYS FOR A MIRROR

These rays are shown in Figure 32-15. The intersection of any two paraxial rays locates the image point of the head. The three principal rays are easier to draw than any of the other rays. Typically, you draw two of the principal rays to locate the image, and then draw the third principal ray as a check to verify the result. Ray diagrams are best drawn with the mirror replaced by a straight line that extends as far as necessary to intercept the rays, as shown in Figure 32-16. Note that the image in this case is inverted and smaller than the object.

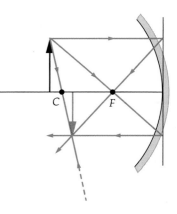

FIGURE 32-16 Ray diagrams are easier to construct if the curved surface is replaced by a plane tangent to the surface at the vertex.

When the object is between the mirror and its focal point, the rays reflected from the mirror do not converge but appear to diverge from a point behind the mirror, as illustrated in Figure 32-17. In this case, the image is virtual and upright (*upright* meaning not inverted relative to the object). For an object between the mirror and the focal point, s is less than $r/2$, so the image distance s' calculated from Equation 32-1 turns out to be negative. We can apply Equations 32-1, 32-2, 32-3, and 32-4 to this case and to convex mirrors if we adopt a convenient sign convention. Whether the mirror is convex or concave, real images can be formed only in front of the mirror, that is, on the same side of the mirror as the reflected light (and the object). Virtual images are formed behind the mirror where there is no actual light from the object. Our sign convention is as follows:

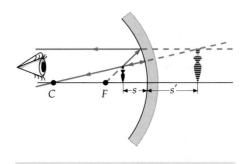

FIGURE 32-17 A virtual image is formed by a concave mirror when the object is inside the focal point. Here the image is located by the radial ray, which is reflected back on itself, and the focal ray, which is reflected parallel to the axis. The two reflected rays appear to diverge from an image point behind the mirror. This image point is found by constructing extensions to the reflected rays.

1. s is positive if the object is on the incident-light side of the mirror.
2. s' is positive if the image is on the reflected-light side of the mirror.
3. r (and f) is positive if the mirror is concave so the center of curvature is on the reflected-light side of the mirror.

SIGN CONVENTIONS FOR REFLECTION

The incident-light side and the reflected-light side are, of course, the same. The parameters s, s', r, and f are all positive if a real object[†] is in front of a concave mirror that forms a real image. A parameter is negative if it does not meet the stated condition for being positive.

The ratio of the image size to the object size is defined as the **lateral magnification** m of the image. From Figure 32-18 and Equation 32-2, we see that the lateral magnification is

$$m = \frac{y'}{y} = -\frac{s'}{s} \qquad\qquad 32\text{-}5$$

LATERAL MAGNIFICATION

A negative magnification, which occurs when both s and s' are positive, indicates that the image is inverted.

For plane mirrors, the radius of curvature is infinite. The focal length given by Equation 32-3 is then also infinite. Equation 32-4 then gives $s' = -s$, indicating that the image is behind the mirror at a distance equal to the object distance. The magnification given by Equation 32-5 is then $+1$, indicating that the image is upright and the same size as the object.

Although the preceding equations, coupled with our sign conventions, are relatively easy to use, we often need to know only the approximate location and magnification of the image and whether it is real or virtual and upright or inverted. This knowledge is usually easiest to obtain by constructing a ray diagram. It is always a good idea to use both the graphical method and the algebraic method to locate an image, so that one method serves as a check on the results of the other.

FIGURE 32-18 Geometry for showing the lateral magnification. Rays from the top of the object at P, upon reflection, intersect at P'; and rays from the bottom of the object at Q intersect at Q', where points P and P' have vertical positions y and y', respectively. The lateral magnification m is given by the ratio y'/y. In accord with Equation 32-2, $y'/y = -s'/s$. The minus sign results from the fact that y'/y is negative when s and s' are both positive. A negative m means the image is inverted.

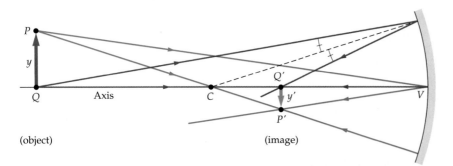

(object)   (image)

---

† An object is real if it is on the same side of the mirror as the incident light.

Convex Mirrors Figure 32-19 shows a ray diagram for an object in front of a convex mirror. The central ray heading toward the center of curvature C is perpendicular to the mirror and is reflected back on itself. The parallel ray is reflected as if it came from the focal point F behind the mirror. The focal ray (not shown) would be drawn toward the focal point and would be reflected parallel to the axis. We can see from the figure that the image is behind the mirror and is therefore virtual. The image is also upright and smaller than the object.

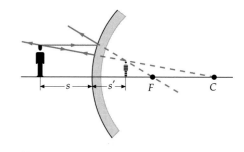

FIGURE 32-19 Ray diagram for an object in front of a convex mirror.

IMAGE IN A CONVEX MIRROR **EXAMPLE 32-2**

An object 2 cm high is 10 cm from a convex mirror with a radius of curvature of 10 cm. (*a*) Locate the image and (*b*) find the height of the image.

PICTURE THE PROBLEM The ray diagram for this problem is the same as shown in Figure 32-19. From this figure, we see that the image is upright, virtual, and smaller than the object. To find the exact location and height of the image, we use the mirror equation, with $s = 10$ cm and $r = -10$ cm.

(*a*) 1. The image distance s' is related to the object distance s and the focal length f by the mirror equation:
$$\frac{1}{s} + \frac{1}{s'} = \frac{1}{f}$$

2. Calculate the focal length of the mirror:
$$f = \tfrac{1}{2}r = \tfrac{1}{2}(-10 \text{ cm}) = -5 \text{ cm}$$

3. Substitute $s = 10$ cm and $f = -5$ cm into the mirror equation to find the image distance:
$$\frac{1}{10 \text{ cm}} + \frac{1}{s'} = \frac{1}{-5 \text{ cm}}$$

4. Solve for s':
$$s' = \boxed{-3.33 \text{ cm}}$$

(*b*) 1. The height of the image is m times the height of the object:
$$y' = my$$

2. Calculate the magnification m:
$$m = -\frac{s'}{s} = -\frac{-3.33 \text{ cm}}{10 \text{ cm}} = +0.333$$

3. Use m to find the height of the image:
$$y' = my = (0.333)(2 \text{ cm}) = \boxed{0.666 \text{ cm}}$$

REMARKS The image distance is negative, indicating a virtual image behind the mirror. The magnification is positive and less than one, indicating that the image is upright and smaller than the object.

EXERCISE Find the image distance and magnification for an object 5 cm away from the mirror in Example 32-2, and draw a ray diagram. (*Answer* $s' = -2.5$ cm, $m = +0.5$; the image is upright, virtual, and reduced in size. The ray diagram is shown in Figure 32-20.)

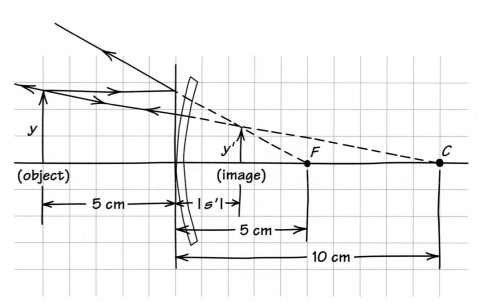

FIGURE 32-20

You have a part-time job at Pleasant Hills Golf Course. The fairway of the 16th hole is horizontal for the first 50 yd and then goes down a not-too-steep hill (Figure 32-21), so the people on the tee cannot see the party in front. To prevent people from driving off the tee into the party in front, a convex mirror is mounted on a pole, enabling golfers on the tee to see if the party in front is out of range. Your boss says that a range finder that works by triangulation could be placed facing the mirror, so the golfers could measure how the image of the party in front is behind the mirror. Then the golfers could be given a chart telling them how far the next party is from the tee. Your boss knows you are taking a physics course, so he asks you to calculate the distance of the image behind the mirror if the next party is 250 yd from the tee. The radius of curvature of the mirror has a magnitude of 20 yd.

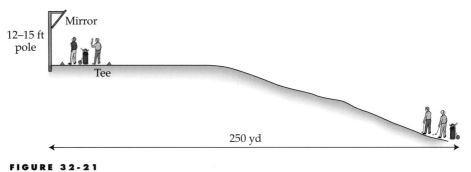

FIGURE 32-21

PICTURE THE PROBLEM The image distance is related to the object distance by the mirror formula, and the focal length of the lens is half the radius of curvature.

1. Use the mirror equation. For a convex mirror, the radius of curvature is negative:

$$\frac{1}{s} + \frac{1}{s'} = \frac{1}{f}$$

and

$$f = \frac{2}{r}$$

so

$$\frac{1}{250 \text{ yd}} + \frac{1}{s'} = \frac{2}{-20 \text{ yd}}$$

2. The image is 9.62 yd behind the mirror:

$$s' = \boxed{-9.62 \text{ yd}}$$

EXERCISE What is the distance to the party in front if the image is 9.75 yd behind the mirror? (*Answer* 390 yd)

(*a*)

(*b*)

(*a*) A convex mirror resting on paper with equally spaced parallel stripes. Note the large number of lines imaged in a small space and the reduction in size and distortion in shape of the image. (*b*) A convex mirror is used for security in a store.

32-2 Lenses

Images Formed by Refraction

One end of a long transparent cylinder is machined and polished to form a convex spherical surface. Figure 32-22 illustrates the formation of an image by refraction at such a surface. Suppose the cylinder is submerged in a transparent liquid with index of refraction n_1, and suppose the cylinder is made of a plastic material with index of refraction n_2, where n_2 is greater than n_1. Again, only paraxial rays converge to one point. An equation relating the image distance to the object distance, the radius of curvature, and the indexes of refraction can be derived by applying the law of refraction (Snell's law) to these rays and using small-angle approximations. The geometry is shown in Figure 32-23. The result is

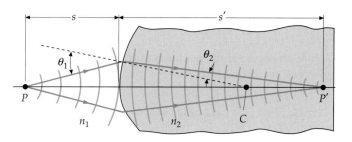

FIGURE 32-22 Image formed by refraction at a spherical surface between two media where the waves move slower in the second medium.

$$\frac{n_1}{s} + \frac{n_2}{s'} = \frac{n_2 - n_1}{r}$$

32-6

REFRACTION AT A SINGLE SURFACE

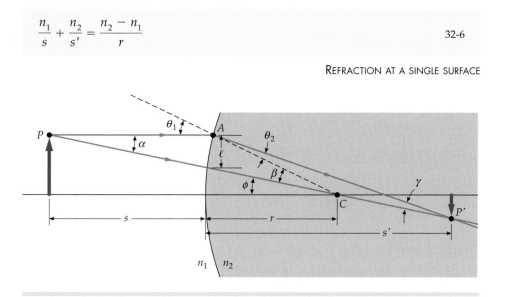

FIGURE 32-23 Geometry for relating the image position to the object position for refraction at a single spherical surface. The angles θ_1 and θ_2 are related by Snell's law of refraction: $n_1 \sin \theta_1 = n_2 \sin \theta_2$. The small-angle approximation $\sin \theta = \theta$ gives $n_1 \theta_1 = n_2 \theta_2$. From triangle ACP', we have $\beta = \theta_2 + \gamma = (n_1/n_2)\theta_1 + \gamma$. We can obtain another relation for θ_1 from triangle PAC: $\theta_1 = \alpha + \beta$. Eliminate θ_1 from these two equations: $n_1\alpha + n_1\beta + n_2\gamma = n_2\beta$. Simplify: $n_1\alpha + n_2\gamma = (n_2 - n_1)\beta$. Using the small-angle approximations $\alpha \approx \ell/s$, $\beta \approx \ell/r$, and $\gamma \approx \ell/s'$ gives Equation 32-6.

In refraction, real images are formed in back of the surface, which we will call the refracted-light side, whereas virtual images occur on the incident-light side, in front of the surface. The sign conventions we use for refraction are similar to those for reflection:

1. s is positive for objects on the incident-light side of the surface.

2. s' is positive for images on the refracted-light side of the surface.

3. r is positive if the center of curvature is on the refracted-light side of the surface.

SIGN CONVENTIONS FOR REFRACTION[†]

[†] The sign convention of choice for advanced work on optical design is the Cartesian sign convention. It can readily be found on the Internet.

We see that parameters s, s', and r are all positive if a real object is in front of a convex refracting surface that forms a real image. A parameter is negative if it does not meet the stated condition for being positive.

MAGNIFICATION BY A REFRACTING SURFACE   **E X A M P L E   3 2 - 4**   **Try It Yourself**

Derive an expression for the magnification $m = y'/y$ of an image formed by a spherical refracting surface.

PICTURE THE PROBLEM The magnification is the ratio of y' to y. Using Figure 32-18 and Figure 32-23 as guides, draw a ray diagram suitable for this derivation. These heights are related to the tangents of the angles θ_1 and θ_2, as shown in Figure 32-24. The angles are related by Snell's law. For paraxial rays, you can use the approximations $\tan\theta \approx \sin\theta \approx \theta$, and $\cos\theta \approx 1$.

Cover the column to the right and try these on your own before looking at the answers.

Steps

1. Using Figure 32-18 and Figure 32-23 as guides, draw a ray diagram suitable for this derivation. This drawing should include an object, a real image, a refracting surface, and an axis. Then draw an incident ray from the top of the object to the intersection of the axis with the refracting surface, and draw the refracted ray to the corresponding image point.

Answers

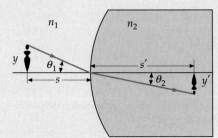

FIGURE 32-24

2. Write expressions for $\tan\theta_1$ and $\tan\theta_2$ in terms of the heights y and $-y'$ and the object and image distances s and s'. (Since y' is negative, use $-y'$, so that $\tan\theta_2$ is positive.)

$$\tan\theta_1 = \frac{y}{s}; \tan\theta_2 = \frac{-y'}{s'}$$

3. Apply the small-angle approximation $\tan\theta \approx \theta$ to your expressions.

$$\theta_1 = \frac{y}{s}; \theta_2 = \frac{-y'}{s'}$$

4. Write Snell's law of refraction relating the angles θ_1 and θ_2 using the small-angle approximation $\sin\theta \approx \theta$.

$$n_1\sin\theta_1 = n_2\sin\theta_2$$
$$n_1\theta_1 = n_2\theta_2$$

5. Substitute the expressions for θ_1 and θ_2 found in step 3.

$$n_1\left(\frac{y}{s}\right) = n_2\left(\frac{-y'}{s'}\right)$$

6. Solve for the magnification $m = y'/y$.

$$m = \frac{y'}{y} = -\frac{n_1 s'}{n_2 s}$$

We see from Example 32-4 that the magnification due to refraction at a spherical surface is

$$m = \frac{y'}{y} = -\frac{n_1 s'}{n_2 s} \qquad\qquad 32\text{-}7$$

IMAGE SEEN FROM A GOLDFISH BOWL
EXAMPLE 32-5

Goldie the goldfish is in a 15-cm-radius spherical bowl of water with an index of refraction of 1.33. Fluffy the cat is sitting on the table with her nose 10 cm from the surface of the bowl (Figure 32-25). The light from Fluffy's nose is refracted by the air–water boundary to form an image. Find (*a*) the image distance and (*b*) the magnification of the image of Fluffy's nose. Neglect any effect of the bowl's thin glass wall.

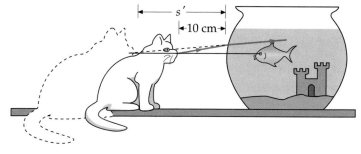

FIGURE 32-25

PICTURE THE PROBLEM We find the image distance *s'* using Equation 32-6 and the magnification using Equation 32-7. Since we are interested in light that goes from Fluffy's nose to the bowl, it follows that the air–water boundary is convex, and that air is the incident-light side of boundary and water is the refracted-light side of boundary. With these identifications, we have $n_1 = 1$, $n_2 = 1.33$, $s = +10$ cm, and $r = +15$ cm.

(*a*) 1. The equation relating the object distance to the image distance is Equation 32-6:

$$\frac{n_1}{s} + \frac{n_2}{s'} = \frac{n_2 - n_1}{r}$$

2. Identify and assign signs to the parameters in the previous step:

$n_1 = 1$, $n_2 = 1.33$, $s = +10$ cm, and $r = +15$ cm

3. Substitute numerical values and solve for *s'*:

$$\frac{1}{10 \text{ cm}} + \frac{1.33}{s'} = \frac{1.33 - 1}{15 \text{ cm}}$$

so

$$s' = \boxed{-17.1 \text{ cm}}$$

(*b*) Substitute numerical values into Equation 32-7 to find the magnification *m*:

$$m = -\frac{n_1 s'}{n_2 s} = -\frac{(1)(-17.1 \text{ cm})}{(1.33)(10 \text{ cm})} = \boxed{1.29}$$

REMARKS Since *s'* is negative, the image is virtual; that is, the image is on the opposite side of the refracting surface as the refracted light, as shown in Figure 32-25. The fish, Goldie, would see Fluffy to be slightly farther away ($|s'| > s$) than she actually is, and larger ($|m| > 1$) than she actually is. That *m* is positive indicates the image is upright.

EXERCISE If Goldie is 7.5 cm from the side of the bowl nearest Fluffy, find (*a*) the location and (*b*) the magnification of Goldie's image, as seen by Fluffy. (*Answer* $n_1 = 1.33$, $n_2 = 1$, $s = 7.5$ cm, $r = -15$ cm; thus, (*a*) $s' = -6.44$ cm and (*b*) $m = 1.14$. Fluffy sees Goldie to be slightly closer and slightly larger than she actually is.)

IMAGE SEEN FROM AN OVERHEAD BRANCH
EXAMPLE 32-6

During the summer months, Goldie the fish spends much of her time in a small pond in her owner's backyard. While enjoying a rest at the bottom of the 1-m deep pond, Goldie is being watched by Fluffy the cat, who is perched on a tree limb 3 m above the surface of the pond. How far below the surface is the image of the fish that Fluffy sees? (The index of refraction of water is 1.33.)

PICTURE THE PROBLEM The surface of the pond is a spherical refracting surface with an infinite radius of curvature. Thus, Equation 32-6 applies. Since the light reaching Fluffy originates in the water, use $n_1 = \frac{4}{3}$ and $n_2 = 1$.

1. Draw a picture of the situation. Label the object distance and the indexes of refraction of the media. Goldie is the object (Figure 32-26):

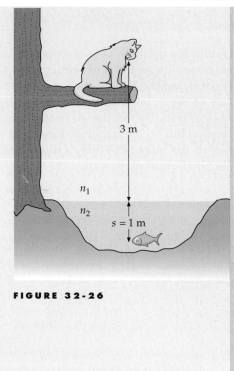

FIGURE 32-26

2. Using Equation 32-6, relate the image position s' to the other relevant parameters:

$$\frac{n_1}{s} + \frac{n_2}{s'} = \frac{n_2 - n_1}{r}$$

3. The refracting surface is flat. Using $r = \infty$, solve for s':

$$s' = -\frac{n_2}{n_1}s$$

4. Using the given values $n_1 = 1.33$, $n_2 = 1$, and $s = 1$ m, substitute to obtain s':

$$s' = -\frac{1}{1.33}(1\text{ m}) = -0.75\text{ m}$$

That the image is negative means that the image is on the side of the surface opposite the refracted light. That is, it is 0.75 m below the surface.

REMARKS (1) This image can be seen at the calculated position only when the object is viewed from directly overhead, or nearly so. From that observation point the rays are paraxial, a condition necessary for Equation 32-6 to be valid. If Fluffy is standing on the edge of the pond, the rays will not satisfy the paraxial approximation and Equation 32-6 will not correctly predict the location of the image. (2) The distance $(n_2/n_1)s$ is called the apparent depth of the submerged object. If $n_2 = 1$, the apparent depth equals s/n_1.

EXERCISE Draw a ray diagram for the image of Goldie, as described in Example 32-6. That is, draw several rays diverging from an object point P on Goldie, and show that after refracting the rays appear to diverge from an image point P' somewhat above the object point (Figure 32-27).

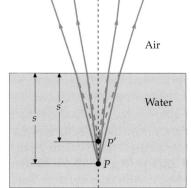

FIGURE 32-27 Ray diagram for the image of an object in water as viewed from directly overhead. The depth of the image is less than the depth of the object.

Thin Lenses

The most important application of Equation 32-6 for refraction at a single surface is finding the position of the image formed by a lens. This is done by considering the refraction at each surface of the lens separately to derive an equation relating the image distance to the object distance, the radius of curvature of each surface of the lens, and the index of refraction of the lens.

We will consider a thin lens of index of refraction n with air on both sides. Let the radii of curvature of the surfaces of the lens be r_1 and r_2. If an object is at a distance s from the first surface (and therefore from the lens), the distance s_1' of the image due to refraction at the first surface can be found using Equation 32-6:

$$\frac{1}{s} + \frac{n}{s_1'} = \frac{n - 1}{r_1}$$

32-8

The light refracted at the first surface is again refracted at the second surface. Figure 32-28 shows the case when the image distance s_1' for the first surface is negative, indicating a virtual image to the left of the surface. Rays in the glass refracted from the first surface diverge as if they came from the image point P_1'. The rays strike the second surface at the same angles as if there were an object at image point P_1'. The image for the first surface therefore becomes the object for the second surface. Since the lens is of negligible thickness, the object distance is equal in magnitude to s_1'. Object distances for objects on the incident-light side of a surface are positive, whereas image distances for images located there are negative. Thus, the object distance for the second surface is $s_2' = -s_1'$.[†] We now write Equation 32-6 for the second surface with $n_1 = n$, $n_2 = 1$, and $s = -s_1'$. The image distance for the second surface is the final image distance s' for the lens:

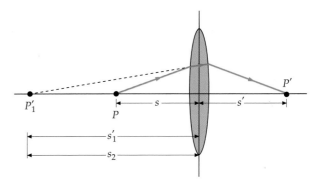

FIGURE 32-28 Refraction occurs at both surfaces of a lens. Here, the refraction at the first surface leads to a virtual image at P_1'. The rays strike the second surface as if they came from P_1'. Image distances are negative when the image is on the incident-light side of the surface, whereas object distances are positive for objects located there. Thus, $s_2 = -s_1'$ is the object distance for the second surface of the lens.

$$\frac{n}{-s_1'} + \frac{1}{s'} = \frac{1 - n}{r_2} \qquad 32\text{-}9$$

We can eliminate the image distance for the first surface s_1' by adding Equations 32-8 and 32-9. We obtain

$$\frac{1}{s} + \frac{1}{s'} = (n - 1)\left(\frac{1}{r_1} - \frac{1}{r_2}\right) \qquad 32\text{-}10$$

Equation 32-10 gives the image distance s' in terms of the object distance s and the properties of the thin lens—r_1, r_2, and n. As with mirrors, the focal length f of a thin lens is defined as the image distance when the object distance is infinite. Setting s equal to infinity and writing f for the image distance s', we obtain

$$\frac{1}{f} = (n - 1)\left(\frac{1}{r_1} - \frac{1}{r_2}\right) \qquad 32\text{-}11$$

LENS-MAKER'S EQUATION

Equation 32-11 is called the **lens-maker's equation;** it gives the focal length of a thin lens in terms of the properties of the lens. Substituting $1/f$ for the right side of Equation 32-10, we obtain

$$\frac{1}{s} + \frac{1}{s'} = \frac{1}{f} \qquad 32\text{-}12$$

THIN-LENS EQUATION

This **thin-lens equation** is the same as the mirror equation (Equation 32-4). Recall, however, that the sign conventions for refraction are somewhat different from those for reflection. For lenses, the image distance s' is positive when the image is on the refracted-light side of the lens, that is, when it is on the side opposite the incident-light side. The sign of the focal length (see Equation 32-11) is determined by the sign convention for a single refracting boundary. That is, r is positive if the center of curvature is on the same side of the surface as the refracted light. For a lens like that shown in Figure 32-28, r_1 is positive and r_2 is negative, so f is positive.

---

† If s_1' were positive, the rays would be converging as they strike the second surface. The object for the second surface would then be a virtual object located to the right of the second surface. This object would be a virtual object. Again, $s_2 = -s_1'$.

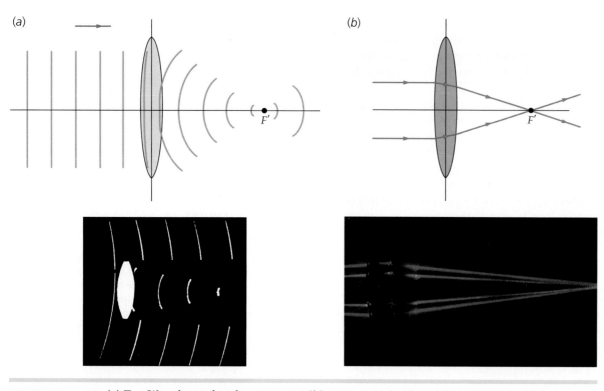

FIGURE 32-29 (a) *Top:* Wavefronts for plane waves striking a converging lens. The central part of the wavefront is retarded more by the lens than the outer part, resulting in a spherical wave that converges at the focal point F'. *Bottom:* Wavefronts passing through a lens, shown by a photographic technique called *light-in-flight recording* that uses a pulsed laser to make a hologram of the wavefronts of light. (b) *Top:* Rays for plane waves striking a converging lens. The rays are bent at each surface and converge at the focal point. *Bottom:* A photograph of rays focused by a converging lens.

Figure 32-29*a* shows the wavefronts of plane waves incident on a double convex lens. The central part of the wavefront strikes the lens first. Since the wave speed in the lens is less than that in air (assuming $n > 1$), the central part of the wavefront lags behind the outer parts, resulting in a spherical wavefront that converges at the focal point F'. The rays for this situation are shown in Figure 32-29*b*. Such a lens is called a **converging lens.** Since its focal length as calculated from Equation 32-11 is positive, it is also called a **positive lens.** Any lens that is thicker in the middle than at the edges is a converging lens (providing that the index of refraction of the lens is greater than that of the surrounding medium). Figure 32-30 shows the wavefronts and rays for plane waves incident on a double concave lens. In this case, the outer part of the wavefronts lag behind the central parts, resulting in outgoing spherical waves that diverge from a focal point on the incident-light side of the lens. The focal length of this lens is negative. Any lens that is thinner in the middle than at the edges is a **diverging,** or **negative,** lens (providing that the index of refraction of the lens is greater than that of the surrounding medium).

FIGURE 32-30 (a) Wavefronts for plane waves striking a diverging lens. Here, the outer part of the wavefront is retarded more than the central part, resulting in a spherical wave that diverges as it moves out, as if it came from the focal point F' to the left of the lens. (b) Rays for plane waves striking the same diverging lens. The rays are bent outward and diverge, as if they came from the focal point F'. (c) A photograph of rays passing through a diverging lens.

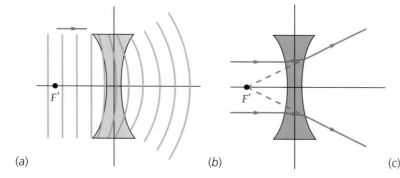

(a)   (b)   (c)

THE LENS-MAKER'S FORMULA   **E X A M P L E   3 2 - 7**

A double convex, thin glass lens with index of refraction $n = 1.5$ has radii of curvature of magnitude 10 cm and 15 cm, as shown in Figure 32-31. Find its focal length.

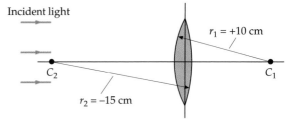

Incident light
$r_1 = +10$ cm
C_2
C_1
$r_2 = -15$ cm

FIGURE 32-31

PICTURE THE PROBLEM We can find the focal length using the lens-maker's equation (Equation 32-11). Here, light is incident on the surface with the smaller radius of curvature. The center of curvature of this surface, C_1, is on the refracted-light side of the lens; thus, $r_1 = +10$ cm. For the second surface, the center of curvature, C_2, is on the incident-light side; therefore, $r_2 = -15$ cm.

Numerical substitution in Equation 32-11 yields the focal length f:

$$\frac{1}{f} = (n - 1)\left(\frac{1}{r_1} - \frac{1}{r_2}\right)$$

$$= (1.5 - 1)\left(\frac{1}{10\ \text{cm}} - \frac{1}{-15\ \text{cm}}\right) = 0.5\left(\frac{1}{6\ \text{cm}}\right)$$

$$f = \boxed{12\ \text{cm}}$$

REMARKS Note that both surfaces tend to converge the light rays; therefore, they both make a positive contribution to the focal length of the lens.

EXERCISE A double convex thin lens has an index of refraction $n = 1.6$ and radii of curvature of equal magnitude. If its focal length is 15 cm, what is the magnitude of the radius of curvature of each surface? (*Answer*   18 cm)

EXERCISE Show that if you reverse the direction of the incoming light for the lens shown in Example 32-7, so that it is incident on the surface with the greater radius of curvature, you get the same result for the focal length.

If parallel light strikes the lens of Example 32-7 from the left, it is focused at a point 12 cm to the right of the lens; whereas if parallel light strikes the lens from the right, it is focused at 12 cm to the left of the lens. Both of these points are focal points of the lens. Using the reversibility property of light rays, we can see that light diverging from a focal point and striking a lens will leave the lens as a parallel beam, as shown in Figure 32-32. In a particular lens problem in which the direction of the incident light is specified, the object point for which light emerges as a parallel beam is called the **first focal point** F, and the point at which parallel light is focused is called the **second focal point** F'. For a positive lens, the first focal point is on the incident-light side and the second focal point is on the refracted-light side. If parallel light is incident on the lens at a small angle with the axis, as in Figure 32-33, it is focused at a point in the **focal plane** a distance f from the lens.

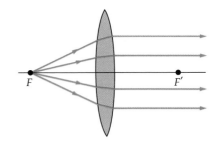

F
F'

FIGURE 32-32 Light rays diverging from the focal point of a positive lens emerge parallel to the axis.

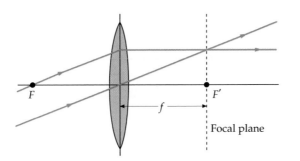

F
F'
f
Focal plane

FIGURE 32-33 Parallel rays incident on the lens at an angle to its axis are focused at a point in the focal plane of the lens.

The reciprocal of the focal length is called the **power of a lens.** When the focal length is expressed in meters, the power is given in reciprocal meters, called **diopters** (D):

$$P = \frac{1}{f}$$

32-13

The power of a lens measures its ability to focus parallel light at a short distance from the lens. The shorter the focal length, the greater the power. For example, a lens with a focal length of 25 cm = 0.25 m has a power of 4 D. A lens with a focal length of 10 cm = 0.10 m has a power of 10 D. Since the focal length of a diverging lens is negative, its power is negative.

POWER OF A LENS

E X A M P L E   3 2 - 8

The lens shown in Figure 32-34 has an index of refraction of 1.5 and radii of curvature of magnitude 10 cm and 13 cm. Find (*a*) its focal length and (*b*) its power.

PICTURE THE PROBLEM For the orientation of the lens relative to the incident light shown in Figure 32-34, the radius of curvature of the first surface is $r_1 = +10$ cm and that of the second surface is $r_2 = +13$ cm.

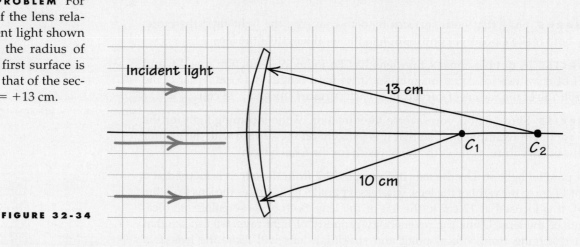

FIGURE 32-34

1. Calculate f from the lens-maker's equation using the given value of n and the values of r_1 and r_2 for the orientation shown:

$$\frac{1}{f} = (n - 1)\left(\frac{1}{r_1} - \frac{1}{r_2}\right)$$

$$= (1.5 - 1)\left(\frac{1}{10 \text{ cm}} - \frac{1}{13 \text{ cm}}\right)$$

$$f = \boxed{86.7 \text{ cm}}$$

2. The power is the reciprocal of the focal length expressed in meters:

$$P = \frac{1}{f} = \frac{1}{0.867 \text{ m}} = \boxed{1.15 \text{ D}}$$

REMARKS We obtain the same result no matter which surface the light strikes first.

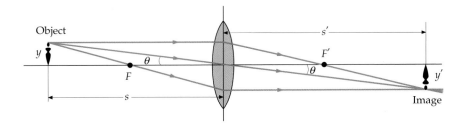

FIGURE 32-35 Ray diagram for a thin converging lens. We assume that all the bending of light takes place at the central plane. The ray through the center is undeflected because the lens surfaces there are parallel and close together.

In laboratory experiments involving lenses, it is usually much easier to measure the focal length than to calculate the focal length from the radii of curvature of the surfaces.

Ray Diagrams for Lenses

As with images formed by mirrors, it is convenient to locate the images of lenses by graphical methods. Figure 32-35 illustrates the graphical method for a thin converging lens. We consider the rays to bend at the plane through the center of the lens. The three principal rays are as follows:

1. The **parallel ray,** drawn parallel to the axis. The emerging ray is directed toward (or away from) the second focal point of the lens.

2. The **central ray,** drawn through the center (the vertex) of the lens. This ray is undeflected. (The faces of the lens are parallel at this point, so the ray emerges in the same direction but displaced slightly. Since the lens is thin, the displacement is negligible.)

3. The **focal ray,** drawn through the first focal point.[†] This ray emerges parallel to the axis.

The weight and bulk of a large-diameter lens can be reduced by constructing the lens from annular segments at different angles so that light from a point is refracted by the segments into a parallel beam. Such an arrangement is called a Fresnel lens. Several Fresnel lenses are used in this lighthouse to produce intense parallel beams of light from a source at the focal point of the lenses. The illuminated surface of an overhead projector is a Fresnel lens.

PRINCIPAL RAYS FOR A THIN LENS

These three rays converge to the image point, as shown in Figure 32-35. In this case, the image is real and inverted. From the figure, we have $\tan \theta = y/s = -y'/s'$. The lateral magnification is then

$$m = \frac{y'}{y} = -\frac{s'}{s}$$

32-14

This expression is the same as the expression for mirrors. Again, a negative magnification indicates that the image is inverted. The ray diagram for a diverging lens is shown in Figure 32-36.

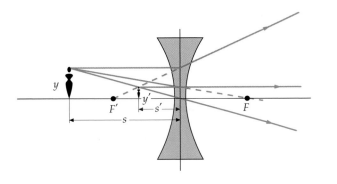

FIGURE 32-36 Ray diagram for a diverging lens. The parallel ray is bent away from the axis, as if it came from the second focal point F'. The ray toward the first focal point F emerges parallel to the axis.

† The focal ray is drawn toward the first focal point for a diverging lens.

IMAGE FORMED BY A LENS   **E X A M P L E   3 2 - 9**

An object 1.2 cm high is placed 4 cm from a double convex lens with a focal length of 12 cm. Locate the image both graphically and algebraically, state whether the image is real or virtual, and find its height. Place an eye on the figure positioned and oriented so as to view the image.

1. Draw the parallel ray. This ray leaves the object parallel to the axis, then is bent by the lens to pass through the second focal point, F' (Figure 32-37):

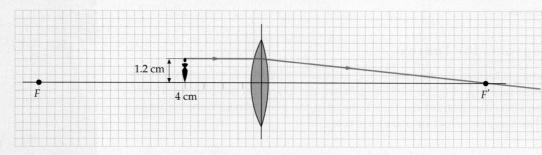

FIGURE 32-37

2. Draw the central ray, which passes undeflected through the center of the lens. Since the two rays are diverging on the refracted-light side, we extend them back to the incident-light side to find the image (Figure 32-38):

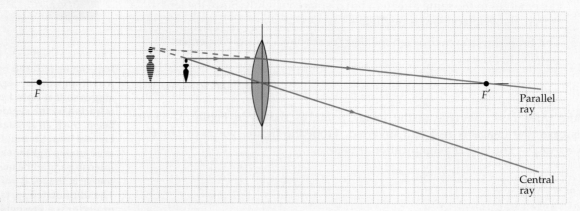

FIGURE 32-38

3. As a check, we also draw the focal ray. This ray leaves the object on a line passing through the first focal point, then emerges parallel to the axis. Note that the image is virtual, upright, and enlarged (Figure 32-39):

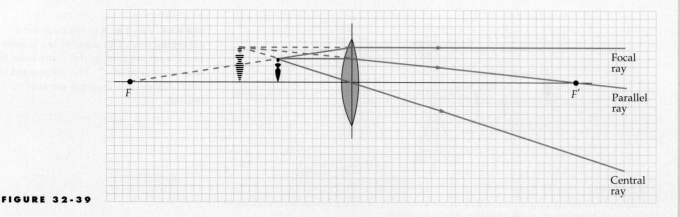

FIGURE 32-39

4. The eye must be positioned so the light from the image enters the eye.

5. We now verify the results of the ray diagram algebraically. First, find the image distance using Equation 32-12:

$$\frac{1}{4 \text{ cm}} + \frac{1}{s'} = \frac{1}{12 \text{ cm}}$$

$$\frac{1}{s'} = \frac{1}{12 \text{ cm}} - \frac{1}{4 \text{ cm}} = -\frac{1}{6 \text{ cm}}$$

$$s' = -6 \text{ cm}$$

6. The height of the image is found from the height of the object and the magnification:

$$h' = mh$$

7. The magnification m is given by Equation 32-14:

$$m = -\frac{s'}{s} = -\frac{-6 \text{ cm}}{4 \text{ cm}} = \boxed{+1.5}$$

8. Using this result we find the height of the image, h':

$$h' = mh = (1.5)(1.2 \text{ cm}) = \boxed{1.8 \text{ cm}}$$

REMARKS Note the agreement between the algebraic and ray diagram results. Algebraically, we find that the image is 6 cm from the lens on the incident-light side (since $s' < 0$); that is, the image is 2 cm to the left of the object. Since $m > 0$, it follows that the image is upright, and because $m > 1$, the image is enlarged. It is good practice to process lens problems both graphically and algebraically and to compare the results.

EXERCISE An object is placed 15 cm from a double convex lens of focal length 10 cm. Find the image distance and the magnification. Draw a ray diagram. Is the image real or virtual? Is the image upright or inverted? (*Answer* $s' = 30$ cm, $m = -2$; real, inverted)

EXERCISE Work the previous exercise for an object placed 5 cm from a lens with a focal length of 10 cm. (*Answer* $s' = -10$ cm, $m = 2$; virtual, upright)

Combinations of Lenses

If we have two or more thin lenses, we can find the final image produced by the system by finding the image distance for the first lens and then using it, along with the distance between the lenses, to find the object distance for the second lens. That is, we consider each image, whether it is real or virtual—and whether it is actually formed or not—as the object for the next lens.

IMAGE FORMED BY A SECOND LENS   **EXAMPLE   32-10**

A second lens of focal length +6 cm is placed 12 cm to the right of the lens in Example 32-9. Locate the final image.

PICTURE THE PROBLEM The principal rays used to locate the image of the first lens will not necessarily be principal rays for the second lens. In this example, however, we have chosen the position of the second lens (Figure 32-40a) so that the parallel ray for the first lens turns out to be the central ray for the second lens. Also, the focal ray for the first lens emerges parallel to the axis and is therefore the parallel ray for the second lens. If additional principal rays for the second lens are needed, we simply draw them from the image formed by the first lens. For example, in Figure 32-40b we added such a ray, drawn from the first image through the first focal point F_2 of the second lens.

FIGURE 32-40

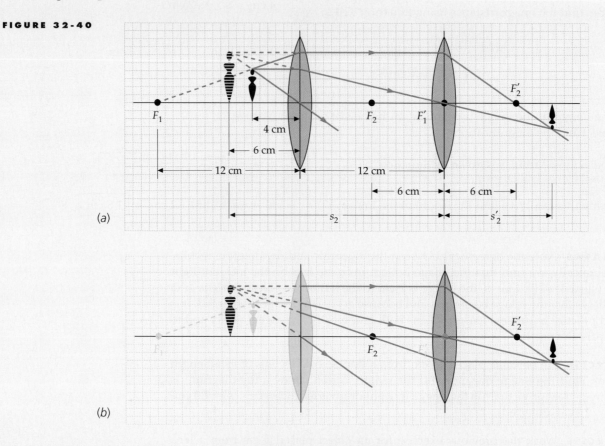

(a)

(b)

Algebraically we use $s_2 = 18$ cm, because the first image is 6 cm to the left of the first lens and therefore 18 cm to the left of the second lens.

Use $s_2 = 18$ cm and $f = 6$ cm to calculate s_2':

$$\frac{1}{s_2} + \frac{1}{s_2'} = \frac{1}{f_2}$$

$$\frac{1}{18 \text{ cm}} + \frac{1}{s_2'} = \frac{1}{6 \text{ cm}}$$

$$s_2' = \boxed{9 \text{ cm}}$$

A COMBINATION OF TWO LENSES   **E X A M P L E   3 2 - 1 1**   **Try It Yourself**

Two lenses, each of focal length 10 cm, are 15 cm apart. Find the final image of an object 15 cm from one of the lenses.

PICTURE THE PROBLEM Use a ray diagram to find the location of the image formed by lens 1. When these rays strike lens 2 they are further refracted, leading to the final image. More accurate results are obtained algebraically using the thin-lens equation for both lens 1 and lens 2.

Cover the column to the right and try these on your own before looking at the answers.

Steps   Answers

1. Draw the (*a*) parallel, (*b*) central, and
   (*c*) focal rays for lens 1 (Figure 32-41). If
   lens 2 did not alter these rays, they
   would form an image at I_1.

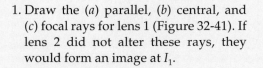

FIGURE 32·41

2. To locate the final image, add three
   principal rays (*d, e,* and *f*) for lens 2.
   The intersection of these rays gives the
   image location (Figure 32-42).

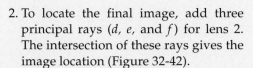

FIGURE 32·42

3. To proceed algebraically, use the thin-lens equation to $s_1' = 30$ cm
   find the image distance s_1' produced by lens 1.

4. For lens 2, the image, I_1 is 15 cm from the lens on the $s_2' = \boxed{6 \text{ cm}}$
   refracted-light side; hence, $s_2 = -15$ cm. Use this to find
   the final image distance s_2'.

REMARKS From the ray diagram we see that the final image is real, inverted,
and slightly reduced.

Compound Lenses

When two thin lenses of focal lengths f_1 and f_2 are placed together, the effective
focal length of the combination f_{eff} is given by

$$\frac{1}{f_{\text{eff}}} = \frac{1}{f_1} + \frac{1}{f_2}$$ 32-15

as is shown in the following Example 32-12. The power of two lenses in contact is
given by

$$P_{\text{eff}} = P_1 + P_2$$ 32-16

Two Lenses in Contact **EXAMPLE 32-12** Try It Yourself

For two lenses very close together, derive the relation $\dfrac{1}{f_{eff}} = \dfrac{1}{f_1} + \dfrac{1}{f_2}$.

PICTURE THE PROBLEM Apply the thin-lens equation to each lens using the fact that the distance between the lenses is zero, so the object distance for the second lens is the negative of the image distance for the first lens.

Cover the column to the right and try these on your own before looking at the answers.

Steps	Answers
1. Write the thin-lens equation for lens 1.	$\dfrac{1}{s} + \dfrac{1}{s_1'} = \dfrac{1}{f_1}$
2. Using $s_2 = -s_1'$, write the thin-lens equation for lens 2.	$\dfrac{1}{-s_1'} + \dfrac{1}{s'} = \dfrac{1}{f_2}$
3. Add your two resulting equations to eliminate s_1'.	$\boxed{\dfrac{1}{s} + \dfrac{1}{s'} = \dfrac{1}{f_1} + \dfrac{1}{f_2} = \dfrac{1}{f_{eff}}}$

*32-3 Aberrations

When all the rays from a point object are not focused at a single image point, the resultant blurring of the image is called **aberration.** Figure 32-43 shows rays from a point source on the axis traversing a thin lens with spherical surfaces. Rays that strike the lens far from the axis are bent much more than are the rays near the axis, with the result that not all the rays are focused at a single point. Instead, the image appears as a circular disk. The **circle of least confusion** is at point C, where the diameter is minimum. This type of aberration in a lens is called **spherical aberration;** it is the same as the spherical aberration of mirrors discussed in Section 32-1. Similar but more complicated aberrations called *coma* (for the comet-shaped image) and *astigmatism* occur when objects are off axis. The aberration in the shape of the image of an extended object that occurs, because the magnification depends on the distance of the object point from the axis, is called **distortion.** We will not discuss these aberrations further, except to point out that they do not arise from any defect in the lens or mirror but instead result from the application of the laws of refraction and reflection to spherical surfaces. These aberrations are not evident in our simple equations, because we used small-angle approximations in the derivation of these equations.

Some aberrations can be eliminated or partially corrected by using nonspherical surfaces for mirrors or lenses, but nonspherical surfaces are usually much more difficult and costly to produce than spherical surfaces. One example of a nonspherical reflecting surface is the parabolic mirror illustrated in Figure 32-44. Rays that are parallel to the axis of a parabolic surface are reflected and focused at a common point, no matter how far the rays are from the axis. Parabolic reflecting surfaces are sometimes used in large astronomical telescopes, which need a large reflecting surface to gather as much light as possible to make the image as intense as possible (reflecting

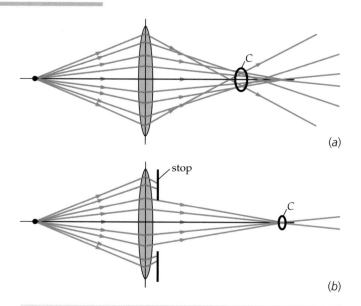

(a)

(b)

FIGURE 32-43 Spherical aberration in a lens. (*a*) Rays from a point object on the axis are not focused at a point. (*b*) Spherical aberration can be reduced by using a stop to block off the outer parts of the lens, but this also reduces the amount of light reaching the image.

telescopes are described in the upcoming optional Section 32-4). Satellite dishes use parabolic surfaces to focus microwaves from communications satellites. A parabolic surface can also be used in a searchlight to produce a parallel beam of light from a small source placed at the focal point of the surface.

An important aberration found with lenses but not found with mirrors is **chromatic aberration,** which is due to variations in the index of refraction with wavelength. From Equation 32-11, we can see that the focal length of a lens depends on its index of refraction and is therefore different for different wavelengths. Since n is slightly greater for blue light than for red light, the focal length for blue light will be shorter than the focal length for red light. Because chromatic aberration does not occur for mirrors, many large telescopes use a large mirror instead of the large, light-gathering (objective) lens.

Chromatic aberration and other aberrations can be partially corrected by using combinations of lenses instead of a single lens. For example, a positive lens and a negative lens of greater focal length can be used together to produce a converging lens system that has much less chromatic aberration than a single lens of the same focal length. The lens of a good camera typically contains six elements to correct the various aberrations that are present.

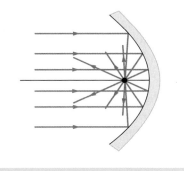

FIGURE 32-44 A parabolic mirror focuses all rays parallel to the axis to a single point with no spherical aberration.

*32-4 Optical Instruments

*The Eye

The optical system of prime importance is the eye, which is shown in Figure 32-45. Light enters the eye through a variable aperture, the pupil. The light is focused by the cornea, with assistance from the lens, on the retina, which has a film of nerve fibers covering the back surface. The retina contains tiny sensing structures called *rods* and *cones,* which detect the light and transmit the information along the optic nerve to the brain. The shape of the crystalline lens can be altered slightly by the action of the ciliary muscle. When the eye is focused on an object far away, the muscle is relaxed and the cornea–lens system has its maximum focal length, about 2.5 cm, which is the distance from the cornea to the retina. When the object is brought closer to the eye, the ciliary muscle increases the curvature of the lens slightly, thereby decreasing its focal length, so that the image is again focused on the retina. This process is called *accommodation.* If the object is too close to the eye, the lens cannot focus the light on the retina and the image is blurred. The closest point for which the lens can focus the image on the retina is called the **near point.** The distance from the eye to the near point varies greatly from one person to another and changes with age. At 10 years, the near point may be as close as 7 cm, whereas at 60 years it may recede to 200 cm because of the loss of flexibility of the lens. The standard value taken for the near point is 25 cm.

If the eye underconverges, resulting in the images being focused behind the retina, the person is said to be farsighted. A farsighted person can see distant objects where little convergence is required, but has trouble seeing close objects. Farsightedness is corrected with a converging (positive) lens (Figure 32-46).

On the other hand, the eye of a nearsighted person overconverges and focuses light from distant objects in front of the retina. A nearsighted person can see nearby objects for which the widely diverging incident rays can be focused on

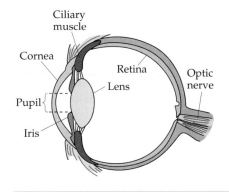

FIGURE 32-45 The human eye. The amount of light entering the eye is controlled by the iris, which regulates the size of the pupil. The lens thickness is controlled by the ciliary muscle. The cornea and lens together focus the image on the retina, which contains approximately 125 million receptors, called rods and cones, and approximately 1 million optic-nerve fibers.

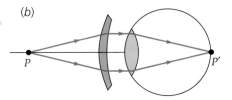

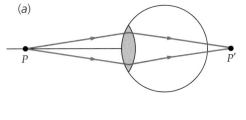

FIGURE 32-46 (*a*) A farsighted eye focuses rays from a nearby object to a point behind the retina. (*b*) A converging lens corrects this defect by bringing the image onto the retina. These diagrams, and those following, are drawn as if all the focusing of the eye is done at the lens; whereas, in fact, the lens and cornea system act more like a spherical refracting surface than a thin lens.

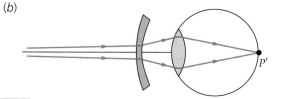

(a)

FIGURE 32-47 (*a*) A nearsighted eye focuses rays from a distant object to a point in front of the retina. (*b*) A diverging lens corrects this defect.

the retina, but has trouble seeing distant objects clearly. Nearsightedness is corrected with a diverging (negative) lens (Figure 32-47).

Another common defect of vision is astigmatism, which is caused by the cornea being not quite spherical but having a different curvature in one plane than in another. This results in a blurring of the image of a point object into a short line. Astigmatism is corrected by glasses using lenses of cylindrical rather than spherical shape.

FOCAL LENGTH OF THE CORNEA–LENS SYSTEM   **EXAMPLE 32-13**

Both a thin lens and a spherical mirror have a focal length given by the formula $\frac{1}{s} + \frac{1}{s'} = \frac{1}{f}$, where f is a constant. Using the same formula, we define the focal length of a spherical refracting surface. However, in this case, the focal length is not constant but depends upon s. By how much does the focal length of the cornea–lens system of the eye change if the object is moved from infinity to the near point at 25 cm? Assume that all the focusing is done at the cornea, and that the distance from the cornea to the retina is 2.5 cm.

PICTURE THE PROBLEM With the object at infinity, the focal length is 2.5 cm. We use the thin-lens equation to calculate the focal length when $s = 25$ cm and $s' = 2.5$ cm.

1. Use the thin-lens equation to calculate f:

$$\frac{1}{f} = \frac{1}{s} + \frac{1}{s'} = \frac{1}{25 \text{ cm}} + \frac{1}{2.5 \text{ cm}}$$

$$= \frac{1}{25 \text{ cm}} + \frac{10}{25 \text{ cm}} = \frac{11}{25 \text{ cm}}$$

so

$$f = 2.27 \text{ cm}$$

2. Subtract the original focal length of 2.5 cm to find the change: $\Delta f = 2.27 \text{ cm} - 2.5 \text{ cm} = \boxed{-0.23 \text{ cm}}$

REMARKS In terms of the power of the cornea–lens system, when the focal length is 2.5 cm = 0.025 m for distant objects, the power is $P = 1/f = 40$ D. When the focal length is 2.27 cm, the power is 44 D.

EXERCISE Find the change in the focal length of the eye when an object originally at 4 m is brought to 40 cm from the eye. (Assume that the distance from the cornea to the retina is 2.5 cm.) (*Answer* −0.13 cm)

The apparent size of an object is determined by the actual size of the image on the retina. The larger the image on the retina, the greater the number of rods and cones activated. From Figure 32-48, we see that the size of the image on the retina is greater when the

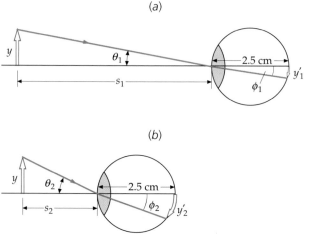

FIGURE 32-48 (*a*) A distant object of height y looks small because the image on the retina is small. (*b*) When the same object is closer, it looks larger because the image on the retina is larger.

object is close than it is when the object is far away. The apparent size of an object is thus greater when it is closer to the eye. The image size is proportional to the angle θ subtended by the object at the eye. For Figure 32-48,

$$\phi \approx \frac{y'}{2.5 \text{ cm}} \quad \text{and} \quad \theta \approx \frac{y}{s} \qquad\qquad 32\text{-}17$$

for small angles. Applying the law of refraction gives $n_{\text{Air}} \sin \theta = n \sin \phi$, where $n_{\text{Air}} = 1.00$ and n is the refractive index inside the eye. For small angles this becomes

$$\theta \approx n\phi \qquad\qquad 32\text{-}18$$

Combining Equations 32-17 and 32-18 gives

$$\frac{y}{s} \approx n \frac{y'}{2.5 \text{ cm}}, \quad \text{or} \quad y' \approx \frac{2.5 \text{ cm}}{n} \frac{y}{s} \qquad\qquad 32\text{-}19$$

The size of the image on the retina is proportional to the size of the object and inversely proportional to its distance from the eye. Since the near point is the closest point to the eye for which a sharp image can be formed on the retina, the distance to the near point is called the *distance of most distinct vision*.

READING GLASSES	**E X A M P L E 3 2 - 1 4**

The near-point distance of a person's eye is 75 cm. With a reading glasses lens placed a negligible distance from the eye, the near-point distance of the lens–eye system is 25 cm. That is, if an object is placed 25 cm in front of the lens, then the lens forms a virtual image of the object a distance 75 cm in front of the lens. (*a*) What power is the reading glasses lens and (*b*) what is the lateral magnification of the image formed by the lens? (*c*) Which produces the bigger image on the retina, (1) the object at the near point of, and viewed by, the unaided eye or (2) the object at the near point of the lens–eye system and viewed through the lens that is immediately in front of the eye?

PICTURE THE PROBLEM   A near-point distance of the lens–eye system of 25 cm means the lens forms a virtual image 75 cm in front of the lens if an object is placed 25 cm in front of the lens. Figure 32-49*a* shows a diagram of an object 25 cm from a converging lens that produces a virtual, upright image at $s' = -75$ cm. Figure 32-49*b* shows the image on the retina formed by the focusing power of the eye.

FIGURE 32-49

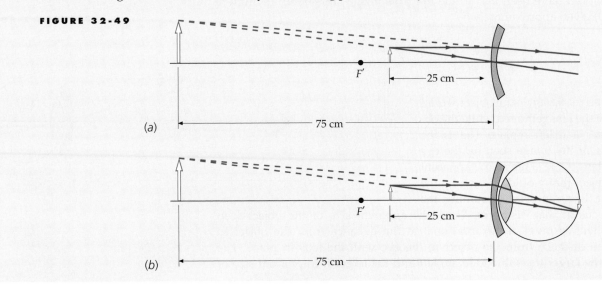

(a)

(b)

1. Use the thin-lens equation with $s = 25$ cm and $s' = -75$ cm to calculate the power, $1/f$:

$$\frac{1}{f} = \frac{1}{s} + \frac{1}{s'} = \frac{1}{25 \text{ cm}} + \frac{1}{-75 \text{ cm}}$$

$$= \frac{2}{75 \text{ cm}} = \frac{2}{0.75 \text{ m}} = \boxed{2.67 \text{ D}}$$

2. Using $m = -s'/s$, find m:

$$m = -\frac{s'}{s} = -\frac{-75 \text{ cm}}{25 \text{ cm}} = \boxed{3}$$

3. In both cases, the rays entering the eye appear to diverge from an image 75 cm in front of the eye. However, with the lens in place, the image there is larger by a factor of 3: $\boxed{\text{Option 2}}$

REMARKS (1) If your near point is 75 cm, you are farsighted. To read a book you must hold it at least 75 cm from your eye to be able to focus on the print. The image of the print on your retina is then very small. The reading glasses lens produces an image also 75 cm from your eye, and this image is three times larger than the actual print. Thus, looking through the lens, the image of the print on the retina is larger by a factor of 3. (2) In this example, the distance from the lens to the eye was negligible. The results are slightly different if this distance is not negligible and is factored into the calculations.

EXERCISE Calculate the power of the eye for which the near point is 75 cm and the cornea–retina distance is 2.5 cm, and calculate the combined power of the lens and eye when they are in contact. Compare this with the power of a lens for which $s' = 2.5$ cm, when $s = 25$ cm. (*Answer* $P_{\text{eye}} = 41.33$ D; $P_c = 41.33$ D + 2.67 D = 44 D; $P = 44$ D)

*The Simple Magnifier

We saw in Example 32-14 that the *apparent* size of an object can be increased by using a converging lens placed next to the eye. A converging lens is called a **simple magnifier** if it is placed next to the eye and if the object is placed closer to the lens than its focal length, as was the case for the lens in Example 32-14. In that example, the lens formed a virtual image at the near point of the eye, the same location that the object must be placed for best viewing by the unaided eye. So, with the lens in place, the magnitude of the image distance $|s'|$ was greater than the object distance s, so the image seen by the eye is magnified by $m = |s'|/s$. If the actual height of the object was y, then the height y' of the image formed by the lens would have been my. To the eye, this image subtended an angle θ (Figure 32-50) given approximately by

FIGURE 32-50

$$\theta = \frac{my}{|s'|} = m\frac{y}{|s'|} = \frac{|s'|}{s}\frac{y}{|s'|} = \frac{y}{s}$$

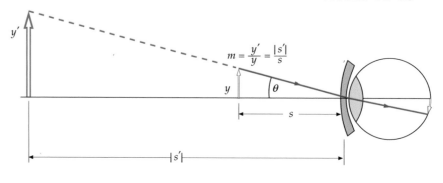

which is the *very same angle* the object would subtend were the lens removed while the object and the eye were left in place. That is, the apparent size of the image seen by the eye through the lens is the same as the apparent size of the object that would be seen by the eye were the lens removed (assuming the eye could focus at that distance). Thus, the apparent size of the object seen through the lens is inversely proportional to the distance from the object to the eye to the distance from the object to the eye with the lens in place. The smaller s is, the larger the subtended angle θ and the larger the apparent size of the object.

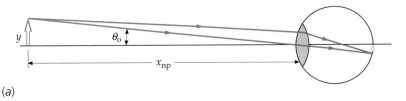

(a)   (b)

FIGURE 32-51 (*a*) An object at the near point subtends an angle θ_0 at the naked eye. (*b*) When the object is at the focal point of the converging lens, the rays emerge from the lens parallel and enter the eye as if they came from an object a very large distance away. The image can thus be viewed at infinity by the relaxed eye. When *f* is less than the near-point distance, the converging lens allows the object to be brought closer to the eye. This increases the angle subtended by the object to θ, thereby increasing the size of the image on the retina.

In Figure 32-51*a*, a small object of height *y* is at the near point of the eye at a distance x_{np}. The angle subtended, θ_0, is given approximately by

$$\theta_0 = \frac{y}{x_{np}}$$

In Figure 32-51*b*, a converging lens of focal length *f* that is smaller than x_{np} is placed a negligible distance in front of the eye, and the object is placed in the focal plane of the lens. The rays emerge from the lens parallel, indicating that the image is located an infinite distance in front of the lens. The parallel rays are focused by the relaxed eye on the retina. The angle subtended by this image is equal to the angle subtended by the object (assuming that the lens is a negligible distance from the eye). The angle subtended by the object is approximately

$$\theta = \frac{y}{f}$$

The ratio θ/θ_0 is called the *angular magnification* or *magnifying power M* of the lens:

$$M = \frac{\theta}{\theta_0} = \frac{x_{np}}{f} \qquad\qquad 32\text{-}20$$

Simple magnifiers are used as eyepieces (called oculars) in microscopes and telescopes to view the image formed by another lens or lens system. To correct aberrations, combinations of lenses that result in a short positive focal length may be used in place of a single lens, but the principle of the simple magnifier is the same.

ANGULAR MAGNIFICATION OF A SIMPLE MAGNIFIER   **EXAMPLE   3 2 - 1 5**   **Try It Yourself**

A person with a near point of 25 cm uses a 40-D lens as a simple magnifier. What angular magnification is obtained?

PICTURE THE PROBLEM The angular magnification is found from the focal length *f* (Equation 32-20), which is the reciprocal of the power.

Cover the column to the right and try these on your own before looking at the answers.

Steps	Answers
1. Calculate the focal length of the lens.	$f = 2.5$ cm
2. Use your result from step 1 and incorporate the result into Equation 32-20 to calculate the angular magnification.	$M = \boxed{10}$

REMARKS Looking through the lens, the object appears 10 times larger because it can be placed at 2.5 cm rather than at 25 cm from the eye, thus increasing the size of the image on the retina tenfold.

EXERCISE What is the magnification in this example if the near point of the person is 30 cm rather than 25 cm? (*Answer* $M = 12$)

*The Compound Microscope

The compound microscope (Figure 32-52) is used to look at very small objects at short distances. In its simplest form, it consists of two converging lenses. The lens nearest the object, called the **objective,** forms a real image of the object. This image is enlarged and inverted. The lens nearest the eye, called the **eyepiece** or **ocular,** is used as a simple magnifier to view the image formed by the objective.

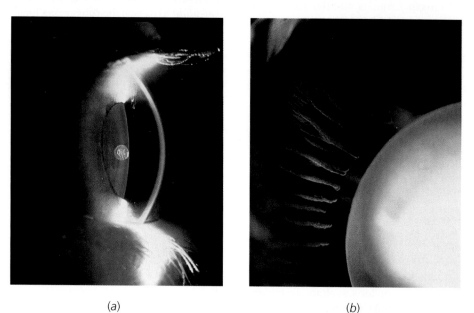

(a)

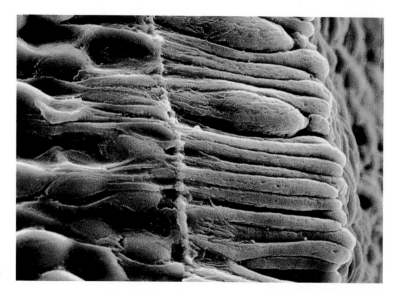

(b)

(*a*) The human eye in profile. (*b*) The lens of the eye is kept in place by the ciliary muscle (shown here in the upper left), which rings the lens. When the ciliary muscle contracts, the lens tends to bulge. The greater lens curvature enables the eye to focus on nearby objects. (*c*) Some of the 120 million rods and 7 million cones in the eye, magnified approximately 5000 times. The rods (the more slender of the two) are more sensitive in dim light, whereas the cones are more sensitive to color. The rods and cones form the bottom layer of the retina and are covered by nerve cells, blood vessels, and supporting cells. Most of the light entering the eye is reflected or absorbed before reaching the rods and cones. The light that does reach them triggers electrical impulses along nerve fibers that ultimately reach the brain. (*d*) A neural net used in the vision system of certain robots. Loosely modeled on the human eye, it contains 1920 sensors.

(c)

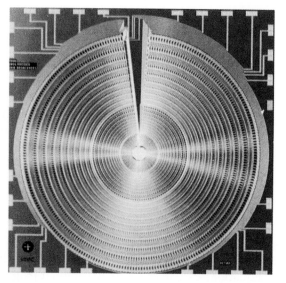

(d)

The eyepiece is placed so that the image formed by the objective falls at the first focal point of the eyepiece. The light from each point on the object thus emerges from the eyepiece as a parallel beam, as if it were coming from a point a great distance in front of the eye. (This is commonly called *viewing the image at infinity.*)

The distance between the second focal point of the objective and the first focal point of the eyepiece is called the **tube length** L. It is typically fixed at approximately 16 cm. The object is placed just outside the first focal point of the objective so that an enlarged image is formed at the first focal point of the eyepiece a distance $L + f_o$ from the objective, where f_o is the focal length of the objective. From Figure 32-52, $\tan \beta = y/f_o = -y'/L$. The lateral magnification of the objective is therefore

$$m_o = \frac{y'}{y} = -\frac{L}{f_o} \qquad\qquad 32\text{-}21$$

The angular magnification of the eyepiece (from Equation 32-20) is

$$M_e = \frac{x_{np}}{f_e}$$

where x_{np} is the near-point distance of the viewer, and f_e is the focal length of the eyepiece. The magnifying power of the compound microscope is the product of the lateral magnification of the objective and the angular magnification of the eyepiece:

$$M = m_o M_e = -\frac{L}{f_o} \frac{x_{np}}{f_e} \qquad\qquad 32\text{-}22$$

MAGNIFYING POWER OF A MICROSCOPE

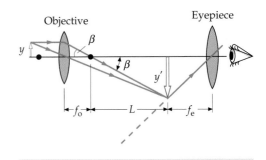

FIGURE 32-52 Schematic diagram of a compound microscope consisting of two positive lenses, the objective of focal length f_o and the ocular, or eyepiece, of focal length f_e. The real image of the object formed by the objective is viewed by the eyepiece, which acts as a simple magnifier. The final image is at infinity.

THE COMPOUND MICROSCOPE   **E X A M P L E   3 2 - 1 6**

A microscope has an objective lens of focal length 1.2 cm and an eyepiece of focal length 2 cm. These lenses are separated by 20 cm. (a) Find the magnifying power if the near point of the viewer is 25 cm. (b) Where should the object be placed if the final image is to be viewed at infinity?

(*a*) 1. The magnifying power is given by Equation 32-22: $M = -\dfrac{L}{f_o} \dfrac{x_{np}}{f_e}$

2. The tube length L is the distance between the lenses minus the focal distances: $L = 20\ \text{cm} - 2\ \text{cm} - 1.2\ \text{cm} = 16.8\ \text{cm}$

3. Substitute this value for L and the given values of x_{np}, f_o, and f_e to calculate M: $M = -\dfrac{L}{f_o} \dfrac{x_{np}}{f_e} = -\dfrac{16.8\ \text{cm}}{1.2\ \text{cm}} \dfrac{25\ \text{cm}}{2\ \text{cm}}$

$$= \boxed{-175}$$

(*b*) 1. Calculate the object distance s in terms of the image distance for the objective s' and the focal length f_o: $\dfrac{1}{s} + \dfrac{1}{s'} = \dfrac{1}{f_o}$

2. From Figure 32-52, the image distance for the image of the objective is $f_o + L$:

$$s' = f_o + L = 1.2 \text{ cm} + 16.8 \text{ cm}$$
$$= 18 \text{ cm}$$

3. Substitute to calculate s:

$$\frac{1}{s} + \frac{1}{18 \text{ cm}} = \frac{1}{1.2 \text{ cm}}$$

$$s = \boxed{1.29 \text{ cm}}$$

REMARKS The object should thus be placed at 1.29 cm from the objective or 0.09 cm outside its first focal point.

*The Telescope

A telescope is used to view objects that are far away and are often large. The telescope works by creating a real image of the object that is much closer than the object. The astronomical telescope, illustrated schematically in Figure 32-53, consists of two positive lenses—an objective lens that forms a real, inverted image and an eyepiece that is used as a simple magnifier to view that image. Because the object is very far away, the image of the objective lies in the focal plane of the objective, and the image distance equals the focal length f_o. The image formed by the objective is much smaller than the object because the object distance is much larger than the focal length of the objective. For example, if we are looking at the moon, the image of the moon formed by the objective is much smaller than the moon itself. The purpose of the objective is not to magnify the object, but to produce an image that is close to us so it can be viewed by the eyepiece. The eyepiece is placed a distance f_e from the image, where f_e is the focal length of the eyepiece, so the final image can be viewed at infinity. Since this image is at the second focal plane of the objective and at the first focal plane of the eyepiece, the objective and eyepiece must be separated by the sum of the focal lengths of the objective and eyepiece, $f_o + f_e$.

The magnifying power of the telescope is the angular magnification θ_e / θ_o, where θ_e is the angle subtended by the final image as viewed through the eyepiece and θ_o is the angle subtended by the object when it is viewed directly by the unaided eye. The angle θ_o is the same as that subtended by the object at the objective shown in Figure 32-53. (The distance from a distant object, such as the moon, to the objective is essentially the same as the distance to the eye.) From this figure, we can see that

$$\tan \theta_o = \frac{y}{s} = -\frac{y'}{f_o} \approx \theta_o$$

where we have used the small-angle approximation $\tan \theta \approx \theta$. The angle θ_e in the figure is that subtended by the image at infinity formed by the eyepiece:

$$\tan \theta_e = \frac{y'}{f_e} \approx \theta_e$$

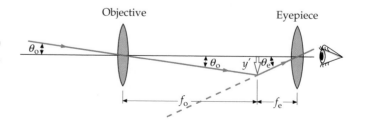

FIGURE 32-53 Schematic diagram of an astronomical telescope. The objective lens forms a real, inverted image of a distant object near its second focal point, which coincides with the first focal point of the eyepiece. The eyepiece serves as a simple magnifier to view the image.

Since y' is negative, θ_e is negative, indicating that the image is inverted. The magnifying power of the telescope is then

$$M = \frac{\theta_e}{\theta_o} = -\frac{f_o}{f_e}$$

32-23

MAGNIFYING POWER OF A TELESCOPE

From Equation 32-23, we can see that a large magnifying power is obtained with an objective of large focal length and an eyepiece of short focal length.

EXERCISE The world's largest refracting telescope is at the Yerkes Observatory of the University of Chicago at Williams Bay, Wisconsin. The telescope's objective has a diameter of 102 cm and a focal length of 19.5 m. The focal length of the eyepiece is 10 cm. What is its magnifying power? (*Answer* -195)

The main consideration with an astronomical telescope is not its magnifying power but its light-gathering power, which depends on the size of the objective. The larger the objective, the brighter the image. Very large lenses without aberrations are difficult to produce. In addition, there are mechanical problems in supporting very large, heavy lenses by their edges. A reflecting telescope (Figure 32-54 and Figure 32-55) uses a concave mirror instead of a lens for its objective. This offers several advantages. For one, a mirror does not produce chromatic aberration. In addition, mechanical support is much simpler, since the mirror weighs far less than a lens of equivalent optical quality, and the mirror can be supported over its entire back surface. In modern earth-based telescopes, the objective mirror consists of several dozen adaptive mirror segments that can be adjusted individually to correct for minute variations in gravitational stress when the telescope is tilted, and to compensate for thermal expansions and contractions and other changes caused by climatic conditions. In addition, they can adjust to nullify the distortions produced by atmospheric fluctuations.

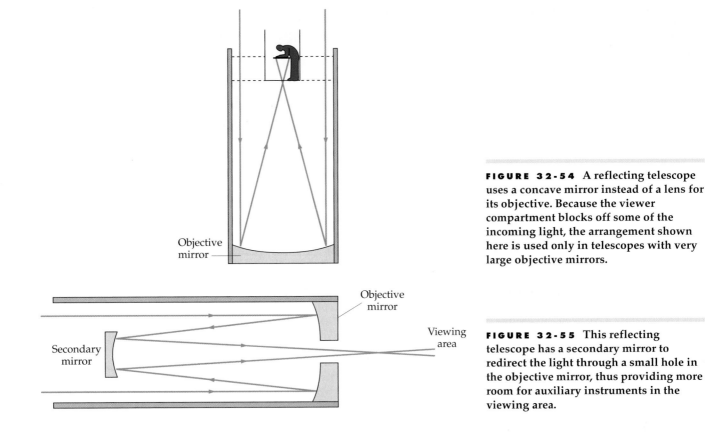

Objective
mirror

Secondary
mirror

Objective
mirror

Viewing
area

FIGURE 32-54 A reflecting telescope uses a concave mirror instead of a lens for its objective. Because the viewer compartment blocks off some of the incoming light, the arrangement shown here is used only in telescopes with very large objective mirrors.

FIGURE 32-55 This reflecting telescope has a secondary mirror to redirect the light through a small hole in the objective mirror, thus providing more room for auxiliary instruments in the viewing area.

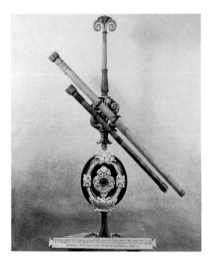

(a)

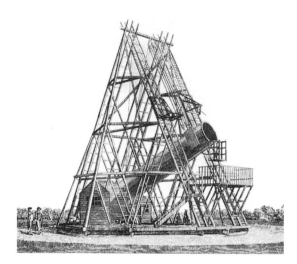

(b)

(c)

(d)

(e)

Astronomy at optical wavelengths began with Galileo approximately 400 years ago. In this century, astronomers began to explore the electromagnetic spectrum at other wavelengths; beginning with radio astronomy in the 1940s, satellite-based X-ray astronomy in the early 1960s, and more recently, ultraviolet, infrared, and gamma-ray astronomy. (*a*) Galileo's seventeenth-century telescope, with which he discovered mountains on the moon, sunspots, Saturn's rings, and the bands and moons of Jupiter. (*b*) An engraving of the reflector telescope built in the 1780s and used by the great astronomer Friedrich Wilhelm Herschel, who was the first to observe galaxies outside our own. (*c*) Because it is difficult to make large, flaw-free lenses, refractor telescopes like this 91.4-cm telescope at Lick Observatory have been superseded in light-gathering power by reflector telescopes. (*d*) The great astronomer Edwin Powell Hubble, who discovered the apparent expansion of the universe, is shown seated in the observer's cage of the 5.08-m Hale reflecting telescope, which is large enough for the observer to sit at the prime focus itself. (*e*) This 10-m optical reflector at the Whipple Observatory in southern Arizona is the largest instrument designed exclusively for use in gamma-ray astronomy. High-energy gamma rays of unknown origin strike the upper atmosphere and create cascades of particles. Among these particles are high-energy electrons that emit Cerenkov radiation observable from the ground. According to one hypothesis, high-energy gamma rays are emitted as matter is accelerated toward ultradense rotating stars called pulsars.

(a)

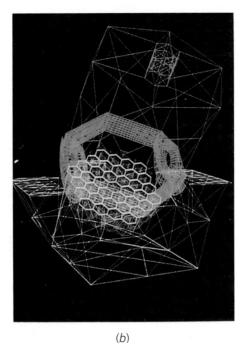

(b)

(c)

(*a*) The Keck Observatory, atop the inactive volcano of Mauna Kea, Hawaii, houses the world's largest optical telescope. The clear, dry air and lack of light pollution make the remote heights of Mauna Kea an ideal site for astronomical observations. (*b*) The Keck telescope is composed of 36 hexagonal mirror segments performing together as if they were a single mirror 10-m wide—roughly twice as large as the largest single-mirror telescope presently in operation. (*c*) Beneath each Keck mirror is a system of computer-controlled sensors and motor-driven actuators that can continuously vary the mirror's shape. These variations, which are sensitive to within 100 nm, enable the system to correct for variations in the alignments of the segments due to minute variations in gravitational stress when the telescope is tilted and to compensate for thermal expansions and contractions and fluctuations caused by gusts of wind on the mountaintop.

The Hubble Space Telescope is high above the atmospheric turbulence that limits the ability of ground-based telescopes to resolve images at optical wavelengths.

Topic	Relevant Equations and Remarks	
1. Virtual and Real Images and Objects		
Images	An image is real if actual light rays converge to each image point. This can occur in front of a mirror, or on the refracted-light side of a thin lens or refracting surface. An image is virtual if only extensions of the actual light rays converge to each image point. This can occur behind a mirror or on the incident-light side of a lens or refracting surface.	
Virtual object	An object is real if actual light rays diverge from each object point. This can occur only on the incident-light side of a mirror, lens, or refracting surface. A real object is either an actual object or a real image. An object is virtual if only extensions of actual light rays diverge from each object point. This can occur only behind a mirror or on the refracted-light side of a lens or refracting surface.	
2. Spherical Mirrors		
Focal length	The focal length is the image distance when the object is at infinity, so the incident light is parallel to the axis:	
Mirror equation (for locating an image)	$$\frac{1}{s} + \frac{1}{s'} = \frac{1}{f}$$	32-4
	where	
	$$f = \frac{r}{2}$$	32-3
Lateral magnification	$$m = \frac{y'}{y} = -\frac{s'}{s}$$	32-5
Ray diagrams	Images can be located by a ray diagram using any two paraxial rays. The parallel, focal, and radial rays are the easiest to draw: 1. The parallel ray, drawn parallel to the axis, is reflected through the focal point. 2. The focal ray, drawn through the focal point, is reflected parallel to the axis. 3. The radial ray, drawn through the center of curvature, strikes the mirror perpendicular to its surface and is thus reflected back on itself.	
Sign conventions for reflection	1. s is positive if the object is on the incident-light side of the mirror. 2. s' is positive if the image is on the reflected-light side of the mirror. 3. r (and f) is positive if the mirror is concave so the center of curvature is on the reflected-light side of the mirror.	
3. Images Formed by Refraction		
Refraction at a single surface	$$\frac{n_1}{s} + \frac{n_2}{s'} = \frac{n_2 - n_1}{r}$$	32-6
	where n_1 is the index of refraction of the medium on the incident-light side of the surface.	
Magnification	$$m = \frac{y'}{y} = -\frac{n_1 s'}{n_2 s}$$	32-7

| Sign conventions for refraction | 1. s is positive for objects on the incident-light side of the surface.
2. s' is positive for images on the refracted-light side of the surface.
3. r is positive if the center of curvature is on the refracted-light side of the surface. | |

4. Thin Lenses

Focal length (lens-maker's equation)	$$\frac{1}{f} = (n-1)\left(\frac{1}{r_1} - \frac{1}{r_2}\right)$$	32-11
	A positive lens ($f > 0$) is a converging lens (like a double convex lens). A negative lens ($f < 0$) is a diverging lens (like a double concave lens).	
First and second focal points	Incident rays parallel to the axis emerge directed either toward or away from the *first focal point F'*. Incident rays directed either toward or away from the *second focal point F* emerge parallel with the axis.	
Power	$$P = \frac{1}{f}$$	32-13
Thin-lens equation (for locating image)	$$\frac{1}{s} + \frac{1}{s'} = \frac{1}{f}$$	32-12
Magnification	$$m = \frac{y'}{y} = -\frac{s'}{s}$$	32-14
Ray diagrams	Images can be located by a ray diagram using any two paraxial rays. The parallel, central, and focal rays are the easiest to draw: 1. The parallel ray, drawn parallel to the axis, emerges directed toward (or away from) the second focal point of the lens. 2. The central ray, drawn through the center of the lens, is not deflected. 3. The focal ray, drawn through (or toward) the first focal point, emerges parallel to the axis.	
Sign conventions for lenses	The sign conventions are the same as for refraction at a spherical surface.	

5. *Aberrations

Blurring of the image of a single object point is called aberration. Spherical aberration results from the fact that a spherical surface focuses only paraxial rays (those that travel close to the axis) at a single point. Nonparaxial rays are focused at nearby points depending on the angle made with the axis. Spherical aberration can be reduced by blocking the rays farthest from the axis. This, of course, reduces the amount of light reaching the image.

Chromatic aberration, which occurs with lenses but not mirrors, results from the variation in the index of refraction with wavelength. Lens aberrations are most commonly reduced by using a series of lens elements.

6. *The Eye

The cornea–lens system of the eye focuses light on the retina, where it is sensed by the rods and cones that send information along the optic nerve to the brain. When the eye is relaxed, the focal length of the cornea–lens system is about 2.5 cm, which is the distance to the retina. When objects are brought near the eye, the lens changes shape to decrease the overall focal length so that the image remains focused on the retina. The closest distance for which the image can be focused on the retina is called the near point, typically about 25 cm. The apparent size of an object depends on the size of the image on the retina. The closer the object, the larger the image on the retina and therefore the larger the apparent size of the object.

7. *The Simple Magnifier

A simple magnifier consists of a lens with a positive focal length that is smaller than the near point.

Magnifying power (angular magnification)	$M = \dfrac{\theta}{\theta_0} = \dfrac{x_{np}}{f}$	32-20

8. *The Compound Microscope

The compound microscope is used to look at very small objects that are nearby. It consists of two converging lenses (or lens systems), an objective, and an ocular or eyepiece. The object to be viewed is placed just outside the focal point of the objective, which forms an enlarged image of the object at the focal plane of the eyepiece. The eyepiece acts as a simple magnifier to view the final image.

Magnifying power (angular magnification)	$M = m_o M_e = -\dfrac{L}{f_o} \dfrac{x_{np}}{f_e}$	32-22

where L is the tube length, the distance between the second focal point of the objective and the first focal point of the eyepiece.

9. *The Telescope

The telescope is used to view objects that are far away. The objective of the telescope forms a real image of the object that is much smaller than the object but much closer. The eyepiece is then used as a simple magnifier to view the image. A reflecting telescope uses a mirror for its objective.

Magnifying power (angular magnification)	$M = \dfrac{\theta_e}{\theta_o} = -\dfrac{f_o}{f_e}$	32-23

PROBLEMS

• Single-concept, single-step, relatively easy

•• Intermediate-level, may require synthesis of concepts

••• Challenging

SSM Solution is in the *Student Solutions Manual*

iSOLVE Problems available on iSOLVE online homework service

iSOLVE ✓ These "Checkpoint" online homework service problems ask students additional questions about their confidence level, and how they arrived at their answer.

In a few problems, you are given more data than you actually need; in a few other problems, you are required to supply data from your general knowledge, outside sources, or informed estimates.

Conceptual Problems

1 • Can a virtual image be photographed?

2 • Suppose each axis of a coordinate system, like the one shown in Figure 32-4, is painted a different color. One photograph is taken of the coordinate system and another is taken of its image in a plane mirror. Is it possible to tell that one of the photographs is of a mirror image, rather than both being photographs of the real coordinate system from different angles?

3 •• iSOLVE True or False

(a) The virtual image formed by a concave mirror is always smaller than the object.

(b) A concave mirror always forms a virtual image.

(c) A convex mirror never forms a real image of a real object.

(d) A concave mirror never forms an enlarged real image of an object.

4 •• SSM Under what condition will a concave mirror produce (a) an upright image, (b) a virtual image, (c) an image smaller than the object, and (d) an image larger than the object?

5 •• Answer Problem 4 for a convex mirror.

6 •• Convex mirrors are often used for rearview mirrors on cars and trucks to give a wide-angle view. Below the mirror is written, "Warning, objects are closer than they appear." Yet, according to a ray diagram, such as the diagram shown in Figure 32-19, the image distance for distant objects is much smaller than the object distance. Why then do they appear more distant?

7 •• As an object is moved from a great distance toward the focal point of a concave mirror, the image moves from (a) a great distance toward the focal point and is always real. (b) the focal point to a great distance from the mirror and is always real. (c) the focal point toward the center of curvature of the mirror and is always real. (d) the focal point to a great distance from the mirror and changes from a real image to a virtual image.

8 • A bird above the water is viewed by a scuba diver submerged beneath the water's surface directly below the bird. Does the bird appear to the diver to be closer to or farther from the surface than it actually is?

9 • **SSM** Under what conditions will the focal length of a thin lens be (a) positive and (b) negative? Consider both the case where the index of refraction of the lens is greater than and less than the surrounding medium.

10 • The focal length of a simple lens is different for different colors of light. Why?

11 •• An object is placed 40 cm from a lens of focal length −10 cm. The image is (a) real, inverted, and diminished. (b) real, inverted, and enlarged. (c) virtual, inverted, and diminished. (d) virtual, upright, and diminished. (e) virtual, upright, and enlarged.

12 •• If a real object is placed just inside the focal point of a converging lens, the image is (a) real, inverted, and enlarged. (b) virtual, upright, and diminished. (c) virtual, upright, and enlarged. (d) real, inverted, and diminished.

13 • Both the eye and the camera work by forming real images, the eye's image forming on the retina and the camera's image forming on the film. Explain the difference between the ways in which these two systems accommodate objects located at different object distances and still keep a focused image.

14 • **SSM** If an object is placed 25 cm from the eye of a farsighted person who does not wear corrective lenses, a sharp image is formed (a) behind the retina, and the corrective lens should be convex. (b) behind the retina, and the corrective lens should be concave. (c) in front of the retina, and the corrective lens should be convex. (d) in front of the retina, and the corrective lens should be concave.

15 •• Myopic (nearsighted) persons sometimes claim to see better under water without corrective lenses. Why? (a) The accommodation of the eye's lens is better under water. (b) Refraction at the water–cornea interface is less than at the air–cornea interface. (c) Refraction at the water–cornea interface is greater than at the air–cornea interface. (d) No reason; the effect is only an illusion and not really true.

16 •• A nearsighted person who wears corrective lenses would like to examine an object at close distance. Identify the correct statement. (a) The corrective lenses give an enlarged image and should be worn while examining the object. (b) The corrective lenses give a reduced image of the object and should be removed. (c) The corrective lenses result in a magnification of unity; it does not matter whether they are worn or removed.

17 • **SSM** The image of a real object formed by a convex mirror (a) is always real and inverted. (b) is always virtual and enlarged. (c) may be real. (d) is always virtual and diminished.

18 • The image of a real object formed by a converging lens (a) is always real and inverted. (b) is always virtual and enlarged. (c) may be real. (d) is always virtual and diminished.

19 • The glass of a converging lens has an index of refraction of 1.6. When the lens is in air, its focal length is 30 cm. If the lens is immersed in water, its focal length will be (a) greater than 30 cm. (b) less than 30 cm. (c) the same as before, 30 cm. (d) negative.

20 •• True or false:
(a) A virtual image cannot be displayed on a screen.
(b) A negative image distance implies that the image is virtual.
(c) All rays parallel to the axis of a spherical mirror are reflected through a single point.
(d) A diverging lens cannot form a real image from a real object.
(e) The image distance for a positive lens is always positive.

21 • **SSM** Explain the following statement: A microscope is an object magnifier, but a telescope is an angle magnifier.

Estimation and Approximation

22 •• The lens-maker's equation contains three design parameters, the index of refraction of the lens and the radius of curvature of its two surfaces. Thus, there are many ways to design a lens with a particular focal length. Use the lens-maker's equation to design three different thin converging lenses, each with a focal length of 27 cm and each made from glass with an index of refraction of 1.6. Draw a sketch of each of your designs.

23 •• Repeat Problem 22, but for a diverging lens of focal length −27 cm.

24 •• **SSM** Estimate the maximum value that could be usefully obtained for the magnifying power of a simple magnifier, using Equation 32-20. (*Hint: Think about the smallest focal length lens that could be made from glass and still be used as a magnifier.*)

Plane Mirrors

25 • The image of the point object P in Figure 32-56 is viewed by an eye, as shown. Draw a bundle of rays from the object that reflect from the mirror and enter the eye. For this object position and mirror, indicate the region of space in which the eye can be positioned and still see the image.

FIGURE 32-56 Problem 25

26 • A person 1.62 m tall wants to be able to see her full image in a plane mirror. (a) What must be the minimum height of the mirror? (b) How far above the floor should the mirror be placed, assuming that the top of the person's head is 15 cm above her eye level? Draw a ray diagram.

27 • **SSM** Two plane mirrors make an angle of 90°. The light from an object point that is arbitrarily positioned in front of the mirrors produces images at three locations. For each image location, draw two rays from the object that, upon one or two reflections, appear to come from the image location.

28 • (a) Two plane mirrors make an angle of 60° with each other. Draw a sketch to show the location of all the images formed of a point object on the bisector of the angle between the mirrors. (b) Repeat for an angle of 120°.

29 •• When two plane mirrors are parallel, such as on opposite walls in a barber shop, multiple images arise because each image in one mirror serves as an object for the other mirror. A point object is placed between parallel mirrors separated by 30 cm. The object is 10 cm in front of the left mirror and 20 cm in front of the right mirror. (*a*) Find the distance from the left mirror to the first four images in that mirror. (*b*) Find the distance from the right mirror to the first four images in that mirror.

Spherical Mirrors

30 •• SSM A concave spherical mirror has a radius of curvature of 24 cm. Draw ray diagrams to locate the image (if one is formed) for an object at a distance of (*a*) 55 cm, (*b*) 24 cm, (*c*) 12 cm, and (*d*) 8 cm from the mirror. For each case, state whether the image is real or virtual; upright or inverted; and enlarged, reduced, or the same size as the object.

31 • Use the mirror equation (Equation 32-4) to locate and describe the images for the object distances and mirror of Problem 30.

32 •• Repeat Problem 30 for a convex mirror with the same radius of curvature.

33 • Use the mirror equation (Equation 32-4) to locate and describe the images for the object distances and convex mirror of Problem 32.

34 • Show that a convex mirror cannot form a real image of a real object, no matter where the object is placed, by showing that s' is always negative for a positive s.

35 • SSM ISOLVE✓ A dentist wants a small mirror that will produce an upright image with a magnification of 5.5 when the mirror is located 2.1 cm from a tooth. (*a*) What should the radius of curvature of the mirror be? (*b*) Should the mirror be concave or convex?

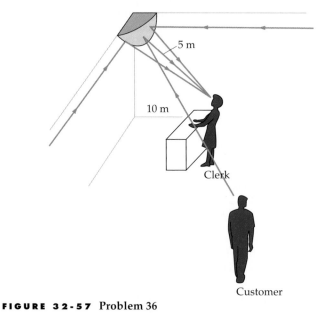

FIGURE 32-57 Problem 36

36 •• ISOLVE✓ Convex mirrors are used in stores to provide a wide angle of surveillance for a reasonable mirror size. The mirror shown in Figure 32-57 allows a clerk 5 m away

from the mirror to survey the entire store. It has a radius of curvature of 1.2 m. (*a*) If a customer is 10 m from the mirror, how far from the mirror surface is his image? (*b*) Is the image in front of or behind the mirror? (*c*) If the customer is 2 m tall, how tall is his image?

37 •• ISOLVE A certain telescope uses a concave spherical mirror of radius 8 m. Find the location and diameter of the image of the moon formed by this mirror. The moon has a diameter of 3.5×10^6 m and is 3.8×10^8 m from the earth.

38 •• A concave spherical mirror has a radius of curvature of 6 cm. A point object is on the axis 9 cm from the mirror. Construct a precise ray diagram showing rays from the object that make angles of 5°, 10°, 30°, and 60° with the axis, which strike the mirror and are reflected back across the axis. (Use a compass to draw the mirror, and use a protractor to measure the angles needed to find the reflected rays.) What is the spread δx of the points where these rays cross the axis?

39 •• SSM A concave mirror has a radius of curvature 6 cm. Draw rays parallel to the axis at 0.5 cm, 1 cm, 2 cm, and 4 cm above the axis, and find the points at which the reflected rays cross the axis. (Use a compass to draw the mirror and a protractor to find the angle of reflection for each ray.) (*a*) What is the spread δx of the points where these rays cross the axis? (*b*) By what percentage could this spread be reduced if the edge of the mirror were blocked off so that parallel rays more than 2 cm from the axis could not strike the mirror?

40 •• ISOLVE An object located 100 cm from a concave mirror forms a real image 75 cm from the mirror. The mirror is then turned around so that its convex side faces the object. The mirror is moved so that the image is now 35 cm behind the mirror. How far was the mirror moved? Was it moved toward the object or away from the object?

41 •• Parallel light from a distant object strikes the large mirror shown in Figure 32-58 at $r = 5$ m and is reflected by the small mirror that is 2 m from the large mirror. The small mirror is actually spherical, not planar as shown. The light is focused at the vertex of the large mirror. (*a*) What is the radius of curvature of the small mirror? (*b*) Is the mirror convex or concave?

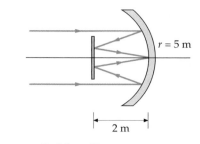

FIGURE 32-58 Problem 41

Images Formed by Refraction

42 • A sheet of paper with writing on it is protected by a thick glass plate having an index of refraction of 1.5. If the plate is 2 cm thick, at what distance beneath the top of the plate does the writing appear when it is viewed from directly overhead?

43 • ☑️✓ A fish is 10 cm from the front surface of a fish bowl of radius 20 cm. (*a*) Where does the fish appear to be to someone viewing the fish from in front of the bowl? (*b*) Where does the fish appear to be when it is 30 cm from the front surface of the bowl?

44 •• SSM A very long glass rod of 3.5-cm diameter has one end ground to a convex spherical surface of radius 7.2 cm. Its index of refraction is 1.5. (*a*) A point object in air is on the axis of the rod 35 cm from the surface. Find the image and state whether the image is real or virtual. Repeat (*b*) for an object 6.5 cm from the surface and (*c*) for an object very far from the surface. Draw a ray diagram for each case.

45 •• At what distance from the glass rod of Problem 44 should the object be placed, so that the light rays in the rod are parallel? Draw a ray diagram for this situation.

46 •• Repeat Problem 44 for a glass rod with a concave hemispherical surface of radius −7.2 cm.

47 •• Repeat Problem 44 when the glass rod and the objects are immersed in water.

48 •• Repeat Problem 44 for a glass rod with a concave hemispherical surface of radius −7.5 cm when the glass rod and the objects are immersed in water.

49 •• SSM ☑️ A glass rod 96 cm long with an index of refraction of 1.6 has its ends ground to convex spherical surfaces of radii 8 cm and 16 cm. A point object is in air on the axis of the rod 20 cm from the end with the 8-cm radius. (*a*) Find the image distance due to refraction at the first surface. (*b*) Find the final image due to refraction at both surfaces. (*c*) Is the final image real or virtual?

50 •• Repeat Problem 49 for a point object in air on the axis of the glass rod 20 cm from the end with the 16-cm radius.

Thin Lenses

51 • The following thin lenses are made of glass with an index of refraction of 1.5. Make a sketch of each lens, and find its focal length in air: (*a*) double convex, $r_1 = 15$ cm and $r_2 = -26$ cm; (*b*) plano-convex, $r_1 = \infty$ and $r_2 = -15$ cm; (*c*) double concave, $r_1 = -15$ cm and $r_2 = +15$ cm; and (*d*) plano-concave, $r_1 = \infty$ and $r_2 = +26$ cm.

52 • Find the focal length of a glass lens of index of refraction 1.62 that has a concave surface with radius of magnitude 100 cm and a convex surface with a radius of magnitude 40 cm.

53 • SSM A double concave lens of index of refraction 1.45 has radii of magnitudes 30 cm and 25 cm. An object is located 80 cm to the left of the lens. Find (*a*) the focal length of the lens, (*b*) the location of the image, and (*c*) the magnification of the image. (*d*) Is the image real or virtual? Is the image upright or inverted?

54 • ☑️ The following thin lenses are made of glass of index of refraction 1.6. Make a sketch of each lens, and find its focal length in air: (*a*) $r_1 = 20$ cm and $r_2 = 10$ cm, (*b*) $r_1 = 10$ cm and $r_2 = 20$ cm, and (*c*) $r_1 = -10$ cm and $r_2 = -20$ cm.

55 • SSM ☑️✓ An object 3 cm high is placed 25 cm in front of a thin lens of power 10 D. Draw a precise ray diagram to find the position and the size of the image, and check your results using the thin-lens equation.

56 • Repeat Problem 55 for an object 1.5 cm high that is placed 20 cm in front of a thin lens of power 10 D.

57 • Repeat Problem 55 for an object 1.5 cm high that is placed 20 cm in front of a thin lens of power −10 D.

58 •• (*a*) What is meant by a negative object distance? How can a negative object distance occur? Find the image distance and the magnification and state whether the image is virtual or real and upright or inverted for a thin lens in air when (*b*) $s = -20$ cm, $f = +20$ cm and (*c*) $s = -10$ cm, $f = -30$ cm. Draw a ray diagram for each of these cases.

59 •• SSM ☑️✓ Two converging lenses, each of focal length 10 cm, are separated by 35 cm. An object is 20 cm to the left of the first lens. (*a*) Find the position of the final image using both a ray diagram and the thin-lens equation. (*b*) Is the image real or virtual? Is the image upright or inverted? (*c*) What is the overall lateral magnification of the image?

60 •• Rework Problem 59 for a second lens that is a diverging lens of focal length −15 cm.

61 •• (*a*) Show that to obtain a magnification of magnitude *m* with a converging thin lens of focal length *f*, the object distance must be given by $s = (m - 1)f/m$. (*b*) A camera lens with a 50-mm focal length is used to take a picture of a person 1.75 m tall. How far from the camera should the person stand so that the image size is 24 mm?

62 •• A converging lens has a focal length of $f = 12$ cm. (*a*) Using a spreadsheet program or graphing calculator, plot the image distance s' as a function of the object distance s, for values of s ranging from $s = 1.1f$ to $s = 10f$. (*b*) On the same graph used in Part (*a*), but using a different *y* axis, plot the magnification of the lens as a function of the object distance s. (*c*) What type of image is produced for this range of object distances, real or virtual, upright or inverted? (*d*) Discuss the significance of any asymptotic limits your graph has.

63 •• A converging lens has a focal length of $f = 12$ cm. (*a*) Using a spreadsheet program or graphing calculator, plot the image distance s' as a function of the object distance s, for values of s ranging from $s = 0.01f$ to $s = 0.9f$. (*b*) On the same graph used in Part (*a*), but using a different *y* axis, plot the magnification of the lens as a function of the object distance s. (*c*) What type of image is produced for this range of object distances, real or virtual, upright or inverted? (*d*) Discuss the significance of any asymptotic limits your graph has.

64 •• SSM An object is 15 cm in front of a positive lens of focal length 15 cm. A second positive lens of focal length 15 cm is 20 cm from the first lens. Find the final image and draw a ray diagram.

65 •• Rework Problem 64 for a second lens with a focal length of −15 cm.

66 ••• In a convenient form of the thin-lens equation used by Newton, the object and image distances are measured from the focal points. Show that if $x = s - f$ and $x' = s' - f$, the thin-lens equation can be written as $xx' = f^2$, and the lateral magnification is given by $m = -x'/f = -f/x$. Indicate x and x' on a sketch of a lens.

67 ••• In *Bessel's method* for finding the focal length f of a lens, an object and a screen are separated by distance D, where $D > 4f$. It is then possible to place the lens at either of two locations, both between the object and the screen, so that there is an image of the object on the screen, in one case magnified and in the other case reduced. Show that if the distance between the two lens locations is given by L, that

$$f = \frac{D^2 - L^2}{4D}$$

(*Hint: Refer to Figure 32-59.*) The two lens locations are such that the object distance with the lens in the one setting is equal to the image distance with the lens in the other setting and vice versa.

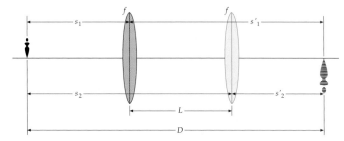

FIGURE 32-59 Problems 67 and 68

68 •• An optician uses *Bessel's method* to find the focal length of a lens, as described in Problem 67. The object-to-image distance was set at 1.7 m. The position of the lens was then adjusted to get a sharp image on the screen. A second image was found when the lens was moved a distance of 72 cm. (*a*) Using the result from Problem 67, find the focal length of the lens. (*b*) What were the two locations of the lens with respect to the object?

69 ••• An object is 17.5 cm to the left of a lens of focal length 8.5 cm. A second lens of focal length -30 cm is 5 cm to the right of the first lens. (*a*) Find the distance between the object and the final image formed by the second lens. (*b*) What is the overall magnification? (*c*) Is the final image real or virtual? Is the final image upright or inverted?

*Aberrations

70 • [SSM] Chromatic aberration is a common defect of (*a*) concave and convex lenses. (*b*) concave lenses only. (*c*) concave and convex mirrors. (*d*) all lenses and mirrors.

71 • True or false:

(*a*) Aberrations occur only for real images.
(*b*) Chromatic aberration does not occur with mirrors.

72 • A double convex lens of radii $r_1 = +10$ cm and $r_2 = -10$ cm is made from glass with indexes of refraction of 1.53 for blue light and 1.47 for red light. Find the focal length of this lens for (*a*) red light and (*b*) blue light.

*The Eye

73 •• [SSM] The Model Eye I: A simple model for the eye is a lens with variable power P located a fixed distance d in front of a screen, with the space between the lens and the screen filled by air. Refer to Figure 32-60. The "eye" can focus for all values of s such that $x_{np} \leq s \leq x_{fp}$. This "eye" is said to be normal if it can focus on very distant objects. (*a*) Show that for a normal "eye," the minimum value of P is

$$P_{min} = \frac{1}{d}$$

(*b*) Show that the maximum value of P is

$$P_{max} = \frac{1}{x_{np}} + \frac{1}{d}$$

(*c*) The difference $A = P_{max} - P_{min}$ is called the accommodation. Find the minimum power and accommodation for a model eye with $d = 2.5$ cm and $x_{np} = 25$ cm.

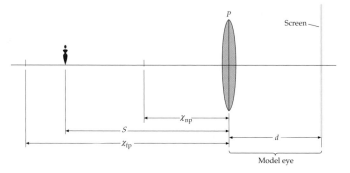

FIGURE 32-60 Problems 73, 74, and 75

74 •• The Model Eye II: In an eye that exhibits nearsightedness, the eye cannot focus on distant objects. Refer to Figure 32-60 and to Problem 73. (*a*) Show that for a nearsighted model eye capable of focusing out to a maximum distance x_{fp}, the minimum value of P is greater than that of a normal eye and is given by

$$P_{min} = \frac{1}{x_{fp}} + \frac{1}{d}$$

(*b*) To correct for nearsightedness, a contact lens may be placed directly in front of the model-eye's lens. What power contact lens would be needed to correct the vision of a nearsighted model eye with $x_{fp} = 50$ cm?

75 •• The Model Eye III: In an eye that exhibits farsightedness, the eye may be able to focus on distant objects but cannot focus on close objects. Refer to Figure 32-60 and to Problem 73. (*a*) Show that for a farsighted model eye capable of focusing only as close as a distance x'_{np}, the maximum value of P is given by

$$P_{max} = \frac{1}{x'_{np}} + \frac{1}{d}$$

(*b*) Show that compared to a model eye capable of focusing as close as a distance x_{np} (where $x_{np} < x'_{np}$), the maximum power of the farsighted lens is too small by

$$\frac{1}{x_{np}} - \frac{1}{x'_{np}}$$

(c) What power contact lens would be needed to correct the vision of a farsighted model eye, with x_{np} = 150 cm, so that the eye may focus on objects as close as 15 cm?

76 • ISOLVE✔ Suppose the eye were designed like a camera with a lens of fixed focal length f = 2.5 cm that could move toward or away from the retina. Approximately how far would the lens have to move to focus the image of an object 25 cm from the eye onto the retina? (*Hint:* Find the distance from the retina to the image behind it for an object at 25 cm.)

77 • ISOLVE✔ Find the change in the focal length of the eye when an object originally at 3 m is brought to 30 cm from the eye.

78 • A farsighted person requires lenses with a power of 1.75 D to read comfortably from a book that is 25 cm from the eye. What is that person's near point without the lenses?

79 • SSM If two point objects close together are to be seen as two distinct objects, the images must fall on the retina on two different cones that are not adjacent. That is, there must be an unactivated cone between them. The separation of the cones is about 1 μm. Model the eye as a uniform 2.5-cm-diameter sphere with a refractive index of 1.34. (a) What is the smallest angle the two points can subtend? (See Figure 32-61.) (b) How close together can two points be if they are 20 m from the eye?

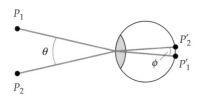

FIGURE 32-61 Problem 79

80 •• A person with a near point of 80 cm needs to read from a computer screen that is 45 cm from her eye. (a) Find the focal length of the lenses in reading glasses that will produce an image of the screen at 80 cm from her eye. (b) What is the power of the lenses?

81 •• A nearsighted person cannot focus clearly on objects that are more distant than 225 cm from her eye. What power lenses are required for her to see distant objects clearly?

82 •• Since the index of refraction of the lens of the eye is not very different from that of the surrounding material, most of the refraction takes place at the cornea, where n changes abruptly from 1.0 in air to approximately 1.4. Assuming the cornea to be a homogeneous sphere with an index of refraction of 1.4, calculate the cornea's radius if it focuses parallel light on the retina a distance 2.5 cm away. Do you expect your result to be larger or smaller than the actual radius of the cornea?

83 •• The near point of a certain person is 80 cm. Reading glasses are prescribed so that he can read a book at 25 cm from his eye. The glasses are 2 cm from the eye. What diopter lens should be used in the glasses?

84 ••• At age 45, a person is fitted for reading glasses of power 2.1 D in order to read at 25 cm. By the time she reaches 55, she discovers herself holding her newspaper at a distance of 40 cm in order to see it clearly with her glasses on. (a) Where was her near point at age 45? (b) Where is her near point at age 55? (c) What power is now required for the lenses of her reading glasses so that she can again read at 25 cm? (Assume the glasses are 2.2 cm from her eyes.)

*The Simple Magnifier

85 • SSM ISOLVE A person with a near-point distance of 30 cm uses a simple magnifier of power 20 D. What is the magnification obtained if the final image is at infinity?

86 • A person with a near-point distance of 25 cm wishes to obtain a magnifying power of 5 with a simple magnifier. What should be the focal length of the lens used?

87 • What is the magnifying power of a lens of focal length 7 cm when the image is viewed at infinity by a person whose near point is at 35 cm?

88 •• A lens of focal length 6 cm is used as a simple magnifier with the image at infinity by one person whose near point is 25 cm and by another person whose near point is 40 cm. What is the effective magnifying power of the lens for each person? Compare the size of the image on the retina when each person looks at the same object with the magnifier.

89 •• A botanist examines a leaf using a convex lens of power 12 D as a simple magnifier. What is the expected angular magnification if (a) the final image is at infinity and (b) the final image is at 25 cm?

90 •• SSM (a) Show that if the final image of a simple magnifier is to be at the near point of the eye rather than at infinity, the angular magnification is given by

$$M = \frac{x_{np}}{f} + 1$$

(b) Find the magnification of a 20-D lens for a person with a near point of 30 cm if the final image is at the near point. Draw a ray diagram for this situation.

91 •• Show that when the image of a simple magnifier is viewed at the near point, the lateral and angular magnification of the magnifier are equal.

*The Microscope

92 •• ISOLVE✔ A microscope objective has a focal length of 17 mm. It forms an image at 16 cm from its second focal point. (a) How far from the objective is the object located? (b) What is the magnifying power for a person whose near point is at 25 cm if the focal length of the eyepiece is 51 mm?

93 •• SSM A microscope has an objective of focal length 8.5 mm and an eyepiece that gives an angular magnification of 10 for a person whose near point is 25 cm. The tube length is 16 cm. (a) What is the lateral magnification of the objective? (b) What is the magnifying power of the microscope?

94 •• ![ISOLVE]✔ A crude, symmetric handheld microscope consists of two converging 20-D lenses fastened in the ends of a tube 30 cm long. (*a*) What is the tube length of this microscope? (*b*) What is the lateral magnification of the objective? (*c*) What is the magnifying power of the microscope? (*d*) How far from the objective should the object be placed?

95 •• [SSM] A compound microscope has an objective lens with a power of 45 D and an eyepiece with a power of 80 D. The lenses are separated by 28 cm. Assuming that the final image is formed 25 cm from the eye, what is the magnifying power?

96 ••• A microscope has a magnifying power of 600, and an eyepiece of angular magnification of 15. The objective lens is 22 cm from the eyepiece. Without making any approximations, calculate (*a*) the focal length of the eyepiece, (*b*) the location of the object so that it is in focus for a normal relaxed eye, and (*c*) the focal length of the objective lens.

*The Telescope

97 • ![ISOLVE]✔ A simple telescope has an objective with a focal length of 100 cm and an eyepiece of focal length 5 cm. It is used to look at the moon, which subtends an angle of about 0.009 rad. (*a*) What is the diameter of the image formed by the objective? (*b*) What angle is subtended by the final image at infinity? (*c*) What is the magnifying power of the telescope?

98 • The objective lens of the refracting telescope at the Yerkes Observatory has a focal length of 19.5 m. When it is used to look at the moon, which subtends an angle of about 0.009 rad, what is the diameter of the image of the moon formed by the objective?

99 •• [SSM] The 200-in (5.1-m) mirror of the reflecting telescope at Mt. Palomar has a focal length of 1.68 m. (*a*) By what factor is the light-gathering power increased over the 40-in (1.016-m) diameter refracting lens of the Yerkes Observatory telescope? (*b*) If the focal length of the eyepiece is 1.25 cm, what is the magnifying power of this telescope?

100 •• ![ISOLVE] An astronomical telescope has a magnifying power of 7. The two lenses are 32 cm apart. Find the focal length of each lens.

101 •• A disadvantage of the astronomical telescope for terrestrial use (e.g., at a football game) is that the image is inverted. A Galilean telescope uses a converging lens as its objective, but a diverging lens as its eyepiece. The image formed by the objective is behind the eyepiece at its focal point so that the final image is virtual, upright, and at infinity. (*a*) Show that the magnifying power is $M = -f_o/f_e$, where f_o is the focal length of the objective and f_e is that of the eyepiece (which is negative). (*b*) Draw a ray diagram to show that the final image is indeed virtual, upright, and at infinity.

102 •• A Galilean telescope (see Problem 101) is designed so that the final image is at the near point, which is 25 cm (rather than at infinity). The focal length of the objective is 100 cm and that of the eyepiece is −5 cm. (*a*) If the object distance is 30 m, where is the image of the objective? (*b*) What is the object distance for the eyepiece so that the final image is at the near point? (*c*) How far apart are the lenses? (*d*) If the object height is 1.5 m, what is the height of the final image? What is the angular magnification?

103 ••• If you look into the wrong end of a telescope, that is, into the objective, you will see distant objects reduced in size. For a refracting telescope with an objective of focal length 2.25 m and an eyepiece of focal length 1.5 cm, by what factor is the angular size of the object reduced?

General Problems

104 • Show that a diverging lens can never form a real image from a real object. (*Hint: Show that s' is always negative.*)

105 • [SSM] A camera uses a positive lens to focus light from an object onto film. Unlike the eye, the camera lens has a fixed focal length, but the lens itself can be moved slightly to vary the image distance to the image on the film. A telephoto lens has a focal length of 200 mm. By how much must the lens move to change from focusing on an object at infinity to an object at a distance of 30 m?

106 • A wide-angle lens of a camera has a focal length of 28 mm. By how much must the lens move to change from focusing on an object at infinity to an object at a distance of 5 m? (See Problem 105.)

107 • A thin converging lens of focal length 10 cm is used to obtain an image that is twice as large as a small object. Find the object and image distances if (*a*) the image is to be upright and (*b*) the image is to be inverted. Draw a ray diagram for each case.

108 •• You are given two converging lenses with focal lengths of 75 mm and 25 mm. (*a*) Show how the lenses should be arranged to form an astronomical telescope. State which lens to use as the objective, which lens to use as the eyepiece, how far apart to place the lenses and, what angular magnification you expect. (*b*) Draw a ray diagram to show how rays from a distant object are magnified by the telescope.

109 •• (*a*) Show how the same two lenses in Problem 108 should be arranged as a compound microscope with a tube length of 160 mm. State which lens to use as the objective, which lens to use as the eyepiece, how far apart to place the lenses, and what overall magnification you expect to get, assuming the user has a near point of 25 cm. (*b*) Draw a ray diagram to show how rays from a close object are magnified into a larger image.

110 •• [SSM] A scuba diver wears a diving mask with a face plate that bulges outward with a radius of curvature of 0.5 m. There is thus a convex spherical surface between the water and the air in the mask. A fish is 2.5 m in front of the diving mask. (*a*) Where does the fish appear to be? (*b*) What is the magnification of the image of the fish?

111 •• ![ISOLVE] A 35-mm camera has a picture size of 24 mm by 36 mm. It is used to take a picture of a person 175-cm tall so that the image just fills the height (24 mm) of the film. How far should the person stand from the camera if the focal length of the lens is 50 mm?

112 •• A 35-mm camera with interchangeable lenses is used to take a picture of a hawk that has a wing span of 2 m. The hawk is 30 m away. What would be the ideal focal length of the lens used so that the image of the wings just fills the width of the film, which is 36 mm?

113 •• An object is placed 12 cm to the left of a lens of focal length 10 cm. A second lens of focal length 12.5 cm is placed 20 cm to the right of the first lens. (a) Find the position of the final image. (b) What is the magnification of the image? (c) Sketch a ray diagram showing the final image.

114 •• (a) Show that if f is the focal length of a thin lens in air, its focal length in water is

$$f' = \frac{n_w(n-1)}{n - n_w}$$

where n_w is the index of refraction of water and n is that of the lens. (b) Calculate the focal length in air and in water of a double concave lens of index of refraction $n = 1.5$ that has radii of magnitude 30 cm and 35 cm.

115 •• **SSM** (a) Find the focal length of a *thick*, double convex lens with an index of refraction of 1.5, a thickness of 4 cm, and radii of +20 cm and −20 cm. (b) Find the focal length of this lens in water.

116 •• A 2-cm-thick layer of water ($n = 1.33$) floats on top of a 4-cm-thick layer of carbon tetrachloride ($n = 1.46$) in a tank. How far below the top surface of the water does the bottom of the tank appear to be to an observer looking from above at normal incidence?

117 •• While sitting in your car, you see a jogger in your side mirror, which is convex with a radius of curvature of magnitude 2 m. The jogger is 5 m from the mirror and is approaching at 3.5 m/s. How fast does the jogger appear to be running when viewed in the mirror?

118 •• In the seventeenth century, Antonie van Leeuwenhoek, the first great microscopist, used simple spherical lenses made first of water droplets and then of glass for his first instruments. He made staggering discoveries with these simple lenses. Consider a glass sphere of radius 2.0 mm with an index of refraction of 1.50. Find the focal length of this lens. (*Hint:* Use the equation for refraction at a single spherical surface to find the image distance for an infinite object distance for the first surface. Then use this image point as the object point for the second surface.)

119 ••• An object is 15 cm to the left of a thin convex lens of focal length 10 cm. A concave mirror of radius 10 cm is 25 cm to the right of the lens. (a) Find the position of the final image formed by the mirror and lens. (b) Is the image real or virtual?

Is the image upright or inverted? (c) On a diagram, show where your eye must be to see this image.

120 ••• **SSM** When a bright light source is placed 30 cm in front of a lens, there is an upright image 7.5 cm from the lens. There is also a faint inverted image 6 cm in front of the lens due to reflection from the front surface of the lens. When the lens is turned around, this weaker, inverted image is 10 cm in front of the lens. Find the index of refraction of the lens.

121 ••• A horizontal concave mirror with radius of curvature of 50 cm holds a layer of water with an index of refraction of 1.33 and a maximum depth of 1 cm. At what height above the mirror must an object be placed so that its image is at the same position as the object?

122 ••• A lens with one concave side with a radius of magnitude 17 cm and one convex side with a radius of magnitude 8 cm has a focal length in air of 27.5 cm. When placed in a liquid with an unknown index of refraction, the focal length increases to 109 cm. What is the index of refraction of the liquid?

123 ••• A glass ball of radius 10 cm has an index of refraction of 1.5. The back half of the ball is silvered so that it acts as a concave mirror (Figure 32-62). Find the position of the final image seen by an eye positioned to the left of the object and ball, for an object at (a) 30 cm and (b) 20 cm to the left of the front surface of the ball.

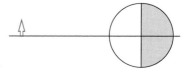

FIGURE 32-62 Problem 123

124 ••• (a) Show that a small change dn in the index of refraction of a lens material produces a small change in the focal length df given approximately by $df/f = -dn/(n-1)$. (b) Use this result to find the focal length of a thin lens for blue light, for which $n = 1.53$, if the focal length for red light, for which $n = 1.47$, is 20 cm.

125 ••• **SSM** The lateral magnification of a spherical mirror or a thin lens is given by $m = -s'/s$. Show that for objects of small horizontal extent, the longitudinal magnification is approximately $-m^2$. (*Hint:* Show that $ds'/ds = -s'^2/s^2$.)

Interference and Diffraction

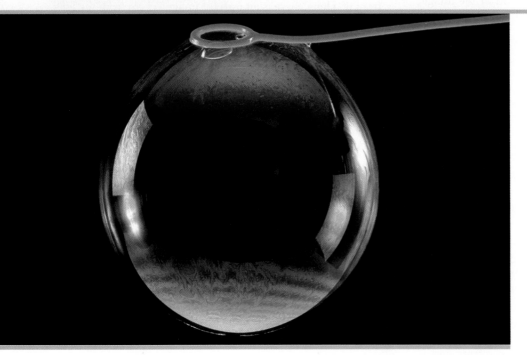

WHITE LIGHT IS REFLECTED OFF A SOAP BUBBLE. WHEN LIGHT OF ONE WAVELENGTH IS INCIDENT ON A THIN SOAP-AND-WATER FILM, LIGHT IS REFLECTED BOTH OFF THE FRONT SURFACE AND OFF THE BACK SURFACE OF THE FILM. IF THE ORDER OF MAGNITUDE OF THE THICKNESS OF THE FILM IS THAT OF THE WAVELENGTH OF THE LIGHT, THE TWO REFLECTED LIGHT WAVES INTERFERE. IF THE TWO REFLECTED WAVES ARE 180° OUT OF PHASE, THE REFLECTED WAVE INTERFERES DESTRUCTIVELY, SO THE NET RESULT IS THAT NO LIGHT IS REFLECTED. IF WHITE LIGHT, WHICH CONTAINS MANY WAVELENGTHS, IS INCIDENT ON THE THIN FILM, THEN THE REFLECTED WAVES WILL INTERFERE DESTRUCTIVELY ONLY FOR CERTAIN WAVELENGTHS, AND FOR OTHER WAVELENGTHS THEY WILL INTERFERE CONSTRUCTIVELY. THIS PROCESS PRODUCES THE COLORED FRINGES THAT YOU SEE IN THE SOAP BUBBLE.

? **Have you ever wondered if the phenomena that produces the bands that you see in the light reflected off a soap bubble has any practical applications? Example 33-2 and Problem 21 reveal how the density of the bands relates to the difference in the thickness of the film for a given distance along the film.**

nterference and diffraction are the important phenomena that distinguish waves from particles.[†] Interference is the combining by superposition of two or more waves that meet at one point in space. Diffraction is the bending of waves around corners that occurs when a portion of a wavefront is cut off by a barrier or obstacle.

➤ **In this chapter, we will see how the pattern of the resulting wave can be calculated by treating each point on the original wavefront as a point source, according to Huygens's principle, and calculating the interference pattern resulting from these sources.**

33-1 Phase Difference and Coherence

When two harmonic waves of the same frequency and wavelength but differing in phase combine, the resultant wave is a harmonic wave whose amplitude depends on the phase difference. If the phase difference is zero or an integer times 360°,

[†] Before you study this chapter, you may wish to review Chapter 15 and Chapter 16, where the general topics of interference and diffraction of waves are first discussed.

the waves are in phase and interfere constructively. The resultant amplitude equals the sum of the individual amplitudes, and the intensity (which is proportional to the square of the amplitude) is maximum. If the phase difference is 180° or any odd integer times 180°, the waves are out of phase and interfere destructively. The resultant amplitude is then the difference between the individual amplitudes, and the intensity is a minimum. If the amplitudes are equal, the maximum intensity is four times that of either source and the minimum intensity is zero.

A phase difference between two waves is often the result of a difference in path length. A path difference of one wavelength produces a phase difference of 360°, which is equivalent to no phase difference at all. A path difference of one-half wavelength produces a 180° phase difference. In general, a path difference of Δr contributes a phase difference δ given by

$$\delta = \frac{\Delta r}{\lambda} 2\pi = \frac{\Delta r}{\lambda} 360° \qquad\qquad 33\text{-}1$$

PHASE DIFFERENCE DUE TO A PATH DIFFERENCE

PHASE DIFFERENCE　　　　　　　　　　　**EXAMPLE　33-1**

(*a*) **What is the minimum path difference that will produce a phase difference of 180° for light of wavelength 800 nm?** (*b*) **What phase difference will that path difference produce in light of wavelength 700 nm?**

PICTURE THE PROBLEM The phase difference is to 360° as the path length difference is to the wavelength.

(a) The phase difference δ is to 360° as the path length difference Δr is to the wavelength λ. We know that $\lambda = 800$ nm and $\delta = 180°$:

$$\frac{\delta}{360°} = \frac{\Delta r}{\lambda}$$

$$\Delta r = \frac{\delta}{360°}\lambda = \frac{180°}{360°}(800 \text{ nm}) = \boxed{400 \text{ nm}}$$

(b) Set $\lambda = 700$ nm, $\Delta r = 400$ nm, and solve for δ:

$$\delta = \frac{\Delta r}{\lambda} 360° = \frac{400 \text{ nm}}{700 \text{ nm}} 360°$$

$$= \boxed{206° = 3.59 \text{ rad}}$$

Another cause of phase difference is the 180° phase change a wave sometimes undergoes upon reflection from a boundary surface. This phase change is analogous to the inversion of a pulse on a string when it reflects from a point where the density suddenly increases, such as when a light string is attached to a heavier string or rope. The inversion of the reflected pulse is equivalent to a phase change of 180° for a sinusoidal wave (which can be thought of as a series of pulses). When light traveling in air strikes the surface of a medium in which light travels more slowly, such as glass or water, there is a 180° phase change in the reflected light. When light is originally traveling in glass or water, there is no phase change in the light reflected from the glass–air or water–air interface. This is analogous to the reflection without inversion of a pulse on a heavy string at a point where the heavy string is attached to a lighter string.

If light traveling in one medium strikes the surface of a medium in which light travels more slowly, there is a 180° phase change in the reflected light.

PHASE DIFFERENCE DUE TO REFLECTION

As we saw in Chapter 16, interference of waves is observed when two or more coherent waves overlap. Interference of overlapping waves from two sources is not observed unless the sources are coherent. Because the light from each source is usually the result of millions of atoms radiating independently, the phase difference between the waves from such sources fluctuates randomly many times per second, so two light sources are usually not coherent. Coherence in optics is often achieved by splitting the light beam from a single source into two or more beams that can then be combined to produce an interference pattern. The light beam can be split by reflecting the light from the two closely spaced surfaces of a thin film (Section 33-2), by diffracting the beam through two small openings or slits in an opaque barrier (Section 33-3), or by using a single point source and its image in a plane mirror for the two sources (Section 33-3). Today, lasers are the most important sources of coherent light in the laboratory.

Light from an ideal monochromatic source is an infinitely long sinusoidal wave, and light from certain lasers approach this ideal. However, light from conventional *monochromatic* sources, such as gas discharge tubes designed for this purpose, consists of packets of sinusoidal light that are only a few million wavelengths long. The light from such a source consists of many such packets, each approximately the same length. The packets have essentially the same wavelength, but the packets differ in phase in a random manner. The length of one of these packets is called the **coherence length** of the light, and the time it takes one of the packets to pass a point in space is the **coherence time.** The light emitted by a gas discharge tube designed to produce monochromatic light has a coherence length of only a few millimeters. By comparison, some highly stable lasers produce light with a coherence length many kilometers long.

33-2 Interference in Thin Films

You have probably noticed the colored bands in a soap bubble or in the film on the surface of oily water. These bands are due to the interference of light reflected from the top and bottom surfaces of the film. The different colors arise because of variations in the thickness of the film, causing interference for different wavelengths at different points.

Consider a thin film of water (such as a small section of a soap bubble) of uniform thickness viewed at small angles with the normal, as shown in Figure 33-1. Part of the light is reflected from the upper air–water interface where it undergoes a 180° phase change. Most of the light enters the film and part of it is reflected by the bottom water–air interface. There is no phase change in this reflected light. If the light is nearly perpendicular to the surfaces, both the ray reflected from the top surface and the ray reflected from the bottom surface can enter the eye at point P in the figure. The path difference between these two rays is $2t$, where t is the thickness of the film. This path difference produces a phase difference of $(2t/\lambda')360°$, where $\lambda' = \lambda/n$ is the wavelength of the light in the film, and n is the index of refraction of the film. The total phase difference between these two rays is thus 180° plus the phase difference due to the path difference. Destructive interference occurs when the path difference $2t$ is zero or a whole number of wavelengths λ' (in the film). Constructive interference occurs when the path difference is an odd number of half-wavelengths.

When a thin film of water lies on a glass surface, as in Figure 33-2, the ray that reflects from the lower water–glass interface also undergoes a 180° phase change, because the index of refraction of glass (approximately 1.50) is greater than that of water (approximately 1.33). Thus, both the rays shown in the figure have undergone a 180° phase change upon reflection. The phase difference between these rays is due solely to the path difference and is given by $\delta = (2t/\lambda')360°$.

When a thin film of varying thickness is viewed with monochromatic light, such as the yellow light from a sodium lamp, alternating bright and dark bands

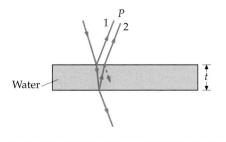

FIGURE 33-1 Light rays reflected from the top and bottom surfaces of a thin film are coherent because both rays come from the same source. If the light is incident almost normally, the two reflected rays will be very close to each other and will produce interference.

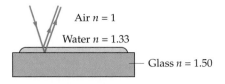

FIGURE 33-2 The interference of light reflected from a thin film of water resting on a glass surface. In this case, both rays undergo a change in phase of 180° upon reflection.

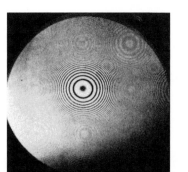

(a)

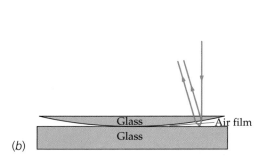

(b)

or lines called **fringes** are observed. The distance between a bright fringe and a dark fringe is that distance over which the film's thickness changes so that the path difference $2t$ is $\lambda'/2$. Figure 33-3*a* shows the interference pattern observed when light is reflected from an air film between a spherical glass surface and a plane glass surface in contact. These circular interference fringes are known as **Newton's rings.** Typical rays reflected at the top and bottom of the air film are shown in Figure 33-3*b*. Near the point of contact of the surfaces, where the path difference between the ray reflected from the upper glass–air interface and the ray reflected from the lower air–glass interface is essentially zero or is at least small compared with the wavelength of light, the interference is perfectly destructive because of the 180° phase shift of the ray reflected from the lower air–glass interface. This central region in Figure 33-3*a* is therefore dark. The first bright fringe occurs at the radius at which the path difference is $\lambda/2$, which contributes a phase difference of 180°. This adds to the phase shift due to reflection to produce a total phase difference of 360°, which is equivalent to a zero phase difference. The second dark region occurs at the radius at which the path difference is λ, and so on.

FIGURE 33-4 The angle θ, which is less than 0.02°, is exaggerated. The incoming and outgoing rays are essentially perpendicular to all air–glass interfaces.

A WEDGE OF AIR **EXAMPLE 33-2**

A wedge-shaped film of air is made by placing a small slip of paper between the edges of two flat pieces of glass, as shown in Figure 33-4. Light of wavelength 500 nm is incident normally on the glass, and interference fringes are observed by reflection. If the angle θ made by the plates is 3×10^{-4} rad, how many dark interference fringes per centimeter are observed?

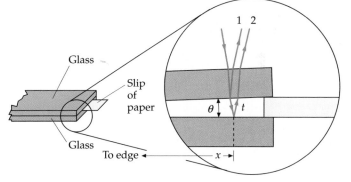

PICTURE THE PROBLEM We find the number of fringes per centimeter by finding the horizontal distance x to the mth fringe and solving for m/x. Because the ray reflected from the bottom plate undergoes a 180° phase shift, the point of contact (where the path difference is zero) will be dark. The first dark fringe after this point occurs when $2t = \lambda'$, where $\lambda' = \lambda$ is the wavelength in the air film, and t is the plate separation at x, as shown in Figure 33-4. Since the angle θ is small, we can use the small-angle approximation $\theta \approx t/x$.

1. The mth dark fringe occurs when the path difference $2t$ equals m wavelengths:

$$2t = m\lambda' = m\lambda$$

$$m = \frac{2t}{\lambda}$$

2. The thickness t is related to the angle θ:

$$\theta = \frac{t}{x}$$

3. Substitute $t = x\theta$ into the equation for m:

$$m = \frac{2x\theta}{\lambda}$$

4. Calculate m/x:

$$\frac{m}{x} = \frac{2\theta}{\lambda} = \frac{2(3 \times 10^{-4})}{5 \times 10^{-7}\text{ m}} = 1200\text{ m}^{-1}$$

$$= \boxed{12\text{ cm}^{-1}}$$

REMARKS We therefore observe 12 dark fringes per centimeter. In practice, the number of fringes per centimeter, which is easy to count, can be used to determine the angle. Note that if the angle of the wedge is increased, the fringes become more closely spaced.

EXERCISE How many dark fringes per centimeter are observed if light of wavelength 650 nm is used? (*Answer* 9.2 cm^{-1})

Figure 33-5*a* shows interference fringes produced by a wedge-shaped air film between two flat glass plates, as in Example 33-2. Plates that produce straight fringes, such as those in Figure 33-5*a*, are said to be **optically flat.** To be optically flat, a surface must be flat to within a small fraction of a wavelength. A similar wedge-shaped air film formed by two ordinary glass plates yields the irregular fringe pattern in Figure 33-5*b*, which indicates that these plates are not optically flat.

One application of interference effects in thin films is in nonreflecting lenses, which are made by covering a lens with a thin film of a material that has an index of refraction of approximately 1.38, which is between that of glass and air. Then the intensities of the light reflected from the top and bottom surfaces of the film are approximately equal, and since both rays undergo a 180° phase change, there is no phase difference between the rays due to reflection. The thickness of the film is chosen to be $\lambda'/4 = \lambda/4n$, where λ is in the middle of the visible spectrum, so that there is a phase change of 180° due to the path difference of $\lambda'/2$. Reflection from the coated surface is thus minimized, whereas transmission through the surface is maximized.

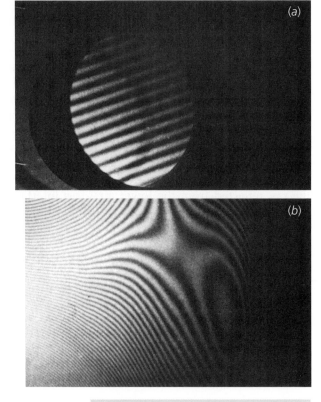

(a)

(b)

FIGURE 33-5 (*a*) Straight-line fringes from a wedge-shaped film of air, like that shown in Figure 33-4. The straightness of the fringes indicates that the glass plates are optically flat. (*b*) Fringes from a wedge-shaped film of air between glass plates that are not optically flat.

33-3 Two-Slit Interference Pattern

Interference patterns of light from two or more sources can be observed only if the sources are coherent. The interference in thin films discussed previously can be observed because the two beams come from the same light source but are separated by reflection. In Thomas Young's famous experiment, in which he demonstrated the wave nature of light, two coherent light sources are produced by illuminating two very narrow parallel slits with a single light source. We saw in Chapter 15 that when a wave encounters a barrier with a very small opening, the opening acts as a point source of waves (Figure 33-6). In Young's experiment, diffraction causes each slit to act as a line source (which is equivalent to a point source in two dimensions). The interference pattern is observed on a screen far from the slits (Figure 33-7*a*). At very large distances from the slits, the lines from

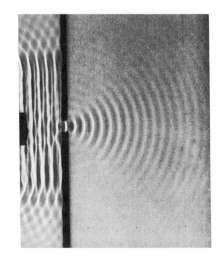

FIGURE 33-6 Plane water waves in a ripple tank encountering a barrier with a small opening. The waves to the right of the barrier are circular waves that are concentric about the opening, just as if there were a point source at the opening.

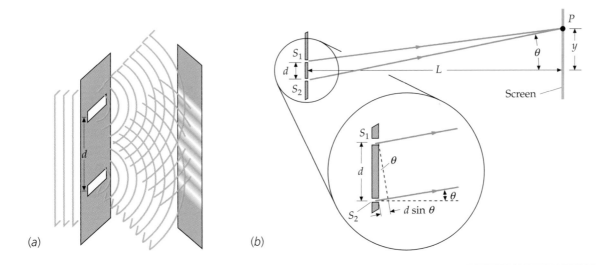

(a)   (b)

FIGURE 33-7 (*a*) Two slits act as coherent sources of light for the observation of interference in Young's experiment. Cylindrical waves from the slits overlap and produce an interference pattern on a screen. (*b*) Geometry for relating the distance *y* measured along the screen to *L* and *θ*. When the screen is very far away compared with the slit separation, the rays from the slits to a point on the screen are approximately parallel, and the path difference between the two rays is *d* sin *θ*.

the two slits to some point P on the screen are approximately parallel, and the path difference is approximately $d \sin \theta$, where d is the separation of the slits, as shown in Figure 33-7b. When the path difference is equal to an integral number of wavelengths, the interference is constructive. We thus have interference maxima at an angle θ_m given by

$$d \sin \theta_m = m\lambda, \qquad m = 0, 1, 2, \ldots \qquad \text{33-2}$$

TWO-SLIT INTERFERENCE MAXIMA

where m is called the **order number.** The interference minima occur at

$$d \sin \theta_m = (m - \tfrac{1}{2})\lambda, \qquad m = 1, 2, 3, \ldots \qquad \text{33-3}$$

TWO-SLIT INTERFERENCE MINIMA

The phase difference δ at a point P is related to the path difference $d \sin \theta$ by

$$\frac{\delta}{2\pi} = \frac{d \sin \theta}{\lambda} \qquad \text{33-4}$$

We can relate the distance y_m measured along the screen from the central point to the mth bright fringe (see Figure 33-7b) to the distance L from the slits to the screen:

$$\tan \theta_m = \frac{y_m}{L}$$

For small angles, $\tan \theta \approx \sin \theta$. Substituting y_m/L for $\sin \theta_m$ in Equation 33-2 and solving for y_m gives

$$y_m = m\frac{\lambda L}{d} \qquad \text{33-5}$$

DISTANCE ON SCREEN TO THE MTH BRIGHT FRINGE

From this result, we see that for small angles the fringes are equally spaced on the screen.

Two narrow slits separated by 1.5 mm are illuminated by yellow light of wavelength 589 nm from a sodium lamp. Find the spacing of the bright fringes observed on a screen 3 m away.

PICTURE THE PROBLEM The distance y_m measured along the screen to the mth bright fringe is given by Equation 33-2, with $L = 3$ m, $d = 1.5$ mm, and $\lambda = 589$ nm.

Cover the column to the right and try these on your own before looking at the answers.

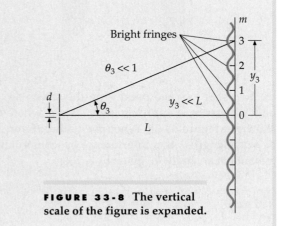

Steps	Answers
1. Make a sketch of the situation (Figure 33-8).	
2. Using the sketch, obtain an expression for the spacing between fringes.	fringe spacing $= \dfrac{y_3}{3}$
3. Apply Equation 33-2 to the $m = 3$ fringe.	$d \sin \theta_3 = 3\lambda$
4. Using trig, relate y_3 and θ_3.	$\sin \theta_3 \approx \tan \theta_3 = \dfrac{y_3}{L}$
5. Substitute into the step 3 result and solve for the fringe spacing.	$\dfrac{y_3}{3} = \boxed{1.18 \text{ mm}}$

FIGURE 33-8 The vertical scale of the figure is expanded.

REMARKS The fringes are uniformly spaced only to the degree that the small-angle approximation is valid. That is, to the degree that $\lambda/d \ll 1$. In this example, $\lambda/d = (589 \text{ nm})/(1.5 \text{ mm}) = 0.0004$.

EXERCISE A point source of light ($\lambda = 589$ nm) is placed 0.4 mm above the surface of a glass mirror. Interference fringes are observed on a screen 6 m away, and the interference is between the light reflected off the front surface of the glass and the light traveling from the source directly to the screen. Find the spacing of the fringes. (*Answer* 4.42 mm)

Calculation of Intensity

To calculate the intensity of the light on the screen at a general point P, we need to add two harmonic wave functions that differ in phase.[†] The wave functions for electromagnetic waves are the electric field vectors. Let E_1 be the electric field at some point P on the screen due to the waves from slit 1, and let E_2 be the electric field at that point due to waves from slit 2. Since the angles of interest are small, we can assume that these fields are parallel. Both electric fields oscillate with the same frequency (they result from a single source that illuminates both slits) and they have the same amplitude. (The path difference is only of the order of a few wavelengths of light at most.) They have a phase difference δ given by Equation 33-4. If we represent these wave functions by

$$E_1 = A_0 \sin \omega t$$

and

$$E_2 = A_0 \sin(\omega t + \delta)$$

the resultant wave function is

---

† We did this in Chapter 16 where we first discussed the superposition of two waves.

$$E = E_1 + E_2 = A_0 \sin \omega t + A_0 \sin(\omega t + \delta)$$
$$= 2A_0 \cos \tfrac{1}{2}\delta \sin(\omega t + \tfrac{1}{2}\delta) \qquad \text{33-6}$$

where we used the identity

$$\sin \alpha + \sin \beta = 2 \cos \tfrac{1}{2}(\alpha - \beta) \sin \tfrac{1}{2}(\alpha + \beta) \qquad \text{33-7}$$

The amplitude of the resultant wave is thus $2A_0 \cos \tfrac{1}{2}\delta$. It has its maximum value of $2A_0$ when the waves are in phase and is zero when they are 180° out of phase. Since the intensity is proportional to the square of the amplitude, the intensity at any point P is

$$I = 4I_0 \cos^2 \tfrac{1}{2}\delta \qquad \text{33-8}$$

INTENSITY IN TERMS OF PHASE DIFFERENCE

where I_0 is the intensity of the light on the screen from either slit separately. The phase angle δ is related to the position on the screen by Equation 33-4.

Figure 33-9a shows the intensity pattern as seen on a screen. A graph of the intensity as a function of $\sin \theta$ is shown in Figure 33-9b. For small θ, this is equivalent to a plot of intensity versus y (since $y = L \tan \theta \approx L \sin \theta$). The intensity I_0 is that from each slit separately. The dashed line in Figure 33-9b shows the average intensity $2I_0$, which is the result of averaging over a distance containing many interference maxima and minima. This is the intensity that would arise from the two sources if they acted independently without interference, that is, if they were not coherent. Then the phase difference between the two sources would fluctuate randomly, so that only the average intensity would be observed.

Figure 33-10 shows another method of producing the two-slit interference pattern, an arrangement known as **Lloyd's mirror.** A single slit is placed at a distance $\tfrac{1}{2}d$ above the plane of a mirror. Light striking the screen directly from the source interferes with the light that is reflected from the mirror. The reflected light can be considered to come from the virtual image of the slit formed by the mirror. Because of the 180° change in phase upon reflection at the mirror, the interference pattern is that of two coherent line sources that differ in phase by 180°. The pattern is the same as that shown in Figure 33-9 for two slits, except that the maxima and minima are interchanged. Constructive interference occurs at points for which the path difference is a half-wavelength or any odd number of half-wavelengths. At these points, the 180° phase difference due to the path difference combines with the 180° phase difference of the sources to produce constructive interference.

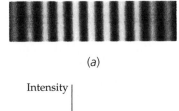

(a)

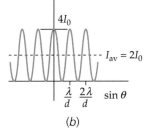

(b)

FIGURE 33-9 (a) The interference pattern observed on a screen far away from the two slits shown in Figure 33-7. (b) Plot of intensity versus $\sin \theta$. The maximum intensity is $4I_0$, where I_0 is the intensity due to each slit separately. The average intensity (dashed line) is $2I_0$.

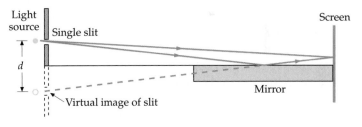

FIGURE 33-10 Lloyd's mirror for producing a two-slit interference pattern. The two sources (the slit and its image) are coherent and are 180° out of phase. The central interference band at the point equidistant from the sources is dark.

33-4 Diffraction Pattern of a Single Slit

In our discussion of the interference patterns produced by two or more slits, we assumed that the slits were very narrow so that we could consider the slits to be line sources of cylindrical waves, which in our two-dimensional diagrams are point sources of circular waves. We could therefore assume that the intensity due to one slit acting alone was the same (I_0) at any point P on the screen, independent of the angle θ made between the ray to point P and the normal line between the slit and the screen. When the slit is not narrow, the intensity on a screen far away is not independent of angle but decreases as the angle increases. Consider a slit of width a. Figure 33-11 shows the intensity pattern on a screen far away from the slit of width a as a function of $\sin \theta$. We can see that the intensity is maximum in the forward direction ($\sin \theta = 0$) and decreases to zero at an angle that depends on the slit width a and the wavelength λ.

Most of the light intensity is concentrated in the broad **central diffraction maximum**, although there are minor secondary maxima bands on either side of the central maximum. The first zeroes in the intensity occur at angles specified by

$$\sin \theta_1 = \lambda / a \qquad \text{33-9}$$

Note that for a given wavelength λ, Equation 33-9 describes how variations in the slit width result in variations in the angular width of the central maximum. If we *increase* the slit width a, the angle θ_1 at which the intensity first becomes zero *decreases*, giving a more narrow central diffraction maximum. Conversely, if we *decrease* the slit width, the angle of the first zero *increases*, giving a wider central diffraction maximum. When a is smaller than λ, then $\sin \theta_1$ would have to exceed 1 to satisfy Equation 33-9. Thus, for a less than λ, there are no points of zero intensity in the pattern, and the slit acts as a line source (a point source in two dimensions) radiating light energy essentially equal in all directions.

Multiplying both sides of Equation 33-9 by $a/2$ gives

$$\tfrac{1}{2}a \sin \theta_1 = \tfrac{1}{2}\lambda \qquad \text{33-10}$$

The quantity $\tfrac{1}{2}a \sin \theta_1$ is the path difference between a light ray leaving the middle of the upper half of the slit and one leaving the middle of the lower half of the slit. We see that the first diffraction *minimum* occurs when these two rays are 180° out of phase, that is, when their path difference equals a half-wavelength. We can understand this result by considering each point on a wavefront to be a point source of light in accordance with Huygens's principle. In Figure 33-12, we have placed a line of dots on the wavefront at the slit to represent these point sources schematically. Suppose, for example, that we have 100 such dots and that we look at an angle θ_1 for which $a \sin \theta_1 = \lambda$. Let us consider the slit to be divided into two halves, with the first 50 sources in the upper half and sources 51 through 100 in the lower half. When the path difference between the middle of the upper half and the middle of the lower half of the slit equals a half-wavelength, the path difference between source 1 (the first source in the upper half) and source 51 (the first source in the lower half) is $\tfrac{1}{2}\lambda$. The waves from these two sources will be out of phase by 180° and will thus cancel. Similarly, waves from the second source in each region (source 2 and source 52) will cancel. Continuing this argument, we can see that the waves from each pair of sources separated by $a/2$ will cancel. Thus, there will be no light energy at this angle. We can extend this argument to the second and third minima in the diffraction pattern of Figure 33-11. At an angle θ_2 where $a \sin \theta_2 = 2\lambda$, we can divide the slit into four regions, two regions for the top half and two regions for the bottom half. Using this same argument, the light intensity from the top half is zero because of the cancellation of pairs of sources, and, similarly, the light intensity from the bottom half is zero. The general expression for the points of zero intensity in the diffraction pattern of a single slit is thus

(a)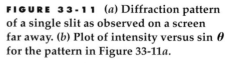

(b)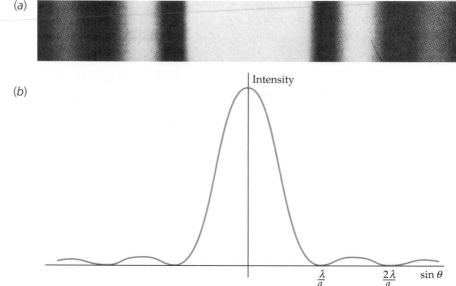

FIGURE 33-11 (a) Diffraction pattern of a single slit as observed on a screen far away. (b) Plot of intensity versus $\sin \theta$ for the pattern in Figure 33-11a.

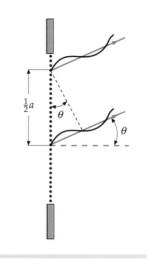

FIGURE 33-12 A single slit is represented by a large number of point sources of equal amplitude. At the first diffraction minimum of a single slit, the waves from each point source in the upper half of the slit 180° out of phase with the wave from the point source a distance $a/2$ lower in the slit. Thus, the interference from each such pair of point sources is destructive.

$$a \sin \theta_m = m\lambda, \qquad m = 1, 2, 3 \dots \qquad\qquad\qquad 33\text{-}11$$

POINTS OF ZERO INTENSITY FOR A SINGLE-SLIT DIFFRACTION PATTERN

Usually, we are just interested in the first occurrence of a minimum in the light intensity because nearly all of the light energy is contained in the central diffraction maximum.

In Figure 33-13, the distance y_1 from the central maximum to the first diffraction minimum is related to the angle θ_1 and the distance L from the slit to the screen by

$$\tan \theta_1 = \frac{y_1}{L}$$

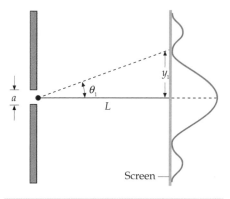

FIGURE 33-13 The distance y_1 measured along the screen from the central maximum to the first diffraction minimum is related to the angle θ_1 by $\tan \theta_1 = y_1/L$, where L is the distance to the screen.

WIDTH OF THE CENTRAL DIFFRACTION MAXIMUM **EXAMPLE 33-4**

In a lecture demonstration of single-slit diffraction, a laser beam of wavelength 700 nm passes through a vertical slit 0.2 mm wide and hits a screen 6 m away. Find the width of the central diffraction maximum on the screen; that is, find the distance between the first minimum on the left and the first minimum on the right of the central maximum.

PICTURE THE PROBLEM Referring to Figure 33-13, the width of the central diffraction maximum is $2y_1$.

1. The half-width of the central maxima y_1 is related to the angle θ_1 by:

$$\tan \theta_1 = \frac{y_1}{L}$$

2. The angle θ_1 is related to the slit width a by Equation 33-11:

$$\sin \theta_1 = \lambda/a$$

3. Solve the step 2 result for θ_1, substitute into the step 1 result, and solve for $2y_1$:

$$2y_1 = 2L \tan \theta_1 = 2L \tan\left(\sin^{-1}\frac{\lambda}{a}\right)$$

$$= 2(6 \text{ m}) \tan\left(\sin^{-1}\frac{700 \times 10^{-9} \text{ m}}{0.0002 \text{ m}}\right)$$

$$= 4.2 \times 10^{-2} \text{ m} = \boxed{4.20 \text{ cm}}$$

REMARKS Since $\sin \theta_1 = \lambda/a = (700 \text{ nm})/(0.2 \text{ mm}) = 0.0035$, we can use the small-angle approximation to evaluate $2y_1$. In this approximation, $\sin \theta_1 = \tan \theta_1$, so $\lambda/a = y_1/L$ and $2y_1 = 2L\lambda/a = 2(6 \text{ m})(700 \text{ nm})/(0.2 \text{ mm}) = 4.20 \text{ cm}$. (This approximate value is in agreement with the exact value to within 0.0006 percent.)

Interference–Diffraction Pattern of Two Slits

When there are two or more slits, the intensity pattern on a screen far away is a combination of the single-slit diffraction pattern and the multiple-slit interference pattern we have studied. Figure 33-14 shows the intensity pattern on a screen far from two slits whose separation d is 10 times the width a of each slit. The pattern is the same as the two-slit pattern with very narrow slits (Figure 33-11) except that it is modulated by the single-slit diffraction pattern; that is, the intensity due to each slit separately is now not constant but decreases with angle, as shown in Figure 33-14b.

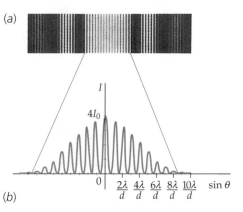

(a)

(b)

FIGURE 33-14 (a) Interference–diffraction pattern for two slits whose separation d is equal to 10 times their width a. The tenth interference maximum on either side of the central interference maximum is missing because it falls at the first diffraction minimum. (b) Plot of intensity versus $\sin \theta$ for the central band of the pattern in Figure 33-14a.

Note that the central diffraction maximum in Figure 33-14 contains 19 interference maxima—the central interference maximum and 9 maxima on either side. The tenth interference maximum on either side of the central one is at the angle θ_{10}, given by $\sin \theta_{10} = 10\lambda/d = \lambda/a$, since $d = 10a$. This coincides with the position of the first diffraction minimum, so this interference maximum is not seen. At these points, the light from the two slits would be in phase and would interfere constructively, but there is no light from either slit because the points are diffraction minima. In general, we can see that if $m = d/a$, the mth interference maximum will fall at the first diffraction minimum. Since the mth fringe is not seen, there will be $m - 1$ fringes on each side of the central fringe for a total of N fringes in the central maximum, where N is given by

$$N = 2(m - 1) + 1 = 2m - 1 \qquad\qquad 33\text{-}12$$

INTERFERENCE AND DIFFRACTION   **E X A M P L E   3 3 - 5**

Two slits of width $a = 0.015$ mm are separated by a distance $d = 0.06$ mm and are illuminated by light of wavelength $\lambda = 650$ nm. How many bright fringes are seen in the central diffraction maximum?

PICTURE THE PROBLEM We need to find the value of m for which the mth interference maximum coincides with the first diffraction minimum. Then there will be $N = 2m - 1$ fringes in the central maximum.

1. Find the angle θ_1 of the first diffraction minimum:
$$\sin \theta_1 = \frac{\lambda}{a} \text{ (first diffraction minimum)}$$

2. Find the angle θ_m of the mth interference maxima:
$$\sin \theta_m = \frac{m\lambda}{d} \text{ (mth interference maxima)}$$

3. Set these angles equal and solve for m:
$$\frac{m\lambda}{d} = \frac{\lambda}{a}$$
$$m = \frac{d}{a} = \frac{0.06 \text{ mm}}{0.015 \text{ mm}} = 4$$

4. The first diffraction minimum coincides with the fourth bright fringe. Therefore, there are 3 bright fringes visible on either side of the central diffraction maximum. These 6 maxima, plus the central interference maximum, combine for a total of 7 bright fringes in the central diffraction maximum:
$$N = \boxed{7 \text{ bright fringes}}$$

---

*33-5   Using Phasors to Add Harmonic Waves

To calculate the interference pattern produced by three, four, or more coherent light sources and to calculate the diffraction pattern of a single slit, we need to combine several harmonic waves of the same frequency that differ in phase. A simple geometric interpretation of harmonic wave functions leads to a method of adding harmonic waves of the same frequency by geometric construction.

Let the wave functions for two waves at some point be $E_1 = A_1 \sin \alpha$ and $E_2 = A_2 \sin(\alpha + \delta)$, where $\alpha = \omega t$. Our problem is then to find the sum:

$$E_1 + E_2 = A_1 \sin \alpha + A_2 \sin(\alpha + \delta)$$

We can represent each wave function by a two-dimensional vector, as shown in Figure 33-15. The geometric method of addition is based on the fact that the y (or x) component of the resultant of two vectors equals the sum of the y (or x) components of the vectors, as illustrated in the figure. The wave function E_1 is represented by the vector $\vec{A}_1$. As the time varies, this vector rotates in the xy plane with angular frequency ω. Such a vector is called a **phasor**. (We encountered phasors in our study of ac circuits in Section 29-4.) The wave function E_2 is the y component of a phasor of magnitude A_2 that makes an angle $\alpha + \delta$ with the x axis. By the laws of vector addition, the sum of these components equals the y component of the resultant phasor $\vec{A}$, as shown in Figure 33-15. The y component of the resultant phasor, $A \sin(\alpha + \delta')$, is a harmonic wave function that is the sum of the two original wave functions:

$$A_1 \sin \alpha + A_2 \sin(\alpha + \delta) = A \sin(\alpha + \delta') \qquad \text{33-13}$$

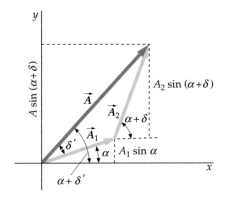

FIGURE 33-15 Phasor representation of wave functions.

where A (the amplitude of the resultant wave) and δ' (the phase of the resultant wave relative to the first wave) are found by adding the phasors representing the waves. As time varies, α varies. The phasors representing the two wave functions and the resultant phasor representing the resultant wave function rotate in space, but their relative positions do not change because they all rotate with the same angular velocity ω.

WAVE SUPERPOSITION USING PHASORS   **EXAMPLE 33-6**   **Try It Yourself**

Use the phasor method of addition to derive Equation 33-16 for the superposition of two waves of the same amplitude.

PICTURE THE PROBLEM Represent the waves $y_1 = A_0 \sin \alpha$ and $y_2 = A_0 \sin(\alpha + \delta)$ by vectors (phasors) of length A_0 making an angle δ with one another. The resultant wave $y_r = A \sin(\alpha + \delta')$ is represented by the sum of these vectors, which form an isosceles triangle, as shown in Figure 33-16.

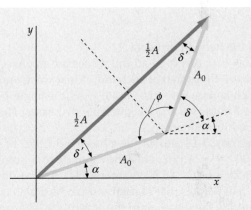

FIGURE 33-16

Cover the column to the right and try these on your own before looking at the answers.

Steps	Answers
1. Find the phase angle δ' in terms of ϕ from the fact that the three angles in the triangle must sum to $180°$.	$\delta' + \delta' + \phi = 180°$
2. Relate ϕ to δ.	$\delta + \phi = 180°$
3. Eliminate ϕ from the step 1 and step 2 results and solve for δ'.	$\delta' = \tfrac{1}{2}\delta$
4. Write $\cos \delta'$ in terms of A and A_0.	$\cos \delta' = \dfrac{\tfrac{1}{2}A}{A_0}$
5. Solve for A in terms of δ.	$A = 2A_0 \cos \delta' = 2A_0 \cos \tfrac{1}{2}\delta$
6. Use your results for A and δ' to write the resultant wave function.	$y_r = A \sin(\alpha + \delta')$
	$= \boxed{2A_0 \cos(\tfrac{1}{2}\delta) \sin(\alpha + \tfrac{1}{2}\delta)}$

EXERCISE Find the resultant wave function of the two waves $E_1 = 4 \sin(\omega t)$ and $E_2 = 3 \sin(\omega t + 90°)$ [*Answer* $E_1 + E_2 = 5 \sin(\omega t + 37°)$]

*The Interference Pattern of Three or More Equally Spaced Sources

We can apply the phasor method of addition to calculate the interference pattern of three or more equally spaced, coherent sources in phase. We are most interested in the interference maxima and minima. Figure 33-17 illustrates the case of three sources. The geometry is the same as for two sources. At a great distance from the sources, the rays from the sources to a point P on the screen are approximately parallel. The path difference between the first and second source is then $d \sin \theta$, as before, and the path difference between the first and third source is $2d \sin \theta$. The wave at point P is the sum of three waves. Let $\alpha = \omega t$ be the phase of the first wave at point P. We thus have the problem of adding three waves of the form

$$E_1 = A_0 \sin \alpha$$

$$E_2 = A_0 \sin(\alpha + \delta)$$

$$E_3 = A_0 \sin(\alpha + 2\delta) \qquad \text{33-14}$$

where

$$\delta = \frac{2\pi}{\lambda} d \sin \theta \approx \frac{2\pi \, yd}{\lambda \, L} \qquad \text{33-15}$$

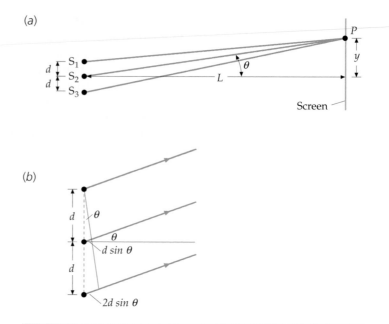

FIGURE 33-17 Geometry for calculating the intensity pattern far away from three equally spaced, coherent sources that are in phase.

as in the two-slit problem.

At $\theta = 0$, $\delta = 0$, so all the waves are in phase. The amplitude of the resultant wave is 3 times that of each individual wave and the intensity is 9 times that due to each source acting separately. As the angle δ increases from $\theta = 0$, the phase angle δ increases and the intensity decreases. The position $\theta = 0$ is thus a position of maximum intensity.

Figure 33-18 shows the phasor addition of three waves for a phase angle $\delta = 30° = \pi/6$ rad. This corresponds to a point P on the screen for which θ is given by $\sin \theta = \lambda\delta/(2\pi d) = \lambda/(12d)$. The resultant amplitude A is considerably less than 3 times that of each source. As the phase angle δ increases, the resultant amplitude decreases until the amplitude is zero at $\delta = 120°$. For this phase difference, the three phasors form an equilateral triangle (Figure 33-19). This first interference minimum for three sources occurs at a smaller phase angle δ (and therefore at a smaller space angle θ) than it does for only two sources (for which the first minimum occurs at $\delta = 180°$). As δ increases from 120°, the resultant amplitude increases, reaching a secondary maximum near $\delta = 180°$. At the phase angle $\delta = 180°$, the amplitude is the same as that from a single source, since the waves from the first two sources cancel each other, leaving only the third. The intensity of the secondary maximum is one-ninth that of the maximum at $\theta = 0$. As δ increases beyond 180°, the amplitude again decreases and is zero at $\delta = 180° + 60° = 240°$. For δ greater than 240°, the amplitude increases and is again 3 times that of each source when $\delta = 360°$. This phase angle corresponds to a path difference of 1 wavelength for the waves from the first two sources and 2 wavelengths for the waves from the first and third sources. Hence, the

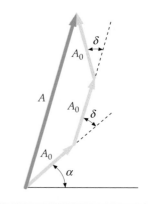

FIGURE 33-18 Phasor diagram for determining the resultant amplitude A due to three waves, each of amplitude A_0, that have phase differences of δ and 2δ due to path differences of $d \sin \theta$ and $2d \sin \theta$. The angle $\alpha = \omega t$ varies with time, but this does not affect the calculation of A.

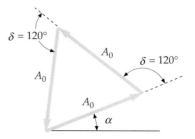

FIGURE 33-19 The resultant amplitude for the waves from three sources is zero when δ is 120°. This interference minimum occurs at a smaller angle θ than does the first minimum for two sources, which occurs when δ is 180°.

three waves are in phase at this point. The largest maxima, called the principal maxima, are at the same positions as for just two sources, which are those points corresponding to the angles θ given by

$$d \sin \theta_{m} = m\lambda, \qquad m = 0, 1, 2, \ldots \qquad \text{33-16}$$

These maxima are stronger and narrower than those for two sources. They occur at points for which the path difference between adjacent sources is zero or an integral number of wavelengths.

These results can be generalized to more than three sources. For four equally spaced sources that are in phase, the principal interference maxima are again given by Equation 33-16, but these maxima are even more intense, they are narrower, and there are two small secondary maxima between each pair of principal maxima. At $\theta = 0$, the intensity is 16 times that due to a single source. The first interference minimum occurs when δ is 90°, as can be seen from the phasor diagram of Figure 33-20. The first secondary maximum is near $\delta = 132°$, leaving only the wave from the fourth source. The intensity of the secondary maximum is approximately one-sixteenth that of the central maximum. There is another minimum at $\delta = 180°$, another secondary maximum near $\delta = 228°$, and another minimum at $\delta = 270°$ before the next principal maximum at $\delta = 360°$.

Figure 33-21 shows the intensity patterns for two, three, and four equally spaced coherent sources. Figure 33-22 shows a graph of I/I_0, where I_0 is the intensity due to each source acting separately. For three sources, there is a very small secondary maximum between each pair of principal maxima, and the principal maxima are sharper and more intense than those due to just two sources. For four sources, there are two small secondary maxima between each pair of principal maxima, and the principal maxima are even more narrow and intense.

From this discussion, we can see that as we increase the number of sources, the intensity becomes more and more concentrated in the principal maxima given by Equation 33-16, and these maxima become narrower. For N sources, the intensity of the principal maxima is N^2 times that due to a single source. The first minimum occurs at a phase angle of $\delta = 360°/N$, for which the N phasors form a closed polygon of N sides. There are $N - 2$ secondary maxima between each pair of principal maxima. These secondary maxima are very weak compared with the principal maxima. As the number of sources is increased, the principal maxima become sharper and more intense, and the intensities of the secondary maxima become negligible compared to those of the principal maxima.

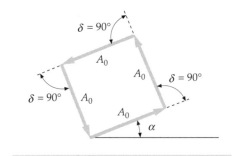

FIGURE 33-20 Phasor diagram for the first minimum for four equally spaced in-phase sources. The amplitude is zero when the phase difference of the waves from adjacent sources is 90°.

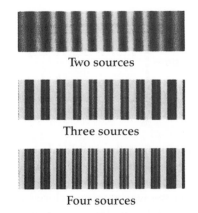

Two sources

Three sources

Four sources

FIGURE 33-21 Intensity patterns for two, three, and four equally spaced coherent sources. There is a secondary maximum between each pair of principal maxima for three sources, and two secondary maxima for four sources.

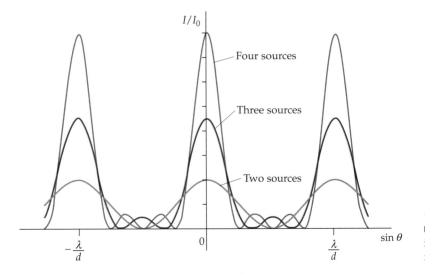

FIGURE 33-22 Plot of relative intensity versus sin θ for two, three, and four equally spaced coherent sources.

*Calculating the Single-Slit Diffraction Pattern

We now use the phasor method for the addition of harmonic waves to calculate the intensity pattern shown in Figure 33-11. We assume that the slit of width a is divided into N equal intervals and that there is a point source of waves at the midpoint of each interval (Figure 33-23). If d is the distance between two adjacent sources and a is the width of the opening, we have $d = a/N$. Since the screen on which we are calculating the intensity is far from the sources, the rays from the sources to a point P on the screen are approximately parallel. The path difference between any two adjacent sources is $d \sin \theta$, and the phase difference δ is related to the path difference by

$$\frac{\delta}{2\pi} = \frac{d \sin \theta}{\lambda}$$

If A_0 is the amplitude due to a single source, the amplitude at the central maximum, where $\theta = 0$ and all the waves are in phase, is $A_{max} = NA_0$ (Figure 33-24).

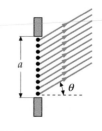

FIGURE 33-23 Diagram for calculating the diffraction pattern far away from a narrow slit. The slit width a is assumed to contain a large number of in-phase point sources separated by a distance d. The rays from these sources to a point far away are approximately parallel. The path difference for the waves from adjacent sources is $d \sin \theta$.

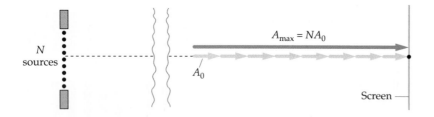

$A_{max} = NA_0$

N sources

A_0

Screen

FIGURE 33-24 A single slit is represented by N sources, each of amplitude A_0. At the central maximum point, where $\theta = 0$, the waves from the sources add in phase, giving a resultant amplitude $A_{max} = NA_0$.

We can find the amplitude at some other point at an angle θ by using the phasor method for the addition of harmonic waves. As in the addition of two, three, or four waves, the intensity is zero at any point where the phasors representing the waves form a closed polygon. In this case, the polygon has N sides (Figure 33-25). At the first minimum, the wave from the first source just below the top of the opening and the wave from the source just below the middle of the opening are 180° out of phase. In this case, the waves from the source near the top of the opening differ from those from the bottom of the opening by nearly 360°. [The phase difference is, in fact, $360° - (360°/N)$.] Thus, if the number of sources is very large, $360°/N$ is negligible and we get complete cancellation if the waves from the first and last sources are out of phase by 360°, corresponding to a path difference of 1 wavelength, in agreement with Equation 33-11.

We will now calculate the amplitude at a general point at which the waves from two adjacent sources differ in phase by δ. Figure 33-26 shows the phasor diagram for the addition of N waves, where the subsequent waves differ in phase from the first wave by $\delta, 2\delta, \ldots, (N - 1)\delta$. When N is very large and δ is very small, the phasor diagram approximates the arc of a circle. The resultant amplitude A is the length of the chord of this arc. We will calculate this resultant amplitude in terms of the phase difference ϕ between the first wave and the last wave. From Figure 33-26, we have

$$\sin \tfrac{1}{2}\phi = \frac{A/2}{r}$$

$\delta = \dfrac{360°}{N}$

FIGURE 33-25 Phasor diagram for calculating the first minimum in the single-slit diffraction pattern. When the waves from the N sources completely cancel, the N phasors form a closed polygon. The phase difference between the waves from adjacent sources is then $\delta = 360°/N$. When N is very large, the waves from the first and last sources are approximately in phase.

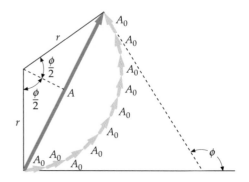

FIGURE 33-26 Phasor diagram for calculating the resultant amplitude due to the waves from N sources in terms of the phase difference ϕ between the wave from the first source just below the top of the slit and the wave from the last source just above the bottom of the slit. When N is very large, the resultant amplitude A is the chord of a circular arc of length $NA_0 = A_{max}$.

or

$$A = 2r \sin \tfrac{1}{2}\phi \qquad\qquad 33\text{-}17$$

where r is the radius of the arc. Since the length of the arc is $A_{max} = NA_0$ and the angle subtended is ϕ, we have

$$\phi = \frac{A_{max}}{r} \qquad\qquad 33\text{-}18$$

or

$$r = \frac{A_{max}}{\phi}$$

Substituting this into Equation 33-17 gives

$$A = \frac{2A_{max}}{\phi} \sin \frac{1}{2}\phi = A_{max} \frac{\sin \tfrac{1}{2}\phi}{\tfrac{1}{2}\phi}$$

Since the amplitude at the center of the central maximum ($\theta = 0$) is A_{max}, the ratio of the intensity at any other point to that at the center of the central maximum is

$$\frac{I}{I_0} = \frac{A^2}{A_{max}^2} = \left(\frac{\sin \tfrac{1}{2}\phi}{\tfrac{1}{2}\phi} \right)^2$$

or

$$I = I_0 \left(\frac{\sin \tfrac{1}{2}\phi}{\tfrac{1}{2}\phi} \right)^2 \qquad\qquad 33\text{-}19$$

INTENSITY FOR A SINGLE-SLIT DIFFRACTION PATTERN

The phase difference ϕ between the first and last waves is related to the path difference $a \sin \theta$ between the top and bottom of the opening by:

$$\frac{\phi}{2\pi} = \frac{a \sin \theta}{\lambda} \qquad\qquad 33\text{-}20$$

Equation 33-19 and Equation 33-20 describe the intensity pattern shown in Figure 33-11. The first minimum occurs at $a \sin \theta = \lambda$, which is the point where the waves from the middle of the upper half and the middle of the lower half of the slit have a path difference of $\lambda/2$ and are 180° out of phase. The second minimum occurs at $a \sin \theta = 2\lambda$, where the waves from the upper half of the upper half of the slit and those from the lower half of the upper half of the slit have a path difference of $\lambda/2$ and are 180° out of phase.

There is a secondary maximum approximately midway between the first and second minima at $a \sin \theta \approx \tfrac{3}{2}\lambda$. Figure 33-27 shows the phasor diagram for determining the approximate intensity of this secondary maximum. The phase difference between the first and last waves is approximately 360° + 180°. The phasors thus complete $1\tfrac{1}{2}$ circles. The resultant amplitude is the diameter of a circle with a circumference that is two-thirds the total length A_{max}. If $C = \tfrac{2}{3}A_{max}$ is the circumference, the diameter A is

$$A = \frac{C}{\pi} = \frac{\tfrac{2}{3}A_{max}}{\pi} = \frac{2}{3\pi} A_{max}$$

and

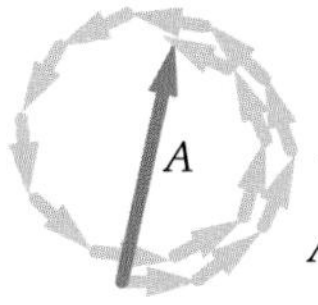

$$\text{Circumference } C = \frac{2}{3} NA_0$$
$$= \frac{2}{3} A_{max} = \pi A$$
$$A = \frac{2}{3\pi} A_{max}$$
$$A^2 = \frac{4}{9\pi^2} A_{max}^2$$

FIGURE 33-27 Phasor diagram for calculating the approximate amplitude of the first secondary maximum of the single-slit diffraction pattern. This secondary maximum occurs near the midpoint between the first and second minima when the N phasors complete $1\tfrac{1}{2}$ circles.

$$A^2 = \frac{4}{9\pi^2} A^2_{max}$$

The intensity at this point is

$$I = \frac{4}{9\pi^2} I_0 \approx \frac{1}{22.2} I_0 \qquad\qquad 33\text{-}21$$

*Calculating the Interference–Diffraction Pattern of Multiple Slits

The intensity of the two-slit interference–diffraction pattern can be calculated from the two-slit pattern (Equation 33-8) with the intensity of each slit (I_0 in that equation) replaced by the diffraction pattern intensity due to each slit, I, given by Equation 33-19. The intensity for the two-slit interference–diffraction pattern is thus

$$I = 4I_0 \left(\frac{\sin\frac{1}{2}\phi}{\frac{1}{2}\phi}\right)^2 \cos^2\frac{1}{2}\delta \qquad\qquad 33\text{-}22$$

INTERFERENCE–DIFFRACTION INTENSITY FOR TWO SLITS

where ϕ is the difference in phase between rays from the top and bottom of each slit, which is related to the width of each slit by

$$\phi = \frac{2\pi}{\lambda} a \sin\theta$$

and δ is the difference in phase between rays from the centers of two adjacent slits, which is related to the slit separation by

$$\delta = \frac{2\pi}{\lambda} d \sin\theta$$

In Equation 33-22, the intensity I_0 is the intensity at $\theta = 0$ due to one slit alone.

FIVE-SLIT INTERFERENCE–DIFFRACTION PATTERN   **E X A M P L E   3 3 - 7**

Find the interference–diffraction intensity pattern for five equally spaced slits, where a is the width of each slit and d is the distance between adjacent slits.

PICTURE THE PROBLEM First, find the interference intensity pattern for the five slits, assuming no angular variations in the intensity due to diffraction. To do this, first construct a phasor diagram to find the amplitude of the resultant wave in an arbitrary direction θ. Intensity is proportional to the square of the amplitude. Next, correct for the variation of intensity with θ by using the single-slit diffraction pattern intensity relation (Equation 33-19 and Equation 33-20).

1. The diffraction pattern intensity I' due to a slit of width a is given by Equation 33-19 and Equation 33-20:

$$I' = I_0 \left(\frac{\sin\frac{1}{2}\phi}{\frac{1}{2}\phi}\right)^2$$

where

$$\phi = \frac{2\pi}{\lambda} a \sin\theta$$

2. The interference pattern intensity I is proportional to the square of the amplitude A of the superposition of the wave functions for the light from the five slits:

$I \propto A^2$

where

$$A \sin(\alpha + \delta') = A_0 \sin \alpha + A_0 \sin(\alpha + \delta) + A_0 \sin(\alpha + 2\delta)$$
$$+ A_0 \sin(\alpha + 3\delta) + A_0 \sin(\alpha + 4\delta)$$

with $\alpha = \omega t$ and $\delta = \dfrac{2\pi}{\lambda} d \sin \theta$

3. To solve for A, we construct a phasor diagram (Figure 33-28). The amplitude A equals the sum of the projections of the individual phasors onto the resultant phasor:

$\delta' = \beta + \delta$

so

$\beta = \delta' - \delta = 2\delta - \delta = \delta$

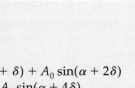

FIGURE 33-28

4. To find δ', we add the exterior angles. The sum of the exterior angles equals 2π. (If you walk the perimeter of a polygon you rotate through the sum of the exterior angles, and you rotate through 2π radians):

$2(\pi - \delta') + 4\delta = 2\pi$

so

$\delta' = 2\delta$

5. Solve for A from the figure:

$A = 2A_0 \cos \delta' + 2A_0 \cos \beta + A_0$

6. Substitute for δ' using the step 4 result, and substitute for β using the relation $\beta = \delta$:

$A = A_0(2 \cos 2\delta + 2 \cos \delta + 1)$

7. Square both sides to relate the intensities. Recall, I' is the intensity from a single slit, and A_0 is the amplitude from a single slit:

$A^2 = A_0^2(2 \cos 2\delta + 2 \cos \delta + 1)^2$

so

$I = I'(2 \cos 2\delta + 2 \cos \delta + 1)^2$

8. Substitute for I' using the step 1 result:

$$I = I_0 \left(\frac{\sin \frac{1}{2}\phi}{\frac{1}{2}\phi} \right)^2 (2 \cos 2\delta + 2 \cos \delta + 1)^2$$

where $\phi = \dfrac{2\pi}{\lambda} a \sin \theta$ and $\delta = \dfrac{2\pi}{\lambda} d \sin \theta$

PLAUSIBILITY CHECK If $\theta = 0$, both $\phi = 0$ and $\delta = 0$. So, for $\theta = 0$, step 5 becomes $A = 5A_0$ and step 8 becomes $I = 5^2 I_0 = 25 I_0$ as expected.

33-6 Fraunhofer and Fresnel Diffraction

Diffraction patterns, like the single-slit pattern shown in Figure 33-11, that are observed at points for which the rays from an aperture or an obstacle are nearly parallel are called **Fraunhofer diffraction patterns.** Fraunhofer patterns can be observed at great distances from the obstacle or the aperture so that the rays reaching any point are approximately parallel, or they can be observed using a lens to focus parallel rays on a viewing screen placed in the focal plane of the lens.

The diffraction pattern observed near an aperture or an obstacle is called a **Fresnel diffraction pattern.** Because the rays from an aperture or an obstacle close to a screen cannot be considered parallel, Fresnel diffraction is much more difficult to analyze. Figure 33-29 illustrates the difference between the Fresnel and the Fraunhofer patterns for a single slit.[†]

Figure 33-30a shows the Fresnel diffraction pattern of an opaque disk. Note the bright spot at the center of the pattern caused by the constructive interference of the light waves diffracted from the edge of the disk. This pattern is of some historical interest. In an attempt to discredit Augustin Fresnel's wave theory of light, Siméon Poisson pointed out that it predicted a bright spot at the center of the shadow, which he assumed was a ridiculous contradiction of fact. However, Fresnel immediately demonstrated experimentally that such a spot does, in fact, exist. This demonstration convinced many doubters of the validity of the wave theory of light. The Fresnel diffraction pattern of a circular aperture is shown in Figure 33-30b. Comparing this with the pattern of the opaque disk in Figure 33-30a, we can see that the two patterns are complements of each other.

Figure 33-31a shows the Fresnel diffraction pattern of a straight edge illuminated by light from a point source. A graph of the intensity versus distance (measured along a line perpendicular to the edge) is shown in Figure 33-31b. The light intensity does not fall abruptly to zero in the geometric shadow, but it decreases rapidly and is negligible within a few wavelengths of the edge. The

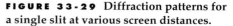
As the screen is moved closer,

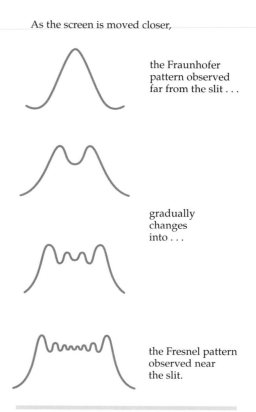

the Fraunhofer pattern observed far from the slit . . .

gradually changes into . . .

the Fresnel pattern observed near the slit.

FIGURE 33-29 Diffraction patterns for a single slit at various screen distances.

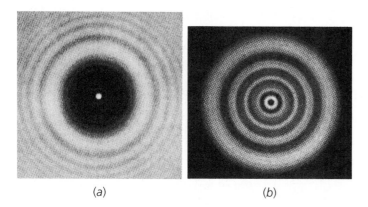

(a)   (b)

FIGURE 33-30 (a) The Fresnel diffraction pattern of an opaque disk. At the center of the shadow, the light waves diffracted from the edge of the disk are in phase and produce a bright spot called the *Poisson spot.* (b) The Fresnel diffraction pattern of a circular aperture. Compare this with Figure 33-30a.

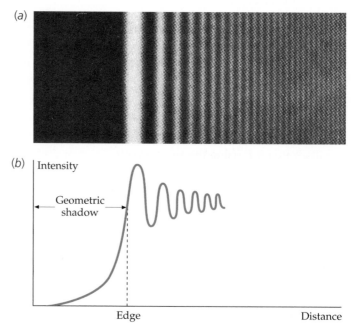

(a)

(b) | Intensity

Geometric shadow

Edge   Distance

FIGURE 33-31 (a) The Fresnel diffraction of a straightedge. (b) A graph of intensity versus distance along a line perpendicular to the edge.

† See Richard E. Haskel, "A Simple Experiment on Fresnel Diffraction," *American Journal of Physics* 38 (1970): 1039.

Fresnel diffraction pattern of a rectangular aperture is shown in Figure 33-32. These patterns cannot be seen with extended light sources like an ordinary lightbulb, because the dark fringes of the pattern produced by light from one point on the source overlap the bright fringes of the pattern produced by light from another point.

33-7 Diffraction and Resolution

Diffraction due to a circular aperture has important implications for the resolution of many optical instruments. Figure 33-33 shows the Fraunhofer diffraction pattern of a circular aperture. The angle θ subtended by the first diffraction minimum is related to the wavelength and the diameter of the opening D by

$$\sin \theta = 1.22 \frac{\lambda}{D} \qquad \text{33-23}$$

Equation 33-23 is similar to Equation 33-9 except for the factor 1.22, which arises from the mathematical analysis, and is similar to the equation for a single slit but more complicated because of the circular geometry. In many applications, the angle θ is small, so $\sin \theta$ can be replaced by θ. The first diffraction minimum is then at an angle θ given by

$$\theta \approx 1.22 \frac{\lambda}{D} \qquad \text{33-24}$$

Figure 33-34 shows two point sources that subtend an angle α at a circular aperture far from the sources. The intensities of the Fraunhofer diffraction pattern are also indicated in this figure. If α is much greater than $1.22\lambda/D$, the sources will be seen as two sources. However, as α is decreased, the overlap of the diffraction patterns increases, and it becomes difficult to distinguish the two sources from one source. At the critical angular separation, α_c, given by

$$\alpha_c = 1.22 \frac{\lambda}{D} \qquad \text{33-25}$$

the first minimum of the diffraction pattern of one source falls on the central maximum of the other source. These objects are said to be just resolved by **Rayleigh's criterion for resolution.** Figure 33-35 shows the diffraction patterns for two sources when α is greater than the critical angle for resolution and when α is just equal to the critical angle for resolution.

Equation 33-25 has many applications. The *resolving power* of an optical instrument, such as a microscope or telescope, is the ability of the instrument to resolve

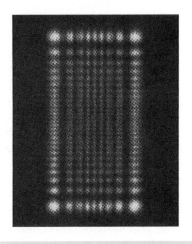

FIGURE 33-32 The Fresnel diffraction pattern of a rectangular aperture.

FIGURE 33-33 The Fraunhofer diffraction pattern of a circular aperture.

(a)

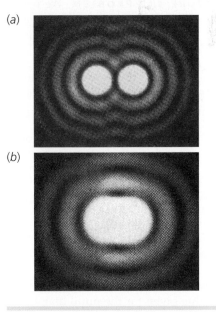

(b)

FIGURE 33-35 The diffraction patterns for a circular aperture and two incoherent point sources when (a) α is much greater than $1.22\lambda/D$ and (b) when α is at the limit of resolution, $\alpha_c = 1.22\lambda/D$.

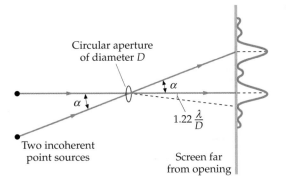

Circular aperture of diameter D

α

α

$1.22 \frac{\lambda}{D}$

Two incoherent point sources

Screen far from opening

FIGURE 33-34 Two distant sources that subtend an angle α. If α is much greater than $1.22\lambda/D$, where λ is the wavelength of light and D is the diameter of the aperture, the diffraction patterns have little overlap and the sources are easily seen as two sources. If α is not much greater than $1.22\lambda/D$, the overlap of the diffraction patterns makes it difficult to distinguish two sources from one.

two objects that are close together. The images of the objects tend to overlap because of diffraction at the entrance aperture of the instrument. We can see from Equation 33-25 that the resolving power can be increased either by increasing the diameter D of the lens (or mirror) or by decreasing the wavelength λ. Astronomical telescopes use large objective lenses or mirrors to increase their resolution as well as to increase their light-gathering power. An array of 27 radio antennas (Figure 33-36) mounted on rails can be configured to form a single telescope with a resolution distance of 36 km (22 mi). In a microscope, a film of transparent oil with index of refraction of approximately 1.55 is sometimes used under the objective to decrease the wavelength of the light ($\lambda' = \lambda/n$). The wavelength can be reduced further by using ultraviolet light and photographic film; however, ordinary glass is opaque to ultraviolet light, so the lenses in an ultraviolet microscope must be made from quartz or fluorite. To obtain very high resolutions, electron microscopes are used—microscopes that use electrons rather than light. The wavelengths of electrons vary inversely with the square root of their kinetic energy and can be made as small as desired.[†]

FIGURE 33-36 The very large array (VLA) of radio antennas is located near Socorro, New Mexico. The 25-m-diameter antennas are mounted on rails, which can be arranged in several configurations, and can be extended over a diameter of 36 km. The data from the antennas are combined electronically, so the array is really a single high-resolution telescope.

---

† The wave properties of electrons are discussed in Chapter 34.

PHYSICS IN THE LIBRARY **EXAMPLE 33-8** **Put It in Context**

While studying in the library, you lean back in your chair and ponder the small holes you notice in the ceiling tiles. You notice that the holes are approximately 5 mm apart. You can clearly see the holes directly above you, about 2 m up, but the tiles far away do not appear to have these holes. You wonder if the reason you cannot see the distant holes is because they are not within the criteria for resolution established by Rayleigh. Is this a feasible explanation for the disappearance of the holes? You notice the holes disappear about 20 m from you.

PICTURE THE PROBLEM We will need to make assumptions about the situation. If we use Equation 33-25, we will need to know the wavelength of light and the aperture diameter. Assuming our pupil is the aperture, we can assume approximately 5 mm for the diameter. (This is the number used in our physics textbook.) The light is probably centered around 500 nm or so.

1. The angular limit for resolution by the eye depends on the ratio of the wavelength and the diameter of the pupil:

$$\theta_c \approx 1.22 \frac{\lambda}{D}$$

2. The angle subtended by two holes depends on their separation distance d and their distance L from your eye:

$$\theta \approx \frac{d}{L}$$

3. Equating the two angles and putting in the numbers gives:

$$\frac{d}{L} \approx 1.22 \frac{\lambda}{D}$$

$$\frac{5 \text{ mm}}{L} \approx 1.22 \frac{500 \text{ nm}}{5 \text{ mm}}$$

4. Solving for L gives:

$$L = 40 \text{ m}$$

5. By a factor of 2, 40 m is too large. However, you are suspect of the value given for the pupil diameter in your physics textbook. You know the pupil is smaller when the light is bright, and the library ceiling is very bright and colored white. An online search for eye pupil diameter soon turns up the information you need. The pupil diameter ranges from 2 to 3 mm up to 7 mm:

> Success. If the pupil diameter is 2.5 mm, the value of L is 20 m.

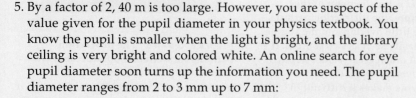

It is instructive to compare the limitation on resolution of the eye due to diffraction, as seen in Example 33-8, with the limitation on resolution due to the separation of the receptors (cones) on the retina. To be seen as two distinct objects, the images of the objects must fall on the retina on two nonadjacent cones. (See Problem 79 in Chapter 32.) Because the retina is about 2.5 cm from the eye lens, the distance y on the retina corresponding to an angular separation of 1.5×10^{-4} rad is found from

$$\alpha_c = 1.5 \times 10^{-4} \text{ rad} = \frac{y}{2.5 \text{ cm}}$$

or

$$y = 3.75 \times 10^{-4} \text{ cm} = 3.75 \times 10^{-6} \text{ m} = 3.75 \ \mu\text{m}$$

The actual separation of the cones in the fovea centralis, where the cones are the most tightly packed, is about 1 μm. Outside this region, they are about 3 μm to 5 μm apart.

*33-8 Diffraction Gratings

A useful tool for measuring the wavelength of light is the **diffraction grating,** which consists of a large number of equally spaced lines or slits on a flat surface. Such a grating can be made by cutting parallel, equally spaced grooves on a glass or metal plate with a precision ruling machine. With a reflection grating, light is reflected from the ridges between the lines or grooves. Phonograph records and compact disks exhibit some of the properties of reflection gratings. In a transmission grating, the light passes through the clear gaps between the rulings. Inexpensive, optically produced plastic gratings with 10,000 or more slits per centimeter are common items. The spacing of the slits in a grating with 10,000 slits per centimeter is $d = (1 \text{ cm})/10{,}000 = 10^{-4}$ cm.

Consider a plane light wave incident normally on a transmission grating (Figure 33-37). Assume that the width of each slit is very small so that it produces a widely diffracted beam. The interference pattern produced on a screen a large distance from the grating is due to a large number of equally spaced light sources. Suppose we have N slits with separation d between adjacent slits. At $\theta = 0$, the light from each slit is in phase with that from all the other slits, so the

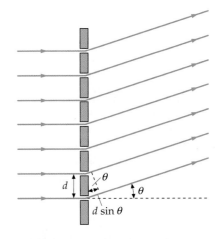

FIGURE 33-37 Light incident normally on a diffraction grating. At an angle θ, the path difference between rays from adjacent slits is $d \sin \theta$.

Compact disks act as reflection gratings.

amplitude of the wave is NA_0, where A_0 is the amplitude from each slit, and the intensity is N^2I_0, where I_0 is the intensity due to a single slit alone. At an angle θ such that $d \sin \theta = \lambda$, the path difference between any two successive slits is λ, so again the light from each slit is in phase with that from all the other slits and the intensity is N^2I_0. The interference maxima are thus at angles θ given by

$$d \sin \theta_m = m\lambda, \qquad m = 0, 1, 2, \ldots \qquad \text{33-26}$$

The position of an interference maximum does not depend on the number of sources, but the more sources there are, the sharper and more intense the maximum will be.

To see that the interference maxima will be sharper when there are many slits, consider the case of N illuminated slits, where N is large ($N \gg 1$). The distance from the first slit to the Nth slit is $(N - 1)d \approx Nd$. When the path difference for the light from the first slit and that from the Nth slit is λ, the resulting intensity will be zero. (We saw this in our discussion of single-slit diffraction.) Since the first and Nth slits are separated by approximately Nd, the intensity will be zero at angle $\theta_{\min}$ given by

$$Nd \sin \theta_{\min} = \lambda$$

so

$$\theta_{\min} \approx \sin \theta_{\min} = \frac{\lambda}{Nd}$$

The width of the interference maximum $2\theta_{\min}$ is thus inversely proportional to N. Therefore, the greater the number of illuminated slits N, the sharper the maximum. Since the intensity in the maximum is proportional to N^2I_0, the intensity in the maximum times the width of the maximum is proportional to NI_0. The intensity times the width is a measure of power per unit length in the maximum.

Figure 33-38a shows a student spectroscope that uses a diffraction grating to analyze light. In student laboratories, the light source is typically a glass tube containing atoms of a gas (e.g., helium or sodium vapor) that are excited by a bombardment of electrons accelerated by high voltage across the tube. The light emitted by such a source contains only certain wavelengths that are characteristic of the atoms in the source. Light from the source passes through a narrow collimating slit and is made parallel by a lens. Parallel light from the lens is incident on the grating. Instead of falling on a screen a large distance away, the parallel light from the grating is focused by a telescope and viewed by the eye. The telescope is mounted on a rotating platform that has been calibrated so that

FIGURE 33-38 (a) A typical student spectroscope. Light from a collimating slit near the source is made parallel by a lens and falls on a grating. The diffracted light is viewed with a telescope at an angle that can be accurately measured. (b) Aerial view of the very large array (VLA) radio telescope in New Mexico. Radio signals from distant galaxies add constructively when Equation 33-26 is satisfied, where d is the distance between two adjacent telescopes.

(a)

(b)

the angle θ can be measured. In the forward direction ($\theta = 0$), the central maximum for all wavelengths is seen. If light of a particular wavelength λ is emitted by the source, the first interference maximum is seen at the angle θ given by Equation 33-26 with $m = 1$. Each wavelength emitted by the source produces a separate image of the collimating slit in the spectroscope called a **spectral line.** The set of lines corresponding to $m = 1$ is called the **first-order spectrum.** The **second-order spectrum** corresponds to $m = 2$ for each wavelength. Higher orders may be seen, providing the angle θ given by Equation 33-26 is less than $90°$. Depending on the wavelengths, the orders may be mixed; that is, the third-order line for one wavelength may occur before the second-order line for another wavelength. If the spacing of the slits in the grating is known, the wavelengths emitted by the source can be determined by measuring the angle θ.

RESOLVING THE SODIUM D LINES　　　　　**EXAMPLE　33-9**

Sodium light is incident on a diffraction grating with 12,000 lines per centimeter. At what angles will the two yellow lines (called the sodium D lines) of wavelengths 589.00 nm and 589.59 nm be seen in the first order?

PICTURE THE PROBLEM Apply $d \sin \theta_m = m\lambda$ to each wavelength, with $m = 1$ and $d = 1 \text{ cm}/12{,}000$.

1. The angle θ_m is given by $d \sin \theta_m = m\lambda$ with $m = 1$:

$$\sin \theta_1 = \frac{\lambda}{d}$$

2. Calculate θ_1 for $\lambda = 589.00$ nm:

$$\theta_1 = \sin^{-1}\left[\frac{589.00 \times 10^{-9} \text{ m}}{(1 \text{ cm}/12{,}000)} \times \left(\frac{100 \text{ cm}}{1 \text{ m}} \right) \right]$$

$$= \boxed{44.98°}$$

3. Repeat the calculation for $\lambda = 589.59$ nm:

$$\theta_1 = \sin^{-1}\left[\frac{589.59 \times 10^{-9} \text{ m}}{(1 \text{ cm}/12{,}000)} \times \left(\frac{100 \text{ cm}}{1 \text{ m}} \right) \right]$$

$$= \boxed{45.03°}$$

REMARKS Note that light of longer wavelength is diffracted through larger angles.

EXERCISE Find the angles for the two yellow lines if the grating has 15,000 lines per centimeter. (*Answer* 62.07° and 62.18°)

An important feature of a spectroscope is its ability to resolve spectral lines of two nearly equal wavelengths λ_1 and λ_2. For example, the two prominent yellow lines in the spectrum of sodium have wavelengths 589.00 and 589.59 nm. These can be seen as two separate wavelengths if their interference maxima do not overlap. According to Rayleigh's criterion for resolution, these wavelengths are resolved if the angular separation of their interference maxima is greater than the angular separation between an interference maximum and the first interference minimum on either side of it. The **resolving power** of a diffraction grating is defined to be $\lambda/|\Delta\lambda|$, where $|\Delta\lambda|$ is the smallest difference between two nearby wavelengths, each approximately equal to λ, that may be resolved. The resolving power is proportional to the number of slits illuminated because the more slits illuminated, the sharper the interference maxima. The resolving power R can be shown to be

$$R = \frac{\lambda}{|\Delta\lambda|} = mN \qquad\qquad 33\text{-}27$$

where N is the number of slits and m is the order number (see Problem 76). We can see from Equation 33-27 that to resolve the two yellow lines in the sodium spectrum the resolving power must be

$$R = \frac{589.00 \text{ nm}}{589.59 \text{ nm} - 589.00 \text{ nm}} = 998$$

Thus, to resolve the two yellow sodium lines in the first order ($m = 1$), we need a grating containing 998 or more slits in the area illuminated by the light.

*Holograms

An interesting application of diffraction gratings is the production of a three-dimensional photograph called a **hologram** (Figure 33-39). In an ordinary photograph, the intensity of reflected light from an object is recorded on a film. When the film is viewed by transmitted light, a two-dimensional image is produced. In a hologram, a beam from a laser is split into two beams, a reference beam and an object beam. The object beam reflects from the object to be photographed and the interference pattern between it, and the reference beam is recorded on a photographic film. This can be done because the laser beam is coherent so that the relative phase difference between the reference beam and the object beam can be kept constant during the exposure. The interference fringes on the film act as a diffraction grating. When the film is illuminated with a laser, a three-dimensional image of the object is produced.

Holograms that you see on credit cards or postage stamps, called rainbow holograms, are more complicated. A horizontal strip of the original hologram is used to make a second hologram. The three-dimensional image can be seen as the viewer moves from side to side, but if viewed with laser light, the image disappears when the viewer's eyes move above or below the slit image. When viewed with white light, the image is seen in different colors as the viewer moves in the vertical direction.

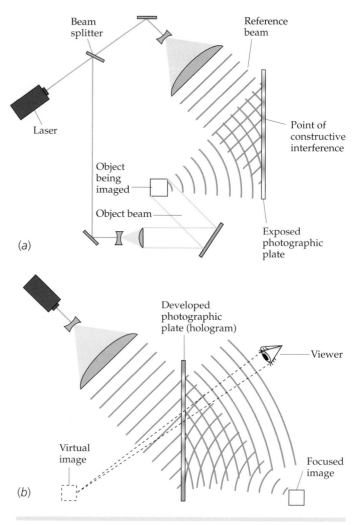

FIGURE 33-39 (*a*) The production of a hologram. The interference pattern produced by the reference beam and object beam is recorded on a photographic film. (*b*) When the film is developed and illuminated by coherent laser light, a three-dimensional image is seen.

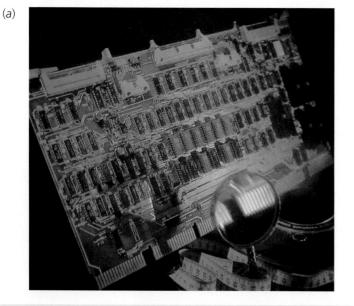

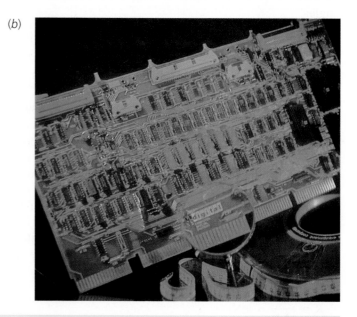

A hologram viewed from two different angles. Note that different parts of the circuit board appear behind the front magnifying lens.

Topic	Relevant Equations and Remarks
1. Interference	Two superposing light waves interfere if their phase difference remains constant for a time long enough to observe. They interfere constructively if their phase difference is zero or an integer times 360°. They interfere destructively if their phase difference is 180° or an odd integer times 180°.
Phase difference due to a path difference	$$\frac{\delta}{2\pi} = \frac{\Delta r}{\lambda} \qquad\qquad\qquad \text{33-1}$$
Phase difference due to reflection	A phase difference of 180° is introduced when a light wave is reflected from a boundary between two media for which the wave speed is greater on the incident-wave side of the boundary.
Thin films	The interference of light waves reflected from the front and back surfaces of a thin film produces interference fringes, commonly observed in soap films or oil films. The difference in phase between the two reflected waves results from the path difference of twice the thickness of the film plus any phase change due to reflection of one or both of the rays.
Two slits	The path difference at an angle θ on a screen far away from two narrow slits separated by a distance d is $d \sin \theta$. If the intensity due to each slit separately is I_0, the intensity at points of constructive interference is $4I_0$, and the intensity at points of destructive interference is zero.
Interference maxima (sources in phase)	$$d \sin \theta_m = m\lambda, \qquad m = 0, 1, 2, \ldots \qquad\qquad \text{33-2}$$
Interference minima (sources 180° out of phase)	$$d \sin \theta_m = (m - \tfrac{1}{2})\lambda, \qquad m = 1, 2, 3, \ldots \qquad\qquad \text{33-3}$$
2. Diffraction	Diffraction occurs whenever a portion of a wavefront is limited by an obstacle or aperture. The intensity of light at any point in space can be computed using Huygens's principle by taking each point on the wavefront to be a point source and computing the resulting interference pattern.
Fraunhofer patterns	Fraunhofer patterns are observed at great distances from the obstacle or aperture so that the rays reaching any point are approximately parallel, or they can be observed using a lens to focus parallel rays on a viewing screen placed in the focal plane of the lens.
Fresnel patterns	Fresnel patterns are observed at points close to the source.
Single slit	When light is incident on a single slit of width a, the intensity pattern on a screen far away shows a broad central diffraction maximum that decreases to zero at an angle θ_1 given by $$\sin \theta_1 = \frac{\lambda}{a} \qquad\qquad\qquad \text{33-9}$$ The width of the central maximum is inversely proportional to the width of the slit. The zeros in the single-slit diffraction pattern occur at angles given by $$a \sin \theta_m = m\lambda, \qquad m = 1, 2, 3, \ldots \qquad\qquad \text{33-11}$$ The maxima on either side of the central maxima have intensities that are much smaller than the intensity of the central maxima.

Two slits	The interference–diffraction pattern of two slits is the two-slit interference pattern modulated by the single-slit diffraction pattern.
Resolution of two sources	When light from two point sources that are close together passes through an aperture, the diffraction patterns of the sources may overlap. If the overlap is too great, the two sources cannot be resolved as two separate sources. When the central diffraction maximum of one source falls at the diffraction minimum of the other source, the two sources are said to be just resolved by Rayleigh's criterion for resolution. For a circular aperture of diameter D, the critical angular separation of two sources for resolution by Rayleigh's criterion is
Rayleigh's criterion	$$\alpha_c = 1.22 \frac{\lambda}{D} \qquad\qquad\qquad 33\text{-}25$$
*Gratings	A diffraction grating consisting of a large number of equally spaced lines or slits is used to measure the wavelength of light emitted by a source. The positions of the mth order interference maxima from a grating are at angles given by $$d \sin \theta_m = m\lambda, \quad m = 0, 1, 2 \ldots \qquad\qquad 33\text{-}26$$ The resolving power of a grating is $$R = \frac{\lambda}{|\Delta\lambda|} = mN \qquad\qquad\qquad 33\text{-}27$$ where N is the number of slits of the grating that are illuminated and m is the order number.
3. *Phasors	Two or more harmonic waves can be added by representing each wave as a two-dimensional vector called a phasor. The phase difference between two harmonic waves is represented as the angle between the phasors.

PROBLEMS

- Single-concept, single-step, relatively easy
- •• Intermediate-level, may require synthesis of concepts
- ••• Challenging
- **SSM** Solution is in the *Student Solutions Manual*
- **iSOLVE** Problems available on iSOLVE online homework service
- **iSOLVE✔** These "Checkpoint" online homework service problems ask students additional questions about their confidence level, and how they arrived at their answer.

In a few problems, you are given more data than you actually need; in a few other problems, you are required to supply data from your general knowledge, outside sources, or informed estimates.

Conceptual Problems

1 • **SSM** When destructive interference occurs, what happens to the energy in the light waves?

2 • Which of the following pairs of light sources are coherent: (*a*) two candles, (*b*) one point source and its image in a plane mirror, (*c*) two pinholes uniformly illuminated by the same point source, (*d*) two headlights of a car, (*e*) two images of a point source due to reflection from the front and back surfaces of a soap film.

3 • The spacing between Newton's rings decreases rapidly as the diameter of the rings increases. Explain qualitatively why this occurs.

4 •• If the angle of a wedge-shaped air film, such as that in Example 33-2, is too large, fringes are not observed. Why?

5 •• Why must a film that is used to observe interference colors be thin?

6 • **SSM** A loop of wire is dipped in soapy water and held so that the soap film is vertical. (*a*) Viewed by reflection with white light, the top of the film appears black. Explain why. (*b*) Below the black region are colored bands. Is the first band red or violet? (*c*) Describe the appearance of the film when it is viewed by *transmitted* light.

7 • As the width of a slit producing a single-slit diffraction pattern is slowly and steadily reduced, how will the diffraction pattern change?

8 • Equation 33-2, which is $d \sin \theta_m = m\lambda$, and Equation 33-11, which is $a \sin \theta_m = m\lambda$, are sometimes confused. For each equation, define the symbols and explain the equation's application.

9 • When a diffraction grating is illuminated with white light, the first-order maximum of green light (*a*) is closer to the central maximum than that of red light. (*b*) is closer to the central maximum than that of blue light. (*c*) overlaps the second-order maximum of red light. (*d*) overlaps the second-order maximum of blue light.

10 • **SSM** A double-slit interference experiment is set up in a chamber that can be evacuated. Using monochromatic light, an interference pattern is observed when the chamber is open to air. As the chamber is evacuated, one will note that (*a*) the interference fringes remain fixed. (*b*) the interference fringes move closer together. (*c*) the interference fringes move farther apart. (*d*) the interference fringes disappear completely.

11 • True or false:

(*a*) When waves interfere destructively, the energy is converted into heat energy.
(*b*) Interference is observed only for waves from coherent sources.
(*c*) In the Fraunhofer diffraction pattern for a single slit, the narrower the slit, the wider the central maximum of the diffraction pattern.
(*d*) A circular aperture can produce both a Fraunhofer diffraction pattern and a Fresnel diffraction pattern.
(*e*) The ability to resolve two point sources depends on the wavelength of the light.

Estimation and Approximation

12 • **SSM** It is claimed that the Great Wall of China is the only human object that can be seen from space with the naked eye. Make an argument in support of this claim based on the resolving power of the human eye. Evaluate the validity of your argument for observers both in low-earth orbit (~ 400 km altitude) and on the moon.

13 • Naturally occuring coronas (brightly colored rings) are sometimes seen around the moon or the sun when viewed through a thin cloud. (Warning: When viewing a sun corona, be sure that the entire sun is blocked by the edge of a building, a tree, or a traffic pole to safeguard your eyes.) These coronas are due to diffraction of light by small water droplets in the cloud. A typical angular diameter for a coronal ring is about 10°. From this, estimate the size of the water droplets in the cloud. Assume that the water droplets can be modeled as opaque disks with the same radius as the droplet, and that the Fraunhofer diffraction pattern from an opaque disk is the same as the pattern from an aperture of the same diameter. (This statement is known as *Babinet's principle*.)

14 • An artificial corona (see Problem 13) can be made by placing a suspension of polystyrene microspheres in water. Polystyrene microspheres are small, uniform spheres made of plastic with an index of refraction of 1.59. Assuming that the water has a refractive index of 1.33, what is the angular diameter of such an artificial corona if 5 μm diameter particles are illuminated by a helium–neon laser with wavelength in air $\lambda = 632.8$ nm?

15 • Coronas (see Problem 13) can be caused by pollen grains, typically of birch or pine. Such grains are irregular in shape, but they can be treated as if they had an average diameter of approximately 25 μm. What is the coronal radius (in degrees) for blue light? What is the coronal radius (in degrees) for red light?

16 •• **SSM** Human hair has a diameter of approximately 70 μm. If we illuminate a hair using a helium–neon laser with wavelength $\lambda = 632.8$ nm and intercept the light scattered from the hair on a screen 10 m away, what will be the separation of the first diffraction peak from the center? (The diffraction pattern of a hair with diameter d is the same as the diffraction pattern of a single slit with width $a = d$.)

Phase Difference and Coherence

17 • **iSOLVE** (*a*) What minimum path difference is needed to introduce a phase shift of 180° in light of wavelength 600 nm? (*b*) What phase shift will that path difference introduce in light of wavelength 800 nm?

18 • **iSOLVE** Light of wavelength 500 nm is incident normally on a film of water 10^{-4} cm thick. The index of refraction of water is 1.33. (*a*) What is the wavelength of the light in the water? (*b*) How many wavelengths are contained in the distance $2t$, where t is the thickness of the film? (*c*) What is the phase difference between the wave reflected from the top of the air–water interface and the wave reflected from the bottom of the water–air interface after it has traveled this distance?

19 •• **SSM** **iSOLVE** ✔ Two coherent microwave sources that produce waves of wavelength 1.5 cm are in the xy plane, one on the y axis at $y = 15$ cm and the other at $x = 3$ cm, $y = 14$ cm. If the sources are in phase, find the difference in phase between the two waves from these sources at the origin.

Interference in Thin Films

20 • A wedge-shaped film of air is made by placing a small slip of paper between the edges of two flat plates of glass. Light of wavelength 700 nm is incident normally on the glass plates, and interference bands are observed by reflection. (*a*) Is the first band near the point of contact of the plates dark or bright? Why? (*b*) If there are five dark bands per centimeter, what is the angle of the wedge?

21 •• **SSM** The diameters of fine fibers can be accurately measured using interference patterns. Two optically flat pieces of glass of length L are arranged with the wire between them, as shown in Figure 33-40. The setup is illuminated by monochromatic light, and the resulting interference fringes are detected. Suppose that $L = 20$ cm and that yellow sodium light ($\lambda \approx 590$ nm) is used for illumination. If 19 bright fringes are seen along this 20-cm distance, what are the limits on the diameter of the wire? (*Hint:* The nineteenth fringe might not be right at the end, but you do not see a twentieth fringe at all.)

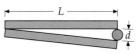

FIGURE 33-40 Problem 21

22 •• [SOLVE] ✓ Light of wavelength 600 nm is used to illuminate two glass plates at normal incidence. The plates are 22 cm in length, touch at one end, and are separated at the other end by a wire of radius 0.025 mm. How many bright fringes appear along the total length of the plates?

23 •• A thin film having an index of refraction of 1.5 is surrounded by air. It is illuminated normally by white light and is viewed by reflection. Analysis of the resulting reflected light shows that the wavelengths 360, 450, and 602 nm are the only missing wavelengths in or near the visible portion of the spectrum. That is, for these wavelengths, there is destructive interference. (a) What is the thickness of the film? (b) What visible wavelengths are brightest in the reflected interference pattern? (c) If this film were resting on glass with an index of refraction of 1.6, what wavelengths in the visible spectrum would be missing from the reflected light?

24 •• [SOLVE] A drop of oil ($n = 1.22$) floats on water ($n = 1.33$). When reflected light is observed from above, as shown in Figure 33-41, what is the thickness of the drop at the point where the second red fringe, counting from the edge of the drop, is observed? Assume red light has a wavelength of 650 nm.

FIGURE 33-41
Problem 24

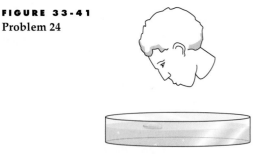

25 •• A film of oil of index of refraction $n = 1.45$ rests on an optically flat piece of glass of index of refraction $n = 1.6$. When illuminated with white light at normal incidence, light of wavelengths 690 nm and 460 nm is predominant in the reflected light. Determine the thickness of the oil film.

26 •• [SSM] [SOLVE] ✓ A film of oil of index of refraction $n = 1.45$ floats on water ($n = 1.33$). When illuminated with white light at normal incidence, light of wavelengths 700 nm and 500 nm is predominant in the reflected light. Determine the thickness of the oil film.

Newton's Rings

27 •• [SSM] A Newton's ring apparatus consists of a plano-convex glass lens with radius of curvature R that rests on a flat glass plate, as shown in Figure 33-42. The thin film is air of variable thickness. The pattern is viewed by reflected light. (a) Show that for a thickness t the condition for a bright (constructive) interference ring is

$$t = \left(m + \frac{1}{2} \right)\frac{\lambda}{2}, \quad m = 0, 1, 2, \ldots$$

(b) Apply the Pythagorean theorem to the triangle of sides r, $R - t$, and hypotenuse R to show that for $t \ll R$, the radius of a fringe is related to t by

$$r = \sqrt{2tR}$$

(c) How would the transmitted pattern look in comparison with the reflected one? (d) Use $R = 10$ m and a lens diameter of 4 cm. How many bright fringes would you see if the apparatus was illuminated by yellow sodium light ($\lambda \approx 590$ nm) and viewed by reflection? (e) What would be the diameter of the sixth bright fringe? (f) If the glass used in the apparatus has an index of refraction $n = 1.5$ and water ($n_w = 1.33$) is placed between the two pieces of glass, what change will take place in the bright-fringe pattern?

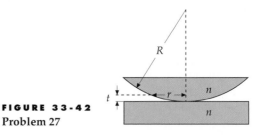

FIGURE 33-42
Problem 27

28 •• [SOLVE] ✓ A plano-convex glass lens of radius of curvature 2.0 m rests on an optically flat glass plate. The arrangement is illuminated from above with monochromatic light of 520-nm wavelength. The indexes of refraction of the lens and plate are 1.6. Determine the radii of the first and second bright fringe in the reflected light. (Use the results from Problem 27b to relate r to t.)

29 •• Suppose that before the lens of Problem 28 is placed on the plate, a film of oil of refractive index 1.82 is deposited on the plate. What will then be the radii of the first and second bright fringes? (Use the results from Problem 27b to relate r to t.)

Two-Slit Interference Pattern

30 • [SSM] Two narrow slits separated by 1 mm are illuminated by light of wavelength 600 nm, and the interference pattern is viewed on a screen 2 m away. Calculate the number of bright fringes per centimeter on the screen.

31 • [SOLVE] ✓ Using a conventional two-slit apparatus with light of wavelength 589 nm, 28 bright fringes per centimeter are observed on a screen 3 m away. What is the slit separation?

32 • Light of wavelength 633 nm from a helium–neon laser is shone normally on a plane containing two slits. The first interference maximum is 82 cm from the central maximum on a screen 12 m away. (a) Find the separation of the slits. (b) How many interference maxima can be observed?

33 •• [SOLVE] Two narrow slits are separated by a distance d. Their interference pattern is to be observed on a screen a large distance L away. (a) Calculate the spacing Δy of the maxima on the screen for light of wavelength 500 nm, when $L = 1$ m and $d = 1$ cm. (b) Would you expect to be able to observe the interference of light on the screen for this situation? (c) How close together should the slits be placed for the maxima to be separated by 1 mm for this wavelength and screen distance?

34 •• Light is incident at an angle ϕ with the normal to a vertical plane containing two slits of separation d (Figure 33-43). Show that the interference maxima are located at angles θ_m given by $\sin\theta_m + \sin\phi = m\lambda/d$.

FIGURE 33-43 Problems 34 and 35

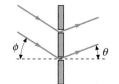

35 •• [SSM] White light falls at an angle of 30° to the normal of a plane containing a pair of slits separated by 2.5 μm. What visible wavelengths give a bright interference maximum in the transmitted light in the direction normal to the plane? (See Problem 34.)

36 •• Two small speakers are separated by a distance of 5 cm, as shown in Figure 33-44. The speakers are driven in phase with a sine wave signal of frequency 10 kHz. A small microphone is placed a distance 1 m away from the speakers on the axis running through the middle of the two speakers, and the microphone is then moved perpendicular to the axis. Where does the microphone record the first minimum and the first maximum of the interference pattern from the speakers? The speed of sound in air is 343 m/s.

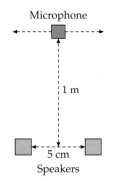

Microphone

1 m

5 cm

Speakers

FIGURE 33-44 Problem 36

Diffraction Pattern of a Single Slit

37 • [ISOLVE] Light of wavelength 600 nm is incident on a long narrow slit. Find the angle of the first diffraction minimum if the width of the slit is (a) 1 mm, (b) 0.1 mm, and (c) 0.01 mm.

38 • [ISOLVE]✓ Plane microwaves are incident on a thin metal sheet with a long, narrow slit of width 5 cm in it. The first diffraction minimum is observed at $\theta = 37°$. What is the wavelength of the microwaves?

39 •• [SSM] Measuring the distance to the moon (lunar ranging) is routinely done by firing short-pulse lasers and measuring the time it takes for the pulses to reflect back from the moon. A pulse is fired from the earth; to send it out, the pulse is expanded so that it fills the aperture of a 6-in-diameter telescope. (a) Assuming the only thing spreading the beam out to be diffraction, how large will the beam be when it reaches the moon, 382,000 km away? (b) The pulse is reflected off a retroreflecting mirror left by the Apollo 11 astronauts. If the diameter of the mirror is 20 in, how large will the beam be when it gets back to the earth? (c) What fraction of the power of the beam is reflected back to the earth? (d) If the beam is refocused on return by the same 6 in telescope, what fraction of the original beam energy is recaptured? Ignore any atmospheric losses.

Interference–Diffraction Pattern of Two Slits

40 • How many interference maxima will be contained in the central diffraction maximum in the interference–diffraction pattern of two slits if the separation d of the slits is 5 times their width a? How many will there be if $d = na$ for any value of n?

41 •• A two-slit Fraunhofer interference–diffraction pattern is observed with light of wavelength 500 nm. The slits have a separation of 0.1 mm and a width of a. (a) Find the width a if the fifth interference maximum is at the same angle as the first diffraction minimum. (b) For this case, how many bright interference fringes will be seen in the central diffraction maximum?

42 •• [ISOLVE]✓ A two-slit Fraunhofer interference–diffraction pattern is observed with light of wavelength 700 nm. The slits have widths of 0.01 mm and are separated by 0.2 mm. How many bright fringes will be seen in the central diffraction maximum?

43 •• [SSM] Suppose that the *central* diffraction maximum for two slits contains 17 interference fringes for some wavelength of light. How many interference fringes would you expect in the first *secondary* diffraction maximum?

44 •• [ISOLVE] Light of wavelength 550 nm illuminates two slits of width 0.03 mm and separation 0.15 mm. (a) How many interference maxima fall within the full width of the central diffraction maximum? (b) What is the ratio of the intensity of the third interference maximum to the side of the centerline (not counting the center interference maximum) to the intensity of the center interference maximum?

*Using Phasors to Add Harmonic Waves

45 • Find the resultant of the two waves $E_1 = 2 \sin \omega t$ and $E_2 = 3 \sin(\omega t + 270°)$.

46 • [SSM] Find the resultant of the two waves $E_1 = 4 \sin \omega t$ and $E_2 = 3 \sin(\omega t + 60°)$.

47 •• At the second secondary maximum of the diffraction pattern of a single slit, the phase difference between the waves from the top and bottom of the slit is approximately 5π. The phasors used to calculate the amplitude at this point complete 2.5 circles. If I_0 is the intensity at the central maximum, find the intensity I at this second secondary maximum.

48 •• (a) Show that the positions of the interference minima on a screen a large distance L away from three equally spaced sources (spacing d, with $d \gg \lambda$) are given approximately by

$$y = \frac{n\lambda L}{3d}, \text{ where } n = 1, 2, 4, 5, 7, 8, 10, \ldots$$

that is, n is not a multiple of 3. (b) For $L = 1$ m, $\lambda = 5 \times 10^{-7}$ m, and $d = 0.1$ mm, calculate the width of the principal interference maxima (the distance between successive minima) for three sources.

49 •• (a) Show that the positions of the interference minima on a screen a large distance L away from four equally spaced sources (spacing d, with $d \gg \lambda$) are given approximately by

$$y = \frac{n\lambda L}{4d}, \text{ where } n = 1, 2, 3, 5, 6, 7, 9, 10, \ldots$$

that is, n is not a multiple of 4. (b) For $L = 2$ m, $\lambda = 6 \times 10^{-7}$ m, and $d = 0.1$ mm, calculate the width of the principal interference maxima (the distance between successive minima) for four sources. Compare this width with that for two sources with the same spacing.

50 •• Light of wavelength 480 nm falls normally on four slits. Each slit is 2 μm wide and is separated from the next slit by 6 μm. (a) Find the angle from the center to the first point of zero intensity of the single-slit diffraction pattern on a distant screen. (b) Find the angles of any bright interference maxima that lie inside the central diffraction maximum. (c) Find the angular spread between the central interference maximum and the first interference minimum on either side of it. (d) Sketch the intensity as a function of angle.

51 ••• Three slits, each separated from its neighbor by 0.06 mm, are illuminated by a coherent light source of wavelength 550 nm. The slits are extremely narrow. A screen is located 2.5 m from the slits. The intensity on the centerline is 0.05 W/m². Consider a location 1.72 cm from the centerline. (a) Draw the phasors, according to the phasor model for the addition of harmonic waves, appropriate for this location. (b) From the phasor diagram, calculate the intensity of light at this location.

52 ••• **SSM** For single-slit diffraction, calculate the first three values of ϕ (the total phase difference between rays from each edge of the slit) that produce subsidiary maxima by (a) using the phasor model and (b) setting $dI/d\phi = 0$, where I is given by Equation 33-19.

Diffraction and Resolution

53 • **SOLVE** ✓ Light of wavelength 700 nm is incident on a pinhole of diameter 0.1 mm. (a) What is the angle between the central maximum and the first diffraction minimum for a Fraunhofer diffraction pattern? (b) What is the distance between the central maximum and the first diffraction minimum on a screen 8 m away?

54 • Two sources of light of wavelength 700 nm are 10 m away from the pinhole of Problem 53. How far apart must the sources be for their diffraction patterns to be resolved by Rayleigh's criterion?

55 • **SSM** Two sources of light of wavelength 700 nm are separated by a horizontal distance x. They are 5 m from a vertical slit of width 0.5 mm. What is the least value of x for which the diffraction pattern of the sources can be resolved by Rayleigh's criterion?

56 • The headlights on a small car are separated by 112 cm. At what maximum distance could you resolve the headlights if the diameter of your pupil is 5 mm and the effective wavelength of the light is 550 nm?

57 • You are told not to shoot until you see the whites of their eyes. If their eyes are separated by 6.5 cm and the diameter of your pupil is 5 mm, at what distance can you resolve the two eyes using light of wavelength 550 nm?

58 •• **SOLVE** ✓ The ceiling of your lecture hall is probably covered with acoustic tile, which has small holes separated by about 6 mm. (a) Using light with a wavelength of 500 nm, how far could you be from this tile and still resolve these holes? The diameter of the pupil of your eye is about 5 mm. (b) Could you resolve these holes better with red light or with violet light?

59 •• **SOLVE** The telescope on Mount Palomar has a diameter of 200 in. Suppose a double star were 4 light-years away. Under ideal conditions, what must be the minimum separation of the two stars for their images to be resolved using light of wavelength 550 nm?

60 •• **SSM** The star Mizar in Ursa Major is a binary system of stars of nearly equal magnitudes. The angular separation between the two stars is 14 seconds of arc. What is the minimum diameter of the pupil that allows resolution of the two stars using light of wavelength 550 nm?

*Diffraction Gratings

61 • **SOLVE** A diffraction grating with 2000 slits per centimeter is used to measure the wavelengths emitted by hydrogen gas. At what angles θ in the first-order spectrum would you expect to find the two violet lines of wavelengths 434 nm and 410 nm?

62 • **SSM** With the diffraction grating used in Problem 61, two other lines in the first-order hydrogen spectrum are found at angles $\theta_1 = 9.72 \times 10^{-2}$ rad and $\theta_2 = 1.32 \times 10^{-1}$ rad. Find the wavelengths of these lines.

63 • Repeat Problem 61 for a diffraction grating with 15,000 slits per centimeter.

64 • **SOLVE** What is the longest wavelength that can be observed in the fifth-order spectrum using a diffraction grating with 4000 slits per centimeter?

65 • The colors of many butterfly wings and beetle carapaces are due to effects of diffraction. The *Morpho* butterfly has structural elements on its wings that effectively act as a diffraction grating with spacing 880 nm. At what angle θ_1 will normally incident blue light of wavelength $\lambda = 440$ nm be diffracted by the *Morpho's* wings?

66 •• A diffraction grating of 2000 slits per centimeter is used to analyze the spectrum of mercury. (a) Find the angular separation in the first-order spectrum of the two lines of wavelength 579 nm and 577 nm. (b) How wide must the beam on the grating be for these lines to be resolved?

67 •• **SSM** A diffraction grating with 4800 lines per centimeter is illuminated at normal incidence with white light (wavelength range of 400 nm to 700 nm). For how many orders can one observe the complete spectrum in the transmitted light? Do any of these orders overlap? If so, describe the overlapping regions.

68 •• **SOLVE** A square diffraction grating with an area of 25 cm² has a resolution of 22,000 in the fourth order. At what angle should you look to see a wavelength of 510 nm in the fourth order?

69 •• Sodium light of wavelength 589 nm falls normally on a 2-cm-square diffraction grating ruled with 4000 lines per centimeter. The Fraunhofer diffraction pattern is projected onto a screen at 1.5 m by a lens of focal length 1.5 m placed immediately in front of the grating. Find (a) the positions of the first two intensity maxima on one side of the central maximum, (b) the width of the central maximum, and (c) the resolution in the first order.

70 •• The spectrum of neon is exceptionally rich in the visible region. Among the many lines are two lines at wavelengths of 519.313 nm and 519.322 nm. If light from a neon discharge tube is normally incident on a transmission grating with 8400 lines per centimeter and the spectrum is observed in second order, what must be the width of the grating that is illuminated, so that these two lines can be resolved?

71 •• **SSM** Mercury has several stable isotopes, among them ¹⁹⁸Hg and ²⁰²Hg. The strong spectral line of mercury, at about 546.07 nm, is a composite of spectral lines from the various mercury isotopes. The wavelengths of this line for ¹⁹⁸Hg and ²⁰²Hg are 546.07532 nm and 546.07355 nm, respectively. What must be the resolving power of a grating capable of resolving these two isotopic lines in the third-order spectrum? If the grating is illuminated over a 2-cm-wide region, what must be the number of lines per centimeter of the grating?

72 •• **SOLVE** A transmission grating is used to study the spectral region extending from 480 nm to 500 nm. The angular spread of this region is 12° in the third order. (a) Find the number of lines per centimeter. (b) How many orders are visible?

73 •• **SOLVE** White light is incident normally on a transmission grating and the spectrum is observed on a screen 8.0 m from the grating. In the second-order spectrum, the separation between light of 520-nm wavelength and 590-nm wavelength is 8.4 cm. (a) Determine the number of lines per centimeter of the grating. (b) What is the separation between these two wavelengths in the first-order spectrum and the third-order spectrum?

74 ••• A diffraction grating has n lines per unit length. Show that the angular separation of two lines of wavelengths λ and $\lambda + \Delta\lambda$ is approximately

$$\Delta\theta = \Delta\lambda / \sqrt{\frac{1}{n^2 m^2} - \lambda^2}$$

where m is the order number.

75 ••• For a diffraction grating in which all the surfaces are normal to the incident radiation, most of the energy goes into the zeroth order, which is useless from a spectroscopic point of view, since in zeroth order all the wavelengths are at 0°. Therefore, modern reflection gratings have shaped, or *blazed*, grooves, as shown in Figure 33-45. This shifts the specular reflection, which contains most of the energy, from the zeroth order to some higher order. (a) Calculate the blaze angle ϕ_m in terms of the groove separation d, the wavelength λ, and the order number m in which specular reflection is to occur for $m = 1, 2, \ldots$. (b) Calculate the proper blaze angle for the specular reflection to occur in the second order for light of wavelength 450 nm incident on a grating with 10,000 lines per centimeter.

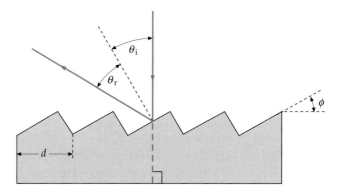

FIGURE 33-45 Problem 75

76 ••• In this problem, you will derive Equation 33-27 for the resolving power of a diffraction grating containing N slits separated by a distance d. To do this, you will calculate the angular separation between the maximum and minimum for some wavelength λ and set it equal to the angular separation of the mth-order maximum for two nearby wavelengths. (a) Show that the phase difference ϕ between the light from two adjacent slits is given by

$$\phi = \frac{2\pi d}{\lambda} \sin\theta$$

(b) Differentiate this expression to show that a small change in angle $d\theta$ results in a change in phase of $d\phi$ given by

$$d\phi = \frac{2\pi d}{\lambda} \cos\theta \, d\theta$$

(c) For N slits, the angular separation between an interference maximum and an interference minimum corresponds to a phase change of $d\phi = 2\pi/N$. Use this to show that the angular separation $d\theta$ between the maximum and minimum for some wavelength λ is given by

$$d\theta = \frac{\lambda}{Nd \cos\theta}$$

(d) The angle of the mth-order interference maximum for wavelength λ is given by Equation 33-26. Compute the differential of each side of this equation to show that angular separation of the mth-order maximum for two nearly equal wavelengths differing by $d\lambda$ is given by

$$d\theta \approx \frac{m \, d\lambda}{d \cos\theta}$$

(e) According to Rayleigh's criterion, two wavelengths will be resolved in the mth order if the angular separation of the wavelengths, given by the Part (d) result, equals the angular separation of the interference maximum and the interference minimum given by the Part (c) result. Use this to derive Equation 33-27 for the resolving power of a grating.

General Problems

77 • [SSM] A long, narrow horizontal slit lies 1 mm above a plane mirror, which is in the horizontal plane. The interference pattern produced by the slit and its image is viewed on a screen 1 m from the slit. The wavelength of the light is 600 nm. (*a*) Find the distance from the mirror to the first maximum. (*b*) How many dark bands per centimeter are seen on the screen?

78 •• A radio telescope is situated at the edge of a lake. The telescope is looking at light from a radio galaxy that is just rising over the horizon. If the height of the antenna is 20 m above the surface of the lake, at what angle above the horizon will the light from the radio galaxy go through its first interference maximum? The wavelength of the radio waves received by the telescope is $\lambda = 20$ cm. Remember that the light has a 180° phase shift on reflection from the water.

79 • [ISOLVE] In a lecture demonstration, a laser beam of wavelength 700 nm passes through a vertical slit 0.5 mm wide and hits a screen 6 m away. Find the horizontal width of the principal diffraction maximum on the screen; that is, find the distance between the first minimum on the left and the first minimum on the right of the central maximum.

80 • What minimum aperture, in millimeters, is required for opera glasses (binoculars) if an observer is to be able to distinguish the soprano's individual eyelashes (separated by 0.5 mm) at an observation distance of 25 m? Assume the effective wavelength of the light to be 550 nm.

81 • The diameter of the aperture of the radio telescope at Arecibo, Puerto Rico, is 300 m. What is the resolving power of the telescope when tuned to detect microwaves of 3.2-cm wavelength?

82 •• [SSM] [ISOLVE] A thin layer of a transparent material with an index of refraction of 1.30 is used as a nonreflective coating on the surface of glass with an index of refraction of 1.50. What should the thickness of the material be for the material to be nonreflecting for light of wavelength 600 nm?

83 •• A *Fabry–Perot interferometer* consists of two parallel, half-silvered mirrors separated by a small distance *a*. Show that when light is incident on the interferometer with an angle of incidence θ, the transmitted light will have maximum intensity when $a = (m\lambda/2) \cos \theta$.

84 •• A mica sheet 1.2 μm thick is suspended in air. In reflected light, there are gaps in the visible spectrum at 421, 474, 542, and 633 nm. Find the index of refraction of the mica sheet.

85 •• A camera lens is made of glass with an index of refraction of 1.6. This lens is coated with a magnesium fluoride film (*n* = 1.38) to enhance its light transmission. This film is to produce zero reflection for light of wavelength 540 nm.

Treat the lens surface as a flat plane and the film as a uniformly thick flat film. (*a*) How thick must the film be to accomplish its objective? (*b*) Would there be destructive interference for any other visible wavelengths? (*c*) By what factor would the reflection for light of wavelengths 400 nm and 700 nm be reduced by this film? Neglect the variation in the reflected light amplitudes from the two surfaces.

86 •• In a pinhole camera, the image is fuzzy because of geometry (rays arrive at the film after passing through different parts of the pinhole) and because of diffraction. As the pinhole is made smaller, the fuzziness due to geometry is reduced, but the fuzziness due to diffraction is increased. The optimum size of the pinhole for the sharpest possible image occurs when the spread due to diffraction equals the spread due to the geometric effects of the pinhole. Estimate the optimum size of the pinhole if the distance from the pinhole to the film is 10 cm and the wavelength of the light is 550 nm.

87 •• [SSM] [ISOLVE]✓ The Impressionist painter Georges Seurat used a technique called *pointillism*, in which his paintings are composed of small, closely spaced dots of pure color, each about 2 mm in diameter. The illusion of the colors blending together smoothly is produced in the eye of the viewer by diffraction effects. Calculate the minimum viewing distance for this effect to work properly. Use the wavelength of visible light that requires the *greatest* distance, so that you are sure the effect will work for *all* visible wavelengths. Assume the pupil of the eye has a diameter of 3 mm.

88 ••• [SSM] A *Jamin refractometer* is a device for measuring or for comparing the indexes of refraction of gases. A beam of monochromatic light is split into two parts, each of which is directed along the axis of a separate cylindrical tube before being recombined into a single beam that is viewed through a telescope. Suppose that each tube is 0.4 m long and that sodium light of wavelength 589 nm is used. Both tubes are initially evacuated, and constructive interference is observed in the center of the field of view. As air is slowly allowed to enter one of the tubes, the central field of view changes to dark and back to bright a total of 198 times. (*a*) What is the index of refraction of air? (*b*) If the fringes can be counted to $\pm$ 0.25 fringe, where one fringe is equivalent to one complete cycle of intensity variation at the center of the field of view, to what accuracy can the index of refraction of air be determined by this experiment?

89 ••• Light of wavelength λ is diffracted through a single slit of width *a*, and the resulting pattern is viewed on a screen a long distance *L* away from the slit. (*a*) Show that the width of the central maximum on the screen is approximately $2L\lambda/a$. (*b*) If a slit of width $2L\lambda/a$ is cut in the screen and is illuminated by light of the same wavelength, show that the width of its central diffraction maximum at the same distance *L* is *a* to the same approximation.

APPENDIX A

SI Units and Conversion Factors

Basic Units

Length	The *meter* (m) is the distance traveled by light in a vacuum in 1/299,792,458 s.
Time	The *second* (s) is the duration of 9,192,631,770 periods of the radiation corresponding to the transition between the two hyperfine levels of the ground state of the ^{133}Cs atom.
Mass	The *kilogram* (kg) is the mass of the international standard body preserved at Sèvres, France.
Current	The *ampere* (A) is that current in two very long parallel wires 1 m apart that gives rise to a magnetic force per unit length of 2×10^{-7} N/m.
Temperature	The *kelvin* (K) is 1/273.16 of the thermodynamic temperature of the triple point of water.
Luminous intensity	The *candela* (cd) is the luminous intensity, in the perpendicular direction, of a surface of area 1/600,000 m^2 of a blackbody at the temperature of freezing platinum at a pressure of 1 atm.

Derived Units

Force	newton (N)	$1\ \text{N} = 1\ \text{kg·m/s}^2$
Work, energy	joule (J)	$1\ \text{J} = 1\ \text{N·m}$
Power	watt (W)	$1\ \text{W} = 1\ \text{J/s}$
Frequency	hertz (Hz)	$1\ \text{Hz} = \text{cy/s}$
Charge	coulomb (C)	$1\ \text{C} = 1\ \text{A·s}$
Potential	volt (V)	$1\ \text{V} = 1\ \text{J/C}$
Resistance	ohm (Ω)	$1\ \Omega = 1\ \text{V/A}$
Capacitance	farad (F)	$1\ \text{F} = 1\ \text{C/V}$
Magnetic field	tesla (T)	$1\ \text{T} = 1\ \text{N/(A·m)}$
Magnetic flux	weber (Wb)	$1\ \text{Wb} = 1\ \text{T·m}^2$
Inductance	henry (H)	$1\ \text{H} = 1\ \text{J/A}^2$

Conversion Factors

Conversion factors are written as equations for simplicity;
relations marked with an asterisk are exact.

Length

1 km = 0.6215 mi

1 mi = 1.609 km

1 m = 1.0936 yd = 3.281 ft = 39.37 in.

*1 in. = 2.54 cm

*1 ft = 12 in. = 30.48 cm

*1 yd = 3 ft = 91.44 cm

1 lightyear = 1 $c \cdot$y = 9.461×10^{15} m

*1 Å = 0.1 nm

Area

*1 m^2 = 10^4 cm^2

1 km^2 = 0.3861 mi^2 = 247.1 acres

*1 $in.^2$ = 6.4516 cm^2

1 ft^2 = 9.29×10^{-2} m^2

1 m^2 = 10.76 ft^2

*1 acre = 43,560 ft^2

1 mi^2 = 640 acres = 2.590 km^2

Volume

*1 m^3 = 10^6 cm^3

*1 L = 1000 cm^3 = 10^{-3} m^3

1 gal = 3.786 L

1 gal = 4 qt = 8 pt = 128 oz = 231 in^3

1 in^3 = 16.39 cm^3

1 ft^3 = 1728 $in.^3$ = 28.32 L
   = 2.832×10^4 cm^3

Time

*1 h = 60 min = 3.6 ks

*1 d = 24 h = 1440 min = 86.4 ks

1 y = 365.24 d = 3.156×10^7 s

Speed

*1 m/s = 3.6 km/h

1 km/h = 0.2778 m/s = 0.6215 mi/h

1 mi/h = 0.4470 m/s = 1.609 km/h

1 mi/h = 1.467 ft/s

Angle and Angular Speed

*π rad = 180°

1 rad = 57.30°

1° = 1.745×10^{-2} rad

1 rev/min = 0.1047 rad/s

1 rad/s = 9.549 rev/min

Mass

*1 kg = 1000 g

*1 tonne = 1000 kg = 1 Mg

1 u = 1.6606×10^{-27} kg

1 kg = 6.022×10^{26} u

1 slug = 14.59 kg

1 kg = 6.852×10^{-2} slug

1 u = 931.50 MeV/c^2

Density

*1 g/cm^3 = 1000 kg/m^3 = 1 kg/L

(1 g/cm^3)g = 62.4 lb/ft^3

Force

1 N = 0.2248 lb = 10^5 dyn

*1 lb = 4.448222 N

(1 kg)g = 2.2046 lb

Pressure

*1 Pa = 1 N/m^2

*1 atm = 101.325 kPa = 1.01325 bars

1 atm = 14.7 $lb/in.^2$ = 760 mmHg
   = 29.9 in.Hg = 33.8 ftH_2O

1 $lb/in.^2$ = 6.895 kPa

1 torr = 1 mmHg = 133.32 Pa

1 bar = 100 kPa

Energy

*1 kW$\cdot$h = 3.6 MJ

*1 cal = 4.1840 J

1 ft$\cdot$lb = 1.356 J = 1.286×10^{-3} Btu

*1 L$\cdot$atm = 101.325 J

1 L$\cdot$atm = 24.217 cal

1 Btu = 778 ft$\cdot$lb = 252 cal = 1054.35 J

1 eV = 1.602×10^{-19} J

1 u$\cdot c^2$ = 931.50 MeV

*1 erg = 10^{-7} J

Power

1 horsepower = 550 ft$\cdot$lb/s = 745.7 W

1 Btu/h = 1.055 kW

1 W = 1.341×10^{-3} horsepower
   = 0.7376 ft$\cdot$lb/s

Magnetic Field

*1 T = 10^4 G

Thermal Conductivity

1 W/(m$\cdot$K) = 6.938 Btu$\cdot$in./(h$\cdot ft^2 \cdot$F°)

1 Btu$\cdot$in./(h$\cdot ft^2 \cdot$F°) = 0.1441 W/(m$\cdot$K)

APPENDIX B

Numerical Data

Terrestrial Data

Free-fall acceleration g	9.80665 m/s^2; 32.1740 ft/s^2
(Standard value at sea level at 45° latitude)[†]	
Standard value	
At sea level, at equator[†]	9.7804 m/s^2
At sea level, at poles[†]	9.8322 m/s^2
Mass of earth M_E	5.98×10^{24} kg
Radius of earth R_E, mean	6.37×10^6 m; 3960 mi
Escape speed $\sqrt{2R_E g}$	1.12×10^4 m/s; 6.95 mi/s
Solar constant[‡]	1.35 kW/m^2
Standard temperature and pressure (STP):	
Temperature	273.15 K
Pressure	101.325 kPa (1.00 atm)
Molar mass of air	28.97 g/mol
Density of air (STP), ρ_{air}	1.293 kg/m^3
Speed of sound (STP)	331 m/s
Heat of fusion of H_2O (0°C, 1 atm)	333.5 kJ/kg
Heat of vaporization of H_2O (100°C, 1 atm)	2.257 MJ/kg.

† Measured relative to the earth's surface.
‡ Average power incident normally on 1 m^2 outside the earth's atmosphere at the mean distance from the earth to the sun.

Astronomical Data[†]

Earth	
Distance to moon[‡]	3.844×10^8 m; 2.389×10^5 mi
Distance to sun, mean[‡]	1.496×10^{11} m; 9.30×10^7 mi; 1.00 AU
Orbital speed, mean	2.98×10^4 m/s
Moon	
Mass	7.35×10^{22} kg
Radius	1.738×10^6 m
Period	27.32 d
Acceleration of gravity at surface	1.62 m/s^2
Sun	
Mass	1.99×10^{30} kg
Radius	6.96×10^8 m

† Additional solar-system data is available from NASA at <http://nssdc.gsfc.nasa.gov/planetary/planetfact.html>.
‡ Center to center.

Physical Constants[†]

Gravitational constant	G	$6.673(10) \times 10^{-11}$ N·m²/kg²
Speed of light	c	$2.997\ 924\ 58 \times 10^8$ m/s
Fundamental charge	e	$1.602\ 1764\ 62(63) \times 10^{-19}$ C
Avogadro's number	N_A	$6.022\ 141\ 99(47) \times 10^{23}$ particles/mol
Gas constant	R	$8.314\ 472(15)$ J/(mol·K)
		$1.987\ 2065(36)$ cal/(mol·K)
		$8.205\ 746(15) \times 10^{-2}$ L·atm/(mol·K)
Boltzmann constant	$k = R/N_A$	$1.380\ 6503(24) \times 10^{-23}$ J/K
		$8.617\ 342(15) \times 10^{-5}$ eV/K
Stefan-Boltzmann constant	$\sigma = (\pi^2/60)k^4/(\hbar^3 c^2)$	$5.670\ 400(40) \times 10^{-8}$ W/(m²k⁴)
Atomic mass constant	$m_u = \frac{1}{12}m(^{12}C)$	$1.660\ 538\ 73(13) \times 10^{-27}$ kg = 1u
Coulomb constant	$k = 1/(4\pi\epsilon_0)$	$8.987\ 551\ 788\ \ldots \times 10^9$ N·m²/C²
Permittivity of free space	ϵ_0	$8.854\ 187\ 817\ \ldots \times 10^{-12}$ C²/(N·m²)
Permeability of free space	μ_0	$4\pi \times 10^{-7}$ N/A²
		$1.256\ 637 \times 10^{-6}$ N/A²
Planck's constant	h	$6.626\ 068\ 76(52) \times 10^{-34}$ J·s
		$4.135\ 667\ 27(16) \times 10^{-15}$ eV·s
	$\hbar = h/2\pi$	$1.054\ 571\ 596(82) \times 10^{-34}$ J·s
		$6.582\ 118\ 89(26) \times 10^{-16}$ eV·s
Mass of electron	m_e	$9.109\ 381\ 88(72) \times 10^{-31}$ kg
		$0.510\ 998\ 902(21)$ MeV/c^2
Mass of proton	m_p	$1.672\ 621\ 58(13) \times 10^{-27}$ kg
		$938.271\ 998(38) \times$ MeV/c^2
Mass of neutron	m_n	$1.674\ 927\ 16(13) \times 10^{-27}$ kg
		$939.565\ 330(38)$ MeV/c^2
Bohr magneton	$m_B = eh/2m_e$	$9.274\ 0008\ 99(37) \times 10^{-24}$ J/T
		$5.788\ 381\ 749(43) \times 10^{-5}$ eV/T
Nuclear magneton	$m_n = eh/2m_p$	$5.050\ 783\ 17(20) \times 10^{-27}$ J/T
		$3.152\ 451\ 238(24) \times 10^{-8}$ eV/T
Magnetic flux quantum	$\phi_0 = h/2e$	$2.067\ 833\ 636(81) \times 10^{-15}$ T·m²
Quantized Hall resistance	$R_K = h/e^2$	$2.581\ 280\ 7572(95) \times 10^4$ Ω
Rydberg constant	R_H	$1.097\ 373\ 156\ 8549(83) \times 10^7$ m⁻¹
Josephson frequency-voltage quotient	$K_J = 2e/h$	$4.835\ 978\ 98(19) \times 10^{14}$ Hz/V
Compton wavelength	$\lambda_C = h/m_e c$	$2.426\ 310\ 215(18) \times 10^{-12}$ m

[†] The values for these and other constants may be found on the Internet at http://physics.nist.gov/cuu/Constants/index.html. The numbers in parentheses represent the uncertainties in the last two digits. (For example, 2.044 43(13) stands for 2.044 43 ± 0.000 13.) Values with without uncertainties are exact, including those values with ellipses (like the value of pi is exactly 3.1415. . .).

For additional data, see the following tables in the text.

APPENDIX C

Periodic Table of Elements

1																	18
1 **H** 1.00797	2											13	14	15	16	17	2 **He** 4.003
3 **Li** 6.941	4 **Be** 9.012											5 **B** 10.81	6 **C** 12.011	7 **N** 14.007	8 **O** 15.9994	9 **F** 19.00	10 **Ne** 20.179
11 **Na** 22.990	12 **Mg** 24.31	3	4	5	6	7	8	9	10	11	12	13 **Al** 26.98	14 **Si** 28.09	15 **P** 30.974	16 **S** 32.064	17 **Cl** 35.453	18 **Ar** 39.948
19 **K** 39.102	20 **Ca** 40.08	21 **Sc** 44.96	22 **Ti** 47.88	23 **V** 50.94	24 **Cr** 52.00	25 **Mn** 54.94	26 **Fe** 55.85	27 **Co** 58.93	28 **Ni** 58.69	29 **Cu** 63.55	30 **Zn** 65.38	31 **Ga** 69.72	32 **Ge** 72.59	33 **As** 74.92	34 **Se** 78.96	35 **Br** 79.90	36 **Kr** 83.80
37 **Rb** 85.47	38 **Sr** 87.62	39 **Y** 88.906	40 **Zr** 91.22	41 **Nb** 92.91	42 **Mo** 95.94	43 **Tc** (98)	44 **Ru** 101.1	45 **Rh** 102.905	46 **Pd** 106.4	47 **Ag** 107.870	48 **Cd** 112.41	49 **In** 114.82	50 **Sn** 118.69	51 **Sb** 121.75	52 **Te** 127.60	53 **I** 126.90	54 **Xe** 131.29
55 **Cs** 132.905	56 **Ba** 137.33	57–71 **Rare Earths**	72 **Hf** 178.49	73 **Ta** 180.95	74 **W** 183.85	75 **Re** 186.2	76 **Os** 190.2	77 **Ir** 192.2	78 **Pt** 195.09	79 **Au** 196.97	80 **Hg** 200.59	81 **Tl** 204.37	82 **Pb** 207.19	83 **Bi** 208.98	84 **Po** (210)	85 **At** (210)	86 **Rn** (222)
87 **Fr** (223)	88 **Ra** (226)	89–103 **Actinides**	104 **Rf** (261)	105 **Ha** (260)	106 (263)	107 (262)	108 (265)	109 (266)									

Rare Earths (Lanthanides)

57 **La** 138.91	58 **Ce** 140.12	59 **Pr** 140.91	60 **Nd** 144.24	61 **Pm** (147)	62 **Sm** 150.36	63 **Eu** 152.0	64 **Gd** 157.25	65 **Tb** 158.92	66 **Dy** 162.50	67 **Ho** 164.93	68 **Er** 167.26	69 **Tm** 168.93	70 **Yb** 173.04	71 **Lu** 174.97

Actinides

89 **Ac** 227.03	90 **Th** 232.04	91 **Pa** 231.04	92 **U** 238.03	93 **Np** 237.05	94 **Pu** (244)	95 **Am** (243)	96 **Cm** (247)	97 **Bk** (247)	98 **Cf** (251)	99 **Es** (252)	100 **Fm** (257)	101 **Md** (258)	102 **No** (259)	103 **Lr** (260)

The 1–18 group designation has been recommended by the International Union of Pure and Applied Chemistry (IUPAC).

Atomic Numbers and Atomic Masses[†]

Name	Symbol	Atomic Number	Mass	Name	Symbol	Atomic Number	Mass
Actinium	Ac	89	227.03	Mercury	Hg	80	200.59
Aluminum	Al	13	26.98	Molybdenum	Mo	42	95.94
Americium	Am	95	(243)	Neodymium	Nd	60	144.24
Antimony	Sb	51	121.75	Neon	Ne	10	20.179
Argon	Ar	18	39.948	Neptunium	Np	93	237.05
Arsenic	As	33	74.92	Nickel	Ni	28	58.69
Astatine	At	85	(210)	Niobium	Nb	41	92.91
Barium	Ba	56	137.3	Nitrogen	N	7	14.007
Berkelium	Bk	97	(247)	Nobelium	No	102	(259)
Beryllium	Be	4	9.012	Osmium	Os	76	190.2
Bismuth	Bi	83	208.98	Oxygen	O	8	15.9994
Boron	B	5	10.81	Palladium	Pd	46	106.4
Bromine	Br	35	79.90	Phosphorus	P	15	30.974
Cadmium	Cd	48	112.41	Platinum	Pt	78	195.09
Calcium	Ca	20	40.08	Plutonium	Pu	94	(244)
Californium	Cf	98	(251)	Polonium	Po	84	(210)
Carbon	C	6	12.011	Potassium	K	19	39.098
Cerium	Ce	58	140.12	Praseodymium	Pr	59	140.91
Cesium	Cs	55	132.905	Promethium	Pm	61	(147)
Chlorine	Cl	17	35.453	Protactinium	Pa	91	231.04
Chromium	Cr	24	52.00	Radium	Ra	88	(226)
Cobalt	Co	27	58.93	Radon	Rn	86	(222)
Copper	Cu	29	63.55	Rhenium	Re	75	186.2
Curium	Cm	96	(247)	Rhodium	Rh	45	102.905
Dysprosium	Dy	66	162.50	Rubidium	Rb	37	85.47
Einsteinium	Es	99	(252)	Ruthenium	Ru	44	101.1
Erbium	Er	68	167.26	Rutherfordium	Rf	104	(261)
Europium	Eu	63	152.0	Samarium	Sm	62	150.36
Fermium	Fm	100	(257)	Scandium	Sc	21	44.96
Fluorine	F	9	19.00	Selenium	Se	34	78.96
Francium	Fr	87	(223)	Silicon	Si	14	28.09
Gadolinium	Gd	64	157.25	Silver	Ag	47	107.870
Gallium	Ga	31	69.72	Sodium	Na	11	22.990
Germanium	Ge	32	72.59	Strontium	Sr	38	87.62
Gold	Au	79	196.97	Sulfur	S	16	32.064
Hafnium	Hf	72	178.49	Tantalum	Ta	73	180.95
Hahnium	Ha	105	(260)	Technetium	Tc	43	(98)
Helium	He	2	4.003	Tellurium	Te	52	127.60
Holmium	Ho	67	164.93	Terbium	Tb	65	158.92
Hydrogen	H	1	1.0079	Thallium	Tl	81	204.37
Indium	In	49	114.82	Thorium	Th	90	232.04
Iodine	I	53	126.90	Thulium	Tm	69	168.93
Iridium	Ir	77	192.2	Tin	Sn	50	118.69
Iron	Fe	26	55.85	Titanium	Ti	22	47.88
Krypton	Kr	36	83.80	Tungsten	W	74	183.85
Lanthanum	La	57	138.91	Uranium	U	92	238.03
Lawrencium	Lr	103	(260)	Vanadium	V	23	50.94
Lead	Pb	82	207.2	Xenon	Xe	54	131.29
Lithium	Li	3	6.941	Ytterbium	Yb	70	173.04
Lutetium	Lu	71	174.97	Yttrium	Y	39	88.906
Magnesium	Mg	12	24.31	Zinc	Zn	30	65.38
Manganese	Mn	25	54.94	Zirconium	Zr	40	91.22
Mendelevium	Md	101	(258)				

[†] More precise values for the atomic masses, along with the uncertainties in the masses, can be found at http://physics.nist.gov/PhysRefData/.

ILLUSTRATION CREDITS

Chapter 26

Opener p. 829 © B. Sidney/TAXI/Getty Images; **p. 832 Figure 26-8 (b)** © 1995 tom pantages; **p. 834 (a)** Larry Langrill; **(b)** © Lawrence Berkeley Laboratory/Science Photo Library; **p. 835 Figure 26-13 (b)** Carl E. Nielsen.

Chapter 27

Opener p. 856 Bob Williamson, Oakland University, Rochester, Michigan; **p. 858 (a and b)** © 1990 Richard Megna/Fundamental Photographs; **p. 860 Figure 27-6 (b)** © 1990 Richard Megna/Fundamental Photographs; **p. 863 Figure 27-10 (c)** © 1990 Richard Megna/Fundamental Photographs; **p. 865** © Bruce Iverson; **p. 866 Figure 27-15 (b)** © 1990 Richard Megna/Fundamental Photographs; **p. 867** Courtesy of F. W. Bell; **p. 869 Figure 27-20 (a)** Photo by Gene Mosca; **p. 873 (a and b)** Courtesy Princeton University Plasma Physics Laboratory; **p. 880 (top)** J. F. Allen, St. Andrews University, Scotland; **(bottom)** Photo by Gene Mosca; **p. 881 (top)** Robert J. Celotta, National Institute of Standards and Technology; **(middle)** © Paul Silverman/Fundamental Photographs, **(bottom a)** Akira Tonomura, Hitachi Advanced Research Library, Hatomaya, Japan; **(bottom b)** © Bruce Iverson; **p. 883 (a)** © 2003 Western Digital Corporation. All rights reserved; **(b)** Tom Chang/IBM Storage Systems Division, San Jose, CA; **p. 885** © Bill Pierce/*Time* Magazines, Inc.

Chapter 28

Opener p. 897 © 1990 Richard Megna/Fundamental Photographs; **p. 915 Figure 28-28 (b)** © Michael Holford, Collection of the Science Museum, London; **p. 922 (top)** A. Leitner/Renselaer Polytechnic Institute; **(bottom)** © Palmer/Kane, Inc./CORBIS; **p. 926 Figures 28-42 (a and b)** Courtesy PASCO Scientific Co.

Chapter 29

Opener p. 922 © Roger Ressmeyer/CORBIS; **p. 936 (a)** Courtesy of U.S. Department of the Interior, Department of Reclamation; **(b)** © Lee Langum/Photo Researchers, Inc.; **p. 956** © George H. Clark Radioana Collection-Archive Center, National Museum of American History; **p. 957 (a)** © Yoav/Phototake; **(b)** © Daniel S. Brody/Stock Boston.

Chapter 30

Opener p. 971 Courtesy of NASA.

Chapter 31

Opener p. 997 © James L. Amos/CORBIS; **p. 998 Figure 31-1** CORBIS-Bettmann; **p. 999 Figure 31-2** CORBIS-Bettmann; **(bottom)** Adapted from Eastman Kodak and Wabash Instrument Corporation; **p. 1002 (a and b)** © 1991 Paul Silverman/Fundamental Photographs; **p. 1007 (a)** © Chuck O'Rear/West Light; **(b)** Courtesy of Ahmed H. Zewail, California Institute of Technology; **(c)** © Chuck O'Rear/West Light; **(d)** © Michael W. Berns/*Scientific American;* **(e)** © 1988 by David Scharf. All rights reserved; **p. 1012 Figure 31-19 (a)** © 1987 Ken Kay/Fundamental Photographs; **(b)** Courtesy Battelle-Northwest Laboratories; **(bottom)** © 1990 Richard Magna/Fundamental Photographs; **p. 1013 Figure 31-20 (b)** © Macduff Everton/CORBIS; **p. 1014** © 1987 Pete Saloutos/The Stock Market; **p. 1015 Figure 31-23 (b)** © 1987 Ken Kay/Fundamental Photographs; **p. 1017 Figure 31-27 (c)** © Ted Horowitz/The Stock Market; **(bottom a)** © Dan Boyd/Courtesy Naval Research Laboratory; **(bottom b)** Courtesy AT&T Archives; **p. 1018 Figure 31-28 (c)** © Robert Greenler; **p. 1019** © David Parker/Science Photo Library/Photo Researchers; **p. 1020 (a)** © Robert Greenler; **(b)** Giovanni DeAmici, NSF, Lawrence Berkeley Laboratory; **p. 1021 (a and b)** Larry Langrill; **p. 1023 (a)** © 1970 Fundamental Photographs; **(b)** 1990 PAR/NYC, Inc./Photo by Elizabeth Algieri; **p. 1025 Figure 31-43 (b)** © 1987 Paul Silverman Photographs; **p. 1026 (a)** Glen A. Izett, U.S. Geological Survey, Denver, Colorado; **(b)** Glen A. Izett, U.S. Geological Survey, Denver, Colorado; **(c)** Dr. Anthony J. Gow/Cold Regions Research and Engineering Laboratory, Hanover, New Hampshire; **(d)** Dr. Anthony J. Gow/Cold Regions Research and Engineering Laboratory, Hanover, New Hampshire; **(e)** © Sepp Seitz/Woodfin Camp and Associates.

Chapter 32

Opener p. 1038 Photo by Gene Mosca; **p. 1039 Figure 32-2** Photo by Demetrios Zangos; **p. 1048 (a and b)** © 1990 Richard Megna/Fundamental Photographs; **p. 1054 Figure 32-29 (a, bottom)** Nils Abramson; **(b, bottom)** © 1974 Fundamental Photographs; **Figure 32.30 (b)** © Fundamental Photographs; **p. 1057** © Bohdan Hrynewych/Stock Boston; **p. 1068 (a, b, and c)** Lennart Nilsson, **(d)** Courtesy IMEC and University of Pennsylvania Department of Electrical Engineering; **p. 1072 (a)** © Scala/Art Resource; **(b)** © Royal Astronomical Society Library; **(c)** Lick Observatory, courtesy of the University of California Regents; **(d)** California Institute of Technology; **(e)** © 1980 Gary Ladd; **p. 1073 (a, b and c)** © California Association for Research in Astronomy; **(bottom)** Courtesy of NASA.

Chapter 33

Opener p. 1084 © Aaron Haupt/Photo Researchers, NY; **p. 1087 Figure 33-3 (a)** Courtesy of Bausch & Lomb; **p. 1088 Figure 33-5 (a and b)** Courtesy T. A. Wiggins; **Figure 33-6** From *PSSC Physics*, 2nd Edition, 1965. D.C. Heath & Co. and Education Development Center, Newton, MA; **p. 1091 Figure 33-9 (a)** Courtesy of Michael Cagnet; **p. 1092 Figure 33-11 (a)** Courtesy of Michael Cagnet; **p. 1093 Figure 33-14 (a)** Courtesy of Michael Cagnet; **p. 1097 Figure 33-21** Courtesy Michael Cagnet; **p. 1102 Figure 33-30 (a and b)** M. Cagnet, M. Fraçon, J.C. Thrierr, *Atlas of Optical Phenomena;* **Figure 33-31**

(a) Courtesy Battelle-Northwest Laboratories; **p. 1103 Figures 33-32, 33-33,** and **33-35 (a** and **b)** Courtesy of Michael Cagnet; **p. 1104 Figure 33-36** Courtesy of National Radio Astronomy Observatory/ Associated Universities, Inc./National Science Foundation. Photographer: Kelly Gatlin. Digital composite: Patricia Smiley; **p. 1105 (bottom)** © Kevin R. Morris/CORBIS; **p. 1106 Figure 33-38 (a)** Clarence Bennett/Oakland University, Rochester, Michigan; **(b)** NRAO/AUI/Science Photo Library/Photo Researchers; **p. 1108 (a** and **b)** © 1981 by Ronald R. Erickson, Hologram by Nicklaus Phillips, 1978, for Digital Equipment Corporation.

ANSWERS

Problem answers are calculated using $g = 9.81$ m/s² unless otherwise specified in the Problem. Differences in the last figure can easily result from differences in rounding the input data and are not important.

Chapter 26

1. *(b)*

3. False

5. The alternating current running through the filament is changing direction every 1/60 s, so in a magnetic field the filament experiences a force that alternates in direction at that frequency.

7. *(a)* True

   (b) True

   (c) True

   (d) False

   (e) True

9. Upward

11. *(a)*

13. From relativity; This is equivalent to the electron moving from right to left at velocity v with the magnet stationary. When the electron is directly over the magnet, the field points directly up, so there is a force directed out of the page on the electron.

15. If only $\vec{F}$ and I are known, one can only conclude that the magnetic field $\vec{B}$ is in the plane perpendicular to $\vec{F}$. The specific direction of $\vec{B}$ is undetermined.

17. *(a)* 177 C/kg

   (b) 53.1 nC

19. *(a)* $-(3.80 \text{ mN})\hat{k}$

   (b) $-(7.51 \text{ mN})\hat{k}$

   (c) 0

   (d) $(7.51 \text{ mN})\hat{j}$

21. 0.962 N

23. 0.621 pN; $\theta_x = 108°$; $\theta_y = 102°$; $\theta_z = 158°$

25. 1.48 A

29. $\vec{B} = (10 \text{ T})\hat{i} + (10 \text{ T})\hat{j} - (15 \text{ T})\hat{k}$

31. *(a)* 87.4 ns

   (b) $6.47 \times 10^7 \text{ m/s}$

   (c) 11.4 MeV

33. *(a)* 142 m

   (b) 2.84 m

35. *(a)* $\dfrac{v_\alpha}{v_p} = \dfrac{1}{2}$

   (b) $\dfrac{K_\alpha}{K_p} = 1$

   (c) $\dfrac{L_\alpha}{L_p} = 2$

39. *(a)* $\phi = 24.0°$; $r_p = 0.492$ m; $v_p = 2.83 \times 10^7 \text{ m/s}$

   (b) $v_d = 1.41 \times 10^7 \text{ m/s}$

41. *(a)* $1.64 \times 10^6 \text{ m/s}$

   (b) 14.0 keV

   (c) 7.66 eV

43. *(a)* 7.34 mm

   (b) 66.2 μT

45. *(a)* 63.3 cm

   (b) 2.60 cm

47. $\Delta t_{58} = 15.8 \ \mu s$; $\Delta t_{60} = 16.3 \ \mu s$

49. *(a)* 21.3 MHz

   (b) 46.0 MeV

   (c) 10.7 MHz; 23.0 MeV

53. *(a)* 0.302 A·m²

   (b) 0.131 N·m

55. *(a)* 0

   (b) $\pm(2.70 \times 10^{-3} \text{ N·m})\hat{j}$

57. $B = \dfrac{mg}{I\pi R}$

59. *(a)*

$\vec{\tau} = (0.840 \text{ N·m})\hat{k}$

   (b)

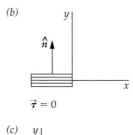

$\vec{\tau} = 0$

   (c)

$\vec{\tau} = 0$

   (d)

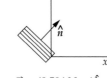

$\vec{\tau} = (0.594 \text{ N·m})\hat{k}$

61. 0.377 A·m²; Into the page.

67. $\mu = \frac{4}{3}\pi\sigma R^4\omega$

69. *(a)* $\tau = \frac{1}{4}\pi\sigma r^4\omega B \sin\theta$

   (b) $\Omega = \dfrac{\pi\sigma r^2 B}{2m}\sin\theta$

71. *(a)* $3.69 \times 10^{-5} \text{ m/s}$

   (b) 1.48 μV

73. 1.02 mV

75. 3.46

77. *(a)* 0.131 μs

   (b) $2.40 \times 10^7 \text{ m/s}$

   (c) 12.0 MeV

81. *(a)* $\vec{B} = -\dfrac{mg}{I\ell}\tan\theta\,\hat{u}_v$

   (b) $g\sin\theta$

83. (a) $\vec{F} = (1.60 \times 10^{-18}\,\text{N})\hat{j}$

    (b) $\vec{E} = (10.0\,\text{V/m})\hat{j}$

    (c) $\Delta V = 20.0\,\text{V}$

85. 5.10 m

Chapter 27

1. (a) The electric forces are repulsive; The magnetic forces are attractive (the two charges moving in the same direction act like two currents in the same direction).

   (b) The electric forces are again repulsive; The magnetic forces are also repulsive.

3.
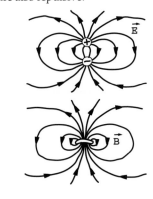

5. (a)

7. (e)

9.
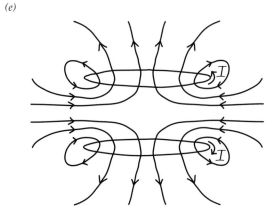

11. (a) True

    (b) True

13. (a) False

    (b) True

    (c) False

    (d) False

    (e) True

15. No. The classical relation between magnetic moment and angular momentum is $\vec{u} = \dfrac{q}{2m}\vec{L}$. Thus, if the angular momentum of the particle is zero, its magnetic moment will also be zero.

17. From Ampère's law, the current enclosed by a closed path within the tube is zero, and from the cylindrical symmetry it follows that $B = 0$ everywhere within the tube.

19. H_2, CO_2, and N_2 are diamagnetic ($\chi_m < 0$); O_2 is paramagnetic ($\chi_m > 0$).

21. 60.0 μT

23. (a) $\vec{B}(0,0) = -(9.00\,\text{pT})\hat{k}$

    (b) $\vec{B}(0,1\,\text{m}) = -(36.0\,\text{pT})\hat{k}$

    (c) $\vec{B}(0,3\,\text{m}) = (36.0\,\text{pT})\hat{k}$

    (d) $\vec{B}(0,4\,\text{m}) = (9.00\,\text{pT})\hat{k}$

25. (a) $\vec{B}(1\,\text{m},3\,\text{m}) = 0$

    (b) $\vec{B}(6\,\text{m},4\,\text{m}) = -(3.56 \times 10^{-23}\,\text{T})\hat{k}$

    (c) $\vec{B}(3\,\text{m},6\,\text{m}) = (4.00 \times 10^{-23}\,\text{T})\hat{k}$

27. $\dfrac{F_B}{F_E} = \dfrac{v^2}{c^2}$

29. $d\vec{B}(3\,\text{m},0,0) = -(9.60\,\text{pT})\hat{i}$

31. (a) $B(0) = 54.5\,\mu$T

    (b) $B(0.01\,\text{m}) = 46.5\,\mu$T

    (c) $B(0.02\,\text{m}) = 31.4\,\mu$T

    (d) $B(0.35\,\text{m}) = 33.9\,$nT

33. (a) 19.1 cm

    (b) 45.3 cm

    (c) 99.5 cm

35. (a)
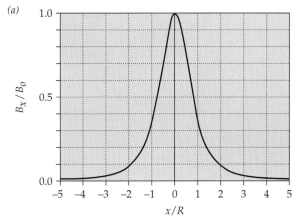

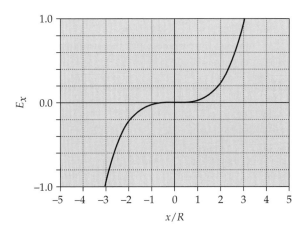

(b)

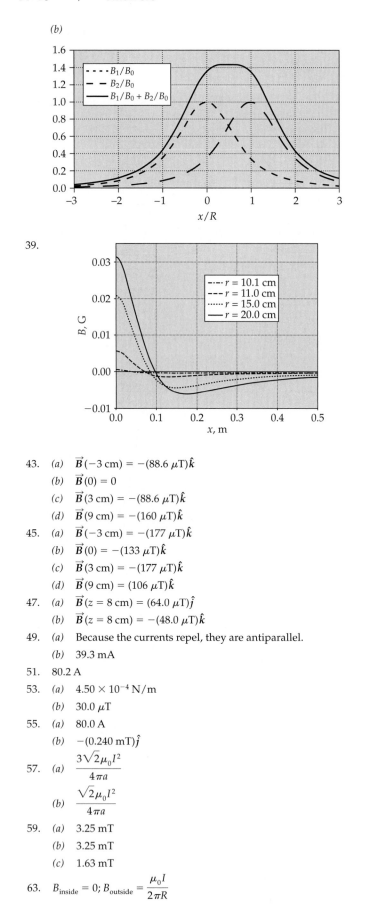

39.

43. (a) $\vec{B}(-3 \text{ cm}) = -(88.6 \ \mu\text{T})\hat{k}$
    (b) $\vec{B}(0) = 0$
    (c) $\vec{B}(3 \text{ cm}) = -(88.6 \ \mu\text{T})\hat{k}$
    (d) $\vec{B}(9 \text{ cm}) = -(160 \ \mu\text{T})\hat{k}$

45. (a) $\vec{B}(-3 \text{ cm}) = -(177 \ \mu\text{T})\hat{k}$
    (b) $\vec{B}(0) = -(133 \ \mu\text{T})\hat{k}$
    (c) $\vec{B}(3 \text{ cm}) = -(177 \ \mu\text{T})\hat{k}$
    (d) $\vec{B}(9 \text{ cm}) = (106 \ \mu\text{T})\hat{k}$

47. (a) $\vec{B}(z = 8 \text{ cm}) = (64.0 \ \mu\text{T})\hat{j}$
    (b) $\vec{B}(z = 8 \text{ cm}) = -(48.0 \ \mu\text{T})\hat{k}$

49. (a) Because the currents repel, they are antiparallel.
    (b) 39.3 mA

51. 80.2 A

53. (a) $4.50 \times 10^{-4} \text{ N/m}$
    (b) 30.0 μT

55. (a) 80.0 A
    (b) $-(0.240 \text{ mT})\hat{j}$

57. (a) $\dfrac{3\sqrt{2}\mu_0 I^2}{4\pi a}$
    (b) $\dfrac{\sqrt{2}\mu_0 I^2}{4\pi a}$

59. (a) 3.25 mT
    (b) 3.25 mT
    (c) 1.63 mT

63. $B_{\text{inside}} = 0$; $B_{\text{outside}} = \dfrac{\mu_0 I}{2\pi R}$

65. (a) $B_{r<R} = \dfrac{\mu_0 I}{2\pi R}$
    (b) $B_{r>R} = 0$

69. (a) $B_{r<a} = 0$
    (b) $B_{a<r<b} = \dfrac{\mu_0 I}{2\pi R}\dfrac{r^2 - a^2}{b^2 - a^2}$
    (c) $B_{r>b} = \dfrac{\mu_0 I}{2\pi R}$

71. (a) 27.3 mT
    (b) 20.0 mT

73. (a) 10.1 mT
    (b) 10.1 mT; 1.52 T

75. (a) 0.544 A/m
    (b) 10.054 A/m

77. -4.00×10^{-5}

79.

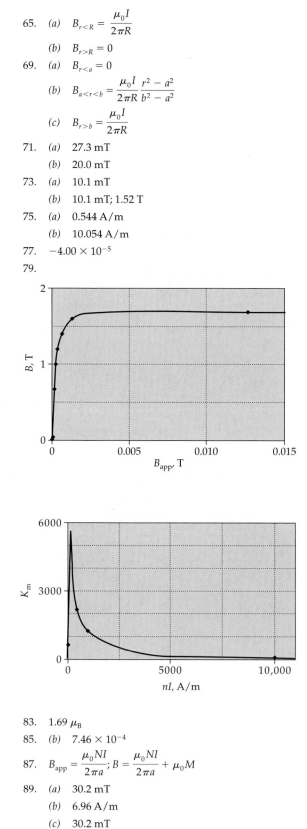

83. 1.69 μ_B

85. (b) 7.46×10^{-4}

87. $B_{\text{app}} = \dfrac{\mu_0 NI}{2\pi a}$; $B = \dfrac{\mu_0 NI}{2\pi a} + \mu_0 M$

89. (a) 30.2 mT
    (b) 6.96 A/m
    (c) 30.2 mT

91. $K_m = 11.75$; $\mu = 1.48 \times 10^{-5}$

93. (a) 12.6 mT

   (b) 1.36×10^6 A/m

   (c) 137

95. (a) 60.3 mT

   (b) 24.0 A

97. (a) 1.42×10^6 A/m

   (b) $K_m = 90.0$; $\mu = 1.13 \times 10^{-4}$ T·m/A; $\chi_m = 89.0$

99. (a) $(8.00 \text{ T/m})r$

   (b) $(3.20 \times 10^{-3} \text{ T·m})\dfrac{1}{r}$

   (c) $(8.00 \times 10^{-6} \text{ T·m})\dfrac{1}{r}$

   (d) Note that the field in the ferromagnetic region is the field that would be produced in a nonmagnetic region by a current of $400I = 1600$ A. The amperian current on the inside of the surface of the ferromagnetic material must therefore be 1600 A − 40 A = 1560 A in the direction of I. On the outside surface, there must then be an amperian current of 1560 A in the opposite direction.

101. $\dfrac{\mu_0 I}{4}\left(\dfrac{1}{R_1} - \dfrac{1}{R_2}\right)\hat{i}$

103. $\dfrac{\mu_0}{2\pi}\dfrac{I}{a}(1 + \sqrt{2})$

105. (a) $\vec{F}_2 = (1.00 \times 10^{-4} \text{ N})\hat{i}$; $\vec{F}_4 = (-0.286 \times 10^{-4} \text{ N})\hat{i}$

   (b) $\vec{F}_{net} = (0.714 \times 10^{-4} \text{ N})\hat{i}$

107. $(7.07\ \mu\text{T})\hat{i}$

109.

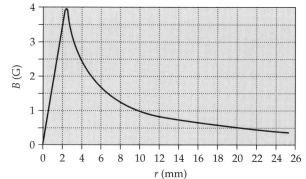

111. $\kappa = 0.246$ N·m/rad; $T = 0.523$ s

115. (a) 5.24×10^{-2} A·m²

   (b) 7.70×10^5 A/m

   (c) 2.31×10^4 A

117. (a) 70.6 A·m²

   (b) 17.7 N·m

119. 3.18 cm

121. (a) 10.0 μT

   (b) 10.0 μT

   (c) 5.00 μT

123. 2.24 A

125. (c) $\dfrac{\mu_0 \omega \sigma}{2}\left(\dfrac{R^2 + 2x^2}{\sqrt{R^2 + x^2}} - 2x\right)$

Chapter 28

1. (d)

3. (a) If the current in B is clockwise, the loops repel one another.

   (b) If the current in B is counterclockwise, the loops attract one another.

5. (a) and (b)

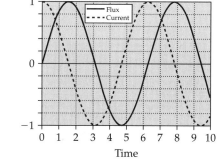

7. The protection is needed because if the current is suddenly interrupted, the resulting emf generated across the inductor due to the large flux change can blow out the inductor. The diode allows the current to flow (in a loop) even when the switch is opened.

9. (a) False

   (b) True

   (c) True

   (d) False

   (e) True

11. The time-varying magnetic field of the magnet sets up eddy currents in the metal tube. The eddy currents establish a magnetic field with a magnetic moment opposite to that of the moving magnet; thus, the magnet is slowed down. If the tube is made of a nonconducting material, there are no eddy currents.

13. (a) 3.14 rad/s

   (b) 1.94 mV

   (c) No. To generate an emf of 1 V, the students would have to rotate the jump rope about 500 times faster.

   (d) The use of multiple strands of lighter wire (so that the composite wire could be rotated at the same angular speed) looped several times around would increase the induced emf.

15. 2.00 kV

17. (a) 0

   (b) 1.37×10^{-5} Wb

   (c) 0

   (d) 1.19×10^{-5} Wb

19. $\pi r^2 B$

21. 6.74×10^{-3} Wb

23. (a) $\mu_0 n I N \pi R_1^2$

   (b) $\mu_0 n I N \pi R_3^2$

25. $\dfrac{\mu_0 I}{4\pi}$

27. (a) 0.314 mV

   (b) 0.785 mA

   (c) 0.247 μW

29. (a)

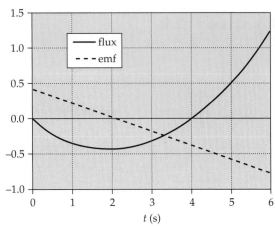

(b) Referring to the graph, we see that the flux is a minimum at $t = 2$ s and that $V = 0$ at this instant.

(c) The flux is zero at $t = 0$ and $t = 4$ s. At these times, $\mathcal{E} = 0.4$ V and -0.4 V, respectively.

31. (a) -1.26 mC

(b) 12.6 mA

(c) 630 mV

33. 79.8 μT

35. (a) $0.693 \dfrac{\mu_0 a}{\pi}$

(b) $4.16 \ \mu\Omega$.

(c) Because the magnetic flux due to I is increasing into the page, the induced current will be in such a direction that its magnetic field will oppose this increase; that is, it will be out of the page. Thus, the induced current is counter-clockwise.

37. 400 m/s

39. (a)

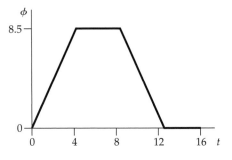

(b)

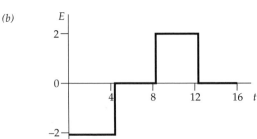

41. (a) $\dfrac{dv}{dt} = \dfrac{B\ell}{mR}(\mathcal{E} - B\ell v)$

(b) $v_t = \dfrac{\mathcal{E}}{B\ell}$

(c) 0

43. $x = \dfrac{mv_0 R}{B^2 \ell^2}$

47. (d)

49. (a) $(15.6 \ \text{T·m})\, v$

(b) 1.61 cm/s

51. (a) $\dfrac{\mu_0 I v b}{2\pi}\left(\dfrac{1}{d + vt} - \dfrac{1}{d + a + vt}\right)$

(b) $\dfrac{\mu_0 I v b}{2\pi}\left(\dfrac{1}{d + vt} - \dfrac{1}{d + a + vt}\right)$

53. (a) $24 \ \text{Wb} + (1600 \ \text{H·A/s})t$

(b) -1.60 kV

55. (a) 6.03 mT

(b) 7.58×10^{-4} Wb

(c) 0.253 mH

(d) -38.0 mV

57. $0; 162 \ \Omega$

63. $\dfrac{\mu_0 I^2}{16\pi}$

65. $0.157 \ \mu$H

67. (a) $I = 0; dI/dt = 25.0 \ \text{A/s}$

(b) $I = 2.27 \ \text{A}; dI/dt = 0.5 \ \text{A/s}$

(c) $I = 7.90 \ \text{A}; dI/dt = 9.20 \ \text{A/s}$

(d) $I = 10.8 \ \text{A}; dI/dt = 3.38 \ \text{A/s}$

69. (a) 44.0 W

(b) 40.4 W

(c) 3.62 W

71. (a) 5.77 s

(b) 28.9 H

73. (a) 3.00 kA/s

(b) 1.50 kA/s

(c) 80.0 mA

(d) 0.123 ms

75. (a) 1.00 A; 0

(b) 100 V; 100 V

(c)

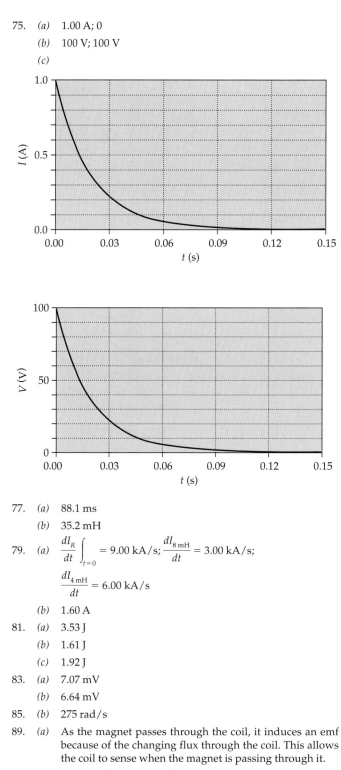

77. (a) 88.1 ms

(b) 35.2 mH

79. (a) $\left.\dfrac{dI_R}{dt}\right|_{t=0} = 9.00 \text{ kA/s}; \dfrac{dI_{8\,mH}}{dt} = 3.00 \text{ kA/s};$

$\dfrac{dI_{4\,mH}}{dt} = 6.00 \text{ kA/s}$

(b) 1.60 A

81. (a) 3.53 J

(b) 1.61 J

(c) 1.92 J

83. (a) 7.07 mV

(b) 6.64 mV

85. (b) 275 rad/s

89. (a) As the magnet passes through the coil, it induces an emf because of the changing flux through the coil. This allows the coil to sense when the magnet is passing through it.

(b) One cannot use a cylinder made of conductive material because eddy currents induced in the material by a falling magnet would slow the magnet.

(c) As the magnet approaches the loop, the flux increases, resulting in the increasing voltage signal. When the magnet is passing the coil, the flux goes from increasing to decreasing, so the induced emf becomes zero and then negative. The time at which the induced emf is zero is the time at which the magnet is at the center of the coil.

(d)

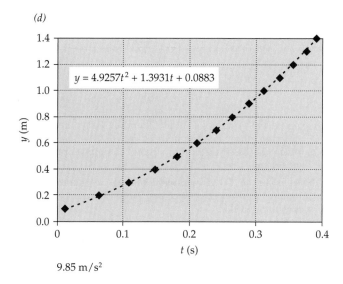

$y = 4.9257t^2 + 1.3931t + 0.0883$

9.85 m/s^2

91. $I(t) = (0.350 \text{ A}) \sin(2 \text{ rad/s})t$

93. (a) $-\tfrac{1}{2}r\mu_0 n I_0 \omega \cos \omega t$

(b) $-\dfrac{\mu_0 n R^2 I_0 \omega}{2r} \cos \omega t$

97. (a) 30.8 N/m

(b) $\dfrac{B y_0 \omega w}{R} \cos \omega t$

(c) $\beta = \dfrac{B^2 w^2}{R}$

(d)

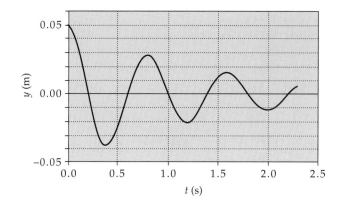

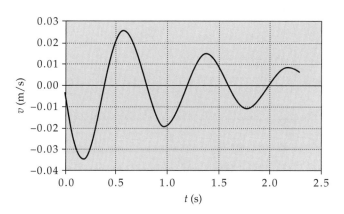

Chapter 29

1. (b)

3. (b)

5. (c)

7. Yes to both questions. While charge is accumulating on the capacitor, the capacitor absorbs power from the generator. When the capacitor is discharging, it supplies power to the generator.

9. To make an *LC* circuit with a small resonance frequency requires a large inductance and large capacitance. Neither is easy to construct.

11. Yes. The power factor is defined to be $\cos \delta = R/Z$ and, because Z is frequency dependent, so is $\cos \delta$.

13. 0

15. True

17. (a) False

 (b) True

19. (a) 39.8 Hz

 (b) 15.1 V

21. (a) 13.6 V

 (b) 486 Hz

23. (a) 0.833 A

 (b) 1.18 A

 (c) 200 W

25. (a) 0.377 Ω

 (b) 3.77 Ω

 (c) 37.7 Ω

27. 1.59 kHz

29. (a) 25.1 mA

 (b) 17.8 mA

31. (a) (0.346 A) cos ωt

 (b) (0.346 A) cos ωt

 (c) (0.344 A) cos(ωt + 0.165 rad)

33. (a) 1.26 ms

 (b) 88.0 mH

35. (a) 2.25 mJ

 (b) 712 Hz

 (c) 0.671 A

37. (a)

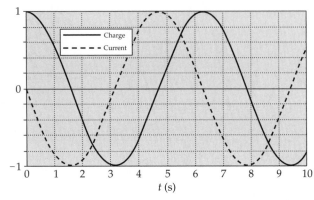

39. 29.2 mH

41. (a) 0.333

 (b) 26.7 Ω

 (c) 0.200 H

 (d) *I* lags $\mathscr{E}$ by 70.5°

43. 0.397

45. (a) $I_{R,\text{rms}} = 6.20$ A; $I_{R_L,\text{rms}} = 2.79$ A; $I_{L,\text{rms}} = 5.52$ A

 (b) $I_{R,\text{rms}} = 3.29$ A; $I_{R_L,\text{rms}} = 2.95$ A; $I_{L,\text{rms}} = 1.47$ A

 (c) 50.3%; 80.5%

47. 60.0 V

49. (b) −90°

 (c) 0

57.

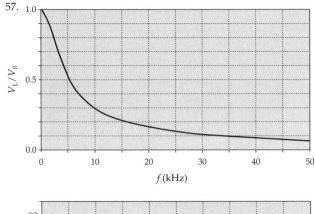

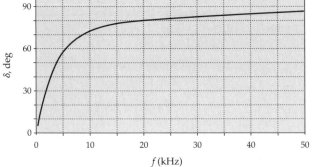

59. $\Delta \omega = R/2L$

61. 170 μF

63. (a) $I = -(18.75 \text{ mA}) \sin\left(1250t + \dfrac{\pi}{4}\right)$

 (b) 22.86 μF

 (c) $U_e = (4.92 \text{ μJ}) \cos^2\left(1250t + \dfrac{\pi}{4}\right)$

 $U_m = (4.92 \text{ μJ}) \sin^2\left(1250t + \dfrac{\pi}{4}\right)$

 $U = U_m + U_e = 4.92$ μJ

65. (a) 5.39×10^{-16} F

 (b) $f(x) = \dfrac{70.0 \text{ MHz}}{\sqrt{1 - (3.96 \text{ m}^{-1})x}}$

67. (a) 0.0444

 (b) 491 rad/s or 509 rad/s

71. (a) 1.13 kHz

 (b) $X_C = 79.6$ Ω; $X_L = 62.8$ Ω

 (c) $Z = 17.5$ Ω; 4.04 A

 (d) $\delta = -73.4°$

73. (a) 14.1
(b) 79.8 Hz
(c) 0.275

75. (a) 10.0 A
(b) 53.1°
(c) 332 μF
(d) 133 V

77. (b) $\delta = -\dfrac{\pi}{2} + \omega RC$

(c) $\delta = \dfrac{\pi}{2} - \dfrac{R}{\omega L}$

79. (a) 80.0 V
(b) 77.5 V
(c) 164 V
(d) 111 V
(e) 181 V

81. 0.933 μF

83. (a)

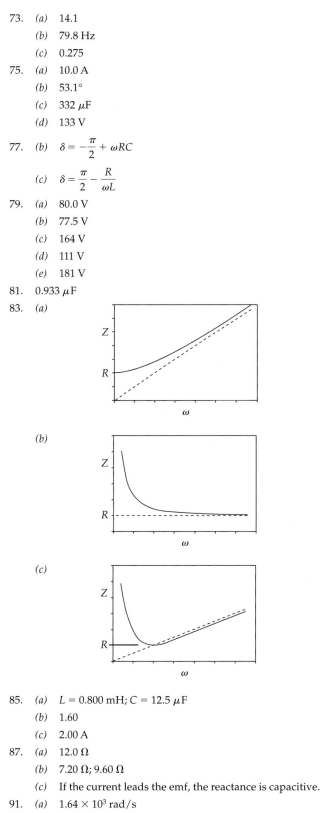

(b)

(c)

85. (a) $L = 0.800$ mH; $C = 12.5\ \mu$F
(b) 1.60
(c) 2.00 A

87. (a) 12.0 Ω
(b) 7.20 Ω; 9.60 Ω
(c) If the current leads the emf, the reactance is capacitive.

91. (a) 1.64×10^3 rad/s
(b) $I_{C,\text{rms}} = 1.39$ A; $\delta_C = -84.4°$
$I_{L,\text{rms}} = 1.41$ A; $\delta_L = 78.5°$
$I_{\text{rms}} = 0.417$ A

93. (a) 0.396 μF
(b) 840 Ω

95. (a) 13.26 kHz
(b) 200 mA; 600 V
(c) $V_L = 433$ V; $V_C = 419$ V

103. (a) 1.67 rad/s
(b) 3.96 W; 7.69 W

105. (a) 1/5
(b) 50.0 A

107. (a) 1.50 A
(b) 19

109. 3333

113. (a) 265 Ω
(b) 2.65 Ω
(c) 2.65 mΩ

115. (a) 12.0 V
(b) 8.49 V

117. (a) $Q_1 = (60\ \mu\text{F})\cos(120\pi t) + 72\ \mu$F;
$Q_2 = (30\ \mu\text{F})\cos(120\pi t) + 36\ \mu$F
(b) $I = -(33.9\ \text{mA})\sin(120\pi t)$
(c) 4.36 mJ
(d) 36.0 μJ

119. $I_{\text{max}} = 1.06$ A; $I_{\text{min}} = -0.0560$ A; $I_{\text{av}} = 0.500$ A; $I_{\text{rms}} = 0.636$ A

121. The inductance acts as a short circuit to the constant voltage source. The current is infinite at all times. Consequently, $I_{\text{max}} = I_{\text{rms}} = \infty$. There is no minimum current.

Chapter 30

1. (a) False
(b) True
(c) True
(d) True
(e) False
(f) True

3. X rays have greater frequencies, whereas light waves have longer wavelengths (see Table 30-1).

5. Consulting Table 30-1, we see that FM radio waves and television waves have wavelengths of the order of a few meters.

7. The dipole antenna should be in the horizontal plane and normal to the line from the transmitter to the receiver.

9. $I = 2.94 \times 10^7$ W/m²; $P = 5.20$ mW

11. (a) $E_{\text{rms}} = 719$ V/m; $B_{\text{rms}} = 2.40\ \mu$T
(b) $P_{\text{av}} = 3.87 \times 10^{26}$ W
(c) $I = 6.36 \times 10^7$ W/m²; $P_r = 0.212$ Pa

13. $F_r = 7.09 \times 10^7$ N; Because the ratio of these forces is 1.65×10^{-14} for the earth and 4.26×10^{-14} for Mars, Mars has the larger ratio. The reason that the ratio is higher for Mars is that the dependence of the radiation pressure on the distance from the sun is the same for both forces (r^{-2}), whereas the dependence on the radii of the planets is different. Radiation pressure varies as R^2, whereas the gravitational force varies as R^3 (assuming that the two planets have the same density, an assumption that is nearly true). Consequently, the ratio of the forces goes as $R^2/R^3 = R^{-1}$. Because Mars is smaller than the earth, the ratio is larger.

15. (a) 3.40×10^{14} V/m·s

    (b) 5.00 A

19. (a) 10.0 A

    (b) 2.26×10^{12} V/m·s

    (c) 7.90×10^{-7} T·m

21. (a) $I = \dfrac{A(0.01 \text{ V/s})}{\rho d} t$

    (b) $I_d = \dfrac{(0.01 \text{ V/s})\epsilon_0 A}{d}$

    (c) $t = \epsilon_0 \rho$

25. (a) 300 m

    (b) 3.00 m

27. 3.00×10^{18} Hz

29. (a) $30.0°$

    (b) 7.07 m

31. $4.14\ \mu\text{W/m}^2$; 5.21×10^{17} photons/(cm²·s)

33. $0.151\ \mu\text{W/m}^2$

35. (a) 283 V/m

    (b) $0.943\ \mu\text{T}$

    (c) $212\ \text{W/m}^2$

    (d) $0.707\ \mu\text{Pa}$

39. (a) 40.0 nN

    (b) 80.0 nN

41. (a) 3.46 V/m; 11.5 nT

    (b) 0.346 V/m; 1.15 nT

    (c) 0.0346 V/m; 0.115 nT

43. $E_{rms} = 75.2$ kV/m; $B_{rms} = 0.251$ mT

45. (a) Positive x direction

    (b) 0.628 m; 477 MHz

    (c) $\vec{E}(x, t) = (194 \text{ V/m}) \cos[10x - (3 \times 10^9)t]\hat{j}$
        $\vec{B}(x, t) = (0.647\ \mu\text{T}) \cos[10x - (3 \times 10^9)t]\hat{k}$

47. 6.10×10^{-3} degrees

49. 3.42 MW/m²

55. The current induced in a loop antenna is proportional to the time-varying magnetic field. For a maximum signal, the antenna's plane should make an angle $\theta = 0°$ with the line from the antenna to the transmitter. For any other angle, the induced current is proportional to $\cos\theta$. The intensity of the signal is therefore proportional to $\cos\theta$.

57. 72.6 nV

59. (a) $I = V_0\left(\dfrac{1}{R}\sin\omega t + \dfrac{\omega\epsilon_0\pi a^2}{d}\cos\omega t\right)$

    (b) $B(r) = \dfrac{\mu_0 V_0}{2\pi r}\left(\dfrac{1}{R}\sin\omega t + \omega\dfrac{r^2}{a^2}\cos\omega t\right)$

    (c) $\delta = \tan^{-1}\left(\dfrac{R\omega\epsilon_0\pi a^2}{d}\right)$

61. (a) $\vec{S} = \dfrac{1}{\mu_0 c}[E_{1,0}^2 \cos^2(k_1 x - \omega_1 t) + 2E_{1,0}E_{2,0}\cos(k_1 x - \omega_1 t)$
        $\times \cos(k_2 x - \omega_2 t + \delta) + E_{2,0}^2 \cos^2(k_2 x - \omega_2 t + \delta)]\hat{i}$

    (b) $\vec{S}_{av} = \dfrac{1}{2\mu_0 c}[E_{1,0}^2 + E_{2,0}^2]\hat{i}$

    (c) $\vec{S} = \dfrac{1}{\mu_0 c}[E_{1,0}^2 \cos^2(k_1 x - \omega_1 t) - E_{2,0}^2 \cos^2(k_2 x + \omega_2 t + \delta)]\hat{i}$

    (d) $\vec{S}_{av} = \dfrac{1}{2\mu_0 c}[E_{1,0}^2 - E_{2,0}^2]\hat{i}$

63. (a) 9.15×10^{-15} T

    (b) $(1.01\ \mu\text{V})\cos(8.80 \times 10^5 \text{ s}^{-1})t$

    (c) $(5.49\ \mu\text{V})\sin(8.80 \times 10^5 \text{ s}^{-1})t$

65. (a) $E = \dfrac{I\rho}{\pi a^2}$

    (b) $B = \dfrac{\mu_0 I}{2\pi a}$

    (c) $\vec{S} = -\dfrac{I^2\rho}{2\pi^2 a^3}\hat{r}$

67. $0.574\ \mu\text{m}$

69. 3.33 mN

Chapter 31

1. The population inversion between the state $E_{2,Ne}$ and the state 1.96 eV below it (see Figure 31-9) is achieved by inelastic collisions between neon atoms and helium atoms excited to the state $E_{2,He}$.

3. The layer of water greatly reduces the light reflected back from the car's headlights, but increases the light reflected by the road of light from the headlights of oncoming cars.

5. The change in atmospheric density results in refraction of the light from the sun, bending it toward the earth. Consequently, the sun can be seen even after it is just below the horizon. Also, the light from the lower portion of the sun is refracted more than the light from the upper portion, so the lower part appears to be slightly higher in the sky. The effect is an apparent flattening of the disk into an ellipse.

7. He takes the path *LES* because the time required to reach the swimmer is the least for this path.

9. (d)

11.

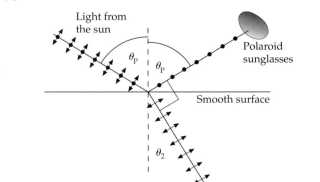

13. (c)

15. In resonance absorption, the molecules respond to the frequency of the light through the Einstein photon relation $E = hf$. Thus, the color appears to be the same in spite of the fact that the wavelength has changed.

17. (a) 2:00 A.M., September 1

    (b) 2:08 A.M., September 1

19. (a) 15.0 mJ

    (b) 5.25×10^{16}

21. (a) 435 nm

    (b) 1210 nm

23. (a) 387.5 nm; $\lambda_{21} = 1138$ nm; $\lambda_{10} = 587.7$ nm

(b) $\lambda_{03} = 285.1$ nm; $\lambda_{32} = 1078$ nm; $\lambda_{21} = 1138$ nm;

$\lambda_{10} = 587.7$ nm; $\lambda_{31} = 553.6$ nm; $\lambda_{20} = 387.5$ nm

25. (a)

27. 5 min, 23 s

29. (a) 20.0 μs

(b) $\Delta t_{\text{reaction}} \approx 10^4 \, \Delta t$

33. $v_{\text{water}} = 2.25 \times 10^8$ m/s; $v_{\text{glass}} = 2.00 \times 10^8$ m/s

35. (a) 50.2°

(b) 38.8°

(c) 26.3°

37. 92.2 percent

41. 62.5°

43. 102 m²

45. 1.30

47. 5.43°

49. (a) 48.7°

(b) Note that θ_2 equals the critical angle for a water–air interface. Therefore, the ray will not leave the water for $\theta_1 \geq 41.8°$.

51. 1.02°

53. (a) 53.1°

(b) 56.3°

55. (a) $\frac{1}{8}I_0$

(b) $\frac{3}{32}I_0$

57. (a) 30.0°

(b) 1.73

59. $I_3 = \frac{1}{8}I_0 \sin^2 2\omega t$

61. 13

63. $I_4 = 0.211 I_0$

65. Right circularly polarized; $\vec{E} = E_0 \sin(kx + \omega t)\hat{j} - E_0 \cos(kx + \omega t)\hat{k}$

67. 35.3°

69. 1.45 m

71. 3.42 m

73. (a) 36.8°

(b) 38.7°

75. (a) -1.00 m

(b) $\theta_i = 26.6°$; $\theta_r = 26.6°$

77. For silicate flint glass: $\theta_p = 58.3°$; for borate flint glass: $\theta_p = 57.5°$; for quartz glass: $\theta_p = 57.0°$; for silicate crown glass: $\theta_p = 56.5°$

79. (b) $\theta_p > \theta_c$

81. (a) 1.33

(b) $\theta_c = 37.2°$

(c) $\theta_2 = 48.8°$

83. (a) $\dfrac{I_t}{I_0} = \left[\dfrac{4n}{(n+1)^2} \right]^{2N}$

(b) 0.783

(c) ≈ 28

85. (a) 24.0°

(b) 4.45 km

(c) $\theta = \tan^{-1}\left(\dfrac{v_{\text{earth}}}{c} \right)$

(d) 2.99×10^8 m/s

Chapter 32

1. Yes. Note that a virtual image is seen because the eye focuses the diverging rays to form a real image on the retina. Similarly, the camera lens can focus the diverging rays onto the film.

3. (a) False

(b) False

(c) True

(d) False

5. A convex mirror always produces a virtual erect image that is smaller than the object. It never produces an enlarged image.

7. (b)

9. (a) The lens will be positive if its index of refraction is greater than that of the surrounding medium, and the lens is thicker in the middle than at the edges. Conversely, if the index of refraction of the lens is less than that of the surrounding medium, the lens will be positive if it is thinner at its center than at the edges.

(b) The lens will be negative if its index of refraction is greater than that of the surrounding medium, and the lens is thinner at the center than at the edges. Conversely, if the index of refraction of the lens is less than that of the surrounding medium, the lens will be negative if it is thicker at the center than at the edges.

11. (d)

13. The eye accommodates by varying the power of a lens located a fixed distance away from the retina. A camera, on the other hand, has a fixed power lens that can move with respect to the film location.

15. (b)

17. (d)

19. (a)

21. Microscopes ordinarily produce images (either the intermediate one produced by the objective or the one viewed through the eyepiece) that are larger than the object being viewed. A telescope, on the other hand, ordinarily produces images that are much reduced compared to the object. The object is normally viewed from a great distance, and the telescope magnifies the angle subtended by the object.

23. Plano-convex lens: $r_1 = -16.2$ cm; $r_2 = \infty$

Biconvex with equal curvature: $r_1 = -32.4$ cm; $r_2 = 32.4$ cm

Biconvex with unequal curvature: $r_1 = 16.2$ cm; $r_2 = 8.10$ cm

25.

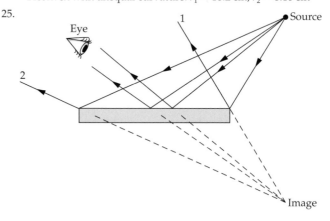

27.

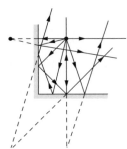

29. (a) The first image in the mirror on the left is 10 cm behind the mirror. The mirror on the right forms an image 20 cm behind that mirror or 50 cm from the left mirror. This image will result in a second image 50 cm behind the left mirror. The first image in the left mirror is 40 cm from the right mirror and forms an image 40 cm behind the right mirror or 70 cm from the left mirror. That image gives an image 70 cm behind the left mirror. The fourth image behind the left mirror is 110 cm behind that mirror.

(b) Proceeding as in Part (a) for the mirror on the right, one finds the location of the images to be 20 cm, 40 cm, 80 cm, and 100 cm behind the right-hand mirror.

31. (a) $s' = 15.8$ cm; $m = -0.316$; Because the image distance is positive and the lateral magnification is less than one and negative, the image is real, inverted, and reduced.

(b) $s' = 24.0$ cm; $m = -1$; Because the image distance is positive and the lateral magnification is one and negative, the image is real, inverted, and the same size as the object.

(c) $s' = \infty$ and there is no image.

(d) $s' = -24.0$ cm; $m = 3$; Because the image distance is negative and the lateral magnification is three and positive, the image is virtual, erect, and three times the size of the object.

33. (a) $s' = -9.85$ cm; $m = 0.179$; Because the image distance is negative and the lateral magnification is less than one in magnitude and positive, the image is virtual, erect, and reduced.

(b) $s' = -8.00$ cm; $m = 0.333$; Because the image distance is negative and the lateral magnification is less than one in magnitude and positive, the image is virtual, erect, and reduced.

(c) $s' = -6.00$ cm; $m = 0.5$; Because the image distance is negative and the lateral magnification is one-half in magnitude and positive, the image is virtual, erect, and half the size of the object.

(d) $s' = -4.80$ cm; $m = 0.600$; Because the image distance is negative and the lateral magnification is less than one and positive, the image is virtual, erect, and six-tenths the size of the object.

35. (a) 5.13 cm

(b) The mirror must be concave. A convex mirror always produces a diminished virtual image.

37. $s' = -4.00$ m; $y' = 3.68$ cm

39. (a)

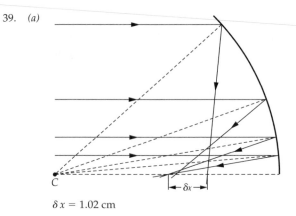

$\delta x = 1.02$ cm

(b) By blocking off the edges of the mirror so that only paraxial rays within 2 cm of the mirror axis are reflected, the spread is reduced by about 83 percent.

41. (a) -1.33 m

(b) Because $f_{small} < 0$, the small mirror is convex.

43. (a) $s' = -8.58$ cm, where the minus sign tells us that the image is 5.58 cm from the front surface of the bowl.

(b) $s' = -35.9$ cm, where the minus sign tells us that the image is 35.9 cm from the front surface of the bowl.

45. 14.4 cm

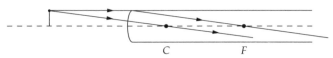

47. (a) $s' = -104$ cm, where the negative image position tells us that the image is 104 cm in front of the surface and is virtual.

(b) $s' = -8.29$ cm, where the minus sign tells us that the image is 8.29 cm in front of the surface and is virtual.

(c) $s' = 63.5$ cm, that is, the image is 63.5 cm behind (to the right of) the surface (at the focal point) and is real.

49. (a) $s' = 64.0$ cm

(b) $s' = -80.0$ cm

(c) The final image is 96 cm − 80 cm = 16 cm from the surface, the radius of the surface is 8 cm and is virtual.

51. (a) 19.0 cm

(b) 30.0 cm

(c) −15.0 cm

(d) −52.0 cm

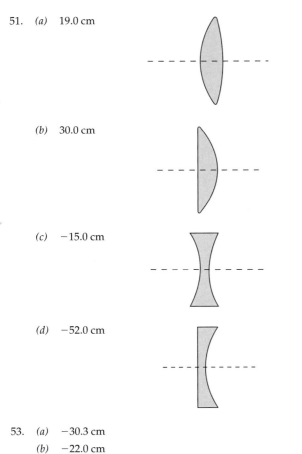

53. (a) −30.3 cm

(b) −22.0 cm

(c) −0.275

(d) Because $s' < 0$ and $m < 0$, the image is virtual and upright.

55.

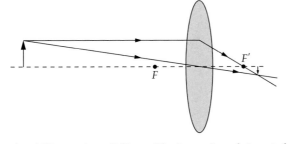

$s' = 16.7$ cm; $y' = -2.00$ cm; The image is real, inverted, and diminished. Because $s' > 0$ and $y' = -2.00$ cm, the image is real, inverted, and diminished in agreement with the ray diagram.

57.

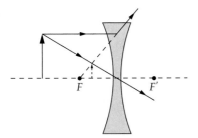

$s' = -6.67$ cm; $y' = 0.500$ cm; The image is virtual, erect, and diminished. Because $s' < 0$ and $y' = 0.500$ cm, the image is virtual, erect, and about one-third the size of the object in agreement with the ray diagram.

59. (a)

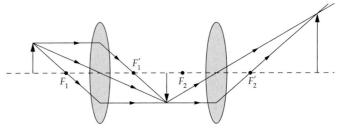

$s'_2 = 30$ cm and the final image is 85.0 cm from the object; $m_2 = -2$

(b) Because $s'_2 > 0$ and $m = m_1 m_2 = 2$, the image is real and upright.

(c) The image magnification is twice the size of the object.

61. (b) 3.70 m

63. (a) and (b)

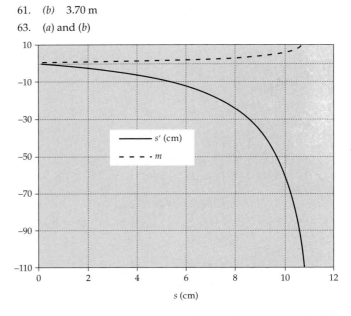

(c) The images are virtual and erect for this range of object distances.

(d) The asymptote of the graph of s' versus s corresponds to the image approaching infinity as the object distance approaches the focal length of the lens. The horizontal asymptote of the graph of m versus s indicates that, as the object moves toward the lens, the height of the image formed by the lens approaches the height of the object.

65. 15.0 cm; The final image is 50 cm from the object, real, inverted, and the same size as the object.

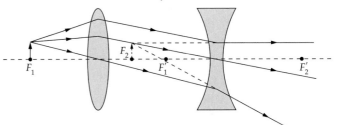

69. (a) 41.2 cm

    (b) −1.53

    (c) Because $m < 0$, the image is inverted. Because $s_2' > 0$, the image is real.

71. (a) False

    (b) True

73. (c) 40.0 D; 4.00 D

75. (c) −7.33 D

77. 1.72 mm

79. (a) 80.0 μrad

    (b) 1.60 mm

81. 0.444 D

83. 3.07 D

85. 6.00

87. 5.00

89. (a) 3.00

    (b) 4.00

93. (a) −1.88

    (b) −18.8

95. −232

97. (a) 9.00 mm

    (b) −20.0; −0.180 rad

99. (a) $P_{\text{Palomar}} = (25.0)P_{\text{Yerkes}}$

    (b) −134

101. (b)

103. −1/150

105. 1.34 mm

107. (a) $s = 5.00$ cm; $s' = -10.0$ cm

    (b) $s = 15.0$ cm; $s' = 30.0$ cm

109. (a) Because the focal lengths appear in the magnification formula as a product, it would appear that it does not matter in which order we use them. The usual arrangement would be to use the shorter focal length lens as the objective, but we get the same magnification in the reverse order. What difference does it make then? It makes no difference in this problem. However, it is generally true that the smaller the focal length of a lens, the smaller its diameter. This condition makes it harder to use the shorter focal length lens, with its smaller diameter, as the eyepiece lens. If we separate the objective and eyepiece lenses by $L + f_e + f_o = 16$ cm $+ 7.5$ cm $+ 2.5$ cm $= 26.0$ cm, the overall magnification will be −21.3.

    (b)

111. 3.70 m

113. (a) 9.52 cm

    (b) −1.19

    (c)

115. (a) 21.3 cm

    (b) 79.2 cm

117. 0.0971 m/s

119. (a) 22.5 cm

    (b) 18.0 cm

    (c)

    To see this image, the eye must be to the left of the image 4.

121. 36.8 cm

123. (a) The final image is 0.8 cm behind the mirror surface.

    (b) The final image is at the mirror surface.

Chapter 33

1. The energy is distributed nonuniformly in space; in some regions the energy is below average (destructive interference), in others it is higher than average (constructive interference).

3. The thickness of the air space between the flat glass and the lens is approximately proportional to the square of d, the diameter of the ring. Consequently, the separation between adjacent rings is proportional to $1/d$.

5. If the film is thick, the various colors (i.e., different wavelengths) will give constructive and destructive interference at that thickness. Consequently, what one observes is the reflected intensity of white light.

7. The first zeros in the intensity occur at angles given by $\sin \theta = \lambda/a$. Hence, decreasing a increases θ, and the diffraction pattern becomes wider.

9. (a)

11. (a) False
    (b) True
    (c) True
    (d) True
    (e) True

13. $3.44 \ \mu m$

15. $\theta_{red} \approx 1.80°$; light, $\theta_{blue} \approx 1.25°$.

17. (a) 300 nm
    (b) 135°

19. 164°

21. $5.46 \ \mu m < d < 5.75 \ \mu m$

23. (a) 600 nm
    (b) From the table, we see that the only wavelengths in the visible spectrum are 720 nm, 514 nm, and 400 nm.
    (c) From the table, we see that the missing wavelengths in the visible spectrum are 720 nm, 514 nm, and 400 nm.

25. 476 nm

27. (c) The transmitted pattern is complementary to the reflected pattern.
    (d) 68 bright fringes
    (e) 1.14 cm
    (f) The wavelength of the light in the film becomes $\lambda_{air}/n = 444$ nm. The separation between fringes is reduced and the number of fringes that will be seen is increased by the factor $n = 1.33$.

29. 0.535 nm; 0.926 nm

31. 4.95 mm

33. (a) $50.0 \ \mu m$
    (b) Not with the unaided eye. The separation is too small to be observed with the naked eye.
    (c) 0.500 mm

35. 625 nm and 417 nm

37. (a) 0.600 mrad
    (b) 6.00 mrad
    (c) 60.0 mrad

39. (a) 1.25 km
    (b) 376 m
    (c) 2.6×10^{-14}

41. (a) $20.0 \ \mu m$
    (b) 9

43. There are eight interference fringes on each side of the central maximum. The secondary diffraction maximum is half as wide as the central one. It follows that it will contain eight interference maxima.

45. $3.61 \sin(\omega t - 56.3°)$

47. $0.0162 I_0$

49. (b) 6.00 mm
    (c) The width for four sources is half the width for two sources.

51. (a)
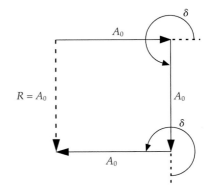
    (b) $5.56 \ mW/m^2$

53. (a) 8.54 mrad
    (b) 6.83 cm

55. 7.00 mm

57. 484 m

59. $5.00 \times 10^9 \ m$

61. 86.9 mrad; 82.1 mrad

63. 40.6°; 38.0°

65. 30.0°

67. Because $m_{max} = 2.98$, one can see the complete spectrum only for $m = 1$ and 2. Because 700 nm $< 2 \times 400$ nm, there is no overlap of the second-order spectrum into the first-order spectrum; however, there is overlap of long wavelengths in the second order with short wavelengths in the third-order spectrum.

69. (a) $y_1 = 0.353$ m; $y_2 = 0.706$ m
    (b) $\Delta y = 88.4 \ \mu m$
    (c) 8000

71. $R = 3.09 \times 10^5$; $n = 5.14 \times 10^4 \ cm^{-1}$

73. (a) $n = 750 \ cm^{-1}$
    (b) 4.20 cm; 12.6 cm

75. (a) $\phi = \sin^{-1}\left(\dfrac{m\lambda}{a}\right)$
    (b) 64.2°

77. (a) 0.150 mm
    (b) $3.33 \times 10^3 \ m^{-1}$

79. 1.68 cm

81. 0.130 mrad

85. (a) 97.8 nm
    (b) No, because 180 nm is not in the visible portion of the spectrum.
    (c) 0.273; 0.124

87. 12.3 m

INDEX

Physical Constants[†]

Atomic mass constant	$m_u = \frac{1}{12}m(^{12}C)$	$1\,u = 1.660\,538\,73(13) \times 10^{-27}\,kg$
Avogadro's number	N_A	$6.022\,141\,99(47) \times 10^{23}\,particles/mol$
Boltzmann constant	$k = R/N_A$	$1.380\,6503(24) \times 10^{-23}\,J/K$ $8.617\,342(15) \times 10^{-5}\,eV/K$
Bohr magneton	$m_B = e\hbar/(2m_e)$	$9.274\,008\,99(37) \times 10^{-24}\,J/T =$ $5.788\,381\,749(43) \times 10^{-5}\,eV/T$
Coulomb constant	$k = 1/(4\pi\epsilon_0)$	$8.987\,551\,788\ldots \times 10^9\,N\cdot m^2/C^2$
Compton wavelength	$\lambda_C = h/(m_e c)$	$2.426\,310\,215(18) \times 10^{-12}\,m$
Fundamental charge	e	$1.602\,176\,462(63) \times 10^{-19}\,C$
Gas constant	R	$8.314\,472(15)\,J/(mol\cdot K) =$ $1.987\,2065(36)\,cal/(mol\cdot K) =$ $8.205\,746(15) \times 10^{-2}\,L\cdot atm/(mol\cdot K)$
Gravitational constant	G	$6.673(10) \times 10^{-11}\,N\cdot m^2/kg^2$
Mass of electron	m_e	$9.109\,381\,88(72) \times 10^{-31}\,kg =$ $0.510\,998\,902(21)\,MeV/c^2$
Mass of proton	m_p	$1.672\,621\,58(13) \times 10^{-27}\,kg =$ $938.271\,998(38)\,MeV/c^2$
Mass of neutron	m_n	$1.674\,927\,16(13) \times 10^{-27}\,kg =$ $939.565\,330(38)\,MeV/c^2$
Permittivity of free space	ϵ_0	$8.854\,187\,817\ldots \times 10^{-12}\,C^2/(N\cdot m^2)$
Permeability of free space	μ_0	$4\pi \times 10^{-7}\,N/A^2$
Planck's constant	h	$6.626\,068\,76(52) \times 10^{-34}\,J\cdot s =$ $4.135\,667\,27(16) \times 10^{-15}\,eV\cdot s$
	$\hbar = h/(2\pi)$	$1.054\,571\,596(82) \times 10^{-34}\,J\cdot s =$ $6.582\,118\,89(26) \times 10^{-16}\,eV\cdot s$
Speed of light	c	$2.997\,924\,58 \times 10^8\,m/s$
Stefan-Boltzmann constant	σ	$5.670\,400(40) \times 10^{-8}\,W/(m^2\cdot K^4)$

† The values for these and other constants can be found in Appendix B as well as on the Internet at http://physics.nist.gov/cuu/Constants/index.html. The numbers in parentheses represent the uncertainties in the last two digits. (For example, 2.044 43(13) stands for 2.044 43 ± 0.000 13.) Values without uncertainties are exact. Values with ellipses are exact (like the number $\pi = 3.1415\ldots$).

Derivatives and Definite Integrals

$$\frac{d}{dx}\sin ax = a\cos ax \qquad \int_0^\infty e^{-ax}\,dx = \frac{1}{a} \qquad \int_0^\infty x^2 e^{-ax^2}\,dx = \frac{1}{4}\sqrt{\frac{\pi}{a^3}}$$

$$\frac{d}{dx}\cos ax = -a\sin ax \qquad \int_0^\infty e^{-ax^2}\,dx = \frac{1}{2}\sqrt{\frac{\pi}{a}} \qquad \int_0^\infty x^3 e^{-ax^2}\,dx = \frac{4}{a^2}$$

$$\frac{d}{dx}e^{ax} = ae^{ax} \qquad \int_0^\infty xe^{-ax^2}\,dx = \frac{2}{a} \qquad \int_0^\infty x^4 e^{-ax^2}\,dx = \frac{3}{8}\sqrt{\frac{\pi}{a^5}}$$

The a in the six integrals is a positive constant.

Vector Products

$$\vec{A} \cdot \vec{B} = AB\cos\theta \qquad \vec{A} \times \vec{B} = AB\sin\theta\,\hat{n} \quad (\hat{n}\text{ obtained using right-hand rule})$$